What You See in the Book	What You Find on TechComm Web

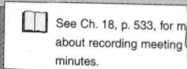

The book icon in the margin directs you to related **discussions elsewhere in the text.**

A book icon connected to a globe icon directs to you **resources on TechComm Web.**

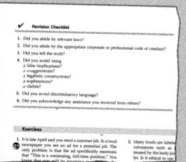

Revision checklists at the end of each chapter summarize the important concepts in the chapter.

Revision checklists are posted with each chapter's materials for quick reference.

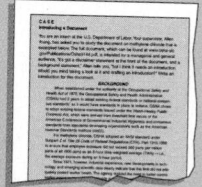

Cases at the end of each chapter present a working-world scenario in which you are asked to solve a problem using your skills as a technical communicator.

A Case of the Month asks you to research a subject on the Internet and respond in various forms of technical communication. The cases from the book are also posted.

An exercise or project with a globe icon calls for **Web research** that will help you **practice the skills** introduced in each chapter.

Flashcards and online quizzes with immediate feedback help you **reinforce and test your understanding** of the material in each chapter.

An exercise or project with a collaboration icon may be assigned to a **collaborative group.**

Downloadable forms for collaboration can be adapted to fit your purpose.

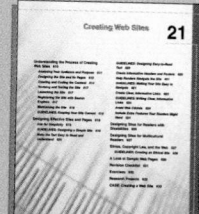

Chapter 21, "Creating Web Sites," includes **advice about designing and coding Web pages.**

A **Web-site design tutorial** shows you how to design a Web site and its pages.

W9-COB-138

Sixth Edition

Technical Communication

Sixth Edition

Technical Communication

Mike Markel
Boise State University

BEDFORD / ST. MARTIN'S
Boston • New York

For Bedford / St. Martin's

Developmental Editor: Leasa Burton
Production Editor: Ara Salibian
Production Supervisors: Donna Peterson, Catherine Hetmansky
Director of Marketing: Karen Melton
Editorial Assistant: Sara Eaton
Production Assistant: Arthur Johnson
Copyeditor: Linda Leet Howe
Text and Cover Design: Anna George
Composition: Monotype Composition Co., Inc.
Printing and Binding: RR Donnelly & Sons Company

President: Charles H. Christensen
Editorial Director: Joan E. Feinberg
Editor in Chief: Karen S. Henry
Director of Editing, Design, and Production: Marcia Cohen
Managing Editor: Elizabeth M. Schaaf

Library of Congress Control Number: 00–103102

Manufactured in the United States of America.

5   4   3   2
f   e   d   c

For information, write: Bedford/St. Martin's, 75 Arlington Street, Boston, MA 02116
(617-399-4000)

ISBN: 0–312–40067–5

Acknowledgments

Figure 1.1: Motorola home page. Source: Motorola Web site <www.motorola.com>. © 1994–2001 Motorola, Inc. All Rights Reserved. Courtesy of Motorola, Inc.

Chapter 1, exercise 2: "The Clean Earth Campaign" Web page. Source: Canon Corporation Web site. Reproduced with permission.

Figure 2.1: "STC Ethical Guidelines for Technical Communicators." Used with permission, from *Code for Communicators,* published by the Society for Technical Communication, Arlington, Virginia.

Acknowledgments and copyrights are continued at the back of the book on pages 746–747, which constitute an extension of the copyright page. It is a violation of the law to reproduce these selections by any means whatsoever without the written permission of the copyright holder.

PREFACE FOR INSTRUCTORS

The principles of good writing have not changed much in the sixteen years since *Technical Communication* was first published in 1984, but almost everything else about technical communication has. The sixth edition of this text reflects the many exciting developments that have occurred in technical communication — and in ways to teach it — in the last few years, most especially the greatly expanded importance of the World Wide Web and electronic communication tools. While maintaining the book's focus on planning, writing, and designing the major types of technical documents, the sixth edition has been thoroughly revised and updated to show the effect of new technologies on the way people produce technical communication.

Overview of the Sixth Edition

Technical Communication is now organized into five parts, highlighting the importance of the writing process in technical communication and giving equal weight to the development of text and visuals in a document.

Part One, "The Technical Communication Environment," provides students with a basic understanding of important topics in technical communication, including ethical and legal considerations, the role of the writing process in planning and developing technical documents, and the practice of collaborating on documents.

Part Two, "Planning the Document," focuses on rhetorical concerns, such as considering an audience and purpose, and communicating persuasively, in addition to the early work of gathering information — both primary and secondary research — and planning the organization of documents.

Part Three, "Developing the Textual Elements," encompasses both drafting and revising text in a document. The part describes techniques for writing definitions and descriptions, improving the coherence of text, improving sentence style, and writing front and back matter.

Part Four, "Developing the Visual Elements," addresses the fundamentals of designing the whole document as well as the individual page. It also includes advice on creating graphics for both print and online documents.

Part Five, "Applications," covers a wide range of types of technical communication: letters, memos, and emails; job-application materials, including print and electronic résumés; proposals; informal reports, such as progress and status reports, trip reports, and meeting minutes; formal reports, including informational, analytical, and recommendations reports; instructions and manuals; Web sites; and oral presentations.

Now a Completely Integrated Book and Companion Web Site

A first among technical communication textbooks, the sixth edition has been thoroughly integrated with its companion Web site. TechComm Web (www.bedfordstmartins.com/techcomm) expands the book's resources in two ways: by providing additional materials for every chapter in the book and by directing students and instructors to the best Web resources available in technical communication. The following features help students use the book and the Web site together:

- *A chart on the inside front cover of the book*, "How to Use This Book and Its Companion Web Site," outlines for students the features that the book and the Web site share.
- *Cross-references in the margins of the book* refer students to the resources available on TechComm Web and on other useful Web sites.
- *The chapter-by-chapter organization on TechComm Web* matches the book's chapters, allowing students to quickly find the information that they need on the Web site.
- *Flashcards and Online Quizzes on TechComm Web* allow students to reinforce and test their understanding of the chapters; these features provide feedback that refers students to the textbook for additional review.

New Design

The sixth edition has been completely redesigned to make the book easier for instructors and students to use. With clean, open pages, the book itself now better illustrates the principles of good technical communication.

- *A new full-color design* uses color for a purpose: to convey information clearly and easily and to show realistic design elements in screen shots and color print documents.
- *Guidelines boxes throughout the book* summarize crucial information that students need to create effective documents. For quick reference, an index of all of the guidelines appears in the back of the book.
- *A new, realistic treatment of sample documents* results in even better models of good design principles.
- *Icons and annotations in the margins* direct students to discussions of related topics in the book, on TechComm Web, and on additional Web sites.
- *A redesigned and reorganized appendix, "Reference Handbook,"* makes editing and documentation advice easier for students to find and use.

New Chapters

Three new chapters have been added to the sixth edition to help students focus on the writing process and rhetorical concerns particular to technical communication.

- *Chapter 3, "Understanding the Writing Process."* Structured according to the stages of the writing process — planning, drafting, and revising — this chapter pays special attention to the role of electronic tools and usability testing. The chapter includes a discussion of the effective use of electronic drafting and revision tools — templates, styles, automated hypertext linking, Web-conversion tools, spell checkers, grammar checkers, and thesauri — explaining their strengths and limitations.

- *Chapter 6, "Communicating Persuasively."* The heart of technical communication is still rhetoric. This new chapter provides a clear and practical introduction to the rhetorical and critical skills students need, focusing on ways to identify an audience's broader goals and craft an effective argument. It also discusses ten common logical fallacies, the role of graphics and design in presenting arguments, and the crucial link between persuasion and ethics. Finally, the chapter includes an analysis of several persuasive arguments, from both print and online sources.

- *Chapter 21, "Creating Web Sites."* The Web is one of the most important media by which organizations communicate with the public, and it is likely to remain so for some time. Students need a basic understanding of the process of creating Web pages and sites. This chapter focuses on the logic of the process — planning and creating the content, and revising, testing, and maintaining the site — as well as the rhetoric of Web pages: designing the text and graphics to meet the needs of the audience and fulfill the writer's purpose. The chapter also discusses designing pages and sites for people with disabilities and for multicultural readers, as well as several important ethical and legal issues. The chapter closes with an analysis of several Web pages.

Expanded Coverage

The book's coverage has been expanded in the following key areas to reflect a wider range of technical documents and new methods of developing them.

- *New clusters of sample documents.* Many of the chapters now conclude by presenting sample documents — both print and Web-based — that demonstrate the principles discussed in the text. These samples and their annotations help students see the principles in action.

- *Research methods.* Chapter 7, "Researching Your Subject," includes a greatly expanded discussion of using the Web as a research tool. With screen shots of actual Web searches, the chapter demonstrates the effective use of search engines and Web-based library catalogs. In addition, it provides new, detailed advice about evaluating Internet sources.

- *Documentation styles.* Included in the appendix for easy reference, the section "Documenting Sources" now covers CBE documentation style in addition to APA and MLA styles. New models for citing electronic sources have been added for all three styles. In addition, the section includes advice on citing sources correctly in order to avoid plagiarism.

- *Ethics and legal considerations.* Chapter 2, "Understanding Ethical and Legal Considerations," now discusses principles that technical writers should follow to ensure that their documents are ethical and abide by relevant laws.

- *Collaboration.* Chapter 4, "Writing Collaboratively," now provides guidelines for conducting meetings effectively to improve the process of collaboration. In addition, the chapter discusses how to use groupware features included in word processors, such as the comment, revision, and highlighting features.

- *Oral presentations.* Chapter 22, "Making Oral Presentations," devotes additional attention to presentation-graphics software, showing students how to use slides to help listeners understand the organization of a presentation, how to use speaker's notes, and how to create handouts.

A Complete Ancillary Package

The ancillary package that accompanies the sixth edition has been expanded to provide a wealth of resources for both new and experienced instructors.

- TechComm Web (www.bedfordstmartins.com/techcomm). In addition to student resources, the book's companion site offers the best Web resources available for instructors in technical communication. Instructors can find everything from password-protected reading quizzes to download and distribute to students, to sample syllabi, in-class activities, and PowerPoint slides that can be adapted for classroom use.

- *Instructor's Resource Manual.* Expanded and thoroughly revised, the resource manual now includes sample course schedules, teaching tips, chapter-by-chapter summaries, and classroom activities; commentaries on the writing and revision exercises in the text; and articles from professional journals on defining technical communication, creating portfolios, evaluating Web sites, and working collaboratively on a research project.

- "Making the Transition from Composition to Technical Communication." A new essay — downloadable from TechComm Web and available in the *Instructor's Resource Manual* — helps TAs and adjuncts apply their knowledge of the composition course to teaching technical communication.

- *Transparency Masters.* The graphics from the text are provided in a convenient form that can be adapted for classroom use. The electronic files of the transparencies may also be downloaded from TechComm Web.

Acknowledgments

All the examples in the book — from single sentences to complete documents — are real. Some were written by my students at Boise State University. Some were written by engineers, scientists, health-care providers, and businesspersons with whom I have worked as a consultant for over twenty-five years. Because much of the information in these documents is proprietary, I have silently changed brand names and other identifying information. I thank these dozens of individuals — students and professionals alike — who have graciously allowed me to reprint their writing. They have been my best teachers.

The sixth edition of *Technical Communication* has benefited greatly from the perceptive observations and helpful suggestions of my fellow instructors throughout the country. Some completed extensive questionnaires about the previous edition; others reviewed the current edition in its draft form. I thank David Berg-Seiter, University of Florida; William Bowers, University of Florida; Erika Derany, University of Florida; Gene Doty, University of Missouri — Rolla; Lise Esch, Trident Technical College; Andrew Flood, University of New Mexico; Sandra L. Friend, East Carolina University; Kristen Hague, University of New Mexico; Kim Kirkpatrick, St. Louis University; Jim Leonhirth, Florida Institute of Technology; Martha Levine, Southwest Missouri State; Martha Mangot, New York Institute of Technology; Kimberly McFetridge, Delaware Technical and Community College; Josephine Jordan Mills, University of Denver; Marriott Nielsen, University of Maryland, Baltimore County; Jennifer Panek, Massachusetts Institute of Technology; Celia Patterson, Pittsburg State University; Tamara Powell, Louisiana Technical University; Cindy Raisor, Texas A&M University; Susan Rode-Perkins, Washington University; Jeffrey A. Schwarz, St. Louis University; Stuart A. Selber, Pennsylvania State University; Carol M. Shehadeh, Florida Institute of Technology; Blake Spence, Bob Jones University; Gregory J. Stratman, University of Missouri — Rolla; Thomas L. Warren, Oklahoma State University; Jacqueline Whipple Walker, University of Florida; and James H. Wilson, Santa Fe Community College.

I also thank the following instructors who contributed their insights and suggestions for the *Instructor's Resource Manual:* Josie Jordon Mills, University of Denver; Christine Mitchell, Southeastern Louisiana University; Tamara Powell, Louisiana Technical University; Cindy Raisor, Texas A&M University; Stuart Selber, Pennsylvania State University; Carol M. H. Shehadeh, Florida Institute of Technology.

I would like to acknowledge two other readers. Kevin S. Wilson of Boise State University, a gifted teacher and editor, has helped me clarify my thinking and simplify my writing in this edition, as he did in previous editions. In addition, he wrote most of the quizzes and classroom activities included on TechComm Web and in the *Instructor's Resource Manual*. The book and site are much improved due to his efforts, and I thank him. John Battalio, also of Boise State University, critiqued the new chapter on Web design, helping me make it clearer and more accurate. In my hallway conversations with John, I learn something new about tech comm and about teaching almost every day. He is an extraordinary colleague.

I have been fortunate, too, to work with a superb team at Bedford/ St. Martin's, led by Leasa Burton, a perceptive, demanding, and supportive editor who has helped me improve the text in many big and small ways. She is responsible for the book's sharper and clearer focus on rhetoric. Anna George and Ara Salibian deserve special thanks as well for their work: Anna, for the much improved design of the text, and Ara, for expertly guiding the manuscript through production. I also want to express my appreciation to Chuck Christensen, Joan Feinberg, Elizabeth Schaaf, and John Amburg for assembling the first-class team that has worked so hard on this edition, including Sara Eaton, Jen Lesar, Denise Wydra, Arthur Johnson, and Ellen Thibault. For me, Bedford/St. Martin's continues to exemplify the highest standards of professionalism in publishing. They have been endlessly encouraging and helpful. I hope they realize the value of their contributions to this book.

I want to thank my colleagues at Boise State University — John Battalio, Jim Frost, Michael Hassett, Rick Leahy, and Theresa Hollenbeck — whose ideas and suggestions have helped me improve the text.

My greatest debt, however, is to my wife, Rita, who over the course of many months and, now, six editions, has helped me say what I mean.

A Final Word

I am more aware than ever before of how much I learn from my students, my fellow instructors, and my colleagues in industry and academia. If you have comments or suggestions for making this a better book, please get in touch with me at the Department of English at Boise State University, Boise, ID 83725. My phone number is (208) 426-3088, or you can send me an email from the Web site: www.bedfordstmartins.com/techcomm. I hope to hear from you.

Mike Markel

BRIEF CONTENTS

CONTENTS

PART THREE
DEVELOPING THE TEXTUAL ELEMENTS   *217*

9. Drafting and Revising Definitions and Descriptions   *219*

PART ONE

THE TECHNICAL COMMUNICATION ENVIRONMENT

Introduction to Technical Communication

1

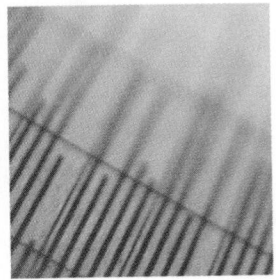

Richard C. Levine, an engineering manager with Bell Northern Research, on his employees' ability to write clearly:

When I see a report or memo that has repeated errors I immediately question the ability and dedication of the person who wrote it. Why didn't they take the time to do it right? Most of the successful engineers I know write clear, well-organized memos and reports. Engineers who can't write well are definitely held back from career advancement. (Qtd. in Beer & McMurrey, 1997, pp. 4–5)

WHAT IS TECHNICAL COMMUNICATION?

See TechComm Web (www.bedfordstmartins.com /techcomm) for additional examples and links related to topics in this chapter.

Producing technical communication involves creating, designing, and transmitting technical information so that people can understand it easily and use it safely, effectively, and efficiently. Most technical communication is produced by people working in or for organizations.

Much of what you read every day — textbooks, phone books, procedures manuals at the office, environmental impact statements, journal articles, Web sites, the owner's manual for your car — is technical communication. The words and the graphics in these documents are meant to be practical. That is, they communicate information to help an audience understand a subject or carry out a task. For example, an introductory biology text helps students understand the fundamentals of plant and animal biology and perform basic experiments. A user's manual for a software program describes how to use the program effectively.

WHO PRODUCES TECHNICAL COMMUNICATION?

Most technical communication is produced by one of two different categories of people:

- *Technical professionals.* Engineers, scientists, businesspeople, and other technically trained individuals do a lot of writing. According to one survey, technical professionals devote at least one-fifth of their time to writing (Anderson, 1985). Engineers can spend as much as 40 percent of their time writing (Beer & McMurrey, 1997). And the percentage of time that technical professionals spend in communicating increases as they advance: supervisors spend 40 percent of their time reading and writing; managers, 50 percent (MIT, 1984). A technical professional is a writer.

- *Technical communicators.* The job of a technical communicator is to create documents, including manuals, proposals, reports, sales literature, Web sites, letters, journal articles, and speeches. Many technical communicators call themselves technical writers (or tech writers) even though the term *technical communicator* more accurately reflects the increasing importance of graphics and the use of other media, such as online documentation. Other terms you will see include *learning product developer, information developer, documentation writer, information architect,* and *information engineer.* In small organizations, one technical communicator might be responsible for all aspects of producing the document, from drafting to production. In large organizations, a technical communicator might work full-time as a writer, editor, graphic artist, designer, Webmaster, or production specialist.

 For more information about technical communicators, see the Society for Technical Communication site (www.stc.org) and the Institute of Scientific and Technical Communicators site (www.istc.org.uk).

Although most technical communicators are employees within organizations, many work as independent contractors — sometimes called *freelancers* — who focus on individual projects that can last a few weeks, months, or years. These independent contractors can work at home, at the client's facility, or both. Still other technical communicators work for companies that offer communication services to other organizations.

THE ROLE OF TECHNICAL COMMUNICATION IN BUSINESS AND INDUSTRY

The working world depends on written communication. Within most modern organizations virtually every action is documented in writing, whether on paper or online. Here are a few examples:

a memo or email requesting information or communicating new policies

a travel report on a conference, a site inspection, or a training session

a set of instructions introducing and explaining a new process or procedure

a proposal intended to persuade management to carry out a project that will address a particular problem

a report to document a completed project

an oral presentation explaining a new policy to employees

In addition, every organization also communicates with other organizations and often with the public. Here are some examples of this kind of communication:

inquiry letters, sales letters, goodwill letters, and claim and adjustment letters to customers, clients, and suppliers

sales and marketing literature for distribution to potential customers

Web sites providing product information and marketing and soliciting job applications

research reports for external organizations

articles for trade and professional journals

TECHNICAL COMMUNICATION AND YOUR CAREER

See Ch. 16 for more about job-application materials.

The technical-communication course you are now taking was probably developed to meet the needs of the working world. In fact, your first step in obtaining a professional-level position is to write two technical documents: an application letter and a résumé. These documents will help an organization decide whether to interview you. At the interview, your oral communication skills will be evaluated along with your other qualifications.

Once you start work, you will write emails, memos, letters, marketing materials, and short reports. You might even be asked to contribute to larger projects, such as proposals and Web sites. During your first few months on the job, your supervisors will be looking at your communication skills as well as your technical abilities.

A survey of engineering managers found that nearly two-thirds thought technical communication "extremely important" as a management tool (Spretnak, 1982). In another survey, more than half of the respondents reported that an ability to write effectively was of critical importance or great importance in their careers (Anderson, 1985).

Job ads in newspapers and professional journals also suggest that the working world values good communication skills:

These job ads mention not only communication skills but also computer skills.

From an organization that manufactures medical instruments:

Design Assurance Engineer. Duties include performing electronic/mechanical product, component, and material qualifications. Requires spreadsheet/word-processing abilities, and excellent written/oral communication skills. BSEE or biology degree preferred.

From a corporation that supplies agricultural products:

Compensation Administrator. The successful candidate will assist in administering the companywide design, development, implementation, and monitoring of the compensation program. Minimum qualifications include 1–2 years of compensation analysis and salary administration experience; experience writing job descriptions, developing salary structures, conducting and participating in salary surveys; broad-based background in the general human resource function with specific knowledge of federal and state regulations affecting compensation systems; demonstrated ability to operate a personal computer and spreadsheet software. Microsoft Excel preferred.

From a corporation that manufactures custom-molded plastic components:
Custom injection molder needs a **process engineer.** Manufacturing knowledge and experience with statistical process control required. Experience with injection molding and ASQC certification a plus. Excellent writing and speaking skills required as well as word-processing and spreadsheet skills.

The steady growth in continuing-education courses in technical and business communication also reflects the demand for effective communicators. The average Fortune 1000 company is spending $300,000 annually on communication training for its employees (Kiggins, 1999). Organizations are paying to send their professionals back to the classroom to improve their technical-communication skills.

The facts of corporate life today are simple: if you can communicate well, you are valuable; if you cannot, you are much less so.

CHARACTERISTICS OF TECHNICAL COMMUNICATION

Technical communication has seven major characteristics.

Technical communication:
Addresses particular readers
Helps readers solve problems
Reflects an organization's goals and culture
Is produced collaboratively
Uses design to increase readability
Consists of words or graphics or both
Is produced using high-tech tools

Addresses Particular Readers

Perhaps the most significant characteristic of technical communication is that it addresses particular readers.

Sometimes you know the reader. For instance, if you are planning to write a proposal addressed to your supervisor, you will think about his or her job responsibilities and the level of detail he or she would be interested in reading as well as more personal factors such as history with the organization, attitudes toward your ideas, and so forth. These are the factors you will need to consider as you decide what kind of document to write, how to structure it, how much detail to include, and what sentence style and vocabulary to use. If you are writing to several different people whose backgrounds and needs vary, you will want to structure the document to make it easy for them to

See Ch. 5, p. 105, for more about addressing a particular audience.

locate and understand the information they seek. If some of your readers are nonnative speakers of English, you will adjust vocabulary and sentence structure to accommodate their needs.

Sometimes, such as when writing a brochure describing one of your company's products, you will not know the reader. In these situations, however, you can establish a profile of your audience. For example, if readers of this brochure are police officers responsible for capital purchases, you might not know their gender or age or any other personal information, but you do know that they share a police background and a common responsibility for approving capital expenditures.

Remember also that although you might direct your writing to particular readers, people you never intended as your audience might read it. Other workers at your organization — especially managers and executives — might see sensitive documents; so might members of the public or the press. You should always avoid writing anything that will embarrass you or your organization if someone for whom it is not intended sees it.

Helps Readers Solve Problems

Technical communication is not meant to express a writer's creativity or to entertain readers; it is intended to help readers learn or do something. For instance, you look at your college's catalog to find out the procedures for registration, to understand the requirements for a major, and to learn what is covered in a particular course. You read your company's employee-benefits manual to help you decide which benefits package you should select. You read them, in other words, because you need information to help you analyze a situation and solve a problem.

Reflects an Organization's Goals and Culture

Technical communication furthers an organization's goals. Consider, for example, a state government department that oversees programs in vocational education. The department submits an annual report to the state legislature describing the purpose of each program, explaining its successes and setbacks, and offering recommendations on how to make the program more effective next time. The department also produces a vast quantity of technical information for the public: flyers, brochures, pamphlets, radio and television advertisements, and course materials such as texts, workbooks, and Web pages. These forms of technical communication help the department secure its funding and reach its audience.

Technical communication also reflects an organization's culture. Some organizations have a rigid hierarchy and way of doing business. Employees are expected to write only to their immediate supervisors and to other employees on their own level, for example, and to use a particular kind of document, such as a memo, and to organize it and format it in a particular way. In other organizations, the culture permits or even encourages employees to communicate much more freely and creatively.

Is Produced Collaboratively

Although you will often work alone in writing email, memos, and letters, you will probably work as part of a team producing more complicated documents. Collaboration can take many forms, from having a colleague review your two-page memo to working with a team of a dozen technical professionals and technical communicators on a 200-page catalog.

See Ch. 4 for more about collaboration.

Collaboration is common in technical communication because no one person has all the information, skills, or time to create a large document. Writers, editors, designers, and production specialists work with subject-matter experts — the various technical professionals — to create a better document than any one of them could have made working alone.

Much of the time you spend on projects is devoted to communicating with your collaborators. Large projects often begin with brainstorming sessions, in which the participants contribute ideas about the content, organization, and style of the document and try to reach consensus. Team members often go off to work on different portions of the project, meeting only periodically for briefings. Team members write progress reports for managers and, at the end of a project, a completion report.

Because technical communication so often involves collaboration, interpersonal skills are essential. You have to be able to listen to people with other views, express yourself clearly and diplomatically, and compromise.

Uses Design to Increase Readability

Technical communicators use design features — typography, spacing, color, special paper, and so forth — to make their documents more effective. Design features have three basic purposes:

See Ch. 13 for more about design.

- *To make the document look attractive and professional.* The reader is more likely to read it and to form a positive impression of your ideas and of you. You are therefore more likely to accomplish your purpose.

- *To help the reader navigate the document.* Because a technical document can be long and complicated, and because most readers want to read only parts of it, design features such as headings, color, or highlighting help readers see where they are and get where they want to be. For Web sites, effective navigation is especially important.

- *To help the reader understand the document.* If, for instance, all first-level headings have one design and all second-level headings another, the reader will recognize this pattern and be better able to follow the main points and subpoints of a discussion. Similarly, if safety warnings appear in a color and size different from the rest of the text, readers will be better able to recognize the importance of the information.

Consists of Words or Graphics or Both

See Ch. 14 for more about graphics.

Most technical documents include words and graphics. Some consist solely of one or the other, but most benefit from incorporating both. In technical communication, graphics help the writer perform five main functions:

- make the document more interesting and appealing to readers
- communicate and reinforce difficult concepts
- communicate instructions and descriptions of objects and processes
- communicate large amounts of quantifiable data
- communicate with nonnative speakers of English

Is Produced Using High-Tech Tools

Every phase of the production of technical documents involves high-tech tools, such as the personal computer. Everyone uses word-processing software, and most technical communicators also rely on graphics software and desktop-publishing software.

Technology has transformed both the processes we use to produce technical communication and the ways we publish it. The growth of the Web is one obvious example. And CD-ROMs and DVDs, with their massive storage

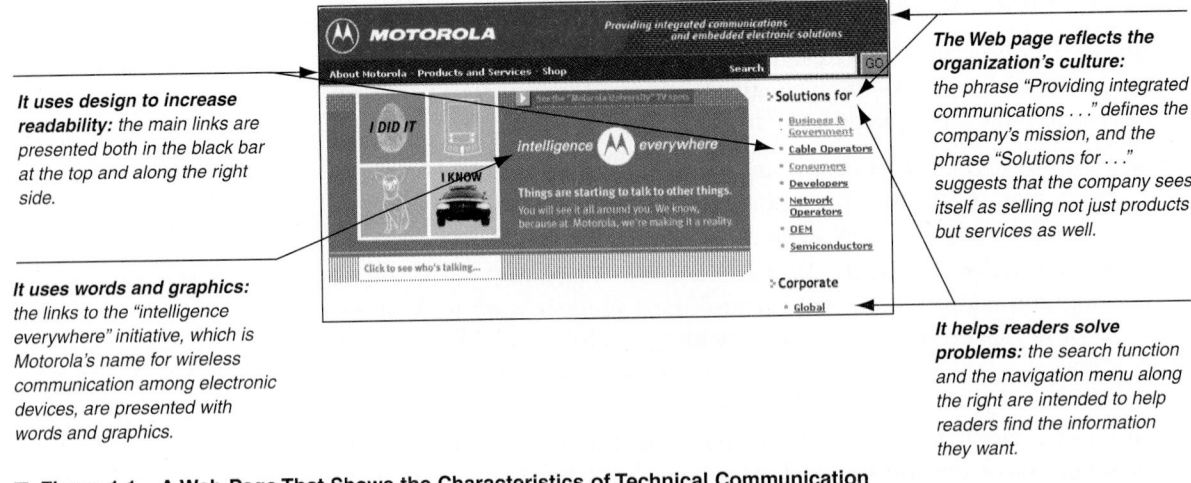

It uses design to increase readability: the main links are presented both in the black bar at the top and along the right side.

It uses words and graphics: the links to the "intelligence everywhere" initiative, which is Motorola's name for wireless communication among electronic devices, are presented with words and graphics.

The Web page reflects the organization's culture: the phrase "Providing integrated communications . . ." defines the company's mission, and the phrase "Solutions for . . ." suggests that the company sees itself as selling not just products but services as well.

It helps readers solve problems: the search function and the navigation menu along the right are intended to help readers find the information they want.

■ **Figure 1.1   A Web Page That Shows the Characteristics of Technical Communication**

*An excellent example of technical communication, this excerpt from a Web page is **addressed to particular readers:** Motorola customers. Like most technical documents, it was **produced collaboratively** (by technical communicators, graphic artists, Web authors, and others), **using high-tech tools,** including a word processor, a graphics program, and a Web editor.*

capabilities, have made it practical to deliver large quantities of information — including text, video, sound, and animation — inexpensively.

As information technology becomes more powerful, easier to use, and less expensive, technical communicators and technical professionals alike continuously upgrade their skills. We are all lifelong learners now.

A LOOK AT SAMPLE DOCUMENTS

The home page from the Motorola Web site (2001) shown in Figure 1.1 is an excellent example of technical communication. It illustrates some of the characteristics of technical communication discussed in this chapter. Figure 1.2 on page 12, from a user's manual for a computer operating system (Microsoft, 1998, p. 47), also illustrates some of the major characteristics of technical communication discussed in this section.

To view Figure 1.1 in context on the Web, follow the links in Chapter 1 on TechComm Web (www.bedfordstmartins .com/techcomm).

Not every example of technical communication is as sophisticated as the Web page in Figure 1.1 or the manual excerpt in Figure 1.2. Figure 1.3 on page 13, an executive summary from a report, also illustrates effective technical communication. An executive summary is addressed to senior managers and focuses on the problem from the executive's perspective.

The writers of this report work for a manufacturer of microchips. A serious problem in the production of microchips is electrostatic discharge (ESD), a static electric charge that can be transmitted by humans to the microchips they are manufacturing. Even a small amount of ESD can damage microchips.

MEASURES OF EXCELLENCE IN TECHNICAL COMMUNICATION

Although every technical document is different, eight measures of excellence characterize all technical communication.

Measures of excellence in technical communication:
Honesty
Clarity
Accuracy
Comprehensiveness
Accessibility
Conciseness
Professional appearance
Correctness

Clear, distinct headings

Marginal glosses

Numbered steps

A screen image

Plenty of white space

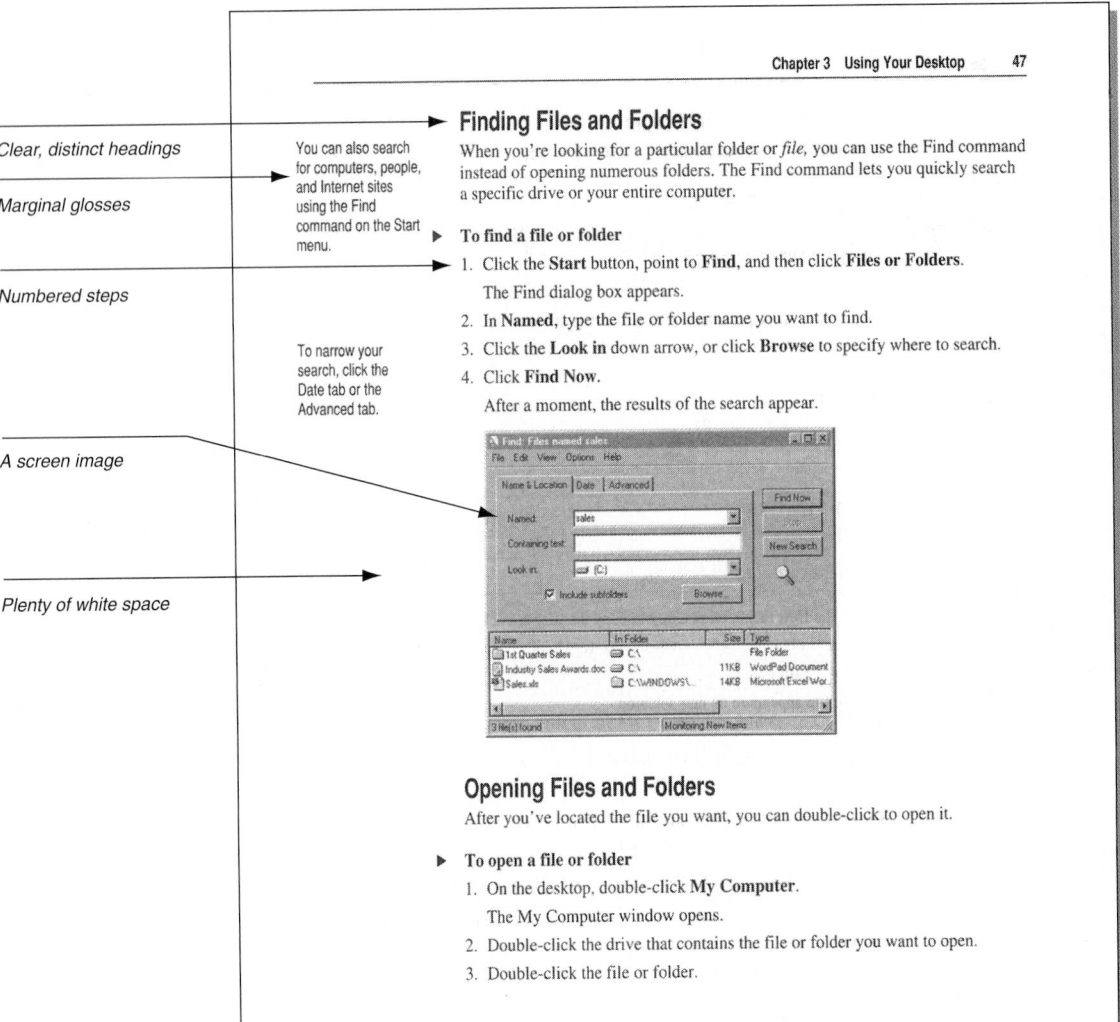

Finding Files and Folders

You can also search for computers, people, and Internet sites using the Find command on the Start menu.

When you're looking for a particular folder or *file,* you can use the Find command instead of opening numerous folders. The Find command lets you quickly search a specific drive or your entire computer.

▶ **To find a file or folder**

1. Click the **Start** button, point to **Find,** and then click **Files or Folders.**

   The Find dialog box appears.

2. In **Named,** type the file or folder name you want to find.

3. Click the **Look in** down arrow, or click **Browse** to specify where to search.

4. Click **Find Now.**

To narrow your search, click the Date tab or the Advanced tab.

   After a moment, the results of the search appear.

Opening Files and Folders

After you've located the file you want, you can double-click to open it.

▶ **To open a file or folder**

1. On the desktop, double-click **My Computer.**

   The My Computer window opens.

2. Double-click the drive that contains the file or folder you want to open.

3. Double-click the file or folder.

■ **Figure 1.2   A Page from a Manual That Shows the Characteristics of Technical Communication**

*This page is **addressed to particular readers** — new users of Windows 98 — to help them learn how to find files and folders. The manual from which this page is taken was **produced collaboratively** by a number of specialists **using high-tech tools,** such as screen-capture software and layout software. A screen image orients readers, while the design of the page, with clear headings, numbered steps, and marginal glosses, helps readers learn the task.*

EXECUTIVE SUMMARY — *should be sans serif* *(handwritten annotation)*

needs headings (handwritten annotation)

Currently, Quantech Technology Inc. uses electrostatic discharge (ESD) booties to minimize the hazards of ESD in its production facilities. Although the booties are as effective as any other technology for minimizing ESD damage, they pose a safety hazard because of poor traction: workers wearing them can slip and fall. Last year, worldwide we had 27 cases of slipping accidents, costing over $300,000 in worker's compensation (Quantech, 1999). A second problem is the cost of replacing worn booties; last year Quantech spent over $4 million on booties.

Because of these two problems, management authorized a study to determine whether Quantech should continue to use ESD booties or switch to another technology. We studied the three available technologies — booties, heel/toe straps, and shoes — according to seven criteria: ESD efficiency, safety, cost, comfort, convenience, training, and stock/availability.

This report documents our research. Our conclusion is that ESD shoes are the best alternative for Quantech. The shoes are as effective as booties and heel/toe straps in minimizing ESD damage, and they provide considerably better traction than booties. Our estimate is that the number of slipping accidents can be reduced by at least two-thirds. In addition, although shoes are more expensive than booties, they have much greater durability. We estimate that switching to shoes will reduce the $4 million outlay by half. In short, we predict an annual savings of $2.2 million by switching from booties to shoes.

We recommend that a further study of the leading ESD shoe manufacturers be conducted to determine which shoe — or choice of shoes — is most appropriate for Quantech employees. If you wish, we would be happy to participate in such a study.

The writers present the problem clearly, in terms of safety and economic loss.

The writers briefly review their methods.

The writers present their conclusions, and then, in the final paragraph, their recommendations.

In the last sentence, the writers offer to work on the next phase of the project if the executive wishes. This polite offer reflects the organization's culture: the writers offer to participate, but leave the decision to the executive.

■ **Figure 1.3   An Executive Summary That Shows Some of the Characteristics of Technical Communication**

This document is low-tech. It contains no graphics and no special design features, and it was produced on a word processor. However, it is effective because it meets the needs of its audience. The executives reading this summary want only a quick overview. They want to know the problems that motivated the study as well as its major findings: conclusions and recommendations.

For more about executive summaries, see Ch. 12, p. 316.

Honesty

See Ch. 2 for more about the ethical and legal aspects of technical communication.

The most important measure of excellence in technical communication is honesty. You have to tell the truth and not mislead the reader.

There are three reasons to be honest as a communicator:

- *It is the right thing to do.* Technical communication is not about using words and pictures to mislead or lie to people; it is about helping people understand how to make wise choices as they use the information available to them in a high-tech culture.
- *Readers can get hurt if you are dishonest.* Misinforming your readers or deliberately omitting important information can have serious consequences. If you lie or mislead in writing about medication dosages, for example, people could become ill or die. If you fail to warn readers about the safety hazards involved in operating a piece of equipment, they could be injured or killed.
- *You and your organization could face serious legal charges if you are dishonest.* If a plaintiff can convince a judge or jury that your document failed to provide honest, appropriate information and that this failure led to a substantial injury or loss, you and your organization are likely to have to pay millions of dollars.

Clarity

Your goal is to produce a document that conveys a single meaning the reader can understand easily. The following directive, written by the British navy (*Technical Communication*, 1990), is an example of what you don't want to do:

> It is necessary for technical reasons that these warheads should be stored upside down, that is, with the top at the bottom and the bottom at the top. In order that there may be no doubt as to which is the top and which is the bottom, for storage purposes, it will be seen that the bottom of each warhead has been labeled with the word TOP.

Technical communication must be clear for two reasons:

- *Unclear technical communication can be dangerous.* A carelessly drafted building code, for example, could tempt contractors to save money by using inferior materials or techniques. More dramatically, the 1986 space-shuttle tragedy might have been prevented if the officials responsible for deciding whether to launch had received clear reports of the safety risks involved that day.
- *Unclear technical communication is expensive.* The average cost of a telephone call to a customer-support center is about $20 (Scholz, 1996). Clear technical communication in the documentation — the instructions that come with the product — can greatly reduce the number of these calls.

Accuracy

Inaccurate writing can cause as many problems as unclear writing. Accuracy seems a simple concept: you must record the facts carefully. If you meant to write *4,000,* double-check your work to make sure that you didn't write *40,000.* The slightest inaccuracy will, at the least, confuse and annoy your readers. A major inaccuracy can be dangerous and expensive.

In another sense, however, accuracy is really a question of ethics. Technical communication must be as objective and free of bias as you can make it. If readers suspect that you are slanting information — by overstating the significance of a particular fact or by omitting important facts — they will doubt the validity of the entire document. Technical communication is effective by virtue of its clarity and organization, but it must also be reasonable, fair, and truthful.

Comprehensiveness

A good technical document provides all the information readers need. It describes the background so that readers who are unfamiliar with the subject can understand it. It contains sufficient detail so that readers can follow the discussion and carry out any required tasks. It refers to supporting materials clearly or includes them as attachments.

Comprehensiveness is crucial because the people who will act on the document need a complete, self-contained discussion in order to use the information safely, effectively, and efficiently. A document also often serves as the official company record of a project, from its inception to its completion.

Here are two examples demonstrating the importance of comprehensiveness:

> A scientific article reporting on an experiment that compares the reaction of a new strain of bacterium to two different compounds will not be considered for publication unless the writer has fully described the methods used in the experiment. Because other scientists should be able to replicate the researcher's methods, every detail must be provided, including the names of the companies from which the researcher obtained all the materials.

> A report recommending that a company open a plant in a new location will be analyzed before management commits itself to such an expensive and important project. The team studying the report will need all the details. If the recommendations are implemented, the company will want a single, complete record in case changes have to be made several months or several years later.

Accessibility

Accessibility refers to the ease with which readers can locate the information they seek. Most technical documents are made up of small, independent sections. Some readers are interested in only one or two sections; others might want to read more. Because few people will pick up a document and read from the first page all the way through, your job is to make the various parts

See Chs. 10 and 13 for more about making documents accessible.

of the document accessible. That is, readers should not be forced to flip through the pages to find the appropriate section. For example, a Web page should include a site map and a clear set of navigation links to help readers understand where they are.

Conciseness

For more about writing concisely, see Ch. 11, p. 291.

To be useful, technical communication must be concise. A longer document is more difficult to use because it takes more of the reader's time. In a sense, conciseness actually works against clarity and comprehensiveness; for a technical explanation to be absolutely clear, it must describe every aspect of the subject in great detail. To balance the conflicting demands of clarity, conciseness, and comprehensiveness, you must make the document just long enough to be clear — given the audience, purpose, and subject — but not a word longer. You can shorten most writing by 10 to 20 percent simply by eliminating unnecessary phrases, choosing short words rather than long ones, and using economical grammatical forms.

The battle for concise writing, however, is often more a matter of psychology than of grammar. Some writers produce long documents to show their readers that they are trying hard. But if the document needs to be short to fulfill its purpose, a writer's job is to figure out how to convey a lot of information in a small space.

Professional Appearance

You start to communicate before anyone reads the first word of the document. If the document looks neat and professional, readers will form a positive impression of both the content and the authors.

Your documents should adhere to the format standards that apply in your organization or your professional field. In addition, your documents should be well designed and neatly printed. For example, a letter should follow one of the traditional letter formats and have generous margins. It should be balanced both vertically and horizontally on the page.

Correctness

Good technical communication observes the conventions of grammar, punctuation, spelling, and usage.

Many of the rules of correctness are clearly important. If you mean to write "The three inspectors — Bill, May, and I — attended the session" but you use commas instead of dashes, your readers might think six people (not three) attended. If you write a sentence with a dangling modifier, such as, "While feeding on the algae, the researchers captured the fish," some readers will have trouble following you — as they puzzle over the image of researchers feeding on algae.

Most of the rules, however, make a difference primarily because readers will judge your thinking on how your writing looks and sounds. Carelessness and grammar errors make your readers doubt the accuracy of your information, or at least lose their concentration. You will still be communicating, but the message won't be the one you had intended. As a result, the document will not achieve its purpose, and readers could well judge you incompetent and unprofessional.

Technical communication is meant to fulfill a mission: to convey information to a particular audience so that they understand something or can carry out a task. To accomplish these goals, it must be honest, clear, accurate, comprehensive, accessible, concise, professional in appearance, and correct.

The rest of this book describes ways to help you say what you want to say.

Exercises

Some of the following exercises ask you to write a memo. See Chapter 15, page 430, for a discussion of memos.

1. Evaluate the following letter from a dentist to all his patients. In a memo to your instructor, explain the strengths and weaknesses of the letter by referring to the measures of excellence described in this chapter.

May 5, 1999

Dear Friends & Patients;

Firstly, I want to take the opportunity to extend my most profound thanks to all of you for your support of my dental practice over these last 10 years. Serving you has been a privilege and I have always endeavored to provide excellent dentistry in an environment where you can feel comfortable.

With increasing busyness in the practice, I have reached a point where I must clarify my goals and define the direction I will be taking in the future. When I started almost 10 years ago, I made a commitment to a practice of excellence, with a special emphasis on quality not quantity. In other words, I wanted to take whatever time was needed to do outstanding dentistry, using the best in materials and techniques, and to put my patients at ease in a non-hurried, congenial environment.

As you may be aware, dentistry, as in other fields of health care, is becoming more and more subject to "management" and outside control. Managed care ensures you receive basic or "average" treatment for a negotiated or reduced fee. It therefore requires that dentists mold their dentistry and their style of practice to a norm dictated by the insurance industry. Unfortunately, this poses a problem for an increasing number of dentists. How can we continue to be a technically progressive practice and improve service to our patients while at the same time managed care is pushing us backwards? As already indicated, I am dedicated to continually improving my level of quality — not remaining stagnant.

With this in mind, I have made the choice to no longer be a "preferred provider" with Phoenix Dental, Blue Cross and

Blue Shield as to do so would require I limit our standard of care to their arbitrary standard of payment. I can no longer do that in good conscience. To remain under their control would mean a continually increasing number of patients I see each day, which translates into longer waits for you, less personal attention and less time to spend on each procedure I perform. Most importantly, decreasing the time I spend with a patient directly effects the quality I can provide. I simply am not going to make that compromise.

How will this effect you? Phoenix Dental, Blue Cross and Blue Shield will still cover a portion of your dental treatment with me as they always have and I will continue to bill your insurance as I always have. None of that will change. In basic services (exams, cleanings) there should be little if any difference in the amounts that will be covered. In others areas of service there may well be some difference. Rest assured, if this is the case, I will work with you in order to obtain the maximum benefits to which you are entitled. Remember, <u>you paid the premium, you are entitled to the same benefit regardless of whether or not we contract</u>.

In the meantime I will remain true to my original goal: to do my very best in every aspect of my service to you and provide great dentistry, not average. I feel I can best do that by not allowing insurance to influence my decision making process and my dentistry. If at any time you feel less than satisfied, please let me know personally. Should you have any additional concerns or questions regarding insurance, feel free to ask at any time.

Sincerly,
Alan Remington, D.D.S.

2. Form small groups to study the following Web page from Canon Corporation (1999). The page describes the company's "Clean Earth Campaign." Meet to discuss which characteristics of technical communication you see in this excerpt. How effective is this excerpt? What changes would you make to improve it? Present your ideas in a memo to your instructor.

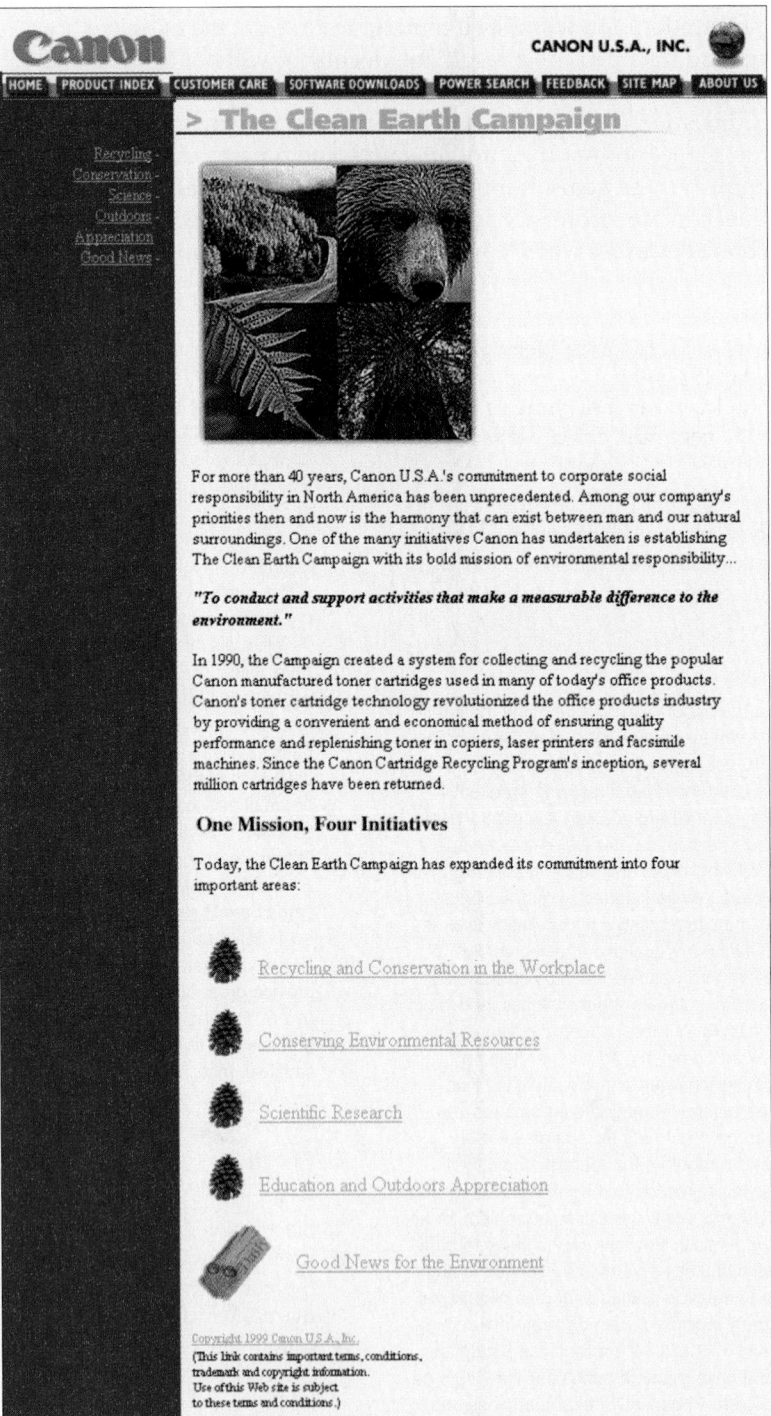

Canon CANON U.S.A., INC.

HOME | PRODUCT INDEX | CUSTOMER CARE | SOFTWARE DOWNLOADS | POWER SEARCH | FEEDBACK | SITE MAP | ABOUT US

> **The Clean Earth Campaign**

Recycling -
Conservation -
Science -
Outdoors -
Appreciation
Good News -

For more than 40 years, Canon U.S.A.'s commitment to corporate social responsibility in North America has been unprecedented. Among our company's priorities then and now is the harmony that can exist between man and our natural surroundings. One of the many initiatives Canon has undertaken is establishing The Clean Earth Campaign with its bold mission of environmental responsibility...

"To conduct and support activities that make a measurable difference to the environment."

In 1990, the Campaign created a system for collecting and recycling the popular Canon manufactured toner cartridges used in many of today's office products. Canon's toner cartridge technology revolutionized the office products industry by providing a convenient and economical method of ensuring quality performance and replenishing toner in copiers, laser printers and facsimile machines. Since the Canon Cartridge Recycling Program's inception, several million cartridges have been returned.

One Mission, Four Initiatives

Today, the Clean Earth Campaign has expanded its commitment into four important areas:

Recycling and Conservation in the Workplace

Conserving Environmental Resources

Scientific Research

Education and Outdoors Appreciation

Good News for the Environment

Copyright 1999 Canon U.S.A., Inc.
(This link contains important terms, conditions, trademark and copyright information.
Use of this Web site is subject to these terms and conditions.)

Research Projects

3. Locate an owner's manual for a consumer product, such as a coffeemaker, bicycle, or hair dryer. In a memo to your instructor, describe and evaluate the manual. To what extent does it meet the measures of excellence discussed in this chapter? In what way does it fall short of these measures? Submit a photocopy of the document (or a representative portion of it) with your memo.

4. Study the job ads in your field from a large newspaper, Web site, or professional journal. To what extent do they suggest that the employers value technical-communication skills? What percentage of the ads explicitly mention communication skills? What kinds of skills are mentioned most often? Are the ads themselves examples of effective technical communication? Write your response in a memo to your instructor.

5. Interview a professional in your major field. What communication skills does this professional use most often? How important are they in the professional's career? What advice does this professional offer you about how to improve your own communication skills? Present your results in a memo to your instructor. See Chapter 7, page 182, for a discussion of interviewing.

6. Locate an example of technical communication on the Web. Describe the aspects of the document that illustrate the characteristics of technical communication discussed in this chapter. Then evaluate the effectiveness of the document using the strategy described in Exercise 1. Write your response in a memo to your instructor. Submit a printout of the document (or a representative portion of it) with your assignment.

CASE
Teaching Technical Communication across Campus

Form groups according to your major. The faculty in your major department have decided to do a thorough review of its curriculum, and they wish to involve students. The department chair has asked your group to determine the extent to which the courses in your major prepare you to communicate effectively. Here are some questions you might focus on:

- What do the professors in the department think is the best approach to helping students learn how to communicate? How do they incorporate technical communication into their courses?

- To what extent do the course descriptions in your department describe the communication skills that are part of a professional's set of skills?

- To what extent do the syllabi for required courses describe the class activities and assignments that call for technical-communication skills?

- What approach is used at other, similar colleges and universities?

Write a memo to the chair of your major department describing and evaluating your department's attempt to teach technical communication. What aspects do you think are effective? How might the department do more to teach students the importance of technical communication to a professional career?

For additional cases, click on Case of the Month and Archive on TechComm Web (www.bedfordstmartins.com/techcomm).

2

Understanding Ethical and Legal Considerations

Brent Gardner, assistant general counsel for Hewlett-Packard Company in Boise, Idaho, credits the company's ethical standards with its performance in the marketplace:

The founders of HP set an expectation that employees act with the highest standards of ethics. They felt it was good business: if you act ethically with your customers there will be a bond of trust established, and they will come back.

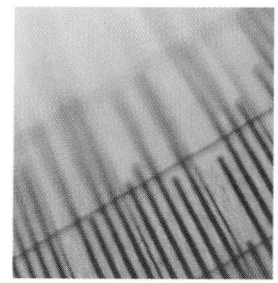

As a technical professional or technical communicator, you need a basic understanding of ethical principles, if only because you are likely to find yourself confronting ethical dilemmas on the job. For instance, you might be asked to write a document in such a way that it could mislead the reader. Should you go along, out of loyalty to your company, or resist because misleading readers might cause harm?

 See TechComm Web (www .bedfordstmartins.com /techcomm) for guidelines boxes, additional examples, and links related to topics in this chapter.

Technical communicators and technical professionals alike also need a basic understanding of several areas of the law related to communication. One such area is liability. In recent years, courts have ruled that instruction manuals, like the products they accompany, can be defective. Companies are now being held legally responsible when their instruction manuals "fail to instruct or warn" customers about the dangers of using the product. In addition to liability, intellectual-property law, which deals with issues of the ownership of information, is becoming extremely important, and the growth of the Web has only complicated what was already a complex legal problem.

Ethical and legal pitfalls can lurk within the words and graphics of many kinds of documents. Here are a few examples:

Proposals. You might be asked to exaggerate or lie about your organization's past accomplishments, pad the résumés of the project personnel, list as project personnel some workers who in fact will not be contributing to the project, or present an unrealistically short work schedule.

Progress reports. You might be asked to describe the project as proceeding smoothly, even though it isn't.

Completion reports. You might be asked to change or leave out data that are inconsistent with the findings presented in the report.

Product information. You might be asked to exaggerate the quality or operating characteristics of products shown in catalogs or manuals or to downplay the hazards of using these products.

Web sites. You might be asked to copy the source code of a competitor's site and then make minor changes so that you are not technically violating the competitor's copyright.

Graphics. You might be asked to hide an item's undesirable features in a photograph by manipulating the photograph electronically. Or you might be tempted to leave out some data in a graph, or even make up data.

This chapter briefly introduces the general principles of ethics and explains the standards commonly followed in deciding ethical questions. It goes on to discuss legal considerations for communicators: copyrights, trademarks, contract law, and liability. Next, it turns to codes of conduct, whistleblowing, and multicultural communication. Finally, it provides a set of principles for communicating technical information ethically.

A BRIEF INTRODUCTION TO ETHICS

Some people think that if an act is legal, it is ethical. Most people, however, believe that ethical standards are more demanding than legal standards. It is legal, for example, to try to sell an expensive life-insurance policy to an impoverished elderly person who has no dependents and therefore no need for such a policy. Yet many people would consider such an attempt unethical.

For many people, ethics is a matter of intuition, what their gut feelings tell them about the rightness or wrongness of an act. Others see ethics in terms of their own religious upbringing or the Golden Rule. Although philosophers cannot agree on what constitutes ethical conduct, most would agree to the following definition: *Ethics is the study of the principles of conduct that apply to an individual or a group.*

The ethicist Manuel G. Velasquez (1998) outlines four moral standards he considers useful in thinking about ethical dilemmas:

- *Rights.* This standard concerns the basic needs and welfare of individuals. Everyone agrees, for example, that people have a right to a reasonably safe workplace.

- *Justice.* This standard concerns how the positive and negative effects of an action or a policy can be distributed fairly among a group. For example, the expense of maintaining a highway should be borne, in part, by people who use that highway. However, since everyone benefits from the highway, it is just that general funds also be used for highway maintenance.

- *Utility.* This standard concerns the positive and negative effects that an action or a policy might have on the general public. For example, if a company is considering closing a plant, utility requires that the company's leaders consider not only the money they would save but also the financial hardship of laid-off workers and the economic effect of the closing on the rest of the community.

- *Care.* This standard concerns the relationships we have with other individuals. We owe care and consideration to all people, but we have greater responsibility to the people in our family, our workplace, and our community.

Although it is best to think about the implications of any serious act in terms of all four standards, these standards can conflict. For instance, from the point of view of care, you might wish to give a promotion to a friend of yours in the company, even though he is not the most deserving candidate. In terms of justice, promoting the friend would be unfair to the other candidates for the promotion. In terms of utility, the promotion would probably not be in the best interests of the organization, although it might be in the best interests of your friend.

When there is a conflict, the standard of rights is usually considered the most important; justice, the second most important; utility, the third most important; and care, the fourth. However, simply ranking the four standards cannot solve all ethical problems. In an individual case, a "less important" standard could outweigh a "more important" one.

For instance, if the power company has to cross your property to repair a transformer on a utility pole, the standard of utility (the need to restore power to all the people affected by the problem) outweighs the standard of rights (your right to private property). The power company is, of course, obligated to respect your rights as much as possible by explaining what it wants to do, trying to accommodate your schedule, and repairing any damage done to your property.

Ethical problems are difficult to resolve because no set rules exist to determine when one standard outweighs another. In the example of the transformer, how many customers have to be deprived of their power before the power company is ethically justified in violating the property owner's right of ownership? Is it a thousand? Five hundred? Ten? Only one? There is no "correct" answer.

YOUR LEGAL OBLIGATIONS

Although most people believe that ethical obligations are more comprehensive and more important than legal obligations, the two sets of obligations are closely related in United States culture. Our ethical values have shaped many of our laws. For this reason, professionals should know the basics of four different bodies of law.

Bodies of law relevant to technical communication:
Copyright law
Trademark law

| Contract law |
| Liability law |

Copyright Law

 See the U.S. Copyright Office site (lcweb.loc.gov /copyright) for more about copyright law.

The Copyright Act of 1976 protects the author of any published or unpublished work — such as printed material, software, and photographs — whether the author is an individual or a corporation. The author is entitled to profit from the sale and distribution of the work in exchange for making the work accessible.

The concept by which work is made accessible is called *fair use*. Under fair-use guidelines, you have the right to use material, without getting permission, for purposes such as criticism, commentary, news reporting, teaching, scholarship, or research. Unfortunately, *fair use* is not a specific legal term but a convention that is interpreted differently in various situations.

GUIDELINES

Determining "Fair Use"

Courts consider four factors in disputes over fair use:

▶ *The purpose and character of the use, especially whether the use is for profit.* If your organization is profit-making, it will be scrutinized more carefully than a nonprofit organization.

▶ *The nature and purpose of the copyrighted work.* When the information you communicate is essential to the public good, as is the case, for example, with medical information, fair use is applied more liberally.

▶ *The amount and substantiality of the portion of the work used.* A 200-word passage would be a small portion of a book but a large portion of a 500-word brochure. Although you will see guidelines stating that 400 words are the maximum you may use, the courts have not always adhered to that number.

▶ *The effect of the use on the potential market for the copyrighted work.* Your use of the work cannot hurt the author's potential to profit from the original work.

Fair use does not apply to graphics: you must obtain written permission to use any graphics.

Writers often treat fair use in different ways depending on whether they are creating an internal document (one that will be published and used only within the organization) or an external document (one that will be published and used by people outside the organization).

In creating internal documents, such as employee manuals, writers are likely to use any material from the existing manual, even if the author of the original manual cannot be identified or no longer works for the organization. This practice is legal under the concept known as *work made for hire.* Anything written on the job by an employee being paid by the organization is the property of the organization. Any revisions to the manual will also be the property of the organization.

However, in creating external documents, such as user's guides, writers are likely to be much more careful about fair use. They will acknowledge the authors and illustrators and refrain from reprinting material unless they are certain that the material was created under work-made-for-hire guidelines.

 The U.S. Copyright Office site (lcweb.loc.gov/copyright) also describes *work made for hire.*

GUIDELINES

Dealing with Copyright Questions

Consider the following advice when using material from another source:

▶ *Abide by the fair-use concept.* Do not rely on excessive amounts of another source's work.

▶ *Seek permission.* Write to the source, stating what portion of the work you wish to use and the publication you wish to use it in. The source is likely to charge you for permission.

▶ *Cite your sources accurately.* Citing your sources fulfills your ethical obligation. It also strengthens your writing by showing the reader the range of your research. Finally, it protects you in case some of the facts or interpretations in your work turn out to be inaccurate: you have not claimed that you were solely responsible for them.

▶ *Discuss authorship questions openly.* The best way to determine the authorship of a document is to discuss the issue openly with everyone who contributed to it. Some contributors might deserve to be listed as authors, whereas others are only credited in an acknowledgment section.

▶ *Seek legal counsel if you have questions.* Copyright questions can be very complicated, especially when you are dealing with Internet documents. Exactly who created a document, or whether the document has been altered without the author's knowledge or permission, is sometimes difficult to determine. In the case of multimedia products, which can require dozens of permissions for a minute of running time, you will have many questions related to permissions. Consult your organization's legal counsel. The fact that you sought counsel shows good faith, an important factor if the copyright question goes to court.

See Appendix, Part A, p. 660 for more about documenting your sources.

For more about copyright issues and the Web, see Ch. 21, p. 627.

Trademark Law

For more on trademarks, see the U.S. Patent and Trademark Office Web site (www.uspto.gov).

Companies use *trademarks* and *registered trademarks* to ensure that the public recognizes the name or logo of their products. Therefore, you need to know the definitions of these two legal terms:

- A *trademark* is a word, phrase, name, or symbol that is identified with a company. The company simply uses the ™ symbol after the product name to claim the design or device as a trademark. Claiming a trademark permits a company to go to state court to prevent other companies from using the trademarked word, phrase, name, or symbol for their own products.

- A *registered trademark* refers to a word, phrase, name, or symbol that the company has registered with the U.S. Patent and Trademark Office. By registering the trademark, the company earns the legal right to use the ® symbol after the product name. Registering a trademark, a process that can take years, ensures much more legal protection in the United States, as well as in other nations.

As a communicator, you are responsible for using the trademark and registered trademark symbols accurately when you refer to a company's products. Unfortunately, doing so is not always easy, because companies often inappropriately claim registered-trademark status for their products.

GUIDELINES

Protecting Trademark

Use the following strategies to protect your client's or employer's trademark:

- ▶ *Distinguish trademarks from other material.* Use boldface, italics, a different type size, or a different color to distinguish the trademarked term.

- ▶ *Use the trademark symbol.* At least once in each document — preferably, the first time — use the appropriate symbol after the name or logo, followed by an asterisk. At the bottom of the page, include a statement such as the following: "*COKE is a registered trademark of the Coca-Cola Company."

- ▶ *Use the trademarked item as an adjective, not as a noun or a verb.* Trademarks can become confused with the generic term they refer to. Xerox, for example, regularly runs ads explaining that you cannot "xerox" anything, even on a Xerox® photocopier; you can only photocopy something. Therefore, use the trademarked item along with the generic term, as in Xerox® photocopier or LaserJet® printer.

▶ *Do not use the plural form or the possessive form of the term.* Doing so reduces the uniqueness of the item and encourages the public to think of the term as generic.

| DOES NOT PROTECT TRADEMARK | take some Kodacolors® |
| PROTECTS TRADEMARK | take some photographs using Kodacolor® film |

| DOES NOT PROTECT TRADEMARK | Kodacolor's® fine quality |
| PROTECTS TRADEMARK | the fine quality of Kodacolor® film |

Contract Law

Contract law deals with agreements between two parties. In most cases, disputes concern whether a product lives up to the manufacturer's claims. These claims are communicated as express warranties or implied warranties.

An *express warranty* is an explicit statement in a piece of product documentation or an oral statement that the product has a particular feature or can perform a particular function. An *implied warranty* is a warranty that is not written or spoken explicitly but rather inferred by the purchaser.

For example, a user's manual for a printer might show sample pages that were produced using the printer. If those sample pages contain sophisticated graphics, the company is setting forth an implied warranty that the printer can print those graphics. Implied warranties also occur in more casual communication, such as letters to customers, or even conversations between salespeople and customers. Therefore, it is important to be careful about what you write or say when you describe or illustrate your company's product.

To protect themselves, most companies include disclaimers with their product information. A *disclaimer* is a statement that the company hopes will limit its liability for the product. Whether the disclaimer in fact *does* limit the company's liability is determined in court on a case-by-case basis. If the court holds that the disclaimer is fair and reasonable, the company wins; if not, it loses. (Many legal disputes do not actually go to trial but are settled in negotiation between the two parties.) Manuals usually include disclaimers such as the following:

> Every effort has been made to supply complete and accurate information. However, Ottorino Publishers assumes no responsibility for the information contained herein or its use.

Software disclaimers often contain statements such as the following:

> Ottorino Software warranties this product to be free of defects at the time of purchase and agrees to replace the product at no cost if it is found to be defective

under normal conditions of use, for a period of ninety (90) days from purchase. However, Ottorino Software assumes no responsibility for incidental or consequential damages.

Liability Law

A product-liability action is "a lawsuit for personal injury, death, property damage, or financial loss caused by a defective product" (Helyar, 1992, p. 126). Liability is an important concern for communicators because courts routinely rule that manufacturers are responsible for providing adequate operating instructions and warning consumers about the risks of using their products.

GUIDELINES

Abiding by Liability Laws

Helyar (1992) summarizes the communicator's obligations, as reflected in recent court rulings, and offers ten guidelines for abiding by liability laws:

▸ *Understand the product and its likely users.* Learn everything possible about the product and its users.

▸ *Describe the product's functions and limitations.* Help potential users determine whether it is the appropriate product to buy. In one court case, the manufacturer was found liable for not stating that its electric smoke alarm does not work during a power outage.

▸ *Instruct users on all aspects of ownership.* Cover assembly, installation, use and storage, testing, maintenance, first aid and emergencies, and disposal of the item.

▸ *Use appropriate words and graphics.* Use common terms, simple sentences, and brief paragraphs. Structure the document logically and include explicit directions. Make graphics clear and easy to understand; where necessary, show people performing tasks. Choose words and create graphics that are appropriate to the education, mechanical ability, manual dexterity, and intelligence of intended users. For products that will be used by children or nonnative speakers of English, include graphics illustrating important information.

▸ *Warn users about the risks of using or misusing the product.* Warn users about the obvious dangers of using the product (such as electrical shock or chemical poisoning) but also warn them about the less-obvious dangers. Describe the cause, extent, and seriousness of the danger. An automobile manufacturer was found guilty for not having warned consumers that parking the car on grass, leaves, or other com-

See Ch. 20, p. 585, for a discussion of *danger, warning,* and *caution.*

bustible material could cause a fire. For particularly dangerous products, explain what the danger is and how to avoid it, and then describe how to use the product safely. Use *mandatory language,* such as *must* and *shall,* rather than the more tentative *might, could,* or *should.* Use the terms *warning* and *caution* appropriately.

▶ *Include warnings along with assertions of safety.* When users read in the product information that a product is safe, they tend to pay less attention to warnings. Therefore, include detailed warnings to balance safety claims.

▶ *Make directions and warnings conspicuous.* Warnings and other safety information must be in large type and easily visible, appear in an appropriate location, and be printed on labels durable enough to withstand ordinary use of the product.

▶ *Make sure that the instructions comply with applicable company standards and local, state, or federal statutes.*

▶ *Perform usability testing on the product (to make sure it is safe and easy to use) and on the instructions (to make sure they are accurate and easy to understand).*

See Ch. 3, p. 60, for a discussion of usability testing.

▶ *Make sure users get the information.* If you discover a problem after the product has been shipped to the retailer, get in touch with the users through direct mail or email if possible or newspaper advertising if not. Automobile-recall notices are one example of how manufacturers contact users.

CODES OF CONDUCT

Between 70 and 90 percent of large corporations have codes of conduct (Murphy, 1995), as do almost all professional societies. Codes of conduct vary greatly from organization to organization. Many are brief, such as the *Ethical Guidelines for Technical Communicators* used by the Society for Technical Communication, reprinted in Figure 2.1 on page 30.

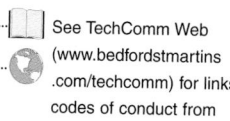
See TechComm Web (www.bedfordstmartins .com/techcomm) for links to codes of conduct from around the world.

An effective code has three major characteristics:

• *It protects the interests of the public rather than the interests of the members of the organization or profession.* For instance, it should condemn unsafe building practices but not advertising, because advertising increases competition and thus lowers prices.

• *It is specific and comprehensive.* If the code merely states that people must not steal, it offers little of value. The code is ineffective if it does not address the ethical offenses most likely to occur in the company or profession, such as bribery in companies that do business overseas.

For a more detailed code, from the Institute of Technical and Scientific Communicators, see www.petecom.demon.co.uk /codecon.htm.

STC ETHICAL GUIDELINES FOR TECHNICAL COMMUNICATORS

As technical communicators, we observe the following guidelines in our professional activities. Their purpose is to help us maintain ethical practices.

Legality
We observe the laws and regulations governing our professional activities in the workplace. We meet the terms and obligations of contracts that we undertake. We ensure that all terms of our contractual agreements are consistent with the *STC Ethical Guidelines*.

Honesty
We seek to promote the public good in our activities. To the best of our ability, we provide truthful and accurate communications. We dedicate ourselves to conciseness, clarity, coherence, and creativity, striving to address the needs of those who use our products. We alert our clients and employers when we believe material is ambiguous. Before using another person's work, we obtain permission. In cases where individuals are credited, we attribute authorship only to those who have made an original, substantive contribution. We do not perform work outside our job scope during hours compensated by clients or employers, except with their permission; nor do we use their facilities, equipment, or supplies without their approval. When we advertise our services, we do so truthfully.

Confidentiality
Respecting the confidentiality of our clients, employers, and professional organizations, we disclose business-sensitive information only with their consent or when legally required. We acquire releases from clients and employers before including their business-sensitive information in our portfolios or before using such material for a different client or employer or for demo purposes.

Quality
With the goal of producing high-quality work, we negotiate realistic, candid agreement on the schedule, budget, and deliverables with clients and employers in the initial project planning stage. When working on the project, we fulfill our negotiated roles in a timely, responsible manner and meet the stated expectations.

Fairness
We respect cultural variety and other aspects of diversity in our clients, employers, development teams, and audiences. We serve the business interests of our clients and employers, as long as such loyalty does not require us to violate the public good. We avoid conflicts of interest in the fulfillment of our professional responsibilities and activities. If we are aware of a conflict of interest, we disclose it to those concerned and obtain their approval before proceeding.

Professionalism
We seek candid evaluations of our professional performance from clients and employers. We also provide candid evaluations of communication products and services. We advance the technical communication profession through our integrity, standards, and performance.

■ **Figure 2.1**
Ethical Guidelines for Technical Communicators (used by permission of the Society for Technical Communication, Arlington, Virginia)

- *It is enforceable.* If it does not stipulate penalties, up to and including dismissal from the company or expulsion from the profession, it is ineffective.

Do codes of conduct really encourage ethical behavior? Sometimes, yes. James Burke, chairman and CEO of Johnson & Johnson, has been praised for withdrawing all Tylenol products from store shelves in 1982 after product tampering killed seven people. Burke stated that withdrawing the product — a move that cost the company $100 million — was the only option because the company's code begins with this sentence: "We believe our first responsibility is to the doctors, nurses, and patients, to mothers, and to all others who use our products and services." He expressed surprise that people would find his decision unusual (De George, 1995, p. 4).

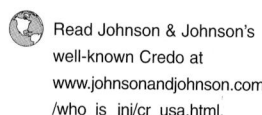 Read Johnson & Johnson's well-known Credo at www.johnsonandjohnson.com /who_is_jnj/cr_usa.html.

Often, however, it is difficult to tell whether codes encourage ethical behavior, as it is impossible to tell how many unethical acts were prevented because people were inspired or frightened by a code of conduct. Many ethicists are skeptical. Because codes must be flexible enough to cover a wide variety of situations, they tend to be so vague that determining whether a person has in fact violated one of their principles is nearly impossible. A survey investigating how chemical engineers make decisions about ethical conflicts revealed that of the 4,318 respondents, fewer than a half-dozen even mentioned codes of conduct in describing their thinking (Bryan, 1992, p. 81).

Codes of conduct are systematically enforced only rarely; many people therefore see them as little more than public-relations tools intended to persuade an organization's employees and members, the general public, and the government that the organization polices its own members. A good-faith effort to establish a code can be valuable, however, because it forces an organization to clarify its own values. Of course, distributing the code might also foster an increased awareness of ethical issues, in itself a positive result.

Figure 2.2 (Texas Instruments, 1999) on page 32 shows Texas Instruments's "TI Ethics Quick Test" and the links to the company's other ethics resources and policies.

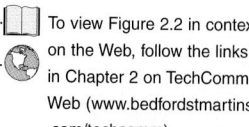 To view Figure 2.2 in context on the Web, follow the links in Chapter 2 on TechComm Web (www.bedfordstmartins .com/techcomm).

WHISTLEBLOWING

Whistleblowing is the practice of going public with information about serious unethical conduct within an organization. For example, an engineer is blowing the whistle in telling a government regulatory agency or a newspaper that quality-control tests on a product the company sells were faked.

In considering whether to blow the whistle, you are choosing between loyalty to your employer and loyalty to your own standards of ethical behavior. Where does loyalty to the employer end and the employee's right to blow the whistle begin? And what should an employee do *before* blowing the whistle?

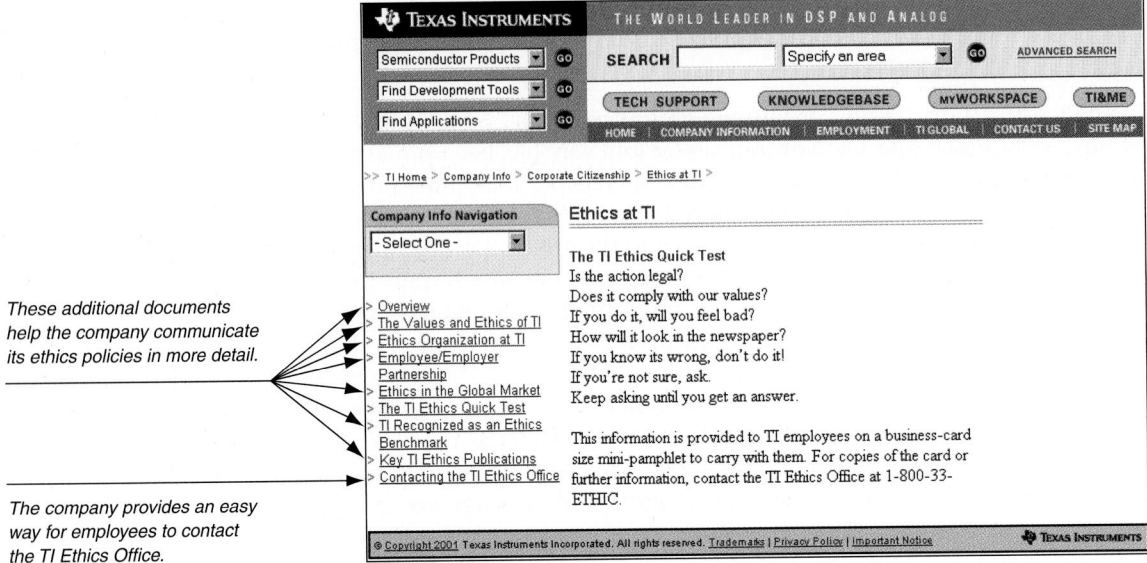

These additional documents help the company communicate its ethics policies in more detail.

The company provides an easy way for employees to contact the TI Ethics Office.

■ **Figure 2.2   Texas Instruments's Ethics Quick Test**

The ethicist Manuel Velasquez (1998, pp. 456–457) argues that whistleblowing is justified if four conditions are satisfied:

- There is strong evidence that the organization is doing something that is hurting or will hurt other parties.
- The employee has made a serious but unsuccessful attempt to prevent the wrongdoing by going through internal channels.
- External whistleblowing is reasonably certain to prevent or stop the wrongdoing.
- The wrongdoing is serious enough to warrant the consequences the whistleblowing will probably have for the employee, his or her family, and any other parties.

But being *justified* in blowing the whistle is not the same as being *obligated* to do so. You are obligated to blow the whistle when you are justified *and* when at least one of the following two conditions is met:

- Your position or professional responsibility calls for you to prevent the wrong. For instance, if you are a professional engineer, your professional code of ethics stipulates that you are to prevent certain kinds of wrongdoing, such as the building of unsafe structures.

- The wrongdoing is sufficiently serious for society, a group, or an individual. Wrongdoing that involves health, safety, or basic economic welfare falls into this category.

A number of organizations encourage employees to bring ethical questions to management rather than to blow the whistle. Among the more common means that companies use are anonymous questionnaires and ombudspersons. An *ombudsperson* is an employee whose job includes bringing ethical grievances to management's attention and serving as an impartial judge.

Whistleblowing remains risky, however, because most companies still have no formal procedures for handling serious ethical questions. Although the federal government and about half the states have laws to protect whistleblowers, these laws are usually ineffective. Organizations can easily penalize the whistleblower through negative performance appraisals, transfers to undesirable locations, or isolation within the company. For this reason, many people feel that an employee who has unsuccessfully tried every method of alerting management to a serious ethical problem would be wise simply to resign rather than to face the professional risks of whistleblowing. Of course, resigning quietly is much less likely to force the organization to remedy the situation.

As many ethicists say, doing the ethical thing does not always advance a person's career.

ETHICS AND MULTICULTURAL COMMUNICATION

Companies face special challenges when they market their products and services to people in other countries and to people in their home countries who come from other cultures. These companies not only have to make their communications understandable and clear to their target audiences, they also have to think about another serious challenge: the ethical norms and practices of the target audience, which may be very different from those of the company.

For instance, what should a company do if it wishes to sell products in a culture in which women are permitted to work as clerks but not as professionals? Should the company revise its product information to remove photographs of female professionals working side by side with male professionals?

The company that does not modify its information risks offending its audience and losing the opportunity to do business. At the same time, the company that tailors its communication to the cultural patterns of its target audience might be reinforcing patterns of discrimination. There is no set formula for figuring out how to deal with this kind of dilemma. You need to consider every situation individually.

The following standard, however, seems reasonable: we are ethically obligated not to reinforce patterns of discrimination in product information. It would be wrong, for example, to include a photograph of a workplace setting that excludes women. Yet we are not obligated to challenge the prevailing prejudice directly by, for example, including a photograph that shows women performing roles they do not normally perform within that culture. Neither are we obligated to portray women wearing clothing or makeup or jewelry that is likely to offend local standards. Although we should avoid the sexist "he," just as we do in writing for a domestic audience, we need not use "he or she" in writing for other cultures. Rather, we should try to find nonsexist ways to avoid the problem altogether, such as those suggested by Maggio (1991). If we can convey the message without risking offense, that is all that is ethically required; it is not necessary to confront the prejudice actively.

For more on avoiding sexist writing, see Ch. 11, p. 295.

This approach is what ethicists such as Thomas Donaldson (1989) call the *moral minimum*. But there is nothing to prevent organizations from adopting a more activist stance. Organizations that actively oppose discrimination are acting admirably.

PRINCIPLES FOR ETHICAL COMMUNICATION

As an employee, you are obligated to further your employer's legitimate aims and to refrain from any activities that run counter to them. You need to be honest, to present information accurately, and to avoid conflicts of interest that pit your own personal goals against the goals of the organization. In addition, you are obligated to help your organization treat its customers fairly by providing safe and effective products or services. The seven principles for ethical communication can help you fulfill these obligations.

Principles for ethical communication:
Abide by relevant laws.
Abide by the appropriate corporate or professional code of conduct.
Tell the truth.
Don't mislead your readers.
Be clear.
Avoid discriminatory language.
Acknowledge assistance from others.

Abide by Relevant Laws

The ethical minimum for any communicator is to abide by the laws that relate to communication. For instance, you must adhere to the laws governing intellectual property.

Here are some examples:

- *Do not plagiarize.* Obtain written permission from the copyright owner when you want to publish copyrighted material, such as graphics you find on the Web.
- *Honor the laws regarding trademarks.* For instance, use the "trademark" symbol (™) and the "registered trademark" symbol (®) properly when you refer to another company's products.
- *Live up to the express and implied warranties on your company's products.*
- *Abide by all laws governing product liability.* Helyar's (1992) guidelines presented on pages 28–29 are a good introduction.

Abide by the Appropriate Corporate or Professional Code of Conduct

Codes often go beyond legal issues to express ethical principles that employees are expected to follow. Principles typically expressed in codes include telling the truth, reporting information accurately, respecting the privacy of others, and avoiding conflicts of interest.

Tell the Truth

Perhaps the simplest and strongest guideline is to express facts honestly and accurately. The Golden Rule in the Christian tradition—treat others as you would like them to treat you — and similar statements in virtually every other religious tradition provide a powerful rationale for telling the truth.

Employees report that sometimes they are asked to lie about their own company's products or about those of their competitors. Obviously, lying — knowingly providing inaccurate information — is unethical. Your responsibility is to resist this pressure, by going over the supervisor's head if necessary.

Telling the truth also means not covering up negative information. For instance, if an information sheet for a portable compact-disc player suggests that it can be used by joggers but does not mention that the bouncing will probably make it skip, the information sheet is misleading.

Don't Mislead Your Readers

A close relative of the lie is the misleading statement. You present the information in a way that invites or even encourages the reader to reach a false conclusion. Although lying is technically different from misleading, a misleading statement is ethically no better.

Avoid these four common kinds of misleading technical communication:

See Ch. 11, p. 286, for a more detailed discussion of misleading writing. See Ch. 14, p. 365, for a discussion of avoiding misleading graphics.

- *False implications.* If you work for SuperBright and write "Use only SuperBright batteries in your new flashlight," you are implying that only that brand will work. If any brand will work, the statement is misleading. Communicators sometimes use clichés such as *user-friendly, ergonomic,* and *state of the art* to make the product sound better than it is. If you make a judgment about the product, back it up with specific and accurate information.

- *Exaggerations.* If you say "Our new Operating System 2500 makes system crashes a thing of the past," but the product in fact only reduces their likelihood, you are exaggerating. Provide the specific technical information on the reduction of crashes. Do not write "We carried out extensive market research" if all you did was make a few phone calls.

- *Legalistic constructions.* It is unethical to write "The 3000X was designed to operate in extreme temperatures, from −40 degrees to 120 degrees Fahrenheit" if the product cannot reliably operate in those temperatures. The fact that the statement might technically be accurate — the product was *designed* to operate in those temperatures — doesn't make it any less misleading.

- *Euphemisms.* If you refer to someone's being fired, say *fired* or *released*, not *granted permanent leave* or *offered an alternative career opportunity.*

Be Clear

See Ch. 11 for more about writing clearly.

Clear writing helps your readers understand your message easily; that is why much of this book is devoted to the issue of clarity. But clarity is also an ethical matter. Your responsibility is to write as clearly as you can to help your audience understand what you are saying.

For instance, if you are writing a product warranty, make it as simple and straightforward as you can. Don't hide behind big words and complicated sentences. Design your documents so that readers can easily locate the information they seek. Use tables of contents, indexes, and other accessing devices to help your readers find what they need.

It is your responsibility to write clearly when you address any audience. When the audience consists of people whose native language is not English, however, you need to be especially careful to be clear.

Avoid Discriminatory Language

For more about discriminatory language, see Ch. 11, p. 295.

Don't use language that discriminates against people because of their sex, religion, ethnicity, race, sexual orientation, or physical or mental abilities.

Acknowledge Assistance from Others

See Ch. 7, p. 160, and Appendix, Part A, for more about citing sources.

Don't suggest that you did all the work yourself if you didn't. Cite your sources and your collaborators accurately and graciously.

✔ **Revision Checklist**

1. Did you abide by relevant laws?
2. Did you abide by the appropriate corporate or professional code of conduct?
3. Did you tell the truth?
4. Did you avoid using
   ❏ false implications?
   ❏ exaggerations?
   ❏ legalistic constructions?
   ❏ euphemisms?
   ❏ clichés?
5. Did you avoid discriminatory language?
6. Did you acknowledge any assistance you received from others?

Exercises

1. It is late April and you need a summer job. In a local newspaper you see an ad for a potential job. The only problem is that the ad specifically mentions that "This is a continuing, full-time position." You know that you will be returning to college in the fall. Is it ethical for you to apply for the job without mentioning this fact? Why or why not? If you feel it is unethical to withhold the information that you plan to return to college in the fall, is there any ethical way you can apply?

2. Many foods are labeled "fat free," yet they contain substances such as monoglycerides, which are treated by the body just like more common forms of fat. Is it ethical to use the phrase "fat free" in such cases? Why or why not?

Research Projects

Some of the following projects ask you to write a memo. See Chapter 15, page 430, for a discussion of memos.

3. Find an article or advertisement in a newspaper or magazine or on the Web that you feel contains untrue or misleading information. Write a memo to your instructor describing the ad and analyzing the unethical techniques. How might the information have been presented more honestly? Include a photocopy or a printout of the ad with your memo.

4. "Graymatters" (onlineethics.org/corp/graymatters /martin.html) is a collection of miniature ethics cases involving typical business scenarios. After each case, you are presented with several alternative responses to the situation. When you select a response, you see an ethical analysis of it. Study several of the cases. Do you agree with the analyses of the ethical options? Why or why not?

5. Study your college's or university's code of conduct for students. Then write a memo to your instructor describing and evaluating it. Consider such questions as the following: How long is the code? How comprehensive is it? Does it provide detailed guidelines or merely make general statements? Where does the code appear? From your experience, does it

appear to be widely publicized, enforced, and adhered to? Are there sources of information on campus that could provide information on how the code is applied?

6. Form small groups. Study the code of conduct of a company or organization in your community. (Many companies and other organizations post their codes on their Web sites.)

   • One group member could study the code to analyze how effectively it states the ideals of the organization, describes proper and improper behavior and practices for employees, and spells out penalties.
   • Another group member could interview the officer who oversees the use of the code in the organization. Who wrote the code? What were the circumstances that led the organization to write it? Is it based on another organization's code? Does this officer of the organization believe the code is effective? Why or why not?
   • A third group member could secure the code of one of the professional groups in the organization's field (search for the code on the Web). For example, if the local organization produces electronic equipment, a professional group would be the Institute for Electrical and Electronics Engineers, Inc. To what extent does the code of the local organization reflect the principles and ideals of the professional group's code?
   • As a team, write a memo to your instructor presenting your findings. Attach the local organization's code to your memo.

7. Form small groups and research an event that involved ethical matters, such as the Chernobyl nuclear accident, the Exxon *Valdez* oil spill, the *Challenger* tragedy, or an incident in which a news organization manipulated a film, video, or photograph. (See the Cases section of the Online Ethics Center for Engineering and Science at onlineethics.org/corp/graymatters/martin.html for dozens of cases.) Each member of the group might investigate one of the following aspects of the case:

   • Study the incident itself. What happened? What factors led to the event, and what were the results immediately after the event?
   • Analyze the event. What were the ethical issues involved, and in what way or ways does the incident represent unethical conduct?
   • Analyze the long-term consequences of the event for the different stakeholders. In the Exxon *Valdez* spill, for instance, what have been the consequences for the environment, for the people who earned their livings from the natural resources, for Exxon, and for the general public?
   • Write a memo to your instructor presenting your findings. Keep in mind that your primary purpose is to analyze the ethical implications of the event, not to describe or document the event itself.

8. Find two Web sites dealing with ethical issues related to the use of computers, such as privacy or "the digital divide": the increasing gap between poor children, who don't have access to computing at home, and other children, who do. (Again, a good place to start the search would be the Online Ethics Center for Engineering and Science at onlineethics.org/corp/graymatters/martin.html.) Evaluate each site in terms of the following questions:

   • How detailed and comprehensive is the site?
   • Does the site attempt to provide a balanced selection of materials, or does it seem to represent an advocacy group whose purpose is to communicate a particular perspective?
   • How effective do you think the site is in helping people understand how to think about ethics and computers?
   • Write a memo to your instructor presenting your findings. Attach printouts of relevant and representative sections from each site.

CASE 1
The Name Game

Crescent Petroleum, an oil refining corporation based in Riyadh, Saudi Arabia, has issued a request for proposals for constructing an intranet that will link its headquarters with its three facilities in the United States and Europe. McNeil Informatics, a networking consulting company, is considering submitting a proposal. Most of the work will be performed at the company headquarters in Riyadh.

Crescent Petroleum was established forty years ago by family members who are related by marriage to the Saudi royal family. At the company headquarters, the support staff and clerical staff include women, most of whom are related to the owners of the company. The professional, managerial, and executive staff is all male, which is traditional in Saudi corporations. Crescent is a large company with revenues in the billions of dollars.

McNeil is a small firm — 12 employees — established two years ago by Denise McNeil, a 29-year-old computer scientist with a master's degree in computer engineering. She is working on her MBA while getting her company off the ground. Her employees include both males and females at all levels. The chief financial officer is female, as are several of the professional staff; the technical writer is male.

Denise McNeil traveled to New York from her headquarters in Pittsburgh to attend a briefing by Crescent. All the representatives from Crescent were middle-aged Saudi men; Denise was the only female among the representatives of the seven companies that attended the briefing. When Denise shook hands with Mr. Fayed, the team leader, he smiled slightly as he mentioned that he did not realize that McNeil Informatics was run by a woman. Denise did not know what to make of his comment, but she got a strong impression that the Crescent representatives felt uncomfortable in her presence. During the break, they drifted off to speak with the men from the other six vendors, leaving Denise to stand awkwardly by herself.

On her flight back to Pittsburgh, Denise McNeil thought about the possibility of gender discrimination but decided to bid for the project anyway, because she believed her company could write a persuasive proposal. McNeil Informatics had done several projects of this type successfully in the last year.

Back at the office, she met with Josh Lipton, the technical writer, to fill him in.

"When you put in the boilerplate about the company, I'd like you to delete the stuff about me founding the company. Don't say that a woman is the president, okay? And when you assemble the résumés of the project team, I'd like you to just use the first initials, not the first names."

"I don't understand, Denise. What's going on?" Josh asked.

 For additional cases, click on Case of the Month and Archive on TechComm Web (www.bedfordstmartins.com /techcomm).

"Well, Crescent looks like an all-male club, very traditional. I'm not sure they would want to hire us if they knew we've got a lot of women at the top," Denise replied.

"You know, Denise, there's another problem."

"Which is?"

"I'm thinking of the lead engineer we used in the other networking projects this year. . . ."

"Mark Feldman," she said, sighing. "What do you think we ought to do?"

"I don't know," Josh said. "I guess we could use another person. Or kind of change his name on the résumé."

"Let me think about this a little bit. I'll get back to you later."

What should Denise do about the fact that the person she wishes to designate as the principal investigator has an ethnic last name that might elicit a prejudiced reaction from Crescent officials? Is Denise's decision to disguise the sex of her employees — and to cover up her own role in founding her company — justified by common sense, or is it giving in to what she perceives as prejudice? Should she assign someone other than Mark Feldman to run the project? Should she tailor his name to disguise his ethnicity? Present your response in a memo to your instructor.

CASE 2
Mysterious Internet Code

John Boorman manages the technical publications department at Santa Barbara Equipment, a company that manufactures equipment used in the semiconductor industry. John supervises four technical communicators, who handle the firm's internal and external communication.

Santa Barbara Equipment is only three years old. It was founded by Eugene Froom, who invented the two major products the company produces. Froom has molded the company in his image. He doesn't delegate authority, and he has no patience with standard business practices. After three years, the company still does not have a policies and procedures manual for new employees, nor is there a code of conduct. If Mr. Froom thinks you are very bright, he will hire you, despite — or even because of — your lack of business experience and sophistication.

The newest member of the technical publications department is Lynn Stone, who was hired largely because of her computer skills. She knows more about computers than anyone else in her department, and in her free time she likes to hang out on the Internet. Lynn is contributing to the department's latest project: online documentation to accompany Santa Barbara's new product. Her job is to help the other technical communicators put the information online. A week before the documentation is to be pressed onto CDs, John Boorman learns through a conversation with Lynn that the software she used in coding the information was downloaded off the Internet. The problem is that one of the executable files of this downloaded software must be included with Santa Barbara's product. Concerned about copyright questions, he asks her to find out everything she can about the software.

Lynn learns that the software was made by a company called Visionary Software, which has no idea how it got onto the Internet. After checking the coding, Visionary's president tells Lynn that the software has been altered. He also tells her that he will refer the matter to Visionary's legal counsel. Lynn reports this information to John Boorman, who communicates it to Eugene Froom. Froom is furious.

In a memo to your instructor, respond to the following questions: What should John Boorman do? What is his obligation in relation to Lynn's conduct? What is his responsibility to Eugene Froom, the company president? To Visionary Software? To Santa Barbara Equipment's customers? In writing your response, be sure to explain your reasoning and evaluate the ethical and practical implications of each recommendation you make.

3 Understanding the Writing Process

Stephanie Rosenbaum, a consultant, on the importance of usability testing to the process of writing large documents:

You're never done with usability testing, just as you're never done with development or support.

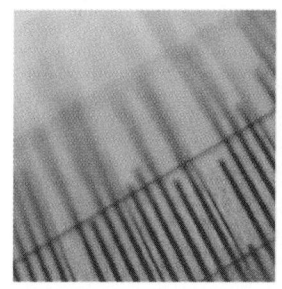

Regardless of the type of document you need to write, the best way to make sure that the document is effective is to divide the process of writing into stages. Planning, drafting, and revising, the three stages that you probably learned about in previous writing courses, apply to technical communication as well. However, everybody's approach to the writing process is somewhat different. You should interpret the information in this chapter as a collection of suggestions or guidelines, not as a rigid set of instructions.

One more point: even though this chapter discusses the writing process as a series of three stages, you will often find yourself doubling back. For example, during the revision stage you might do more research or draft whole new sections of the document. Doubling back is not a sign that you didn't carry out the process in the right way. It means only that you're still thinking about your subject and developing your ideas.

This chapter discusses the three stages of the writing process, concentrating on techniques and tools used by people who write technical documents. The focus of the chapter is on an individual working alone to create a document. Many kinds of technical documents, however, are created by two or more people working together. This technique, called *collaborative writing*, is the subject of Chapter 4.

See TechComm Web (www .bedfordstmartins.com /techcomm) for guidelines boxes and additional links related to topics in this chapter.

PLANNING

Students are sometimes surprised to learn that the planning phase of writing a document can take more than a third of the total time spent on the project. Planning is critically important for every document, from an email to a book-length instruction manual.

Planning involves:		
Analyzing your audience		
Analyzing your purpose		
Generating ideas about your topic		
Researching additional information		
Organizing and outlining your document		
Devising a schedule and a budget		

Analyzing Your Audience

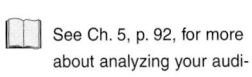

See Ch. 5, p. 92, for more about analyzing your audience.

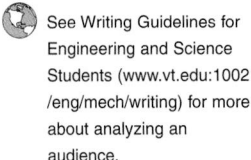

See Writing Guidelines for Engineering and Science Students (www.vt.edu:10021 /eng/mech/writing) for more about analyzing an audience.

As discussed in Chapter 1, technical communication addresses a particular audience. The first step in writing a document is to analyze your audience.

Sometimes you can talk with your audience before you start to create the document. These conversations can help you learn what your readers already know about the topic, what they want to know, and how they would like the information presented. You can test out drafts, making changes as you go.

Often, however, you cannot work with your audience while writing the document. In these cases, you need to learn everything you can about your audience in advance. You can start by determining whether your most important readers are experts, technicians, managers, or general readers. Knowing this information helps you determine the scope, organization, and style that will work best. Then, for each of your most important readers, you can try to answer the following three questions:

- *Who is your reader?* Consider such factors as education, job experience and responsibilities, cultural characteristics, and personal preferences.
- *What are your reader's attitudes and expectations?* Consider the reader's attitudes toward the topic and your message, as well as the reader's expectations about the kind of document you will be presenting.
- *Why and how will the reader use your document?* Think about the physical environment in which the document will be used, the way the reader will read the document, and the reader's skill in reading.

Consider how these questions would apply in a typical situation. The insurance company that administers your employee benefits is planning to raise premiums because your employees have a higher-than-average rate of absences and health-related claims. Executives in your company are considering implementing a health-promotion program to encourage employees to get in shape, improve their diets, and stop smoking. They have asked you to investigate these programs to learn how they work so they can decide whether the company should implement one of them.

Who are your readers, and what are their attitudes and expectations? In this case, your readers — the executives — want a clear, detailed report that helps them determine whether an available program will help solve the problem.

Why and how will your readers use your document? Your readers have not yet made up their minds. They will rely on the report to provide the facts and the economic analyses to help them decide what to do. There are no special requirements for the report; it does not need to be presented in a special format or printed on special paper.

Analyzing Your Purpose

The next step in planning is to analyze your purpose. You cannot start to write until you can state the purpose (or purposes) of the document. Ask yourself these questions: After your readers have read your document, what do you want them to know or to do? What beliefs or attitudes do you want them to hold? A statement of purpose might be as simple as this: "The purpose of this report is to recommend whether we should adopt a health-promotion program." Although the statement of purpose might not appear in this form in the final document, you need to state it clearly during the planning stage. A clear understanding of your purpose helps you stay on track as you gather and organize your information and draft your document.

See Ch. 5, p. 110, for more about analyzing your purpose.

Generating Ideas about Your Topic

With at least a preliminary understanding of your audience and your purpose, you can begin to generate ideas about your topic. Generating ideas is a way to start mapping out the information you will need to include in the document, where to put it, and what additional information may be required.

First, find out what you already know about the topic by using any of the following techniques:

- *Asking journalistic questions.* The journalistic questions are the *who, what, when, where, why,* and *how* of the topic. If, for instance, you are writing about health-promotion programs, ask the following questions:

  - *Who* administers these programs? *Who* uses them?
  - *What* is a health-promotion program?
  - *When* do employees use the programs? *When* do companies introduce them?
  - *Where* do employees go to use the programs? *Where* are such programs used now? *Where* will they be used in the future?
  - *Why* are health-promotion programs used? *What* advantages do they offer?
  - *How* do health-promotion programs work? *How* are they administered? *How* do employees sign up for them?

 See Paradigm Online Writing Assistant (www.powa.org) for an excellent discussion of journalistic questions and other techniques for generating ideas.

Answering these basic questions will help you discover what you already know about the topic and what additional information you need to acquire.

- *Brainstorming.* When you brainstorm, you spend 10 or 15 minutes listing ideas about your subject. List them as quickly as you can, using short phrases, not sentences. For instance, the first five items in your brainstorming list might look like this:

  – definition of health-promotion programs
  – the need for the programs
  – the problems they help solve
  – how employees react to them
  – insurance companies like them

  As the word *brainstorming* suggests, you are trying to inspire thinking, not to impose order on it. You will probably skip from one idea to another, and some ideas will be subsets of larger ideas.

  Brainstorming frees your mind, allowing you to think of as many ideas as possible that relate to your subject. When you construct an outline, you will find that some of the items you have listed probably do not belong in the document. Just toss them out. The advantage of brainstorming is that it is often the most effective and efficient way to generate and catalog ideas that might be important to the document.

- *Freewriting.* When you freewrite, you write without plans or restrictions. If your topic is the technology for mixing oil and water in power plants, you write whatever comes into your mind about it. You don't prepare an outline, stop to think about sentence construction, or consult a reference book. You just make the pen or the cursor move. Although the text you create seldom becomes a part of the final document, freewriting can help you determine what you do and do not understand. And one phrase or sentence might spark an important idea. Here is a sample freewriting text produced in about five minutes. The writer was working for a company considering a health-promotion program.

  > Insurance rates have gone up over 10% a year for last 5 years. Other problem: the loss of employees to preventable health problems. Check out health-promotion programs. How much do they cost? Are they administered by us or by a subcontractor? How about reduced-rate memberships at health clubs? Will we have further liability if someone gets hurt in the programs? How much would we save — in premiums and reduced health problems? Is there an organization that rates these programs? First we have to find out how many employees would be interested. Maybe the insurance company could help us with some of this information.

  This freewriting text, a series of incomplete thoughts, contains more questions than answers. Yet the writer has started to work out some of the problems her company would face and some possible solutions to research. She has also started to think about her own problem: determining what kind of information her readers will need.

- *Talking with someone.* Discussing your topic with someone is an excellent way to find out what you already know about it and to generate new ideas. Have the person ask questions of you as you speak. If you start with your main idea — "Our company is thinking about instituting some sort of health-promotion program for its employees" — the person could ask why. You might respond, "Because our health-insurance premiums are too high and because too many of our employees suffer from preventable health problems." The person might then ask, "How high are the premiums? What is the rate of increase?" or "What percentage of your employees get sick? What is an acceptable rate? Is the rate changing?" As the person asks questions, you will quickly get a sense of how clearly you understand your topic, what additional information you will need, and what questions your readers will want answered. You will also find yourself making new connections from one idea to another. Forcing yourself to put your ideas into sentences for someone else helps you clarify the information.

- *Clustering.* Write your main idea or question in the middle of the page. Then draw a circle around it. Write second-level ideas around the main idea. Then add third-level ideas around the second-level ideas. Figure 3.1 on page 48 is a cluster for the report on health-promotion programs.

- *Branching.* There is only one major difference between clustering and branching: in the cluster format, the movement from larger to smaller ideas is from the center to the perimeter; in the branching format, it is from top to bottom. Figure 3.2 on page 49 is a branching diagram.

    As you continue to think about your topic, you will extend and revise your sketch. You might want to rethink the sketch if, for example, one of the branches contains many more sub-branches than the others. A short branch might signal that you need more research, or it might just be a smaller idea that is fine as it is.

Researching Additional Information

Once you have a good idea of what you already know about your topic, you need to develop a strategy for obtaining the rest of the information you will need. You will find out what other people have already learned and published by reading reference books, scholarly books, and articles in the library. And you will find useful information on the Internet. You will evaluate all the information you have gathered, testing it for accuracy and relevance. In addition, you might compile new information by interviewing experts, distributing surveys and questionnaires, making observations, and conducting experiments.

See Ch. 7 for more about conducting research.

    Think about the best sources for the information you need. Here is how you might research several of the questions about health-promotion programs:

- *What health-related claims are our employees filing?* You meet with the human-resources department in your company to get the facts: the number and kinds of claims, trends over the last five or ten years, and so forth.

The writer has started with the journalistic questions. Notice that she doesn't yet know how to answer the question "Who?" She has to go back to her sources to find out who actually administers these kinds of programs.

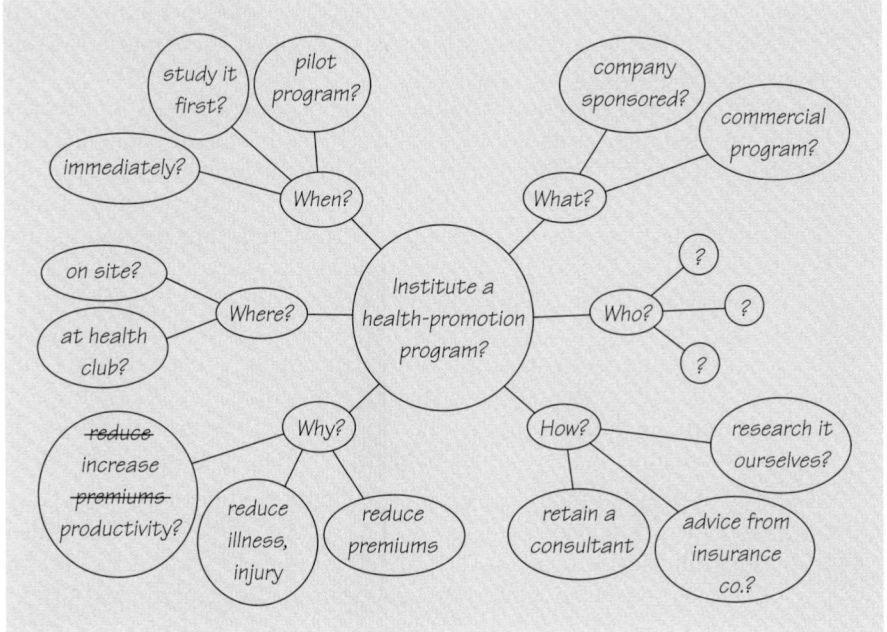

■ **Figure 3.1 Clustering**

And you meet with the insurance company to see how your company's employees stack up against employees of other similar organizations.

- *What kinds of health-promotion programs are there?* Your insurance carrier can probably direct you to many sources of information on different programs. In addition, you will search different databases and the Internet. Vendors will be happy to send you information on their services, because they want you to purchase those services.

- *What are the strengths and weaknesses of the available programs?* Some programs will be easy and inexpensive to implement and administer because they are simple: offering discounts to local health clubs, for instance. Other programs will be more complicated, involving hiring employees and building facilities on site. You could visit health clubs and companies that have implemented more complex programs to see firsthand how these programs work.

Organizing and Outlining Your Document

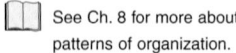 See Ch. 8 for more about patterns of organization.

When you have found and evaluated the additional information, you need to start thinking about how to organize your document. Then you will make an outline to help you write the draft.

Each document will have its own structure, but there are a number of organizational patterns you can adapt to your own situation so that you don't

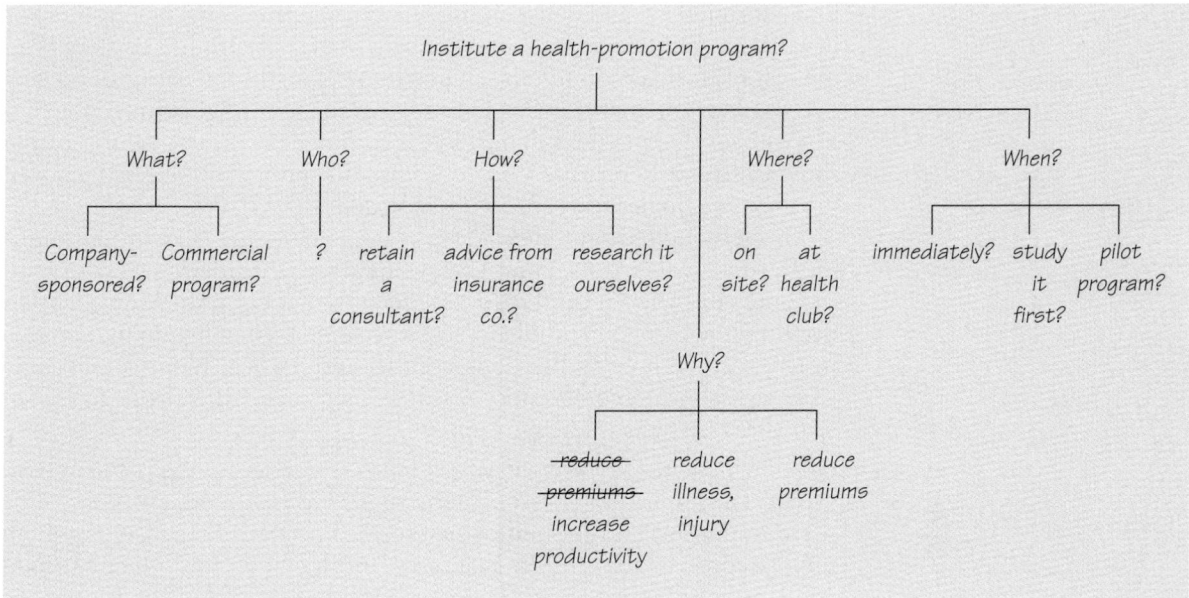

■ **Figure 3.2   Branching**

need to start from scratch. For instance, you are probably familiar with the comparison-and-contrast pattern. In your report on health-promotion programs, comparison and contrast might be an effective way to organize a discussion of the different programs. The cause-and-effect pattern might be a good way to structure a discussion of the effects that implementing a program can be expected to produce. The problem-methods-solution pattern might be an effective structure for the whole document: the problem that led to the study; the methods you carried out; and the results, conclusions, and recommendations of your investigation.

At this point, your organization is only tentative. When you start to draft, you might find that the organizational pattern you have chosen isn't working well and needs to be revised. Or you might find that you need further information that doesn't fit into the organizational pattern you have chosen. Don't let these problems discourage you. They are part of the writing process, and they happen to all writers.

Once you have settled on at least a tentative plan, write an outline. An outline will help you stay on track as you draft the document. To keep your statement of purpose clearly in mind as you work, you might want to write it at the top of your page before you begin your outline.

Making an outline involves five main tasks:

 See Online Technical Writing (www.io.com/~hcexres /tcm1603/acchtml/acctoc .html) for more about outlining.

- *Grouping similar items.* If you used branching or clustering to generate ideas, you have already grouped similar items. But if you used brainstorming, you have not. Start by looking at the first item on your brainstorming list and determining what major category it belongs to. For instance, it could be an item that belongs in the background section. Scan the rest of the list, noting other items that belong in the background section. Link them. Then go back to the top of the list and determine the logical category of the next item. Link it to other items that belong in the same category. Repeat this process until all the items have either been grouped into logical categories or discarded.

Even though the outline is incomplete, some writers like to start drafting the document at this point. They want to see what they have to say, and the only sure way to do that is to start to write. Once they have started drafting, they feel better able to sequence items. They juggle paragraphs instead of juggling outline items. Other writers feel more comfortable working from a more refined outline. For them, the next step is to sequence the items in each group.

- *Ordering the items in the groups.* In some cases, ordering the items in each group is a simple matter — one item might precede another in time or be more important and it should therefore come first. Likewise, one item might represent the problem, another the methods for solving it, and another the solution.

- *Organizing the groups.* To organize the groups, follow the same logical processes you used to organize the items within each group. Often your readers expect a particular structure. For instance, the body of a formal report begins with an introduction followed by a detailed discussion and the major findings. If your material does not lend itself to a traditional organizational pattern, you'll need to come up with one that your readers will find clear and easy to follow.

- *Avoiding common logical problems.* As you refine your outline, avoid *faulty coordination* and *faulty subordination*. Faulty coordination occurs when a writer equates items that are not of equal value or not at the same level of generality. Here is a portion of an outline with faulty coordination followed by a corrected version.

Faulty coordination	*Proper coordination*
Common household tools	Common household tools
• screwdrivers	• screwdrivers
• drills	• drills
• claw hammers	• hammers
• ball peen hammers	— claw hammers
	— ball peen hammers

Faulty subordination occurs when an item is made a subunit of a unit to which it does not belong.

Faulty subordination	Proper subordination
Power sources for lawnmowers	Power sources for lawnmowers
• manual	• manual
• gasoline	• gasoline
• electric	• electric
• riding mowers	

"Riding mowers" is out of place because it is a type of lawnmower, not a power source. Whether it belongs in the outline at all is another question, but it certainly doesn't belong here.

A second kind of faulty subordination occurs when only one subunit is listed. The solution is to incorporate the single subunit into the unit. In the following example, item 1.1 has been deleted and restated as item 1.

Single subunit	Proper subordination
Three types of sound-reproduction systems:	Three types of sound-reproduction systems:
1. records	1. phonograph records
1.1. phonograph records	2. tapes
2. tapes	2.1. cassette
2.1. cassette	2.2. open reel
2.2. open reel	3. compact discs
3. compact discs	

- *Choosing an outline format.* If you are creating an outline purely for your own use, you don't have to worry about its format. However, if your outline will be read by someone else, you should use one of the two standard outline formats: the traditional alphanumeric format or the decimal format.

Traditional outline	Decimal outline
I.	1.
A.	1.1
1.	1.1.1
2.	1.1.2
B.	1.2
1.	1.2.1
a.	1.2.2
b.	2.
(1)	2.1 etc.
(2)	
(a)	
(b)	
2. etc.	

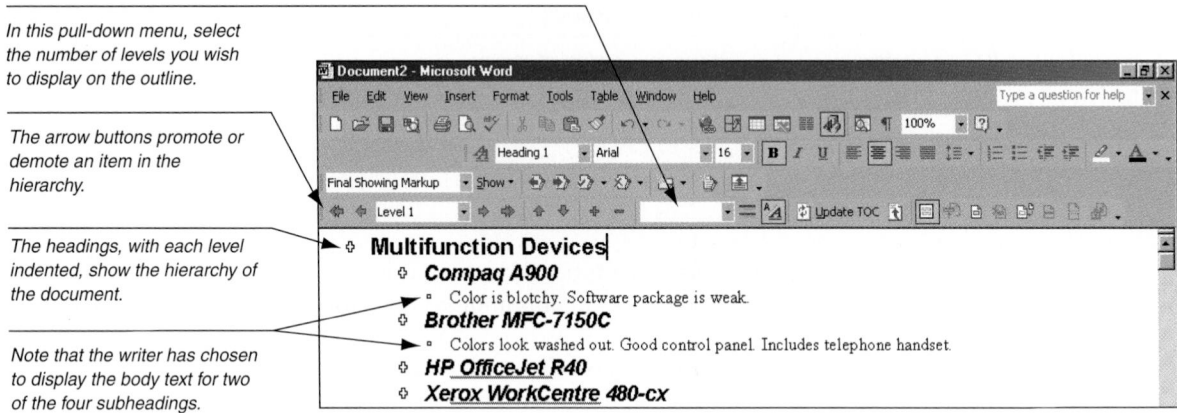

In this pull-down menu, select
the number of levels you wish
to display on the outline.

The arrow buttons promote or
demote an item in the
hierarchy.

The headings, with each level
indented, show the hierarchy of
the document.

Note that the writer has chosen
to display the body text for two
of the four subheadings.

■ **Figure 3.3   The Outline View**

An advantage of the decimal system is that it is simple to use and understand. You don't have to remember how to represent the different heading levels. In the same way, your readers can easily see what level they are reading. Use whichever system helps you stay organized.

As you draft your outline, consider using the *outline view* on your word processor. Figure 3.3 shows the main features of the outline view.

Devising a Schedule and a Budget

During the planning stage, you will also need to decide when you will provide various portions of the document to your readers and how much you can spend on the project.

For the project on health-promotion programs, for instance, your audience might have a strict deadline in mind that will enable them to decide what to do before the new fiscal year begins in two months. In addition, your readers might want to see a progress report, submitted halfway through the project, to tell them whether the project is proceeding on schedule, whether you will in fact be providing the information you thought you would, and whether you need any additional resources. Making a schedule is often a collaborative process: you meet with your main readers, who tell you when they need the information, and you estimate how long the different tasks will take.

See Ch. 18, p. 524, for more
on progress reports.

You also need to set up a budget for the project. In addition to the time you will need to do the study, you might incur additional expenses. For example, you might need to travel to visit companies that have implemented different kinds of health-promotion programs. You might need to conduct specialized database searches, create and distribute questionnaires to employees, or conduct interviews at remote locations. Some projects call for *usability testing* — evaluating potential users as they try out a system or a document. This testing costs money and needs to be included in your budget.

DRAFTING

Some writers start with a detailed outline. Others use only a brief, very general one. Still others work from sketches of graphics. When you have at least a preliminary outline, it is time to start drafting. Some writers like to draft right on the outline on their screens. Others prefer to print a paper copy of their outline and place it on the desk next to their keyboard.

 Purdue University's OWL (owl.english.purdue.edu) has many instructional handouts covering all aspects of the writing process.

GUIDELINES

Drafting Effectively

Try the following techniques when you begin to draft or when you get stuck in the middle of drafting.

▶ *Get comfortable.* Choose a good chair set at the right height for the keyboard and adjust the light so that it doesn't reflect off the screen.

▶ *Start with the easiest topics.* Instead of starting at the beginning of the document, begin with the section you most want to write about.

▶ *Draft quickly.* Many students draft slowly, hoping their first draft will be good enough to serve as a final draft. It won't be. Just try to make your fingers keep up with your brain. Turn the phrases from your outline into paragraphs.

▶ *Don't stop to get more information or to revise.* Set a timer and draft for an hour or two without stopping. When you come to an item that you don't understand or that requires more research, just skip to the next item. Don't worry about sentence structure or spelling. Your goal is to create a long rough draft.

▶ *Try invisible writing.* Darken the screen so that you can look only at your hardcopy outline or the keyboard. You won't stop typing so often because you will be less tempted to revise what you have just written.

▶ *Stop in the middle of a section.* When you stop, do so in the middle of a paragraph, or even in the middle of a sentence. When you start again, you will find it easy to conclude the idea you were working on. This technique will help you avoid writer's block, the mental paralysis that can set in when you stare at a blank page.

You are probably already familiar with the ways in which word-processing software makes your job as a writer easier. You can draft quickly and revise easily. You can move chunks of text around to achieve the most effective organization of your document. And you can copy text from one document to another.

This section discusses how to use three more-advanced tools as you draft: templates, styles, and automated hypertext linking and Web-conversion tools.

Using Templates

Templates are preformatted designs for different kinds of documents, such as business letters, memos, newsletters, and reports. Templates incorporate all the different design specifications for the document, including typeface, type size, margins, and spacing. Once you select a template, you just type in the information; the document is already designed. Using templates, however, can lead to three problems:

See Ch. 13 for more about design and Ch. 14 for more about graphics.

- *Templates do not always reflect the best design principles.* For instance, most letter and memo templates default to 10-point rather than 12-point type, which is easier on the eyes of most readers. Templates for graphics often violate principles of clear, effective communication.

- *Readers get tired of seeing the same designs over and over.* If you choose a memo template from Microsoft Word, for instance, your memos will look like those of many other writers. Modifying the templates, or creating your own designs, can give your documents a fresher look.

- *Templates cannot help you answer the important questions about your document.* Memo templates can help you format the to-from-subject information at the top, saving you a minute or two. But they cannot help you figure out how to organize and write the memo. Sometimes templates can even send you the wrong message. For example, résumé templates in word processors present a set of headings that might work better for some job applicants than for others. Templates cannot help you think about how best to present your own credentials to your audience.

The more you rely on templates, the less likely you are to learn how to use the software to make your documents look professional.

Using Styles

See Ch. 12, p. 314, for an explanation of using styles to set up a table of contents.

Styles are like small templates in that they apply to the design of smaller elements, such as headings and body text, but they are much more valuable.

Like templates, styles save you time. As you draft your document, you don't need to stop to add all the formatting each time, for example, you want to designate an item as a first-level heading. When you finish drafting, you simply put your cursor in the text you want to be a first-level heading and,

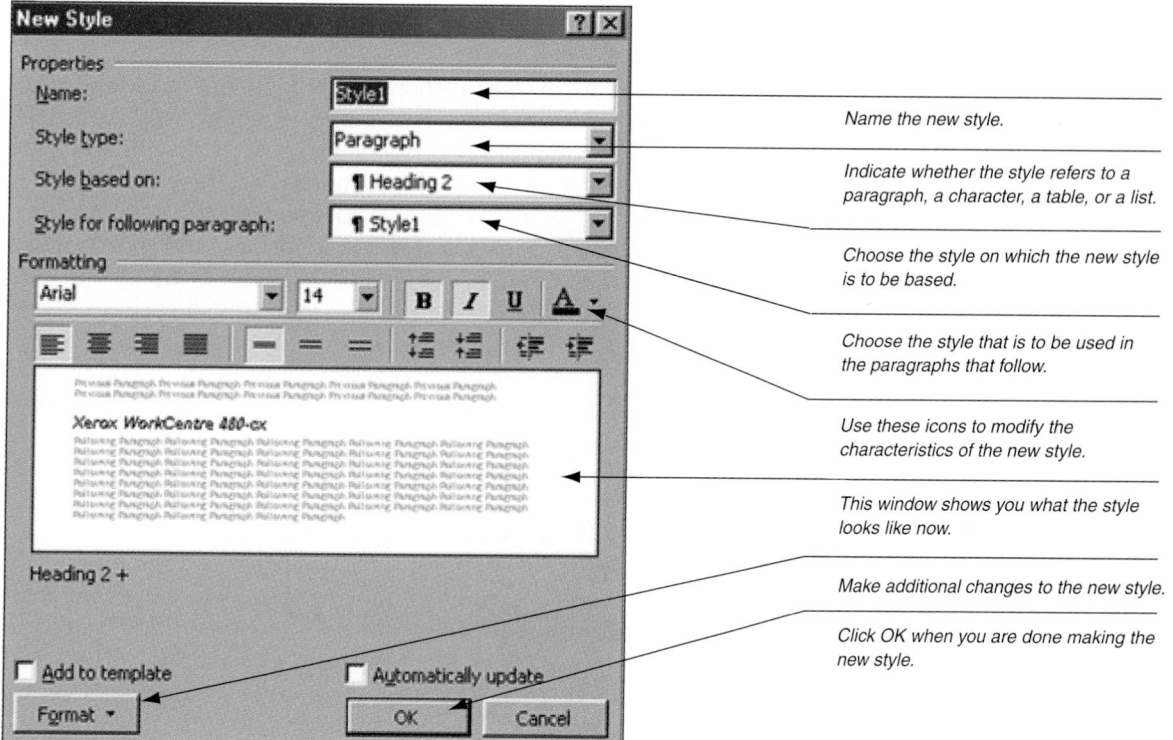

The figure shows the "New Style" dialog box with the following labeled annotations:

Properties

Name: Style1 — *Name the new style.*

Style type: Paragraph — *Indicate whether the style refers to a paragraph, a character, a table, or a list.*

Style based on: ¶ Heading 2 — *Choose the style on which the new style is to be based.*

Style for following paragraph: ¶ Style1 — *Choose the style that is to be used in the paragraphs that follow.*

Formatting

Arial   14   **B** *I* U A — *Use these icons to modify the characteristics of the new style.*

Previous Paragraph Previous Paragraph Previous Paragraph Previous Paragraph Previous Paragraph Previous Paragraph Previous Paragraph Previous Paragraph Previous Paragraph

Xerox WorkCentre 480-cx

Following Paragraph Following Paragraph Following Paragraph Following Paragraph Following Paragraph... — *This window shows you what the style looks like now.*

Heading 2 +

□ Add to template   □ Automatically update

Format ▾ — *Make additional changes to the new style.*

OK   Cancel — *Click OK when you are done making the new style.*

■ **Figure 3.4   Modifying a Style**

This window appears when you modify a style.

using a pull-down menu, select that style. The text automatically incorporates all the specifications of a first-level heading.

Styles also help ensure consistency, because you don't have to change four or five technical specifications every time you want to change an item's style. And if you decide to modify a style — by adding italics to a heading, for instance — you change it only once; the software automatically changes every instance of the style in the document. Styles are particularly useful as you create collaborative documents, because they make it easier for different collaborators to achieve a consistent look.

Figure 3.4 shows the window you use to change the specifications of the styles used in Microsoft® Word 2002.

Using Automated Hypertext Linking and Web-Conversion Tools

Word-processing software today is designed to make it easier to print documents in the traditional way or to put them on the Web. Automated hypertext

See Ch. 21, p. 614, for more about Web-conversion tools and browsers.

linking and Web-conversion tools help you cut down on the time needed to produce documents for the two media.

Automated hypertext linking is a function that recognizes Web addresses and formats them so that they are active links. This way, your readers will see the Web address underlined as a link and can click it to launch a Web browser and go to that site — provided, of course, that they are reading the document online. If your document isn't meant to be read online, disable automated hypertext linking.

A Web-conversion tool adds HTML codes to a document so that it can be viewed through a Web browser on the Internet or on an intranet (a private network). Using this tool, you can save *much* of the work involved in changing word-processed files into HTML files. However, Web-conversion tools still cannot handle every kind of design or effect you wish to create. If you want to create a complex Web page — and make sure people using any Web browser can view it — you still have to learn HTML coding.

REVISING

See Robin A. Cormier's excellent essay (1997) on revising: www.eeicom.com /eye/qc-lead.html.

Revising is the process of making sure that the document says what you want it to say and says it professionally. Everyone uses a different technique for revising, but the important point is that you have a technique. You cannot hope to simply read through your document, waiting for the problems to leap off the page. Some of them might, but most won't.

Revise a document by:
Studying the document by yourself
Using revision software
Seeking help from someone else
Testing readers as they use the document

Studying the Document by Yourself

The first step in revision involves spending time reading and rereading the document. Many writers go through their documents several times: once for overall meaning and clarity, once for organization and logical development, and once for correctness. Experiment to learn what system works best for you.

GUIDELINES

Revising by Yourself

Try the following strategies for revising your drafts before you show them to your instructor, other students, or co-workers.

- *Let it sit.* Set the document aside for a while, overnight if possible. This way, you gain some distance from your work and are better able to approach it more as your readers will.

- *Read it aloud.* Doing this helps you hear awkward phrases and recognize poorly developed ideas, illogical reasoning, or missing evidence.

- *Use checklists.* Take checklists such as those in this book and modify them to include the points your instructors have made about your writing.

- *Review a printout of your draft.* You can revise effectively right on the screen, but be sure to print a copy of your draft. You'll be able to spot more problems on a printout than on a screen because a printer provides a much sharper image than a screen, and you'll see the draft as your readers will.

See the *Writer's Handbook* in the Grammar and Style section of the University of Wisconsin Writing Center (www.wisc.edu/writing) for examples of checklists.

Using Revision Software

Word processors include three major revising tools: spell checkers, grammar checkers, and thesauri. These tools can be valuable, but you need to know their limitations.

Spell Checkers

A *spell checker* alerts you when it sees a word that isn't in its dictionary. That word might be misspelled, but it might also be a correctly spelled word that simply isn't in the spell checker's dictionary. You can add the word to your dictionary so that the spell checker will recognize it in the future. Some spell checkers work as you type; that is, as soon as you have typed a word that is not in the dictionary, they highlight it or make a beeping sound. Consider turning off this function so that you can concentrate on drafting, then fix the spelling later, as you revise.

You are probably aware of the danger of relying on a spell checker: it cannot tell whether you have used the correct word. If you have typed "We need too dozen test tubes," for example, the spell checker won't see a problem.

Grammar Checkers

A *grammar checker* can help you identify and fix potential grammatical and stylistic problems, such as wordiness, subject/verb agreement, and double negatives. Many grammar checkers identify abstract words and suggest more specific ones or point out sexist terms and provide nonsexist alternatives.

Figure 3.5 shows a screen illustrating how grammar checkers work. As you can see, they can have serious drawbacks. For this reason, many writing teachers advise that, unless you are a capable and experienced writer who knows when to ignore bad advice, you shouldn't use a grammar checker at all.

Thesauri

A *thesaurus* (plural *thesauri*) lists synonyms or related words. An electronic thesaurus has the same strengths and weaknesses as a printed one: if you are looking for a word but can't quite think of it, the thesaurus will help you remember it. But the listed terms might not be related to the key term closely enough to function as synonyms. Unless you are aware of the shades of difference, you might substitute an inappropriate word. For example, the entries for the word *famous* in one electronic thesaurus include *infamous* and *notorious*. If you use either as a synonym for *famous*, you could embarrass yourself badly.

The revision tools discussed in this section cannot replace a careful reading by you and by other people. These tools don't understand your subject, your audience, and your purpose. They cannot identify unclear explanations, con-

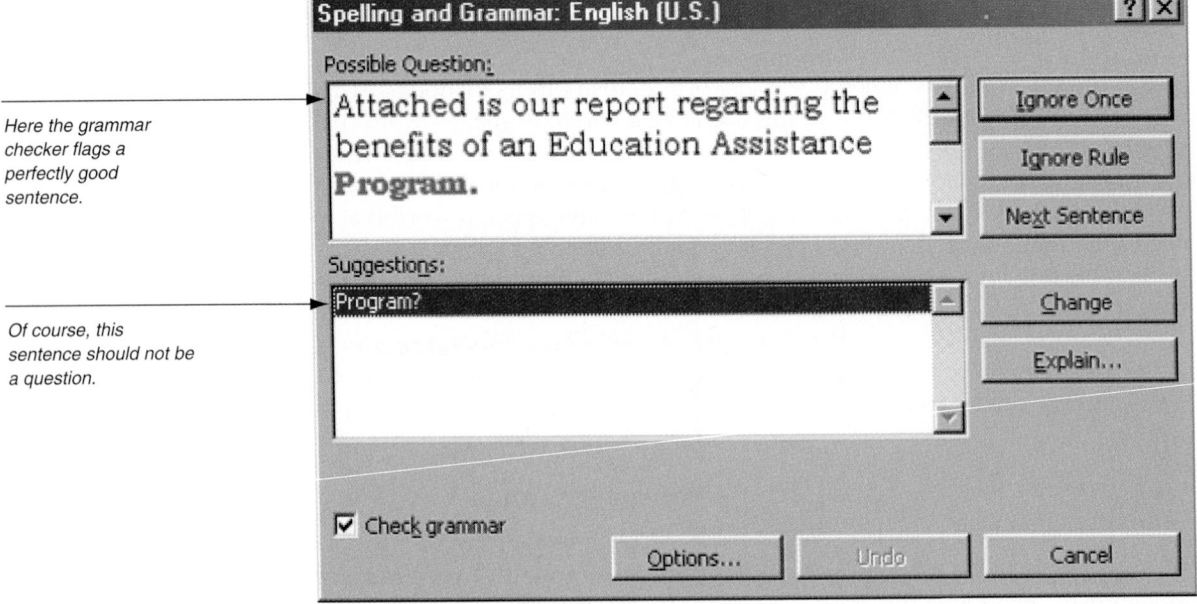

Here the grammar checker flags a perfectly good sentence.

Of course, this sentence should not be a question.

■ **Figure 3.5   Grammar Checker**

traditions, inaccurate data, inappropriate tone, and so forth. Use the tools, but don't rely on them. Revise your document yourself, and then get help from someone you trust.

Seeking Help from Someone Else

Once you have done what you can by yourself, enlist a colleague to help you revise. For technical documents, it is best to turn to two kinds of people for help:

- *Subject-matter experts.* If, for instance, you have written an analysis of alternative fuels for automobiles, you could ask an automotive expert to review it. Important documents are routinely reviewed by technical experts, lawyers, and marketing specialists before being released to the public.

- *People who are like those in the targeted audience.* People who fit the profile of the eventual readers can help you see problems you didn't notice while revising by yourself. People more knowledgeable about the subject than your eventual readers might not be sufficiently critical because they will understand the document even if it isn't clear to other, less-knowledgeable readers.

When you ask someone for help, provide specific instructions. Tell the person as much as you can about your audience and purpose and about the tasks you still have to complete before you finish the document. Then tell the person what kinds of problems you want him or her to look for; better yet, write them down. Here's an example:

Bob —

Thanks a lot for agreeing to look at this draft. I'm particularly interested in the following points:

- Is the background clear? Do you understand the relationship between the insurance company's decision and our study?
- Does the description of each of the three major programs seem proportional, or do I seem to go into more detail on one or more of them?
- Does the draft of the questionnaire on p. 17 work? See any problems? I'm particularly concerned about leading questions or questions that could logically have more than one valid response.

Don't worry about copy editing: I'm going to go over it carefully later. Please get back to me with questions.

As I mentioned on the phone, I'd love to get your response by Thursday, March 14.

Regards,
Ann

Writing a note such as this increases the chances that your colleagues will be able to point out your document's strong points, unclear passages, and sections that need to be expanded, deleted, or revised. In addition, the note reduces the chance that your reader will waste time working on something you don't need.

Testing Readers as They Use the Document

Although analyzing a document can provide useful information, it cannot tell you how well the document will work when it gets into the hands of the people who will use it. To find that out, you must test readers as they use the document. If the document is missing important information or contains unclear or inaccurate information, testing is likely to reveal the problem.

Usability testing is the process of performing experiments with people who represent real users to see how well they understand a document and how easily they can use it. In most cases, instructions and manuals are the types of documents that undergo usability testing. This section covers five topics:

- the goals of usability testing
- the basic concepts of usability testing
- preparing for a usability test
- conducting a usability test
- interpreting and reporting the data from a usability test

The information in this section is based on Rubin (1994) and Dumas & Redish (1993).

The Goals of Usability Testing

For an article describing one organization's usability testing program, see www.oclc.org /oclc/new/n229/ulab.html.

The two main goals of usability testing are to improve the safety of the company's product and to save money. An effective document reduces the risk that a customer will be injured in assembling or using the product it accompanies. As discussed in Chapter 2, courts are now finding companies guilty of publishing "defective" instructions and manuals. Beyond issues of safety, usability testing can save the company money in five ways:

- helping the company understand and exploit the product's competitive advantages
- reducing the number of service calls and customer-support phone lines
- reducing the number of updates to the product
- improving related products
- helping increase customer satisfaction with the product

The Basic Concepts of Usability Testing

There are four basic principles of usability testing:

- *Usability testing permeates product development.* Usability testing involves testing the document rigorously and often to make sure it works and is easy to use. The organization devotes substantial resources to training and equipping the testers and enabling the designers to use the information generated by the tests.

- *Usability testing involves teamwork.* A usability test involves the collaborative efforts of people with different areas of expertise, including product experts, usability experts, technical communicators, and videocam operators.

- *Usability testing involves studying real users as they use the product.* A company always learns important information from real users that it would not learn from people in the organization.

- *Usability testing involves setting measurable goals and determining whether the product meets them.* Usability testing involves determining in advance what the user is supposed to be able to do. For instance, in testing a help system for a word-processing program, the testers might decide that the user should be able to find the section on saving a file and carry out that task successfully in less than two minutes.

Preparing for a Usability Test

Usability testing requires careful planning over a period of weeks or months. According to Kantner (1994), planning accounts for one-half to three-quarters of the time devoted to testing. Eight main tasks must be accomplished in planning a usability test:

- *Understand your users' needs.* Companies frequently conduct *focus groups*, groups of people brought together for a few hours to talk about a product or an issue. Companies also test existing products (either their own or a competitor's), have experts review the product, and conduct on-site interviews and observations of real users in the workplace to observe them doing their work and to interview them.

- *Determine the purpose of the test.* You can test an idea before the product is even designed, to see if people understand it and like it. Or you can test a prototype to see if it is easy to use, or a finished product to see if there are any last-minute changes you can make to improve it.

- *Staff the test team.* Extensive programs in usability testing involve many specialists, each doing one (and only one) job. In many companies, however, two or three people do all the tasks.

- *Set up the test environment.* A basic test environment includes a room for the test participant, equipped with VCRs to record the test, and another room for the test observers.

See Ch. 17 for information on proposals.

- *Develop a test plan.* A *test plan* is a document that describes and justifies what the testers plan to do. The test plan is a proposal, a statement requesting approval and resources.

- *Select participants.* Testers recruit participants who match the profile of the intended users. Generally, it is best not to use company employees, who might know more about the product than a real user and thereby bias the test.

- *Prepare the test materials.* Most tests require legal forms, an orientation script to help the participant understand the purpose of the test, background questionnaires, instructions for the participant to follow in the test, and a log for the testers to record data during the test.

- *Conduct a pilot test.* A *pilot test* is a usability test for the usability test. A pilot test can uncover problems with the equipment, the document being tested, the test materials, and the test design.

Conducting a Usability Test

Although every tester makes mistakes every time, the goal is to minimize the number and severity of the mistakes so that the company can learn as much as possible from each test. There are three important aspects of conducting the test:

- *Staying organized.* Make a checklist and a schedule for the test day. The schedule should cover everything from before the first participant arrives until after the last one leaves, including packing up the equipment in preparation for the next day.

- *Interacting with the participant.* A popular technique for eliciting useful information from the participant is called a *think-aloud protocol.* In this technique, the participant says aloud what he or she is thinking: "I guess I'm supposed to press ENTER here, but I'm not sure because the manual didn't say to do it." If the participant feels comfortable thinking aloud, the information can prove invaluable. However, many people become very uncomfortable doing so.

  A second important aspect of interacting with the participant is to make sure not to influence the participant unintentionally. When testers ask a participant a question, they try not to reveal the answer they want. They do not say "Well, that part of the test was pretty simple, wasn't it?" Dumas & Redish (1993, p. 298) recommend neutral phrasing, such as "How was it performing that procedure?" or "Did you find that procedure easy or difficult?"

- *Debriefing the participant.* After the test, testers usually have questions about the participant's actions. For this reason, they debrief — that is, interview — the participant.

Interpreting and Reporting the Data from a Usability Test

After a usability test, testers have a great deal of data, including notes, questionnaires, and videotapes. The three steps in turning that data into useful information are the following:

- *Tabulate the data.* Testers gather all the information, both quantitative and qualitative, and, when appropriate, perform statistical procedures on it. The quantitative data include *performance* measures, such as how long it took a participant to complete a task, and *attitude* measures, such as how easy the participant found it to perform the task. The qualitative data include nonnumerical information, such as notes taken by the test team and quotations from participants during or after the test.

- *Analyze the data.* Testers analyze the information to understand it. Naturally, they want to concentrate on the most important problems revealed in the test and to determine the severity and the frequency of each problem.

- *Report the data.* Writing a clear, comprehensive report often leads the testers to insights they might not have achieved otherwise.

Usability testing might seem like an extremely expensive and difficult undertaking, and it certainly can be frustrating. But testers who are methodical, open-minded, and curious about how people use the document find that usability testing is the least expensive and most effective way to improve a document's quality.

✔ **Revision Checklist**

1. In planning the document, did you
   ❏ analyze your audience?
   ❏ analyze your purpose?

2. In generating ideas about your topic, did you consider
   ❏ asking journalistic questions?
   ❏ brainstorming?
   ❏ freewriting?
   ❏ talking with someone?
   ❏ clustering?
   ❏ branching?

3. In organizing and creating an outline, did you
   ❏ consider generic patterns, such as comparison and contrast?
   ❏ group similar items?
   ❏ order the items within the groups?
   ❏ organize the groups?
   ❏ avoid common logical problems such as faulty coordination and faulty subordination?

❏ choose an appropriate outline format?
❏ use the outline view in your word processor?

4. In drafting, did you
❏ first make yourself comfortable?
❏ start with the easiest topics?
❏ draft quickly?
❏ keep going when you needed more information?
❏ stop in the middle of a section?
❏ use the styles on your word processor?

5. In revising your document by yourself, did you
❏ let it sit?
❏ read it aloud?
❏ use checklists?

6. In revising the document, did you
❏ use the spell checker and proofread for spelling errors?
❏ use the grammar checker carefully?
❏ use the thesaurus carefully?

7. Did you revise the document by obtaining help from
❏ appropriate subject-matter experts?
❏ people similar to the eventual readers?

8. In revising the document, did you test readers as they used the document if possible?

Exercises

1. Read the online help about using the outline view in your word processor. Make a file with five headings, each of which has a sentence of body text. Practice using the outline feature to do the following tasks:

   a. change a level-one heading to a level-two heading
   b. move the first heading on your outline to the end of the document
   c. hide the body text that goes with one of the headings

2. Read a current article about science or technology from an online newspaper, such as the *New York Times* (www.nytimes.com) or *USA Today* (www.usatoday.com). List the six journalistic questions answered in the article.

3. Create a brief brainstorming list for a report on one of the following topics:

- the need for a technical-communication course tailored to students in your major
- the need for computer-skills training for students in your major
- how to use the Internet to look for a job
- a problem on campus, such as a shortage of parking or inexpensive off-campus housing

Do a 10-minute freewrite on the topic that you selected and then create a cluster sketch or a branching sketch on it.

4. In each of the following excerpts from outlines, identify the logical problem. Then, revise each excerpt to fix the problem.

   a. Arguments for requiring that all students take a computer-skills course:
   • Students would be better prepared for their course work.

- Students would be better prepared for professional employment.
- Students come to college with different levels of computer skills.

b. Problems with inexpensive PCs:
- insufficient multimedia capabilities
  – poor audio quality
  – poor graphics
- insufficient memory
  – not enough RAM
- skimpy documentation
- skimpy software
  – no antivirus software
  – cheap office suites
  – no entertainment software

c. Types of computer printers:
- dot-matrix printers
- laser printers
- ink-jet printers
- color laser printers

5. Your word processor probably contains a number of templates for such documents as letters, memos, faxes, and résumés. Evaluate one of these templates. Is it clear and professional looking? Does it present an effective design for all users or only for some? What changes would you make to the template to improve it?

6. Create a new style by revising one of the styles included in your word processor. Practice changing several of the style's specifications, such as typeface, type size, and indentation. Give your style a new name.

7. Based on instructors' comments on your papers in this course and in previous courses, create a checklist of at least five points that can help you identify potential problems as you revise.

8. Type the following paragraph into your word processor and then use the spell checker and grammar checker to analyze it. Which problems in the passage do the checkers identify? Do the checkers offer useful suggestions for fixing the problems? Do they miss any of the problems in the paragraph?

> The Japanse have a almost mystical feeling for they language. People who speak Japanese are distinct by race and by birth. Native Japanese peoples learn *kokugo*, or national language. On the other hand, foreigners learn *nihingo*, of Japanese language. Leaning nihingo mean that you are racially distinct from the language and culture of native-born Japanese.

Research Project

9. Form groups of three or four for this project. Read one of the two books on usability testing cited in this chapter (see Rubin, as well as Dumas & Redish, in the references section, p. 733). Using students in your class as test participants, carry out an informal usability test on the help instructions on your word processor for performing a particular task. For example, test the help instructions on how to save a file, how to create a new style, how to use a template, or how to number the pages in a document.

C A S E
The Writing Process Online

This case is best for groups of three or four. Assume that your technical-communication instructor, who also teaches first-year writing courses, has asked your group to make a document to help first-year college students learn the writing process. Specifically, your instructor would like a one-page handout that describes the top ten Internet sites explaining the writing process: planning, drafting, and revising. Your instructor suggests that you start your search at Purdue University's Online Writing Laboratory (owl.english .purdue.edu). Following the links from the Purdue site, study a number of sites, noting their strengths. Then create the one-page handout for your instructor.

Writing Collaboratively

4

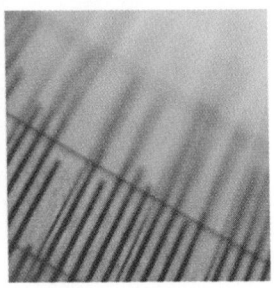

A technical communicator, quoted in Ede and Lunsford (1990, p. 66), comments on the value of knowing how to collaborate effectively:

No one here loses a job because of incompetence; they lose jobs if they can't work with others.

 See TechComm Web (www .bedfordstmartins.com /techcomm) for guidelines boxes and additional links related to topics in this chapter.

Collaborative writing — people working together to create a document — is common in organizations. As documents and the techniques and tools used to produce them become more complex, the amount of collaboration is likely to continue to increase. One survey found that 73.5 percent of 200 college-educated businesspeople collaborate to produce about a quarter of their documents (Faigley & Miller, 1982, p. 567). Another survey found that 87 percent of 520 professionals collaborate at least some of the time (Ede & Lunsford, 1990, p. 20). A third study of more than 400 professionals found that they often write collaboratively (Couture & Rymer, 1989, p. 78).

People collaborate in writing everything from memos to books. The longer, the more complex, or the more important a document, the more likely it is to be written collaboratively. Proposals, reports, manuals, corporate annual reports, and Web sites are often created collaboratively.

Collaboration exists in all kinds of organizations. In some organizations, managers assign tasks for employees to complete. In other organizations, employees have more say in devising their own projects and forming their own collaborative groups (Killingsworth & Jones, 1989).

The chapter begins by describing different patterns of collaboration and the strengths and weaknesses of the technique. It goes on to discuss tasks you want to accomplish in your first group meeting, as well as tips for conducting efficient face-to-face meetings and for communicating diplomatically. It then explains how to critique a group member's draft and how to use groupware and other communication media when collaborators can't work face-to-face. Finally, the chapter discusses the role of gender differences and multiculturalism in collaboration.

Figure 4.1 shows three basic patterns of collaboration.

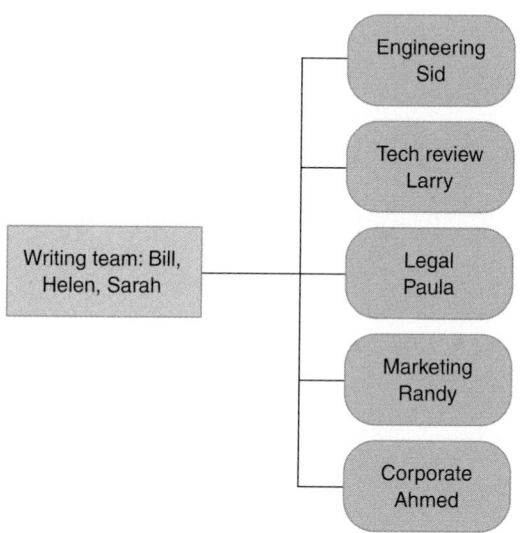

Collaboration based on job specialty. *On this team, an engineer is the subject-matter expert, the person in charge of contributing all the technical information; other professionals are in charge of their own specialties. The writing team writes, edits, and designs the document.*

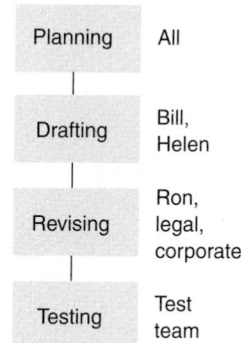

Collaboration based on the stages of the writing process. *Group members collaborate during the planning stage by sharing ideas about the document's content, organization, and style, then establish a production schedule and an evaluation program. They collaborate less during the drafting stage, because drafting collaboratively is much more time consuming than drafting individually. During the revising phase, they return to collaboration.*

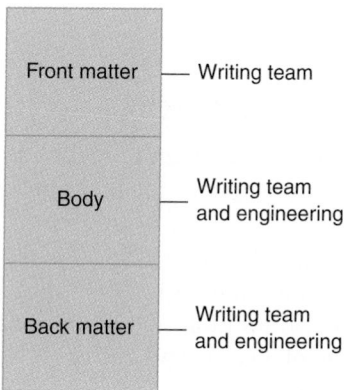

Collaboration based on the section of the document. *One person takes responsibility for one section of the document, another person does another section, and so forth. This pattern is common for large projects with separate sections, such as proposals.*

■ **Figure 4.1**
Patterns of Collaboration

ADVANTAGES AND DISADVANTAGES OF COLLABORATION

Ede and Lunsford's study (1990, p. 50) found that 58 percent of writers considered collaborative writing to be very productive or productive, whereas 42 percent found it not very productive or not at all productive. As a way to create documents, collaboration has both advantages and disadvantages.

Advantages of Collaboration

Writers who collaborate can create a better document and improve the way an organization functions:

- *Collaboration draws on a greater knowledge base.* The more people involved in the writing, the more information the team can draw on. This explains why a collaborative document can often be more comprehensive and more accurate than a single-author document.
- *Collaboration draws on a greater skills base.* No one person can be an expert manager, writer, editor, graphic artist, and production person.
- *Collaboration provides a better idea of how the audience will read the document.* Each group member acts as an audience, asking more questions and suggesting more improvements than one person could writing alone.
- *Collaboration improves communication among employees.* Group members share a goal and therefore get to know one another and learn about each other's jobs, responsibilities, and frustrations.
- *Collaboration helps acclimate new employees to an organization.* Through collaboration, new employees learn how to get things done in the organization: which people to see, what forms to fill out, and so forth. Collaboration also teaches the company's values, such as the importance of ethical conduct and the willingness to work hard and sacrifice for an important initiative.

Disadvantages of Collaboration

Collaboration can also have important disadvantages:

- *Collaboration takes more time than individual writing.* It takes longer to collaborate simply because of the time needed for the collaborators to communicate with each other.
- *Collaboration can lead to groupthink.* When group members place a higher value on getting along than on thinking critically about the subject, they are more prone to *groupthink*. Groupthink, which promotes conformity, can result in an inferior document, because no one wants to cause a scene by asking tough questions about the project.

- *Collaboration can yield a disjointed document.* For example, sections written by different people can contradict each other or contain unintended duplication. And the more people involved, the greater the variation of style in everything from design to punctuation. To prevent these variations, writers need to plan and edit the document carefully.

- *Collaboration can lead to inequitable workloads.* No matter how hard the project leader tries to prevent it, some people will end up doing more work than others.

- *Collaboration can reduce collaborators' motivation to work hard on the document.* The smaller the role a person plays in the project, the less motivated he or she is to make the extra effort.

- *Collaboration can lead to interpersonal conflict.* People can disagree about the best way to create the document or about the document itself. Such disagreements can hurt the working relationship not only during the project but long after.

CONDUCTING MEETINGS

Collaboration requires conducting meetings. If the final document is to succeed, these meetings must be conducted effectively and efficiently.

Conducting meetings involves:
Setting your group's agenda
Conducting efficient face-to-face meetings
Communicating diplomatically
Critiquing a group member's draft

Setting Your Group's Agenda

It's important to get your group off to a smooth start. In the first meeting, start working to define your group's agenda.

GUIDELINES

Setting Your Agenda

▶ *Define the group's task.* What document, or "deliverable," will your group submit? Every group member has to agree, for example, that your task is to revise your company's employee manual by April 10 and that the revision must be no longer than 200 pages. You also need to reach consensus on the more conceptual aspects of the task, including clearly defining the audience, purpose, and scope of the document.

▶ *Choose a group leader.* This person serves as the link between the group and management. (In a school setting, the group leader represents the group in communicating with the instructor.) The group leader also keeps the group on track, leads the meetings, and coordinates communication among group members.

▶ *Define tasks for each group member.* As shown in Figure 4.1 on p. 69, the division of labor can follow different patterns. Naturally, all group members will participate to some extent in each phase of the project, and each group member will review the document at every stage. However, each member will have chief responsibility for a task to which he or she is best suited.

▶ *Establish working procedures.* Members need answers to the following questions — in writing, if possible — before the work of the group proceeds:

  – When and where do we meet?
  – What procedures will we follow in the meetings?
  – How — and how often — are we to communicate with other members of the group, including the group leader?

▶ *Establish a procedure for resolving conflict.* Disagreements about the project are inevitable, but they can also be very valuable. Give every member a chance to express his or her ideas fully, find areas of agreement, and try to resolve the conflict by vote.

See Chs. 10 and 11 for more about establishing a writing style.

▶ *Create a style sheet.* If all group members use a similar style in drafting, the document will need less revision. Discuss as many aspects of style as you can: use of headings and lists, paragraph style and length, level of formality, and so forth. You will probably need to continue this discussion in a follow-up meeting.

Figure 4.2 shows a work schedule form.

▶ *Establish a work schedule.* Starting with the date the document is due, work backward to create a schedule for the entire project. For example, to submit a proposal on February 10, you must complete an outline by January 25, a draft by February 1, and a revision by February 8. Each of these dates is called a *milestone*.

Figure 4.3 on page 74 shows a form that group members can use to evaluate other group members. Figure 4.4 on page 75 shows a self-evaluation form.

▶ *Create evaluation materials.* Group members have a right to know how their participation will be evaluated. In schools, students are likely to evaluate themselves and other group members. In the working world, however, managers are more likely to do the evaluations.

Name of Project:

Principal Reader:

Other Readers:

Group Members:

Milestones	Responsible Member	Status	Date
Deliver Document			
Proofread Document			
Send Document to Print Shop			
Complete Revision			
Review Draft Elements			
Assemble Draft			
Establish Tasks			

Progress Reports	Responsible Member	Status	Date
Progress Report 3			
Progress Report 2			
Progress Report 1			

Meetings	Agenda	Location	Date	Time
Meeting 3				
Meeting 2				
Meeting 1				

Notes

Notice that sometimes milestones are presented in reverse chronological order; the delivery date milestone, for instance, comes first. On other forms, items are presented in normal chronological order.

The form includes spaces for listing the person responsible for each milestone and progress report and for stating the progress of each milestone and progress report.

For printable versions of Figures 4.2, 4.3, and 4.4, click on Forms for Technical Communication on TechComm Web (www.bedfordstmartins.com/techcomm).

■ **Figure 4.2**
Work Schedule Form

Your name: _____

Title of the project: _____

Date: _____

Instructions

Use this form to evaluate the other members of your group. Write the name of each group member in one of the columns, then assign a score of 0 to 10 (0 being the lowest grade, 10 the highest) to each group member for each criterion. Then total the scores for each member. Because each group member has different strengths and weaknesses, the scores you assign will differ. On the back of this sheet, write down any comments you wish to make.

Criteria	Group Members			
1. Regularly attends meetings	1._____	1._____	1._____	1._____
2. Is prepared at the meetings	2._____	2._____	2._____	2._____
3. Meets deadlines	3._____	3._____	3._____	3._____
4. Contributes good ideas in meetings	4._____	4._____	4._____	4._____
5. Contributes ideas diplomatically	5._____	5._____	5._____	5._____
6. Submits high-quality work	6._____	6._____	6._____	6._____
7. Listens to other members	7._____	7._____	7._____	7._____
8. Shows respect for other members	8._____	8._____	8._____	8._____
9. Helps to reduce conflict	9._____	9._____	9._____	9._____
10. Your overall assessment of this person's contribution	10._____	10._____	10._____	10._____
Total Points	_____	_____	_____	_____

■ **Figure 4.3**
Evaluation Form

Your name: _____   Date: _____
Title of the project: _____

Instructions

On this form, record and evaluate your own involvement in this project. In the Log section, record the activities you performed as an individual and you performed as part of the group. For all activities, record the date and the number of hours you spent. In the Evaluation section, write two brief statements, one about aspects of your contribution you think were successful and one about the aspects you want to improve.

Log Individual Activities	Date	Number of Hours
Activities as Part of Group	Date	Number of Hours

Evaluation
Aspects of My Participation That Were Successful

Aspects of My Participation That I Want to Improve in the Future

■ **Figure 4.4**
Self-Evaluation Form

Conducting Efficient Face-to-Face Meetings

Human communication is largely nonverbal. That is, people communicate not only through the words they speak but through the tone, rate, and volume of their speech. They also communicate through body language and in many other ways. For this reason, face-to-face discussions provide the most information about what a person is actually thinking and feeling — and the best opportunity for group members to learn to understand one another.

For an excellent discussion of how to conduct meetings, see Matson (1996) at www.fastcompany.com /online/02/meetings.html.

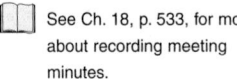

See Ch. 18, p. 533, for more about recording meeting minutes.

> **GUIDELINES**
>
> ### Conducting Efficient Meetings
>
> ▶ *Arrive on time.* If you know you will have to miss a meeting, notify the group leader as soon as possible.
>
> ▶ *Stick to an agenda.* Create the agenda beforehand so that everyone can come to the meeting prepared. Don't stray too far from the agenda. If you find you need to discuss an important point that isn't on the agenda, schedule another meeting.
>
> ▶ *Record the important decisions made at the meeting.* One group member should serve as secretary by making a record of the meeting focusing on any decisions the group makes and any tasks to be carried out after the meeting.
>
> ▶ *Summarize your accomplishments and make sure every member understands what his or her assignment is.* The group leader should formally close the meeting by summarizing the progress the group has made in the meeting and stating the tasks each group member is to work on before the next meeting. If possible, the secretary should give each group member this informal set of meeting minutes.

Communicating Diplomatically

Because collaborating on an important project is stressful, it can lead to interpersonal conflict. People can become frustrated and angry with each other because of personality clashes or because of disputes about the process of collaborating or about the substance of the project itself. But for the project to succeed, group members have to be able to work together productively. When you speak in a group meeting, you want to appear helpful, not critical or overbearing.

> **GUIDELINES**
>
> ### Communicating Diplomatically
>
> ▶ *Listen carefully.* Set aside any preconceptions based on the speaker's age, race, appearance, or sex. Maintain eye contact with the speaker,

paying attention to the ideas, not to the words themselves. Examine the speaker's use of evidence and logic. Compare the speaker's message with what you already know about the subject. Think of questions you might ask later.

▶ *Let the speaker finish.* Don't interrupt.

▶ *Give everyone a chance to speak.*

▶ *Avoid personal remarks and insults.* Be tolerant and respectful of other people's views and working methods. Doing so is right, both ethically and practically. It is ethical to treat people respectfully, as you want them to treat you. And it is practical: if you anger people, they will go out of their way to oppose you.

▶ *Don't overstate your position.* A modest qualifier such as "I think" or "it seems to me" is an effective signal to your listeners that you understand and acknowledge the possibility that everyone may not share your point of view.

OVERBEARING My plan is a sure thing; there's no way we're not going to kill Allied next quarter.

DIPLOMATIC I think this plan has a good chance of success: we're playing off our strengths and Allied's weaknesses.

In the diplomatic version, notice the speaker's decision to call it "this plan" rather than "my plan."

▶ *Don't get emotionally attached to your own ideas.* When you meet opposition, try to understand why other group members don't agree with you. Digging in is usually unwise — unless the matter is a serious one of principle — because, although you may be right and everyone else wrong, it's not likely.

▶ *Ask pertinent questions.* Don't be afraid that you will appear ignorant; the brightest people constantly try to understand what they hear and to connect it to other ideas. Asking pertinent questions also helps other group members by encouraging them to examine what they hear.

▶ *Pay attention to nonverbal communication.* People communicate through tone of voice, facial expression, and body language. Bob might *say*, for example, that he understands a point, but his facial expression might reveal that he doesn't. If a group member's body language suggests confusion, ask him or her about it. A direct question is likely to elicit a statement that will help the group clarify its discussion.

Critiquing a Group Member's Draft

In collaborating, group members often critique drafts written by other group members. Knowing how to offer criticism without offending the writer is a valuable skill.

GUIDELINES

Critiquing a Draft

▶ *Start with a positive comment.* Even if the draft is weak, begin with a positive statement: "You've obviously put a lot of work into this, Joanne. Thanks." Or, "This is a really good start. Thanks, Joanne."

▶ *Discuss the larger issues first.* Begin with the organization and development of the draft, the use of logic and evidence, and the design and use of graphics. Then work on paragraph development, and then on sentence-level matters and word choice. Leave editing and proofreading until the end of the process.

▶ *Talk about the writing, not the writer.*

| RUDE | You don't explain clearly why this criterion is relevant. |
| BETTER | I'm having trouble understanding how this criterion relates to the topic. |

▶ *Focus on the group's document, not on the group member's draft.* Your goal is to improve the quality of the document you are working on, not to evaluate the writer or the draft. Offer suggestions about how to make the communication more effective.

| RUDE | Why didn't you include the price comparisons here, like you said you would? |
| BETTER | I wonder if the report would be stronger if we include the price comparisons here. |

In the better version, the speaker is focusing on the group's goal — to create an effective report — rather than on the writer's failure to include some information. Also, the speaker qualifies his recommendation by saying, "I wonder if. . . ." This approach sounds constructive rather than boastful or annoyed.

USING GROUPWARE AND OTHER COMMUNICATION MEDIA

Electronic media are useful collaborative tools for two reasons. First, face-to-face meetings are not always possible or convenient. Electronic media enable people to communicate *asynchronously*. That is, one person can send an email to another early in the day, and the recipient can read it later that day, even after work.

Second, electronic communication is digital. Group members can store and revise comments and drafts, incorporating them as the document develops. Face-to-face meetings are excellent for communicating ideas and reaching consensus, but unless someone takes notes and then turns those notes into an electronic file, the information can be lost to group members.

This section examines three important skills used in communicating electronically.

Communicating electronically may require:
Using the comment, revision, and highlighting features on a word processor
Using email to send files
Using groupware

Using the Comment, Revision, and Highlighting Features on a Word Processor

Word processors offer three powerful features you will find useful in collaborative work.

The *comment feature* lets a reader add electronic comments to a writer's electronic file. The text that has been commented on appears as highlighted material. When the writer moves a mouse over this highlighted material, the comment pops up. These comments can also be printed out. Figure 4.5 illustrates the comment feature.

The *revision feature* lets a reader mark up a text by deleting, revising, and adding words while allowing the writer of the text to keep track of the suggested changes.

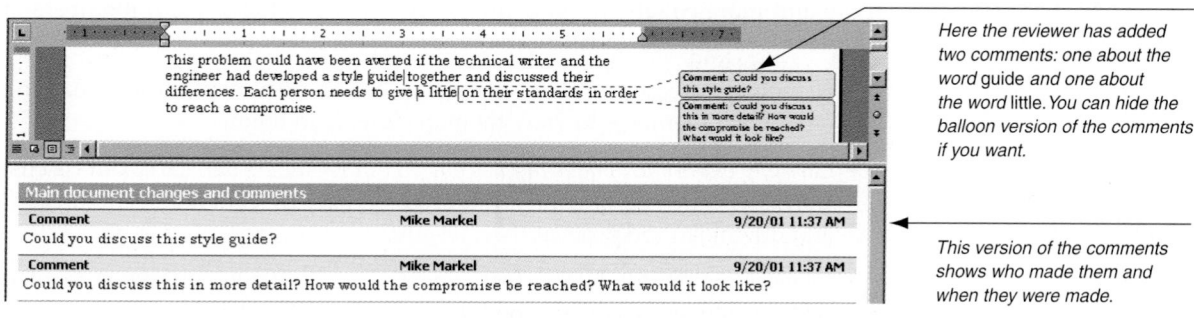

Here the reviewer has added two comments: one about the word guide and one about the word little. You can hide the balloon version of the comments if you want.

This version of the comments shows who made them and when they were made.

■ **Figure 4.5   The Comment Feature on a Word Processor**

> This problem could have been averted if the technical writer and the engineer had developed a style guide together and discussed their differences. Both needed to ~~give a little~~ compromise on their standards in order to reach an agreement.

In this passage, the reader is recommending that the writer substitute the word "compromise" for the phrase "give a little." (The vertical rule in the left margin, called a *change bar*, indicates that a change has been made on that line.) The writer can accept or reject any of the reader's suggestions. The revision feature allows different readers to recommend changes and the writer to know which reader made which recommendation.

The *highlighting feature* lets a reader use one of about a dozen "highlighting pens" to call the writer's attention to a particular passage.

> This problem could have been averted if the technical writer and the engineer had developed a style guide together and discussed their differences. Both needed to compromise on their standards in order to reach an agreement.

Each reader can use a different color, or a single reader can use one color to signal one kind of comment and another color to signal another kind. The writer can remove the highlighting with a mouse click.

Using Email to Send Files

Most email software lets you "attach" a file to an email message. This means that you can easily send a formatted file, such as a document written in WordPerfect or a spreadsheet in Excel, to anyone around the world. A recipient who has the software in which the document was saved will be able to open it.

Using Groupware

Groupware, a category of software that helps people work together, lets people at the same or different locations plan, draft, revise, and track a document. You may already be familiar with groupware programs such as Lotus Notes™ or Microsoft NetMeeting™.

 For a tutorial on NetMeeting's collaborative tools, see www.microsoft.com /netmeeting/demo/demo _plain.htm.

Manufacturers of office suites such as Microsoft and Corel are building more and more collaboration features into their products. Team members at different locations can perform five important collaborative activities:

- *Sharing files.* Team members can post files to a document library, enabling other team members to view them or download them.

- *Carrying out asynchronous discussions.* Team members can carry out discussions by posting comments to a discussion list. All team members can read and download the posts at their leisure.

- *Commenting on documents.* Team members can attach comments to files, without actually changing the files.

- *Distributing announcements.* Team members can post announcements, such as reminders about upcoming deadlines or revisions of the team's schedule.

- *Creating automated change notifications.* Team members can sign up to be notified by email when a document has been changed.

Figure 4.6 shows a discussion list (About.com, 2001).

Several groupware programs also offer two addition tools:

- *Whiteboards.* Whiteboard software lets people at different locations draw on the screen as if they were all in a room with a whiteboard. Anything drawn on one screen is displayed immediately on every screen. The image can be printed or saved as a file. Figure 4.7 shows the whiteboard screen from Microsoft's NetMeeting.

- *Videoconferencing.* Standard videocams can be connected to computers, enabling people to see each other as they talk. Figure 4.8 (PictureTel, 1998) shows a videoconference.

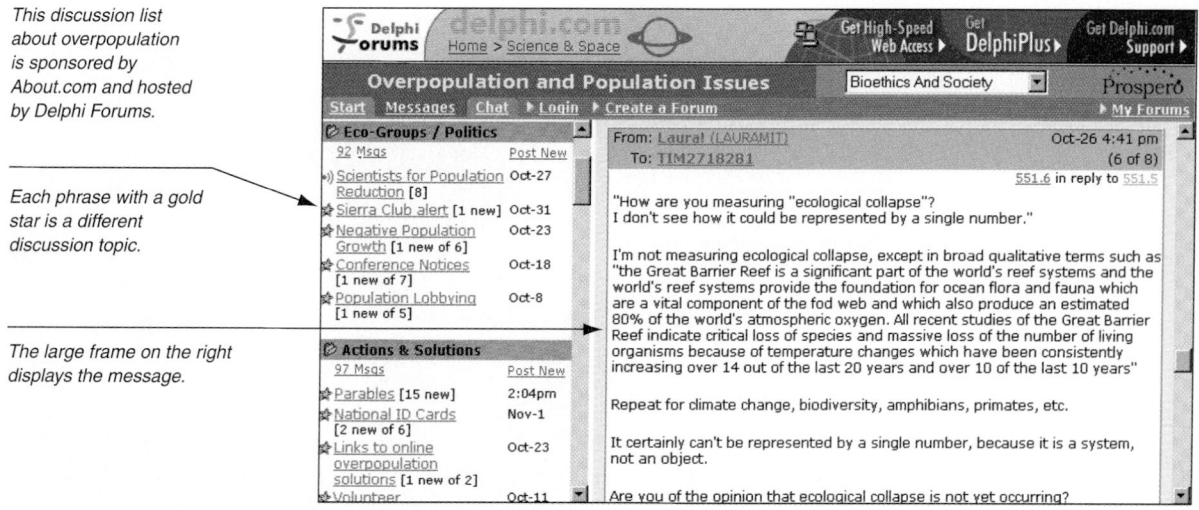

■ **Figure 4.6   A Discussion List**

A discussion list is a Web-based asynchronous forum that lets users post messages.

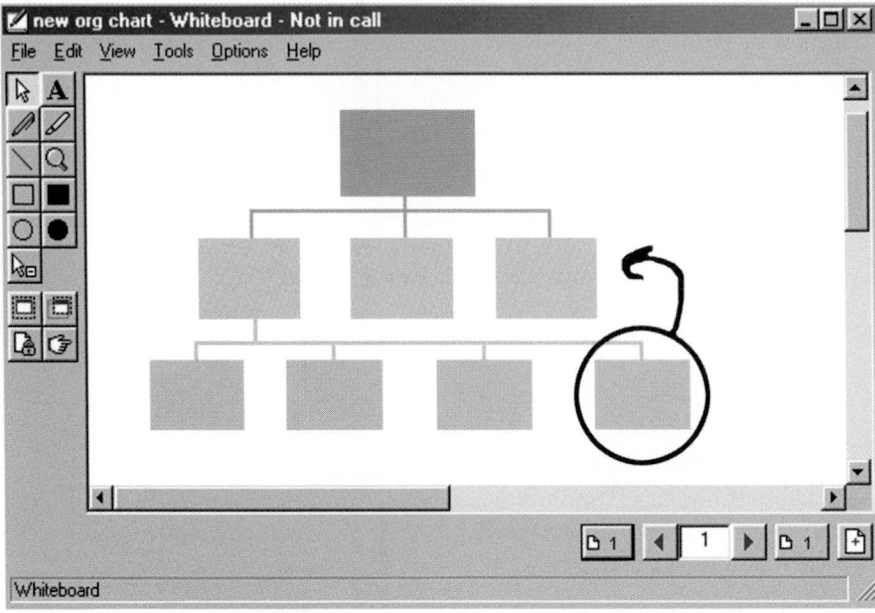

■ **Figure 4.7   A Whiteboard Screen**

All the people participating in the whiteboard session see the same screen. Whenever anyone changes anything on his or her screen, all participants see the change on their screens.

The camera mounted on top of the monitor sends a video image to the other group of participants. More sophisticated videoconferencing systems use room-mounted cameras that can capture the image of everyone in a room.

Most videoconferencing systems let participants display additional "windows" on the screen, such as computer files or desktops.

■ **Figure 4.8   A Videoconference**

GENDER AND COLLABORATION

Effective collaboration involves two related challenges: maintaining the group as a productive, friendly working unit and accomplishing the task. Scholars of gender and collaboration see these two challenges as representing the feminine and the masculine perspectives.

Any discussion of gender studies must begin with a qualifier: when we talk about gender, we are generalizing; we are not talking about particular people. In other words, the differences in behavior between two men or between two women are likely to be greater than the differences between men and women in general.

The differences in how the sexes communicate and work in groups have been traced to every culture's traditional family structure. As the primary caregivers, women have learned to value nurturing, connection, growth, and cooperation; as the primary breadwinners, men have learned to value separateness, competition, debate, and even conflict (Chodorow, 1978).

For decades, scholars have studied the speech differences between women and men. Women tend to use more qualifiers and "tag" questions, such as "Don't you think?" (Tannen, 1990). Some scholars, however, suggest that women might be using these patterns because it is expected of them, and that they use them mainly in groups that include men (McMillan, Clifton, McGrath, & Gale, 1977). Many experts caution against using qualifiers and tag questions, which can suggest subservience and powerlessness.

The study of women and men in collaborative groups is a new research subject, but as might be expected, women appear to value consensus and relationships, to show more empathy, and to demonstrate superior listening skills (Borisoff & Merrill, 1987). Men appear to be more competitive and more likely to assume leadership roles.

Although many questions remain unanswered, it seems clear that women's communication patterns are more focused on maintaining the group, and men's on completing the task. For instance, women may talk more about topics unrelated to the task (Duin, Jorn, & DeBower, 1991), but this talk is central to maintaining group coherence, which is necessary if the group is to accomplish its task.

Scholars of gender recommend that all professionals strive to achieve an androgynous mix of the skills and aptitudes commonly associated with both women and men. According to William Eddy (1983, p. 278):

> Traditional male traits of task focus, objectivity, confrontation, and control are clearly important in many situations. . . . But to build and lead groups that attain effectiveness and viability by fully utilizing their human resources, you also need some of the traits traditionally thought of as female. . . . It is not surprising, when you think about it, why an androgynous combination of skills is best.

For an excellent review of the literature, see Lay (1994).

MULTICULTURALISM AND COLLABORATION

For more about multiculturalism, see Ch. 5, p. 107.

Most collaborative groups in industry and in the classroom include people from other cultures. The challenge for all group members is to understand the ways in which cultural differences can affect group behavior.

Some 70 percent of the world's cultures value the family, the community, and the corporation more highly than the individual (Thiederman, 1991). As Bosley (1993) points out, this fact suggests that people from these cultures would be excellent participants in collaborative projects. But this conclusion is not necessarily accurate.

Collaborative groups in the United States are usually based on the premise that the individual should speak up. However, people from other cultures often find it difficult to assert themselves in such collaborative groups. In a study of college students from other cultures in collaborative groups, Bosley (1993, p. 57) describes several kinds of behavior that differ from those of students from the United States. Students from other cultures

- may be unwilling to respond with a definite "no"
- may be reluctant to admit when they are confused or to ask for clarification
- may avoid criticizing others
- may avoid initiating new tasks or performing creatively

Even the most benign gesture of friendship on the part of a U.S. student can cause confusion. If a U.S. student casually asks a Japanese student about her major and the courses she is taking, the Japanese student might find the question too personal but consider it perfectly appropriate to talk about her family and her religious beliefs (Lustig & Koester, 1993, p. 234).

This very brief discussion of multiculturalism is meant only to give you a sense of the range and variety of cultural differences you are likely to encounter. Although there is no substitute for immersing yourself in another culture, it is usually impractical. The best alternative is to try to remain open to encounters with people from other cultures without jumping to conclusions about what their actions might or might not mean.

A good first step is to read a full-length discussion of multiculturalism, including any or all of these four respected books:

Hall, E. T., & Reed Hall, M. (1990). *Understanding Cultural Differences.* Yarmouth, ME: Intercultural Press.

Hofstede, G. (1991). *Culture and Organizations: Software of the Mind.* New York: McGraw-Hill.

Trompenaars, F. (1993). *Riding the Waves of Culture: Understanding Cultural Diversity in Business.* London: Nicholas Brealey.

Victor, D. A. (1992). *International Business Communication.* New York: HarperCollins.

✔   **Collaborator's Checklist**

1. In your first group meeting, did you
   ❑ define the group's task?
   ❑ choose a group leader?
   ❑ define tasks for each group member?
   ❑ establish working procedures?
   ❑ establish a procedure for resolving conflict?
   ❑ create a style sheet?
   ❑ establish a work schedule?
   ❑ create evaluation materials?

2. To conduct efficient face-to-face meetings, do you
   ❑ arrive on time?
   ❑ stick to an agenda?
   ❑ make sure that a group member records important decisions made at the meeting?
   ❑ make sure that a group member summarizes your accomplishments and that every member understands what his or her assignment is?

3. To communicate diplomatically, do you
   ❑ listen carefully?
   ❑ let the speaker finish?
   ❑ let others talk?
   ❑ avoid personal remarks and insults?
   ❑ avoid overstating your position?
   ❑ avoid getting emotionally attached to your own ideas?
   ❑ ask pertinent questions?
   ❑ pay attention to body language?

4. In critiquing a group member's draft, do you
   ❑ start with a positive comment?
   ❑ discuss the larger issues first?
   ❑ talk about the writing, not the writer?
   ❑ focus on the group's document, not on the group member's draft?

5. If appropriate, do you
   ❑ use the comment, revision, and highlighting features on a word processor?
   ❑ use email to send files?
   ❑ use groupware?

Exercises

1. Experiment with the comment, revision, and high-lighting features on your word processor. Using online help if necessary, learn how to make, revise, and delete comments; make, undo, and accept revisions; and add and delete highlights.

2. Using a search engine, find email shareware or free-ware on the Internet. Download the software and install it on your computer at home. Learn how to use the feature that lets you send attached files.

3. See whether your Internet browser has any of the groupware functions discussed in this chapter, such as synchronous chat, document revision, or white-boards. Install the function and use it with other members of your group. Print out a document you have created using the groupware.

Research Projects

Some of the following exercises ask you to conduct an interview, and some ask you to write a memo. See Chapter 7, page 171, for a discussion of interviewing and Chapter 15, page 430, for a discussion of memos.

4. Interview a professor in your department who has published a co-written article. Ask this professor to describe the collaborative techniques the co-writers used in writing the article and to evaluate the strengths and weaknesses of these techniques. Write a memo to your instructor that briefly sum-marizes the professor's main points and then ana-lyze the information you have gathered. In what ways and to what extent do the professor's insights confirm the discussion about collaboration in this chapter?

5. Interview a technical communicator or technical professional in your community who has partici-pated in a collaborative project. Ask this person to describe the collaborative process and to evaluate its strengths and weaknesses. Write a memo to your instructor that briefly summarizes the technical communicator's main points and analyzes the information you have gathered. In what ways and to what extent do the technical communicator's insights confirm the discussion about collaboration in this chapter?

6. Form small groups. Each member of your group is to attend the same lecture or view the same inter-view show on television (such as *The NewsHour with Jim Lehrer* on the Public Broadcasting System or one of the Sunday morning news shows) and

take notes. Afterward, meet with the other mem-bers of the group to discuss what you have learned. Concentrate on the following questions:

- What were the central points established in the lecture or interview?
- Were the points made clearly or did you have trouble understanding them?
- What techniques might the speaker have used to improve his or her ability to convey the informa-tion?
- Do you think your gender or culture affected how you listened to the lecture or interview?
- What suggestions can you make that might help you improve your listening skills?

Write a memo to your instructor presenting your findings.

7. Form groups and perform a collaborative project like the one described in Exercise 6. Each of you should keep a log of your experiences covering such topics as the following:

- What procedures were used to assign tasks?
- What procedures were used to establish policies for meetings and submitting work?
- How helpful were your contributions — and those of the other group members — both in meetings and in submitting work?
- How effective were you and the other members in establishing a cooperative and efficient envi-ronment for working?
- Did conflicts arise and, if so, how did you and the other group members try to resolve the situation?

How effective were these efforts? How might they have been improved?

Meet as a group and discuss your logs. Then write a memo to your instructor presenting your findings.

8. This project requires a basic understanding of the Web (see Chapters 7 and 21). Your college or university wishes to update its Web site to include a section called "For Prospective International Students." Along with members of your group, first determine whether your school already has information of particular interest to prospective international students. If it does, write a memo to your instructor describing and evaluating the information. Is it accurate? Comprehensive? Clear? Useful? What kinds of information should be added to the site to make it more effective?

If the school's site does not have this information, perform the following tasks:

- *Plan.* What kind of information should it include? Does some of this information already exist, or does it all have to be created from scratch? For example, can you create a link to information on how to obtain a student visa, or does this information not exist on the Web? Write an outline of the main topics that should be covered.
- *Draft.* Write the following sections: "Where to Live on or Near Campus," "Social Activities on or Near Campus," and "If English Is Not Your Native Language." What graphics could you include? Are they already available? What other sites should you link to for these three sections?

In a memo, present your suggestions to your instructor.

CASE
The Reluctant Collaborator

Some of the projects in your technical communication course are to be completed collaboratively. The instructor has assigned you to a group; the other members of the group are Allison and Ken. Allison is a senior in mechanical engineering; Ken is a junior in computer science. You are a senior in accounting. Along with your collaborative assignments, each member of the group is to submit two forms presented in this chapter (Figures 4.3 and 4.4): the evaluation form for the other members of the group and the self-evaluation form. In addition, each member is invited to submit any notes, drafts, or other documents to show the instructor the work he or she contributed to the project.

At your first group meeting, you choose a leader: Allison. As a senior in mechanical engineering, she has had a lot of experience working collaboratively and is happy to take on the task of supervising your group's work. The first meeting, to discuss topics for a research proposal, is scheduled for 7:00 the next night in Conference Room 3 in the library.

It is 7:15, the night of the meeting, and Ken has not arrived. Neither Allison nor you had heard from Ken before the meeting with word that he would be late. At 7:30, you and Allison decide to phone Ken. There is no answer.

You and Allison are angry that Ken hasn't arrived or sent a friend to tell the group that he will be late or will have to cancel. The two of you are unwilling to start work on the project without Ken. You decide to cancel the meeting. Allison tells you she will call him from home later that evening.

 For additional cases, click on Case of the Month and Archive on TechComm Web (www.bedfordstmartins.com /techcomm).

At about 9:00 you get a call from Allison. She reached Ken a few minutes before. He said he had car trouble that night and apologized. Allison asks you if the group could meet the next afternoon at 3:00 in the library. Allison tells you that Ken agreed to her request that each group member bring a list of three possible topics to the meeting. You agree and write the time in your appointment book.

The next day, at 3:15, Ken rushes into the room. "Sorry I'm late," he says with a smile. You and Allison give him a cold look. Allison begins the discussion.

"Okay, we each agreed to bring three topics to the meeting today. Ken, what have you got?"

"That was for today? I'm really sorry. I just blew that off."

Allison turns to you, and you read off your three topics. Then she reads hers. The three of you agree to select one of Allison's topics: a feasibility study of software used in creating engineering drawings. You set a time for the next meeting and decide on an agenda.

As you conclude the meeting, Allison asks you if you could stay a minute. Ken leaves.

"I'm not happy with what I see developing with Ken," she says. "We're going to be slowed down if he doesn't come to the meetings or prepare. What do you think we should do?"

In a memo to your instructor, respond to the following questions: How should you respond to Allison? Should you merely hope that Ken starts to participate more responsibly? Is there some way to delegate tasks that will motivate Ken to participate more actively? Should you and Allison go ahead with the project, letting Ken participate when he chooses to? Should Allison talk with him? Should both of you talk with him together? Should you go to your instructor? See Chapter 15, p. 430, for a discussion of memos.

PART TWO
PLANNING THE DOCUMENT

Analyzing Your Audience and Purpose

5

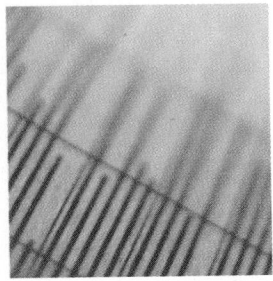

George Rimolower, a translator, comments on the need to consider the cultural background of your audience as you plan your communication:

Recently, McDonald's printed take-out bags decorated with flags from around the world. On the bag was additional text translated into the different languages of the countries the flags depicted. Had McDonald's taken the time to consult with a cultural expert, they would have avoided committing the major faux pas of printing bags that contained a Saudi flag with holy Koran scripture. This was extremely offensive to the Saudis. To print the Saudi flag, which contains holy scripture, on a disposable bag that would wind up in a garbage heap was sacrilegious. (Personal communication with author)

The content and form of every technical document you write are determined by the writing situation: your audience and purpose. Understanding the writing situation helps you devise a strategy to meet your readers' needs — and your own.

Audience and purpose are not unique to technical communication. Most everyday communication relies on both of these elements equally. When a classified advertisement describes a job, the writing situation of the advertiser is clear:

AUDIENCE   prospective applicants

PURPOSE   to describe the job opening so that qualified persons will apply

Once you have defined the two basic elements of your writing situation, you must analyze each of them before deciding what to say and how to say it.

This chapter begins with a discussion of how to understand the basic categories of readers — experts, technicians, managers, and general readers. Then it explains how to analyze the individual characteristics of your readers, discusses how to address multiple audiences and multicultural audiences, and finally, explains how to create a clear statement of purpose for your document.

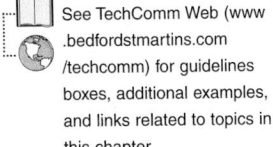

See TechComm Web (www .bedfordstmartins.com /techcomm) for guidelines boxes, additional examples, and links related to topics in this chapter.

ANALYZING AN AUDIENCE

In her book *Dynamics in Document Design*, Karen Schriver (1997) describes three approaches to analyzing an audience:

• *Thinking*. If you are writing an employee-benefit manual, you think about the similar manuals you have used, noting their strengths and weaknesses.

You study samples of these manuals, noting what works and what doesn't. You brainstorm with your collaborators on the project, trying to think of good ideas.

- *Listening.* You go to people who already have the current manual and ask for their opinions. Do they know they have the manual? Do they use it? What do they like or dislike about it? You distribute questionnaires or perform usability testing. As you start to create the new manual, you show potential readers drafts and ask for critiques.

For more on questionnaires, see Ch. 7, p. 174. For more on usability testing, see Ch. 3, p. 60.

- *Classifying.* By classifying readers, you can decide what kind of information to include, how to organize it, what kind of writing style to use, and so forth. You classify your readers into categories on the basis of their knowledge of your subject.

As Schriver points out, it is best to use all three approaches. The following discussion describes in more detail how to classify your audience and use the information you gain to help you plan the document.

IDENTIFYING PRIMARY AND SECONDARY AUDIENCES

In a college course, you have some idea of the professor's expectations for written assignments because he or she has given you guidelines. In the working world, however, you probably won't have explicit guidelines, and you will often write to people you have never met. In cases such as these, start by classifying your readers into two categories:

- A *primary audience* of people who have a direct role in responding to your document. They might be readers who use the information you provide in doing their jobs. They might evaluate and revise your document, or they might act on your recommendations. An executive who decides whether to authorize building a new production facility is a primary reader. So is the treasurer, who has to determine whether the organization can pay for it.

- A *secondary audience* of people who need to stay abreast of developments in the organization, such as salespeople who want to know where a new facility will be located, what products it will produce, and when it will be operational. A secondary audience will not directly act on or respond to your document.

Naturally, the needs of your primary audience are more central than those of your secondary audience. If, for example, several members of your primary audience need to know the financial aspects of the project, you should provide that information prominently. But if only members of a secondary audience will need that information, you should probably put it in a less prominent part of the document.

The next step in classifying your audience is to determine what basic categories they might belong to.

BASIC CATEGORIES OF READERS

See Online Technical Writing (www.io.com/~hcexres /tcm1603/acchtml/acctoc .html) for more about audience analysis.

Although each person is unique, try to classify your readers on the basis of their knowledge of your subject. In general, every reader can be classified into one of four categories in the following list.

Categories of readers:
The expert
The technician
The manager
The general reader

Of course, these categories are only generalizations, but most people fit into one or perhaps two of them. Ellen DeSalvo, for example, a Ph.D. in materials engineering, would be an expert in that particular field; she might also be the manager of the materials group at her company.

The Expert

The expert is a highly trained individual with an extensive theoretical and practical understanding of the subject. Often an expert carries out basic or applied research and communicates these research findings. Here are some examples of experts:

a *physician* trying to understand the AIDS virus who delivers papers at professional conferences and writes research articles for scholarly journals

an *engineer* who is trying to devise a simpler, less expensive test for structural flaws in composite materials

a *forester* who is trying to plan a strategy for dealing with the threat of forest fires during a season of low rainfall

In short, almost everyone with a postgraduate degree — and many people with an undergraduate degree in a technical field — is an expert in one area.

Because experts share a curiosity about their subject and a detailed understanding of its theory, they usually have no trouble understanding technical concepts, formulas, graphics, and vocabulary. Therefore, when you write for experts, you can get right to the details of the technical subject without spending time on the fundamentals. In addition, most experts are comfortable with long sentences, if the sentences are well constructed and no longer than necessary.

Figure 5.1, an excerpt from a description of the Mars Climate Orbiter (Jet Propulsion Lab, 1997b), which crashed into Mars in 1999, illustrates the needs and interests of the expert reader.

The Technician

The technician takes the expert's ideas and turns them into real products and procedures. The technician fabricates, operates, maintains, and repairs mechanisms, and sometimes teaches other people how to operate them. An engineer having a problem with an industrial laser will talk the situation over with a technician. After they agree on a possible cause of the problem and a way to fix it, the technician will go to work.

Like experts, technicians are very interested in their subject, but they know less about theory. They work with their heads and their hands. Technicians have a wide variety of educational backgrounds. Some have a high-school education, while others have attended trade schools or earned an associate's degree or even a bachelor's degree. When you write for technicians, keep in mind that they do not want complex theoretical discussions.

The Mars Climate Orbiter mission aims to provide information about the cycles of water, carbon dioxide, and dust on Mars. The orbiter will study the planet's weather for one year, acquiring data to help scientists better understand the Martian climate. Two instruments aboard the orbiter, the Pressure Modulator Infrared Radiometer (PMIRR) and the Mars Color Imager (MARCI), will collect this data. PMIRR will observe the global distribution and time variation of temperature, pressure, dust, water vapor, and condensates in the Martian atmosphere. MARCI will observe Martian atmospheric processes at global scale and study details of the interaction of the atmosphere with the surface at a variety of scales in both space and time. In addition to the science payload, the orbiter spacecraft will provide an on-orbit data relay capability for future U.S. and/or international surface stations. Below is a list of the primary scientific goals of the mission:

- Observe the Martian climate from a 400 km near circular, near polar mapping orbit.
- Examine general atmospheric circulation patterns and how they affect atmospheric transport and climate change.
- Derive information about atmospheric winds from global temperature observations.
- Observe atmospheric dust to better understand the seasonal dust cycle, including initiation, spreading, and dissipation of global-scale dust storms.
- Examine features on the Martian surface that can provide information about climatic evolution.
- Gain detailed information about interactions occurring between the atmosphere and surface at many scales of space and time.

The writer uses technical vocabulary to describe the purpose of the instruments.

The writer uses metric units, which still are uncommon among general readers in the United States.

All the items in the bulleted list refer to technical concepts that only experts in this field would understand.

■ **Figure 5.1   Writing Addressed to an Expert Audience**

This passage, from the Jet Propulsion Lab Web site, discusses advanced concepts of interest to astronomers.

They want to finish a task safely, effectively, and quickly. Therefore, they need schematic diagrams, parts lists, and step-by-step instructions. Most technicians prefer short or medium-length sentences and common vocabulary, especially in documents such as step-by-step instructions.

Figure 5.2, an excerpt from the Jet Propulsion Lab Web site (1997a), illustrates the needs and interests of the technician.

The Manager

The manager is harder to define than the technical person, because *manager* describes what a person does rather than what a person knows. A manager makes sure an organization operates smoothly and efficiently. For instance, the manager of the procurement department at a manufacturing plant sees that raw materials are purchased and delivered on time so that production will not be interrupted.

Upper-level managers, known as executives, address longer-range concerns. They foresee problems years ahead by considering questions such as the following:

Writing addressed to technicians often includes very little text. This excerpt about the Mars Climate Orbiter flight system consists of a diagram and a set of specifications.

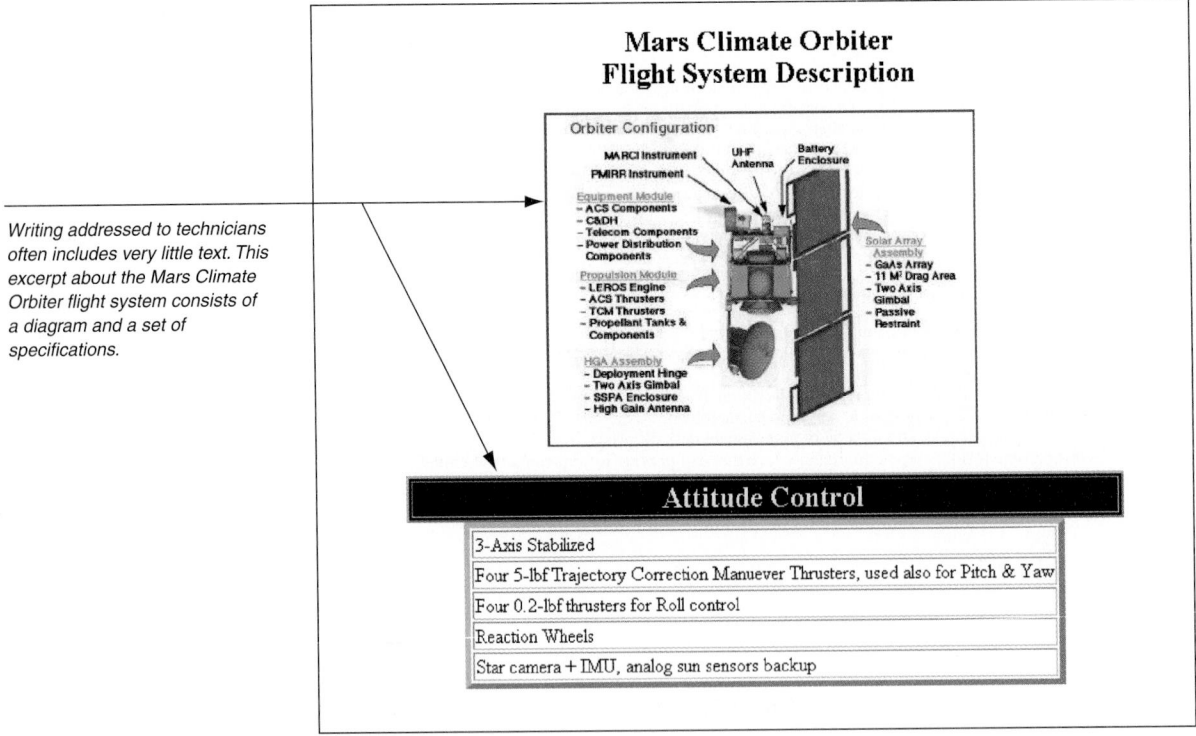

■ **Figure 5.2   Writing Addressed to Technicians**

- Is current technology at the company becoming obsolete?
- What are the newest technologies?
- How expensive are they?
- How much would they disrupt operations if they were adopted?
- What other plans would have to be postponed or dropped altogether?
- When would the conversion start to pay for itself?
- What has been the experience of other companies that have adopted these new technologies?

Executives are concerned with these and dozens of other broad questions that go beyond day-to-day managerial concerns.

Managers might or might not have a technical background. Although generalizing about the typical manager's educational background is difficult, identifying the manager's needs is not: managers want to know the bottom line. They have to get a job done on schedule; they don't have time to consider theory in the way an expert does. Rather, managers must juggle constraints — financial, personnel, time, and informational — and make logical and reasonable decisions quickly. And they have to communicate with their own supervisors.

In writing directed to a manager, try to determine his or her technical background and then choose an appropriate vocabulary and sentence length. Focus on practical information. For example, if you are a police officer describing a new product to the chief of police, you might begin with some theoretical background but then concentrate on the product's capabilities and its advantages over competing products.

If you know that your reader will take the information that you've provided and use it in a document addressed to higher-level managers or executives, make your reader's job easier. Include an executive summary (see Ch. 12, p. 316) and use frequent headings (Ch. 10, p. 252) to highlight the major points you are making. Ask your reader if there is an organizational pattern, a format, or a strategy for writing the document that will help him or her in using your document as source material.

Figure 5.3 on page 98, an overview of the Mars Climate Orbiter mission (Jet Propulsion Lab, 1999a), illustrates the interest, goals, and needs of the manager.

The General Reader

Often your writing will address the general reader, sometimes called the *layperson*. A nuclear scientist reading about economics is a general reader, as is a historian reading about astronomy.

The layperson reads out of curiosity or self-interest. The typical article in the magazine supplement of the Sunday newspaper — on attempts to increase the populations of endangered species in zoos, for example — will attract the

This passage focuses on the overall goals and milestones of the program, not its details.

The National Aeronautics and Space Administration (NASA) has initiated a long-term systematic program of Mars exploration, the Mars Surveyor Program (MSP). The highest priority scientific objectives of this program are to:

- search for evidence of past or present life
- understand the climate and volatile history of Mars
- assess the nature and inventory of resources on Mars

The common thread of these objectives is water: past and present sources and sinks; exchanges between subsurface, surface and atmospheric reservoirs; and the change of volatiles over time.

Notice the focus on "scientific return": achieving good scientific results at a reasonable cost.

The goal of the Mars Surveyor Program is to carry out low-cost missions, each of which provides important, focused, scientific return, and which will in sum constitute a major element of the scientific exploration of Mars. A series of lander and orbiter spacecraft are being launched at each favorable Mars launch opportunity, which occur approximately every 26 months. In 1997 the Mars Surveyor Program launched the Mars Global Surveyor (MGS), which together with the launch of the Discovery Program's Mars Pathfinder lander, initiated the new era of Mars exploration.

In the 1998–1999 launch opportunity, the Mars Surveyor Program successfully launched the following spacecraft, which are now in flight to Mars:

- Mars Climate Orbiter (MCO) launched on December 11, 1998
- Mars Polar Lander (MPL) launched on January 3, 1999

■ **Figure 5.3   Writing Addressed to a Manager**

This excerpt discusses the technical aspects of the program only in very general terms.

general reader's attention if it seems interesting and well written. The general reader may also seek specific information that will bring direct benefits: someone interested in buying a house might read articles on new methods of alternative financing.

In writing for a general audience, use simple vocabulary and relatively short sentences in discussing subjects that might be confusing. Translate jargon into standard English and use analogies and examples to clarify your discussion. Include the human angle: how the situation affects people. Present any special background — historical or ethical, for example — so that your reader can follow the discussion easily. Concentrate on the implications of this information for the general reader. For example, in discussing a new substance that removes graffiti from buildings, focus on its effectiveness and cost, not on its chemical composition.

Figure 5.4, an excerpt from a general discussion of the Mars Climate Orbiter mission (Jet Propulsion Lab, 1999b), illustrates the interests and needs of the general reader.

Why Explore Mars?

After Earth, Mars is the planet with the most hospitable climate in the solar system. So hospitable that it may once have harbored primitive, bacterialike life. Outflow channels and other geologic features provide ample evidence that billions of years ago liquid water flowed on the surface of Mars. Although liquid water may still exist deep below the surface of Mars, currently the temperature is too low and the atmosphere too thin for liquid water to exist at the surface.

What caused the change in Mar's climate? Were the conditions necessary for life to originate ever present on Mars? Could there be bacteria in the subsurface alive today? These are the questions that lead us to explore Mars. The climate of Mars has obviously cooled dramatically. By studying the reasons for climate change on Mars, which lacks the complications of oceans, a biosphere, and industrial contaminants, we may begin to understand the forces driving climate change on Earth. As we begin to explore the universe and search for planets in other solar systems, we must first ask the question "Did life occur on another planet in our own solar system?" and "What are the minimal conditions necessary for the formation of life?"

What Are We Looking For?

The planet Mars landed in the middle of immense public attention on July 4, 1997, when Mars Pathfinder touched down on a windswept, rock-laden ancient flood plain. Two months later, Mars Global Surveyor went into orbit, sending back pictures of towering volcanoes and gaping chasms at resolutions never before seen.

In December 1998 and January 1999, another orbiter and lander were launched to Mars. And every 26 months over the next decade, when the alignment of Earth and Mars are suitable for launches, still more robotic spacecraft will join them at the red planet.

These spacecraft carry varied payloads, ranging from cameras and other sensors to rovers and robotic arms. Some of them have their roots in different NASA programs of science or technology development. But they all have the goal of understanding Mars better, primarily by delving into its geology, climate and history.

Notice the very informal style in the second sentence: a sentence fragment is used for dramatic emphasis. The writer presents a very general answer to the question in the heading.

Questions — in the paragraphs and in the two headings — introduce the main points of the discussion, helping general readers focus on them.

The chronological organization of this passage makes the information easy for general readers to understand.

■ **Figure 5.4   Writing Addressed to the General Reader**

GUIDELINES

Writing for the Basic Categories of Readers

These guidelines summarize the discussion of how to write for the basic categories of readers.

Audience	Reasons for Reading	Guidelines for Writing
Expert	To gain an understanding of the theory and its implications.	Include theory, technical vocabulary, formulas, and sophisticated graphics.

Technician	To gain a hands-on understanding of how something works or how to carry out a task.	Include graphics. Use common words, short sentences, and short paragraphs. Avoid excessive theory.
Manager	To learn the bottom-line facts to aid in making decisions.	Focus on managerial implications, not technical details. Use short sentences and simple vocabulary. Put details in appendices.
General Reader	To satisfy curiosity and for self-interest.	Use short sentences and paragraphs, human appeal, and an informal tone.

INDIVIDUAL CHARACTERISTICS OF READERS

Knowing that a reader is, for example, a manager tells you something about what that person is likely to want to see in a communication. But it doesn't tell you whether that person is a native speaker of English, whether she is receptive or hostile to your message, or whether she needs to use information from your document in a document she is writing. In other words, being able to put a reader into a basic category doesn't tell you anything about that reader as an individual. It is therefore a good idea to try to find out as much as you can about your reader's individual characteristics.

> **Determine individual characteristics by asking:**
>
> Who is your reader?
>
> What are your reader's attitudes and expectations?
>
> Why and how will your reader use your document?

Who Is Your Reader?

In thinking about who your reader is, consider six specific factors:

- *The reader's education.* Think not only about the person's degree but also about when the person earned the degree. A civil engineer who earned a B.S. in 1979 has a much different background from the person who earned the same degree in 1999. Also, don't forget to consider any formal and informal course work the person has completed while on the job.

  Knowing your readers' educational backgrounds helps you determine how much supporting material to provide, what level of vocabulary to use, what kind of sentence structure and length to use, what types of graphics to include, and whether to provide such formal elements as a glossary or an executive summary.

Discovering your readers' educational backgrounds is not easy. You cannot ask people to send you their current résumés. But you can try to learn as much as possible in conversation with your colleagues.

- *The reader's professional experience.* Although education in school and on the job is significant, a person's professional experience is equally important. For example, a nurse with a decade of experience might have represented her hospital on a community committee to encourage citizens to give blood, might have worked with the hospital administration to choose vehicles for the emergency medical staff, and might have contributed to the planning for the hospital's new delivery room. In short, her range of experience might have provided several areas of competence or expertise.

- *The reader's job responsibility.* Consider the major job responsibility of your primary reader and how your document will help that person accomplish it. For example, you are writing a feasibility study of several means of cooling the air for a new office building and you know that your reader — an upper-level manager — will have to worry about electricity or other utility costs over the lifetime of the cooling system. Therefore, in your report you need to explain how you are estimating utility costs in the future.

- *The reader's personal characteristics.* The reader's age might tell you something important about how he or she will read and interpret your document. A senior manager at age sixty may not be as interested in tomorrow's technology as a thirty-year-old manager is. Does your reader have any other personal characteristics you should consider, such as impaired vision, that would affect the way you write and design your document?

- *The reader's personal preferences.* Most people have their own individual tastes and biases, some reasonable, some not. One person might hate to see the first-person pronoun *I* in technical documents. Another might find the word *interface* distracting when the writer isn't discussing computers. A good way to learn a person's preferences is to read that person's own documents. You should try to accommodate as many of your readers' preferences as you can. Sometimes you can't, of course, either because the preferences contradict one another or because the special demands of the subject won't permit you to accommodate them. But try to avoid alienating or distracting your reader.

- *The reader's cultural characteristics.* What you know about your reader's cultural characteristics can help you appeal to his or her interests and avoid confusing or offending. As discussed later in this chapter (p. 107), cultural characteristics can affect virtually every aspect of a reader's comprehension of a document and perception of the writer.

What Are Your Reader's Attitudes and Expectations?

In thinking about your reader's attitudes and expectations, consider these four factors:

- *Your reader's attitude toward you.* Most people will like you because you are hardworking, intelligent, and cooperative. Some won't. If a reader's animosity toward you is irrational or unrelated to the current project, you can do little other than to try to repair the damage. You can try to earn that person's respect and trust by meeting him or her on some neutral ground, perhaps by discussing other, less volatile projects or some shared interest, such as gardening, skiing, or science-fiction novels.

- *Your reader's attitude toward the subject.* If possible, discuss the subject thoroughly with your primary readers to determine whether they are positive, neutral, or negative toward it. Here are some basic strategies for responding to different attitudes:

If . . .	Try this . . .
Your reader is neutral or positively inclined toward your subject	• Write the document so that it responds to the reader's needs; make sure that vocabulary, level of detail, organization, and style are appropriate.
Your reader is hostile to the subject or your approach to it	• Try to find out what the objections are and then answer them directly. Explain clearly why the objections are either not valid or less important than the benefits. • Organize the document so that your actual recommendation follows your explanation of the benefits. This strategy encourages the hostile reader to understand your argument rather than to reject it out of hand. • Avoid describing the subject as a dispute. Seek areas of agreement and concede points. Avoid trying to persuade; people don't like to be persuaded, because it threatens their ego. Instead, suggest that there are new facts that need to be considered. People are less reluctant to change their minds when they realize that there are additional facts to be considered.
Your reader was instrumental in creating the policy or procedure that you are arguing is ineffective	• Be diplomatic in discussing the shortcomings of the present system, but be especially careful if there is any chance you might offend one of your readers. People can be very defensive about their ideas, and sometimes they are most defensive about their worst ideas. You don't want someone to dig in and do everything to oppose you just because you were tactless. When you address such an audience, don't write "The present system for logging customer orders is completely ineffective." Instead, write "While the present system has worked well for many years, new developments in electronic processing of orders might enable us to improve logging speed and reduce errors substantially."

- *Your reader's expectations about the subject.* Think about how your reader expects to see the information treated in terms of scope, pattern of organization, and amount of detail. If you can match your primary readers' expectations about the subject *and* meet your own objectives, you increase your chances of connecting with them.

- *Your reader's expectations about the document.* If your reader expects to see the information presented as a memo, use a memo unless some other format would clearly work better. If your reader expects a report to include an executive summary, include one.

Why and How Will Your Reader Use Your Document?

In thinking about how your reader will use your document, consider the following four factors:

- *Your reader's reasons for reading your document.* Just as you have a purpose in writing, your reader has a purpose in reading. Does the reader need to use the document to carry out a task? To learn the answer to a specific question? To understand the broad outlines of the subject? The more specifically you can answer such questions, the more likely it is that your document will meet your reader's needs.

- *The way your reader will read your document.* Will he or she
    - file it?
    - skim it?
    - read only a portion of it?
    - study it carefully?
    - modify it and submit it to another reader?
    - attempt to implement recommendations?
    - use it to perform a test or carry out a procedure?
    - use it as a source document for another document?

  If only one of fifteen readers will study the document for detailed information, you must provide it. But because you don't want the other fourteen people to have to wade through it, put this information in an appendix. If you know that your reader wants to use your status report as raw material for a report to a higher-level reader, try to write a report that requires little rewriting; you might use the reader's own writing style and put the file on the network so that your report can be merged with the new document without requiring retyping.

- *Your reader's reading skills.* If your document confuses or intimidates your reader, you will not communicate your message. Your assessment of your reader's reading skills should influence every aspect of your thinking. First you need to consider whether you should be writing at all, or whether it would be better to use videotapes, for example, or an oral presentation or computer-based training. If you decide to produce a written document, you need to consider whether your reader can understand how to use the type of document you have selected, handle the level of detail you will present, and understand your graphics, sentence structure, and vocabulary.

For more about designing a document for use in different environments, see Ch. 13, p. 334.

* *The physical environment in which your reader will read your document.* Technical documents are often formatted in a special way or constructed of special materials in order to improve their effectiveness in particular physical settings. A user's manual for a computer system might be bound so that it can lie flat on a table for a long period. Documents used in poorly lit places might be printed in larger-than-normal type. Some documents might be used on ships, on aircraft, or in garages, where they might be exposed to wind, salt water, and grease. You might have to use special waterproof bindings, oil-resistant or laminated paper, coded colors, or nonstandard-sized paper.

GUIDELINES

Identifying Individual Characteristics of Readers

Use the specific questions in the following table to identify your readers' individual characteristics.

General Questions	Specific Questions
Who is your reader?	• What is the reader's education? • What is the reader's professional experience? • What is the reader's job responsibility? • What are the reader's personal characteristics? • What are the reader's personal preferences? • What are the reader's cultural characteristics?
What are your reader's attitudes and expectations?	• What is the reader's attitude toward you? • What is the reader's attitude toward the subject? • What are the reader's expectations about the subject? • What are the reader's expectations about the document?
Why and how will your reader use your document?	• Why is the reader reading your document? • How will the reader read your document? • What are the reader's reading skills? • What is the physical environment in which the reader will read your document?

WRITING FOR A MULTIPLE AUDIENCE

Like the Mars Climate Orbiter Web site excerpts presented earlier, many documents of more than a few pages are addressed to a multiple audience, that is, to more than one reader. In many cases, this multiple audience will consist of people with widely different backgrounds. Some might be experts or technicians, others might be managers, and still others might be general readers.

See Chs. 10 and 15, pp. 252 and 435, for more about writing to a multiple audience.

If you think your document will have a number of readers, consider making it *modular:* break it up into components addressed to different kinds of readers. A modular report might contain an executive summary for the managers who don't have the time, knowledge, or desire to read the whole report. It might also contain a full technical discussion for expert readers, an implementation schedule for technicians, and a financial plan in an appendix for budget officers.

Figure 5.5 illustrates the concept of a modular report.

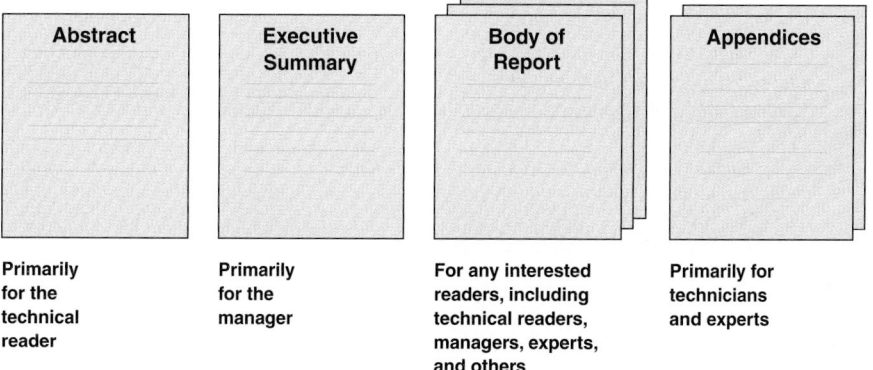

Abstract	Executive Summary	Body of Report	Appendices
Primarily for the technical reader	Primarily for the manager	For any interested readers, including technical readers, managers, experts, and others	Primarily for technicians and experts

■ Figure 5.5
A Modular Report

THE AUDIENCE PROFILE SHEET

To help you analyze your audience, you might draw up an audience profile sheet that you can fill out for each primary and secondary reader (assuming, of course, that there are just a few).

Assume that you work in the drafting department of an architectural engineering firm. You know that the company's computer-assisted design (CAD) equipment is out of date and that recent CAD technology would make it easier and faster for the draftspeople to do their work. You want to persuade your company to authorize the purchase of a CAD workstation that costs about $4,000. Your primary reader is Harry Becker, manager of the Drafting/Design Department. Figure 5.6 is an audience profile sheet for Harry Becker.

For a version of this form that you can print out or save as a word-processing file, click on Forms for Technical Communication on TechComm Web (www.bedfordstmartins.com/techcomm).

AUDIENCE PROFILE SHEET

Reader's name: Harry Becker

Reader's job title: Manager, Drafting and Design Department

Kind of reader: Primary _X_ Secondary ___

Education: B.S., Architectural Engineering, Northwestern, 1989. CAD/CAM Short Course, 1989; Motivating Your Employees Seminar, 1991; Writing on the Job Short Course, 1994

Professional Experience: Worked for two years in a small architecture firm. Started here 12 years ago as a draftsperson. Worked his way up to Assistant Manager, then Manager. Instrumental in the Wilson project, particularly in coordinating personnel and equipment.

Job Responsibility: Supervises a staff of 12 draftspersons. Approves or denies all requests for capital expenditures over $2,000 coming from his department. Works with employees to help them make the best case for the purchase. After approving or denying the request, forwards it to Tina Buterbaugh, Manager, Finance Dept., who maintains all capital expenditure records.

Personal Characteristics: N/A

Personal Preferences: Likes straightforward documents, lots of evidence, clear structure. Dislikes complicated documents full of jargon.

Cultural Characteristics: Nothing unusual.

Attitude toward the Writer: No problems.

Attitude toward the Subject: He understands and approves of my argument.

Expectations about the Subject: Expects to see a clear argument with financial data and detailed comparisons of available systems.

Expectations about the Document: Expects to see a report, with an executive summary, of about 10 pages.

Reasons for Reading the Document: To offer suggestions and eventually approve or deny the request.

Way of Reading the Document:
Skim it ____ Study it _X_ Read a portion of it ____ Which portion?
Modify it and submit it to another reader ____
Attempt to implement recommendations ____
Use it to perform a task or carry out a procedure ____
Use it to create another document ____
Other ____ Explain.

Reading Skill: Excellent

Reader's Physical Environment: N/A

■ **Figure 5.6**
An Audience Profile Sheet

You should modify this form to meet your own needs and those of your organization.

UNDERSTANDING MULTICULTURALISM

Our society and our workforce are becoming increasingly diverse, both culturally and linguistically, and businesses are relying more on exports. As a result, technical communicators and technical professionals often find it necessary to communicate with nonnative speakers of English in the United States and abroad and with speakers of other languages who read texts translated from English into their own languages.

 See Nancy Hoft's site (www.world-ready.com) for bibliographies about writing for multicultural audiences.

The economy of the United States depends on international trade. In 1998, the United States exported almost one trillion dollars' worth of goods and services (U.S. Bureau, 2000, p. 791). In the same year, direct investment abroad by U.S. companies totaled $980 billion (U.S. Bureau, 2000, p. 785). Exports are responsible for four out of five new jobs in the United States (Lustig & Koester, 1999). In addition, the population of the United States itself is truly multicultural. Each year, the United States admits more than half a million immigrants (U.S. Bureau, 2000, p. 10). In 2000, one in ten U.S. residents was foreign born (U.S. Bureau, 2000, p. 55).

Understanding Cultural Variables

Effective communication requires an understanding of cultural variables, those aspects of culture that differ from one culture to another.

First, what is culture? *Culture* refers to the beliefs, attitudes, and values that motivate people's behavior. Culture, in other words, is how people do things. Unless you have lived in a different culture, you are probably unaware of the power and pervasiveness of cultural patterns. We tend to take the cultural patterns of our native culture for granted, sometimes to our own detriment. In fact, more than half of U.S. businesspeople on long overseas assignments return home early because of their inability to adapt to a new culture (Ferraro, 1990).

Communicating effectively with people from another culture requires an understanding of cultural variables. If you are writing a user's manual to accompany a product you want to sell in another culture, you need to answer a number of important questions. For example, what language or languages should you use? What cultural icons should you refer to or avoid? What colors? What designs or graphics? Answering these questions is not an exact science, but it does require that you learn as much as you can about the culture of those with whom you want to communicate.

A brief example: an American manufacturer of deodorant launched an advertising campaign in Japan in which a cute octopus applied the firm's product under each of its eight arms. But the campaign failed. In Japan, an octopus has legs, not arms (Bathon, 1999).

In *International Technical Communication*, Nancy L. Hoft (1995) describes seven major cultural variables that communicators need to research:

- *Political.* This category includes trade issues and legal issues (for example, some countries forbid imports of certain foods or chemicals) and laws concerning such topics as intellectual property and product safety and liability.

- *Economic.* You need to understand the level of economic development in other cultures before you can effectively communicate with members of those cultures. For instance, in many cultures, only a very few people can afford personal computers and other expensive electronic goods in their homes.

- *Social.* This category covers many issues, including these three:
  - *Gender.* In most Western cultures, women play a much greater role in the workplace than they do in many Middle Eastern and Asian cultures.
  - *Time.* Western cultures often view time as a linear concept; once the time has passed, it is gone. Many other cultures see time as a cycle; a meeting that extends two hours beyond its scheduled length is no problem (Limaye & Victor, 1991).
  - *Business customs.* Forms of greeting, business dress, and gift giving differ from culture to culture. Customs related to the level of formality used in addressing a person also differ, as do customs related to nonverbal communication (such as hand gestures and the space that people maintain between themselves and others).

- *Religious.* Religious differences can affect diet, attitudes toward individual colors, style of dress, holidays, hours of business, and, of course, attitudes about the role and importance of business in daily life.

- *Educational.* In the United States, some forty million people are only marginally literate. In other cultures, that rate can be much higher or much lower. In some cultures, classroom learning with a teacher is considered the most acceptable way to study; in others, people are encouraged to study on their own.

- *Technological.* If you sell computer-related products, you need to know whether your readers have the hardware, the software, and the technological infrastructure to use your products.

- *Linguistic.* In some nations where English is not the first language, many people understand and speak English; in others, almost nobody does. In many cultures, text directionality — the orientation of text on a page and in a book — is not from left to right. The following section will discuss some of the major aspects of technical communication that are affected by linguistic variables.

Considering Cultural Variables as You Write

Many cultural variables have a direct effect on a culture's technical communication. In *high-context* cultures such as that of Japan, the main point of a document is communicated implicitly rather than explicitly. The reader is left to infer the details from the context and the tone. In *low-context* cultures, such as that of the United States, the main point is stated explicitly and fully supported by details. Similarly, the typical Asian professional is highly group oriented and rarely presents a personal viewpoint or uses the pronoun *I*. By contrast, the typical U.S. professional is highly individualistic and often uses the pronoun *I*. Cultural differences extend to the nature of evidence itself. In Japan, tradition, authority, and group consensus are persuasive forms of evidence. In most Western countries, empirical and testable data are more persuasive.

In addition to these basic differences, there are dozens of other more superficial differences you need to be aware of. For instance, the United States is the only major country in the world that has not adopted the metric system. Americans also use periods to separate whole numbers from decimals, and commas to separate thousands from hundreds. Much of the rest of the world reverses this usage.

United States: 3,425.6

Europe: 3.425,6

In the United States, we use a format for writing out and abbreviating dates that is different from that of most other cultures:

United States: March 2, 2001 3/2/01

Europe: 2 March 2001 2/3/01

Japan: 2 March 2001 01/2/3

No brief discussion of even the major areas of difference among cultures can answer questions about how to write for a particular multicultural audience. All it can do is point out areas you need to investigate. Do your homework on the culture of your readers. Read everything you can about the society. As you plan the document, seek assistance from someone native to the culture, who can help you prevent blunders that might confuse or offend your readers.

GUIDELINES

Writing for Multicultural Readers

The following guidelines will help you communicate more effectively with multicultural readers:

▶ *Limit your vocabulary.* Every word should have only one meaning, as called for in Simplified English and in other basic-English languages.

For more on Simplified English, see Ch. 11, p. 297.

▶ *Keep sentences short.* There is no magic number, but try to stay within a range of 20 to 25 words.

▶ *Define abbreviations and acronyms in a glossary.* Don't assume that your readers know what a GFI — ground fault interrupter — is, because the abbreviation is derived from English vocabulary and word order.

For more on glossaries, see Ch. 12, p. 320.

> ▶ *Avoid jargon unless you know your readers are familiar with it.* For instance, your readers might not know what a *graphical user interface* is.
>
> ▶ *Avoid idioms.* Because idioms are culture specific, it is wise to avoid them. If you tell your Japanese readers that your company plans to put on a "full court press," most likely they will be confused.
>
> ▶ *Use the active voice whenever possible.* The active voice — in which the actor is the grammatical subject of the sentence — is easier for non-native speakers of English to understand than the passive voice.
>
> ▶ *Be careful with graphics.* The garbage-can icon on the Macintosh computer has often been cited as a graphic that does not translate well, because garbage cans have different shapes and can be made of different materials in other countries.
>
> ▶ *Be sure someone from the culture reviews your document.* Even if you have had help in planning the document, have it reviewed before you publish and distribute it.

For more on voice, see Ch. 11, p. 286.

For more on graphics, see Ch. 14.

DETERMINING YOUR PURPOSE

Once you have identified and analyzed your audience, it is time to examine your purpose in writing. Ask yourself this: "What do I want this document to accomplish?" When your readers have finished reading what you have written, what do you want them to *know* or *believe*? What do you want them to *do*? Your writing should help your readers carry out a task, understand a concept, or hold a particular belief.

In defining your purpose, think of a verb that represents it. (Sometimes, of course, you have several purposes.) The following list of examples is divided into two categories — verbs used to communicate information to your readers and verbs used to convince them to accept a particular point of view:

Communicating Verbs		*Convincing Verbs*
to describe	to authorize	to assess
to explain	to define	to request
to inform	to summarize	to propose
to illustrate		to recommend
to review		to forecast
to outline		to evaluate

This classification is not absolute. For example, *to review* could in some cases be a *convincing verb* rather than a *communicating verb*: one writer's review of a complicated situation might be very different from another's.

Here are a few examples of how you can use these verbs to clarify the purpose of your document (the verbs are italicized):

This report *describes* the research project intended to determine the effectiveness of the new waste-treatment filter.

This report *reviews* the progress in the first six months of the heat-dissipation study.

This letter *authorizes* the purchase of six new PCs for the Jenkintown facility.

This memo *recommends* that we create the Web site as soon as possible.

Sometimes your real purpose differs from your expressed purpose. For instance, if your real purpose is to persuade your reader to lease a new computer system rather than purchase it, you might phrase the purpose this way: *to explain the advantages of leasing over purchasing.* As mentioned earlier, many readers don't want to be *persuaded* but are willing to learn new facts or ideas.

GAINING MANAGEMENT'S APPROVAL

After you have analyzed your audience and purpose, consider gaining the approval of management before you proceed. You needn't seek approval for every kind of project, of course. Sometimes you simply do your research and write your document. But the larger and more complex the project and the document, the more sense it makes to be sure that you are on the right track before you invest too much time and effort.

For example, imagine that you are planning a CAD equipment project. You already have a good understanding of your audience and purpose, and a general outline is taking shape in your mind. Before you actually start to write an outline or gather the information you will need, however, spend another 10 or 15 minutes making sure your primary reader agrees with your thinking. You don't want to waste days or even weeks working on a document that won't fulfill its purpose. If you have misunderstood what your supervisor wants, it is far easier to fix the problem at this early stage.

Your statement can serve another purpose as well: if you want your reader's views on which of two strategies to pursue, you can describe each one and ask your reader to state a preference.

What should this statement look like? It doesn't matter. You can write an email or a memo, as long as you clearly and briefly state what you are trying to do. Here is an example of the statement you might submit to your boss about the CAD equipment.

Harry:

Please tell me if you think this is a good approach for the proposal on CAD equipment.

The purpose of the memo

Outright purchase of the complete system will cost more than $1,000, so you would have to approve it and send it on for Tina's approval. (I'll provide leasing costs

A statement of the audience for the proposal

A statement of the purpose, followed by early statements of the scope of the document.

as well.) I want to show that our CAD hardware and software are badly out of date and need to be replaced. I'll be thorough in recommending new equipment, with independent evaluations in the literature, as well as product demonstrations. The proposal should specify what the current equipment is costing us and show how much we can save by buying the recommended system.

I'll call you later today to get your reaction before I begin researching what's available.

Renu

In composing this statement, the writer drew on her audience profile sheets of the two principal readers. She describes a logical, rational plan for proposing the equipment purchase.

Once you have received your primary reader's approval, you can feel confident about starting to gather information.

    Revision Checklist

Following is a checklist for analyzing your audience and purpose. Remember that your document might be read by one person, several people, a large group, or several groups with various needs.

1. In analyzing your audience, did you consider the following questions about your most important readers?
   ❑ What is your reader's educational background?
   ❑ What is your reader's professional experience?
   ❑ What is your reader's job responsibility?
   ❑ What are your reader's personal characteristics?
   ❑ What are your reader's personal preferences?
   ❑ What are your reader's cultural characteristics?
   ❑ What is your reader's attitude toward you?
   ❑ What is your reader's attitude toward the subject?
   ❑ What are your reader's expectations about the subject?
   ❑ What are your reader's expectations about the document?
   ❑ Why is your reader reading your document?
   ❑ How will your reader read your document?
   ❑ What is your reader's reading skill?
   ❑ What is the physical environment in which your reader will read your document?

2. Did you fill out an audience profile sheet for your primary and secondary audiences that considers
   ❑ education?
   ❑ professional experience?
   ❑ personal preferences?
   ❑ attitudes toward you and your subject?
   ❑ cultural characteristics?
   ❑ use of the document?
   ❑ physical environment?

3. In planning to write for an audience from another culture, did you consider the following cultural variables:
   ❑ political?
   ❑ economic?
   ❑ social?
   ❑ religious?
   ❑ educational?
   ❑ technological?
   ❑ linguistic?

4. In writing for a multicultural audience, did you
   ❑ limit your vocabulary?
   ❑ keep sentences short?
   ❑ define abbreviations and acronyms in a glossary?
   ❑ avoid jargon unless you know that your readers are familiar with it?
   ❑ avoid idioms?
   ❑ use the active voice whenever possible?
   ❑ use graphics carefully?
   ❑ have the document reviewed by someone from the culture?

5. Did you consider your purpose in writing and express it in the form of a verb or verbs?

Exercises

1. Choose a 200-word passage from a technical article addressed to an expert audience, one related to your major course of study. (You can find a technical article on the Web by using a directory search engine, such as Yahoo!, selecting a subject area such as "science," then selecting "journals." In addition, many federal government agencies publish technical articles and reports on the Web.) Rewrite the passage so that it is clear and interesting to the general reader. Submit the original passage along with your revision.

2. Fill out an audience profile sheet about yourself, as if you were the reader of your own writing. Then fill out an audience profile sheet about your instructor. What are the major differences between the two profiles? What more do you need to know about your instructor?

3. The following passage is an advertisement from a translation service. Revise the passage to make it more appropriate for a multicultural audience.

   If your technical documents have to meet the needs of a global market but you find that most translation houses are swamped by the huge volume, fail to accommodate the various languages you require, or fail to make your deadlines, where do you turn?

   Well, your search is over. Translations, Inc. provides comprehensive translations in addition to full-service documentation publishing.

   We utilize ultrasophisticated translation programs that can translate a page in a blink of an eye. Then our crack linguists comb each document to give it that personalized touch.

   No job too large! No schedule too tight! Give us a call today!

Research Projects

Some of the following projects ask you to write a memo. See Chapter 15, page 430, for a discussion of memos.

4. Audience is your primary consideration in many types of nontechnical writing. Choose a one- or two-page magazine advertisement or Web site for an economy car, such as a Geo, and one for a luxury car, such as a Mercedes. In a memo to your instructor, contrast the audiences for the two ads according to age, sex, economic means, hobbies, interests, and leisure activities. In contrasting the two audiences, consider the explicit information in the ads — the writing, as well as the implicit information — hidden persuaders, such as background scenery, color, lighting, angles, and the situation portrayed by any people photographed. Keep in mind that your purpose is to contrast, not merely to describe the content of the ad or its design. Submit color photocopies or the original ads from the magazines or sites along with your memo.

5. Interview a student from another culture. If necessary, consult the international student office, the English department, or the English as a Second Language office. Choose one of the topics discussed in this chapter, such as gender roles, notions of time, or document standards, and write a memo to your instructor comparing and contrasting your culture's view with that of the other culture. See Chapter 7, page 171, for a discussion of interviewing.

6. Locate a fund-raising letter from a charitable organization, such as the American Cancer Society, the March of Dimes, or the Special Olympics. (You can also find fund-raising appeals on these organizations' Web sites.) Write a memo to your instructor describing the tactics used in the letter to persuade you to give money to the cause. Which tactics work well, and which do not? Do not limit your analysis to the argument made by the words themselves; consider also the design of the letter, the appearance of the type, and the type of paper used. Analyze any graphics in the letter and any other materials included in the envelope.

7. Form small groups for this project on audience. Choose two articles on the same subject, one from a general-audience periodical such as *Time* or *Newsweek* and one from a technical journal such as

Science, Journal of Visual Languages and Computing, Management Science, or *Journal of Microcomputer Applications.* (You can also find suitable articles on the Web.) Working *alone,* each member of the group should study *each article,* focusing on the following questions:

- What is the likely background of the article's audience? Does the article require that the reader have specialized knowledge?
- What is the author's purpose? In other words, what is the article intended to accomplish?
- How do the differences in audience and purpose affect the following elements in the article?
  - scope and organization
  - sentence length and structure
  - vocabulary
  - number and type of graphics
  - references within the articles and at the end
  - headings and white space

Meet and discuss your analyses of the two articles. Do you agree more in your analyses of the general-audience article or of the technical article? What are the main areas of difference in your analyses? What do you think accounts for these differences? Write a memo to your instructor that does the following:

- compares and contrasts the two articles in terms of the authors' assessment of the writing situation
- describes and analyzes the differences in each group member's analysis

Submit the two articles (or photocopies) along with your memo. For more on comparison and contrast, see Chapter 8, page 193.

8. Form small groups and study two Web sites that advertise competing products. For instance, you might choose the Web sites of two car manufacturers, two television shows, or two music publishers. Have each person in the group, working *alone,* compare and contrast the two sites according to these three criteria:

- the kind of information they provide: hard, technical information or more emotional information
- the use of multimedia such as animation, sound, or video

- the amount of interactivity they invite, that is, the extent to which you can participate in activities while you visit the site

After each person has separately studied the sites and taken notes about the three points, come together as a group. Each person should share his or her findings and then discuss the differences as a group. Which aspects of these sites cause the most difference in group members' reactions? Which aspects seemed to elicit the most consistent reactions? In a brief memo to your instructor, describe and analyze how the two sites were perceived by the different members of the group.

CASE 1
Writing an Information Booklet

Form small groups for this case on writing for a multicultural audience. The student-exchange program at your college wishes to issue information booklets about each of the countries in which students from your school study. Research one country and write a 2,000-word guide explaining the culture for students preparing to study there. Begin by studying the Central Intelligence Agency's fact book about the country (www.odci.gov/cia /publications) and any materials the student-exchange program can provide. (Be sure to cite your sources properly.) One group member might concentrate on the country itself: the land, the culture, and the language. Another might concentrate on the city in which the college or university is located. A third might focus on the college or university itself: its policies, its student makeup, its opportunities, the expectations for students in that country, and so forth. Do not overlook Web sites in researching the country, the city, and the school.

 For additional cases, click on Case of the Month and Archive on TechComm Web (www.bedfordstmartins.com /techcomm).

CASE 2
Making a Question-and-Answer Sheet

You are an assistant to Gilbert F. Casellas, Chair of the U.S. Equal Employment Opportunity Commission (EEOC). The EEOC has published a number of fact sheets, such as the one on sexual harassment printed here. Mr. Casellas tells you he has received a number of requests for a question-and-answer sheet on sexual harassment that can be posted in workplaces. "Would you mind taking care of this for me? Make sure it is easy to understand, even for nonnative speakers. And be sure you use examples so it doesn't sound like legalese." Drawing on the fact sheet, write a one-page question-and-answer sheet that can be distributed to companies and other organizations requesting it. (For more information on the EEOC, visit its site at www.eeoc.gov.)

This fact sheet is in the Case of the Month and Archive, under Ch. 5, on TechComm Web (www.bedfordstmartins.com/techcomm).

FACTS ABOUT SEXUAL HARASSMENT

Sexual harassment is a form of sex discrimination that violates Title VII of the Civil Rights Act of 1964.

Unwelcome sexual advances, requests for sexual favors, and other verbal or physical conduct of a sexual nature constitute sexual harassment when submission to or rejection of this conduct explicitly or implicitly affects an individual's employment, unreasonably interferes with an individual's work performance, or creates an intimidating, hostile, or offensive work environment.

Sexual harassment can occur in a variety of circumstances, including but not limited to the following:

- The victim as well as the harasser may be a woman or a man. The victim does not have to be of the opposite sex.
- The harasser can be the victim's supervisor, an agent of the employer, a supervisor in another area, a co-worker, or a non-employee.
- The victim does not have to be the person harassed but could be anyone affected by the offensive conduct.
- Unlawful sexual harassment may occur without economic injury to or discharge of the victim.
- The harasser's conduct must be unwelcome.

It is helpful for the victim to directly inform the harasser that the conduct is unwelcome and must stop. The victim should use any employer complaint mechanism or grievance system available.

When investigating allegations of sexual harassment, EEOC looks at the whole record: the circumstances, such as the nature of the sexual advances, and the context in which the alleged incidents occurred. A determination on the allegations is made from the facts on a case-by-case basis.

Prevention is the best tool to eliminate sexual harassment in the workplace. Employers are encouraged to take steps necessary to prevent sexual harassment from occurring. They should clearly communicate to employees that sexual harassment will not be tolerated. They can do so by establishing an effective complaint or grievance process and taking immediate and appropriate action when an employee complains.

Communicating Persuasively

6

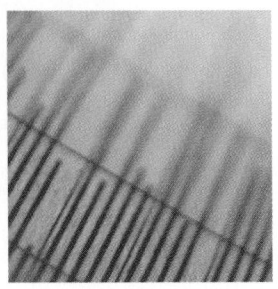

According to Katherine J. Mayberry (1999, p. 160) of the Rochester Institute of Technology, image and reality coincide in persuasive writing:

In successful arguments, writers project an image of intelligence . . . and trustworthiness. There is nothing false or superficial about this kind of image: these qualities cannot be created out of thin air; they must be true reflections of the writer and thus are developed over time and through experience.

 See TechComm Web (www .bedfordstmartins.com /techcomm) for guidelines boxes, additional examples, and links related to topics in this chapter.

As the discussion of purpose in Chapter 5 stated, any communication has at least one and perhaps two main purposes: to communicate information and to convince readers to accept a particular point of view. Both tasks — communicating and convincing — involve persuasion, the subject of this chapter.

In technical communication, your job is often to convince a reader of a point of view — about what factors caused a situation, for example, or what a company ought to do to solve a problem. If you are lucky, you will be reinforcing a viewpoint the reader already has. Sometimes, however, you want the reader either to see the situation in a different light or to change his or her viewpoint completely. An effective presentation, one that convinces the reader, involves the techniques of persuasion discussed in this chapter.

Communicating information also involves persuasion, although in a less obvious way. When you communicate information, you want your reader to trust that your facts are accurate and appropriate. If your document is clear, comprehensive, accessible, correct, and professional in appearance, chances are good that your reader will accept your statements. If the document lacks one or more of these characteristics, however, the reader might begin to doubt your facts.

Whether you are presenting a viewpoint or communicating information, you are presenting an argument. An *argument* is simply an arrangement of facts and judgments about some aspect of the world.

The chapter begins by discussing how to plan a persuasive argument. Next, it explains techniques for crafting a persuasive argument, including using appropriate evidence, considering opposing views, and deciding where to place the claim. It then discusses how to avoid logical fallacies and present an appropriate persona, and finally, the relationship between persuasion and ethics, as well as the challenge of multicultural audiences.

CONSIDERING THE PURPOSE OF YOUR ARGUMENT

An argument can be as short as a sentence or as long as a multivolume report. It can take many forms, including oral communication. And it can discuss almost any kind of issue. Here are some examples:

from a description of a construction site:

Features A, B, and C characterize the site.

from a study of why a competitor is outselling your company:

Company X's dominance can be attributed to the following four major factors: A, B, C, and D.

from a feasibility study considering four courses of action:

Alternative A is better than alternatives B, C, and D.

from a set of instructions for performing a task:

The best and safest way to perform the task is to complete task A, then task B, and so on.

To develop an effective argument, you must understand your audience's broader goals and work within constraints.

Understanding Your Audience's Broader Goals

Chapter 5 discussed your primary readers — who they are, what they think about you and your subject, and why and how they will use your document. By answering these questions you can better plan the scope, organization, design, and style of the document.

 See Business Communication: Managing Information and Relationships (http://spider.hcob .wmich.edu/bis/faculty /bowman/mir.html#Contents) for an excellent discussion of persuasion.

In addition to seeking information about particular readers, however, you need to think about your audience's broader goals. Certainly, most people want their company to prosper, yet most people are also concerned about their own welfare and interests within the company. Your argument is more likely to be effective if it responds to three goals that most people share: security, ego reinforcement, and self-actualization.

Security

People prefer safe actions to controversial ones, especially if the controversial ones threaten their status, power, or livelihood. People oppose an action if they fear that it might hurt them personally, even if it seems likely to help the organization. For obvious reasons, then, those who stand to lose their jobs will resist an argument that their division be eliminated, even if there are many valid reasons to support the argument. Another aspect of security is workload; most people resist an argument that calls for them to work more. This resistance is not necessarily a sign of laziness; additional work may be unreasonable because people want to spend time with their families and pursue other interests.

Ego Reinforcement

People like to be praised for their hard work and their successes. Where appropriate in technical documents, be generous with your praise. Acknowledge colleagues who contributed to a project or who thought up good ideas. Similarly, people hate being humiliated in public. Therefore, in your documents you should allow people to save face. Avoid criticizing their actions or positions and speculating about their motivations. Instead, present your argument as a response to the company's present and future needs. Look ahead, not back, and be diplomatic.

Self-Actualization

People want to develop and grow on the job as well as in their personal lives. In an obvious way, this means that they want to learn new skills and assume new duties. This desire for self-actualization is reflected in more subtle ways in efforts to improve how the organization treats its employees and customers, relates to the community, and coexists with the environment. Your arguments will be more persuasive if you can show how the action you recommend will help your organization become an industry leader, for example, or help needy people in your city or reduce environmental pollution. We want to be associated with — and contribute to — organizations that are good at what they do and that help us become better people.

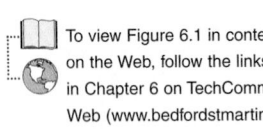 To view Figure 6.1 in context on the Web, follow the links in Chapter 6 on TechComm Web (www.bedfordstmartins .com/techcomm).

Figure 6.1 (Microsoft, 1999), from the employment section on the Microsoft Web site, profiles an employee the company believes reflects the personality and character of those who work at Microsoft.

Working within Constraints

In planning a persuasive document, you need to pay attention to the constraints that shape the environment in which you will be working, because they will affect what you can accomplish. For an argument to be persuasive, it must make sense within those constraints. As a student writer, you routinely work within constraints: the amount of information you can gather for a paper, the required length and format, the due date, and so forth. In business, industry, and government, similar constraints also play a role.

Ethical Constraints

You do your work within a complex web of rights and responsibilities, but your greatest responsibility is always to your own sense of what constitutes ethical behavior. Being asked to lie or mislead in a document can directly chal-

According to Blackburn, the most critical thing you can do to succeed at Microsoft is to focus on what you believe is important. "It's easy to get overwhelmed with everything going on around you, so first and foremost you need to stay focused," he says. "What you focus on needs to be something you firmly believe in so that you'll pursue it and defend your mission with a passion."

Here the writer refers to the goal of believing strongly in what you do at work in order to appeal to applicants who want to be committed to what they do.

He also thinks that one of the biggest differences between his co-workers at Microsoft and those at his former workplace is the sense of ownership and the impact on the business that everyone shares. "We hire people who tend to become personally attached to the products and the success of the company, and the environment really perpetuates this," Blackburn says. "It's much easier to motivate a team at Microsoft than any other company I've been in."

Ambitious people like "a sense of ownership" and making an "impact."

One of the most difficult parts of Blackburn's job is staying ahead of the people who report to him. "Because there are so many smart people and technology is changing so quickly, it is a constant challenge to keep up," he says. "I hope that's because I hire such great people!"

Blackburn praises his intelligent co-workers.

To succeed, Blackburn uses time management tactics and allocates a specific amount of time to education. "My personal goal is to spend at least 20 percent of my time learning new things through formal and informal methods."

Blackburn refers to setting aside time to grow in the job, showing that the company takes an interest in its employees' personal goals.

Microsoft provides myriad ways for people to blow off a little steam, and Blackburn says that popular activities include hitting the athletic club during lunch, jogging around campus on the wooded trails, wakeboarding on Lake Sammamish (five minutes from campus), or scheduling a match at the sand volleyball courts.

"The Northwest has almost unlimited activities you could participate in before, during, or after a workday, so finding something to do to release energy is not a problem," he says. "As for work-related activities, most teams have Friday afternoon social events where people can get together and vent, brag about their accomplishments for the week, or make plans for the weekend. Microsoft people definitely fit the 'work hard/play hard' model."

Who wouldn't want to work at Microsoft?

■ **Figure 6.1   Appealing to an Audience's Broader Goals**

By profiling employee Patrick Blackburn in the employment section of the Microsoft Web site, the writer hopes to encourage like-minded people to apply for positions with the company.

lenge your ethical standards. In most cases, you do have options when you feel you are being asked to act unethically. Some organizations and professional communities have a published code of conduct. In addition, many large companies have hired ombudspersons whose job it is to help employees resolve ethical conflicts and other problems within the organization. If you think you are being asked to act unethically, consider the issue carefully and then take advantage of the resources available to you.

Legal Constraints

You are obligated to abide by all applicable laws on labor practices, environmental issues, fair trade, consumer rights, and so forth. If you think the action recommended by your supervisor has potential legal implications for you or

For more on ethical and legal constraints, see Ch. 2.

your company, meet with your organization's legal counsel and, if necessary, attorneys outside the organization.

Political Constraints

Choose your battles. You can reasonably hope to achieve some goals, but not all. Don't spend your energy and credibility on a losing cause. If you know that your proposal would help the company but also know that upper management disagrees with you or that the company can't afford to act on your proposal, don't be a martyr to the cause. Try to figure out what you might achieve through other means, or scale back the idea so that it is affordable. Two big exceptions to this rule are matters of ethics and matters of safety. As discussed in Chapter 2, under certain circumstances, compromise is unacceptable.

Informational Constraints

You might also face constraints on the kind or the amount of information you are able to use in the document. Most often, the information you need is not available. You might want to recommend that your organization buy a piece of equipment, for example, but you can't find objective evidence that it will do the job. You have located advertising brochures and testimonials from satisfied users, but you can't find reports on the kinds of controlled tests that would convince a skeptical reader.

What do you do? You tell the truth. You state exactly what the situation is, weighing the available evidence and carefully noting what is missing. Your most important credential on the job is credibility; you will lose it if you unintentionally suggest that your evidence is better than it really is. In the same way, you don't want your readers to think that you don't realize your information is incomplete. They will doubt your technical knowledge.

Personnel Constraints

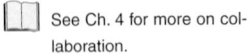 See Ch. 4 for more on collaboration.

Much technical communication is written collaboratively by teams of writers, editors, production specialists, and subject-matter experts. One constraint you will face in working on large projects is lack of access to as many collaborators as you need. In such cases, present a clear and persuasive proposal for hiring the personnel you need, but don't be surprised if you have to make do with fewer people than you want. Everyone will have to work harder and perform some tasks for which they are less than ideally suited.

Financial Constraints

Financial restraints are related to personnel constraints: if you had unlimited funds, you could hire all the personnel you need. But financial constraints can affect other kinds of resources, too: you might not be able to produce as many copies of the document as you want, or you might need to settle for black-and-white printing instead of full color.

Time Constraints

Many managers allot too little time for their employees to do a good job in writing all the necessary documents. Most working professionals resign themselves to taking work home in the evenings or on weekends. Some people come in early to do their writing before things get hectic.

The first piece of necessary information, of course, is the document's deadline. Sometimes a document will have several intermediate deadlines. Once you know when a document (or a part of it) is due, draw up a schedule. Keep in mind that tasks almost always take longer than estimated. People call in sick, information is lost or delayed in the mail — whatever can go wrong will go wrong, sooner or later. And when you collaborate, the number of potential problems increases dramatically, because when one person is delayed, others may lack the necessary information to proceed, causing a log jam.

See Ch. 3, p. 52, for more on scheduling.

Format and Tone Constraints

You will be expected to work within one further set of constraints:

- *Format.* Format constraints are limitations on the size, shape, or style of a document. For example, all tables and figures must be presented at the end of the report, or the recipients of a memo must be listed in alphabetical order. If you are writing to someone in your own organization, follow the format constraints described in the company style guide, if there is one, or check similar documents to see what other writers have done. You should also ask more experienced co-workers for their advice. If you are writing to someone outside your organization, learn what you can about that organization's preferences.

- *Tone.* You already know that you write in one way to an organizational superior and in another way to a peer or a subordinate, just as you speak differently to these people. When addressing superiors, use a formal, polite tone. When addressing peers or subordinates, use a less formal tone but be equally polite. Politeness is common courtesy and common sense: you should not bully people or be rude.

CRAFTING A PERSUASIVE ARGUMENT

Persuasion is important to many kinds of technical communication, whether you wish to affect a reader's attitude or merely present information clearly.

To craft a persuasive argument:
Identify the elements of a persuasive argument.
Use the right kinds of evidence.
Consider opposing viewpoints.
Decide where to present the claim.

Identifying the Elements of Your Argument

A persuasive argument has three main elements:

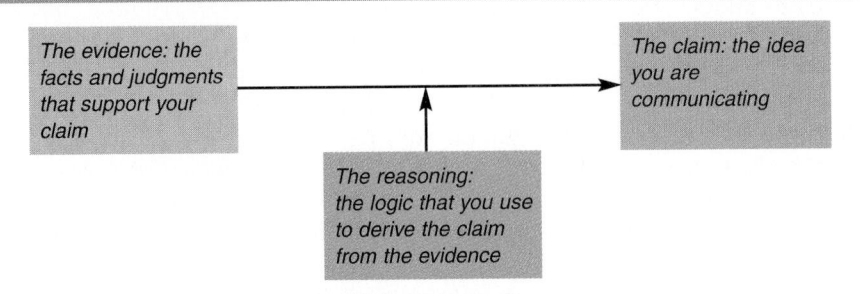

The *claim* is the idea you wish to prove or make compelling: the conclusion you want your readers to accept and, if appropriate, to act on. For example, your claim might be that your company should institute flextime, a scheduling approach that gives employees some flexibility in when they begin and end their workdays. You want your readers to agree with this idea and to take the next steps toward instituting flextime in your company.

The *evidence* is the information you want your readers to consider as they read your discussion. For the argument about flextime, the evidence might include the following:

- The turnover rate of our female employees is double that of our male employees.
- At exit interviews, more than 40 percent of our female employees under the age of 38 state that they quit because they cannot work and also be home for their school-age children.
- Replacing a staff-level employee costs us about one-half the employee's annual salary; replacing a professional-level employee, a whole year's salary.
- Other companies have found that flextime significantly decreases the turnover rate among female employees under the age of 38.
- Other companies have found that flextime has additional benefits and introduces no significant problems.

The *reasoning* is the logic you use to connect the evidence to your claim. In the discussion of flextime, the reasoning involves three links:

- Flextime appears to have reduced the problem of high turnover among younger female employees at other companies.
- Our company is similar to these other companies.
- If flextime has proven helpful at other companies, it is likely to prove helpful at our company.

Using the Right Kinds of Evidence

Every argument calls for a different kind of evidence, but people most often react favorably to four kinds of evidence:

- *"Commonsense" arguments.* In this phrase, *commonsense* has a special meaning: "most people would think that. . . ." The following sentence presents a commonsense argument that binge drinking is harmful:

  > Binge drinking cannot be good for you; it makes you sick while you are doing it and for quite a few hours afterward.

  A commonsense argument appeals to a person's understanding and experience of how the world works. It says, "I don't have hard evidence to support my conclusion, but it stands to reason that. . . ." In this case, the argument is that behavior the body rejects is probably not healthy. If your audience's commonsense matches yours, your argument is likely to be persuasive.

- *Numerical data.* Numerical data — statistics — are generally more persuasive than commonsense arguments.

  > Statistics drawn from campus crime reports (McClellan, 2000) covering over 7,000 rapes reported at 1,672 campuses from 1994–1997 suggest that binge drinking by women is associated with increased rates of rape and sexual assault. Compared with female students who do not binge drink, female binge drinkers are twice as likely as their non-binge-drinking peers to be the victims of sexual assault and rape.

  Notice that the writer states the sample size, and that it is sufficiently large to be persuasive. If the sample size were small — fewer than one hundred, for example — the claim would be much less persuasive. (The discussion of logical fallacies later in this chapter explains such *hasty generalizations*.)

  In addition, you need a representative sample. If you are reporting on computerphobia among students, you need to include a representative percentage of male and female students, as well as students of traditional and nontraditional age.

- *Examples.* An example makes an abstract point more concrete and therefore more vivid and memorable.

  > John M. began to binge drink when he was a junior in high school. Three or four nights a week he and his friends would get together and each drink 6 to 12 beers. Because they were underage and wanted to hide their activities from their parents, they did their drinking away from home — in alleys, in parking lots, at the school athletic fields. One night, as they were walking back from the athletic fields . . .

  Examples are often used in conjunction with numerical data. The example provides a memorable incident, and the numerical data show that the illustration is part of a pattern, not a fluke or a coincidence. The writer using the example of a teenage boy hit by a car when he is binge drinking needs numerical data to prove that a significant number of binge drinkers suffer similar fates; otherwise, the illustration, although powerful, doesn't support the argument.

See Ch. 7, p. 167, for advice on evaluating information from the Internet.

- *Expert testimony.* A message from an expert is more persuasive than the same message from someone without credentials. A well-researched article on binge drinking written by a respected public-health official and published in a reputable medical journal is likely to be persuasive. When you make arguments, you will often cite expert testimony from published sources, Web sites, or interviews you have conducted.

Considering Opposing Viewpoints

When you present an argument, you need to address opposing points of view. If you are proposing to your supervisor that your company study whether to institute flextime and you are aware that several of your important readers oppose flextime because they think it causes new problems — decreased car-pooling opportunities and increased utility bills, for example — you need to consider their objections.

If you don't address opposing views, people who disagree with you will not pay attention to your valid arguments; they will simply conclude that your proposal is flawed because it doesn't address the problems associated with flextime. Regardless of the relative merits of your argument and those of your opponents, you will be in a defensive position because you didn't consider the opposing views.

In meeting the skeptical or hostile reader's possible objections to your case, you can use several different tactics:

- *The opposing argument is based on illogical reasoning or on inaccurate or incomplete facts.* If you can show that the opposing argument has a serious problem, your argument looks stronger because you appear to have thought about the problem carefully. For instance, you might be able to counter the argument that flextime increases utility bills by citing unbiased research studies showing that utility bills do not increase when a company uses flextime.

- *The opposing argument is valid but less powerful than your own.* If you can show that the opposing argument makes sense but is outweighed by your own argument, you look stronger because you appear to be a fair-minded person who understands that reality is complicated. If, for instance, you can show that only 3 percent of your employees use carpooling, and that three-quarters of these employees favor flextime anyway because of its other advantages, you strengthen your argument by removing an opposing argument.

- *There might be a way to reconcile the two arguments.* If an opposition argument is not invalid or clearly inferior to your own, you can express a willingness to study the situation thoroughly to try to find a solution that incorporates the best from each argument. For example, if flextime might cause serious problems for your company's numerous carpoolers, you could propose a trial period. During this period, you would study several ways to help employees find other carpooling opportunities with fellow employees or employees from other organizations. If the organization cannot solve the problem, or if most of the employees prefer the old system, you will switch back to it. This proposal can remove much of the threat posed by your ideas. If your argument turns out to be ill advised, you will admit it, and no permanent harm will result.

When you address an opposing argument, use a gracious, understated tone. Your purpose is to focus the reader's attention on the argument, not to embarrass or discredit your opposition. If your comment is hostile, dismissive, or overstated, you will undermine your own credibility.

In any particular document, there is no one best place to address opposing arguments. In general, however, if you know that important readers hold opposing views, you should address opposing arguments relatively early in your argument. Your goal is to show *all* your readers that you are a fair-minded person who has thought long and hard about the subject, and that your argument is stronger than the opposing arguments. But you don't want to humiliate your opponents, or they will make it a matter of principle to continue opposing you.

Figure 6.2 on page 128, based on excerpts from an article titled "Poverty Accounts for Gap in IQ Scores between Blacks and Whites" (1996), shows two authors confronting an opposing viewpoint.

Deciding Where to Present the Claim

In most cases, the best place to state your claim is at the start of the argument. Then provide the evidence and, if appropriate, the reasoning. This structure is similar to that of a general essay, in which you present the thesis in the first paragraph, then support it in the body of the essay. This structure is also similar to that of the typical paragraph, in which you begin with the topic sentence, then support it in the body of the paragraph.

Sometimes, however, it is more effective to withhold the claim until after you have presented the evidence and the reasoning. This indirect structure works best if you think a substantial number of your readers oppose your claim. If you present your claim right away, these readers might become alienated and stop paying attention. You want a chance to present your evidence and your reasoning without causing this kind of undesirable reaction.

The difference between the claim-first and the claim-delayed approach can be seen in responses you might have received if you had applied for admission to a number of different colleges. If the college accepted you, you

The first three paragraphs summarize the main claim of the article: that environmental conditions, not race, explain the IQ gap between blacks and whites.

Contrary to *The Bell Curve* findings, we have found that poverty and early learning opportunities — not race — account for the gap in IQ scores between blacks and whites. . . . Adjustments for socioeconomic conditions almost completely eliminate differences in IQ scores between black and white children.

As in many other studies, the black children in the study had IQ scores a full 15 points lower than their white counterparts. Poverty alone accounted for 52 percent of that difference, cutting it to 7 points. Controlling for the children's home environment reduced the difference by another 28 percent, to a statistically insignificant 3 points — in essence, eliminating the gap altogether.

The writers explain the methods of the study, commenting on how these methods enabled them to probe the question more fully than previous researchers did. This argument is effective largely because of the authors' careful use of evidence (the statistics) and clear reasoning.

The study includes data from birth to age five on 800 black and white children who were born premature and with a low birth weight. Collected from eight health care sites around the country, it is the only data set that combines high-quality measurement of developmental outcomes (i.e., full-scale IQ tests) with longitudinal data on family economic status, neighborhood conditions, family structure, and home environment. Because the study looks at very young children, the subjects' IQ measures cannot be attributed to such nonfamily influences as schooling or work.

The longitudinal data allowed us to measure persistent poverty — found to be a key factor in the IQ differences. Many children have transitory experiences with poverty. For black children, poverty is likely to be much more persistent. Of the black children in the study, 40 percent lived in persistent poverty, compared to 5 percent of white children. The study also takes into account how impoverished neighborhood conditions and environmental influences can affect even children not living in poverty. Black families are more likely to live in poor neighborhoods, whether or not they are poor themselves. Almost one half of all black children whose families were not poor resided in poor neighborhoods, compared with less than 10 percent of white children. . . .

The writers introduce the opposing view, as presented in The Bell Curve.

Debate over what causes the IQ gap has been highly charged since the 1994 publication of *The Bell Curve*, by Richard Herrnstein and Charles Murray, who view the difference as genetic and impossible to change.

Notice the use of the word "assertion," an understated criticism of the methods of the authors of The Bell Curve.

The *Bell Curve* hypothesis does not depend on any direct evidence, but rather on its authors' assertion that social and economic factors cannot explain it. Because the typical black ranks at the 15th percentile of the white IQ distribution, say Herrnstein and Murray, black socioeconomic status (SES) can only explain the ranking if, on average, it is well below than the 15th percentile of white SES ranking.

Saying that the new study "calls into question" the findings of The Bell Curve *projects an image of careful professionalism. You always want your readers to see that your evidence is at least as strong as you claim it is.*

The *Bell Curve* authors claim no such SES inequality exists, and this is the point that our study calls into question. Most studies of socioeconomic status do not consider such obvious factors as family income or neighborhood conditions, and those that do fail to account for the degree of persistent poverty.

■ **Figure 6.2   Confronting Opposing Viewpoints**

knew it as early as the first line, which often begins "Congratulations" or "We are pleased to tell you. . . ." If, however, the letter began "We have now completed the very difficult process of selecting from among the many excellent applicants . . . ," you knew you would see the word *regret* a few paragraphs down the page.

AVOIDING LOGICAL FALLACIES

Logical fallacies can undercut the persuasiveness of your writing. Table 6.1 explains some of the most common logical fallacies.

▨ Table 6.1 Common Logical Fallacies

Fallacy	Explanation	Example and comment
Ad hominem, also called *argument against the speaker*	Argument against the writer, not the writer's argument.	"Of course Matthew favors buying more computers — he's crazy about computers." The fact that Matthew loves computers doesn't necessarily mean that his argument for buying more computers is unwise.
Argument from ignorance	A claim is true because it has never been proven false, or false because it has never been proven true.	"Nobody has ever proven that global warming is occurring. Therefore, global warming is a myth." The fact that a concept has not yet been proven does not necessarily mean that it is false. Perhaps the measurement techniques are insufficiently precise or not yet available.
Appeal to pity	An argument based on emotion, not reasons.	"We shouldn't sell the Ridgeway division. It's been part of the company for over forty years." The fact that the division has long been a part of the company is not in itself a good reason to retain it.
Argument from authority	An argument that a claim is valid because the person making the claim is an authority.	"According to Dr. Smith, global warming is definitely a fact." Even if Dr. Smith is a recognized authority in this field, saying that global warming is a fact is not valid unless you present a valid argument to support it.

 See Writing Guidelines for Engineering and Science Students (www.vt.edu:10021 /eng/mech/writing/) for exercises on logical fallacies.

Fallacy	Explanation	Example and comment
Circular argument, also called *begging the question*	An argument that assumes what it is attempting to prove.	"Compaq is more successful than its competitors because of its consistently high sales." Because "more successful" means roughly the same thing as achieving "consistently high sales," this statement says only that Compaq outsells its competitors. The writer needs to explain *why* Compaq outsells its competitors and is therefore more successful.
Either-or argument	An argument that poses only two alternatives when in fact there might be more.	"If we don't start selling our products online, we're going to be out of business within a year." This statement does not explain why these are the only two alternatives. The company might improve its sales by taking measures other than selling online.
Ad populum argument, also called the *bandwagon argument*	An argument that a claim is valid because many people think it is or act as if it is.	"Our four major competitors have started selling online. We should too." The fact that our competitors are selling online is not in itself an argument that we should.
Hasty generalization, sometimes called *inadequate sampling*	An argument that draws conclusions on the basis of an insufficient number of cases.	"The new Gull is an unreliable car. Two of my friends own Gulls, and both have had reliability problems." Before reaching any valid conclusions, you would have to study a much larger sample and compare your findings with those for other cars in the Gull's class.
Post-hoc reasoning (the complete phrase is *post hoc, ergo propter hoc*)	An argument that claims that, because A precedes B, A caused B.	"There must be something wrong with the new circuit breaker in the office. Ever since we had it installed, the air conditioners haven't worked right." The air conditioners' malfunctioning might be caused by the circuit breaker, but the malfunctioning might have other causes.

Fallacy	Explanation	Example and comment
Oversimplifying	An argument that omits important information in establishing a causal link.	"The way to solve the balance-of-trade problem is to improve the quality of the products we produce."
		Although improving quality is important, international trade balances are determined by many factors, including tariffs and currency rates, and therefore cannot be explained by simple cause-and-effect reasoning.

PRESENTING YOURSELF EFFECTIVELY

Just as testimony from an expert is more persuasive than the same testimony from a nonexpert, an argument from a person who appears professional is more persuasive than the same argument from a person who appears unprofessional.

The following paragraph shows how a writer can demonstrate the qualities of cooperativeness, moderation, fair-mindedness, and modesty:

This plan is certainly not perfect. For one thing, it calls for a greater up-front investment than we had anticipated. And the return-on-investment through the first three quarters is likely to fall short of our initial goals. However, I think this plan is the best of the three alternatives for the following reasons. . . . Therefore, I recommend that we begin planning immediately to implement the plan. I am confident that this plan will enable us to enter the flat-screen market successfully, building on our fine reputation for high-quality advanced electronics.

In the first three sentences, the writer acknowledges the problems with his recommendation.

The use of "I think" adds an attractive modesty; the recommendation might be unwise.

The recommendation itself is moderate; the writer does not claim that the plan will save the world.

In the last two sentences, the writer shows a spirit of cooperativeness by focusing on the company's goals.

GUIDELINES

Creating a Professional Persona

Your *persona* is how you appear to your readers. Demonstrating the following characteristics will help you establish an attractive professional persona:

▶ *Cooperativeness.* Make clear that your goal is to solve a problem, not advance your own interests.

▶ *Moderation.* Be moderate in your judgments. The problem you are describing will not likely spell doom for your organization, and the solution you propose will not solve all the company's problems.

▶ *Fair-mindedness.* Acknowledge the strengths of opposing points of view, even as you offer counterarguments.

▶ *Modesty.* If you fail to acknowledge that you don't know everything, someone else will be sure to volunteer that insight.

USING GRAPHICS AND DESIGN AS PERSUASIVE ELEMENTS

Graphics and design elements are fundamental in communicating persuasively because they help you convey both technical data and nontechnical information.

Figure 6.3 (Gateway Inc., 1999), for example, shows a typical combination of verbal and visual techniques used to make a persuasive argument.

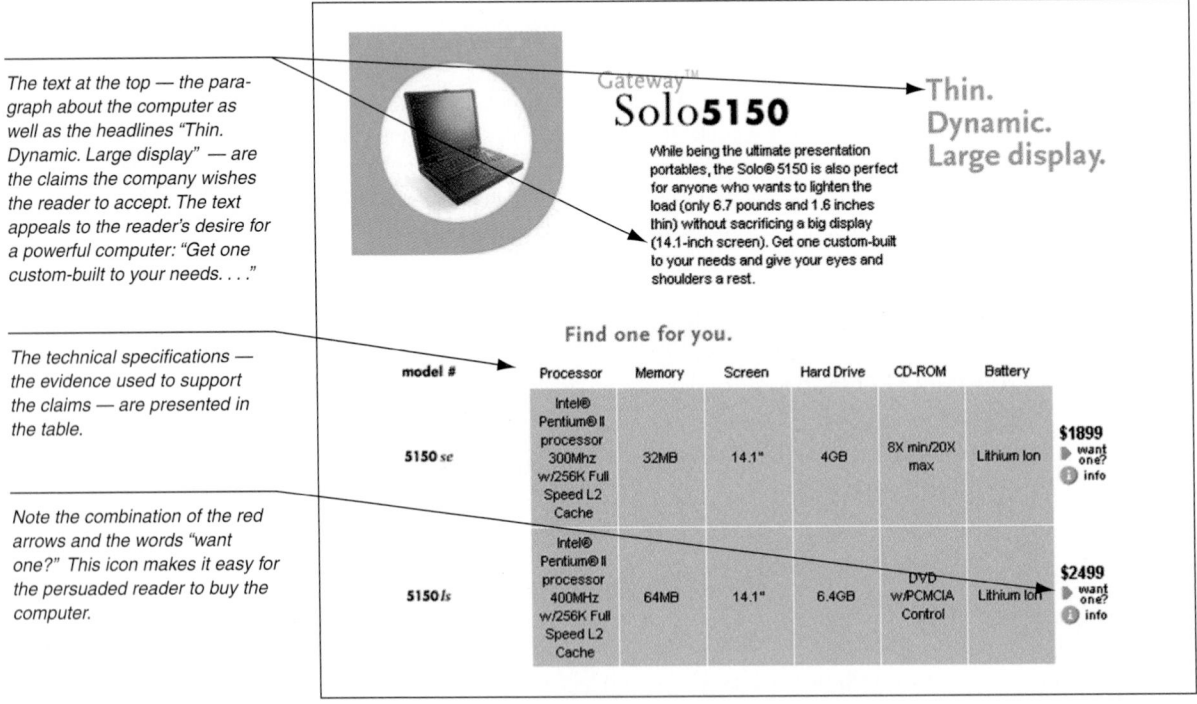

■ Figure 6.3   Verbal and Visual Techniques in Persuasion

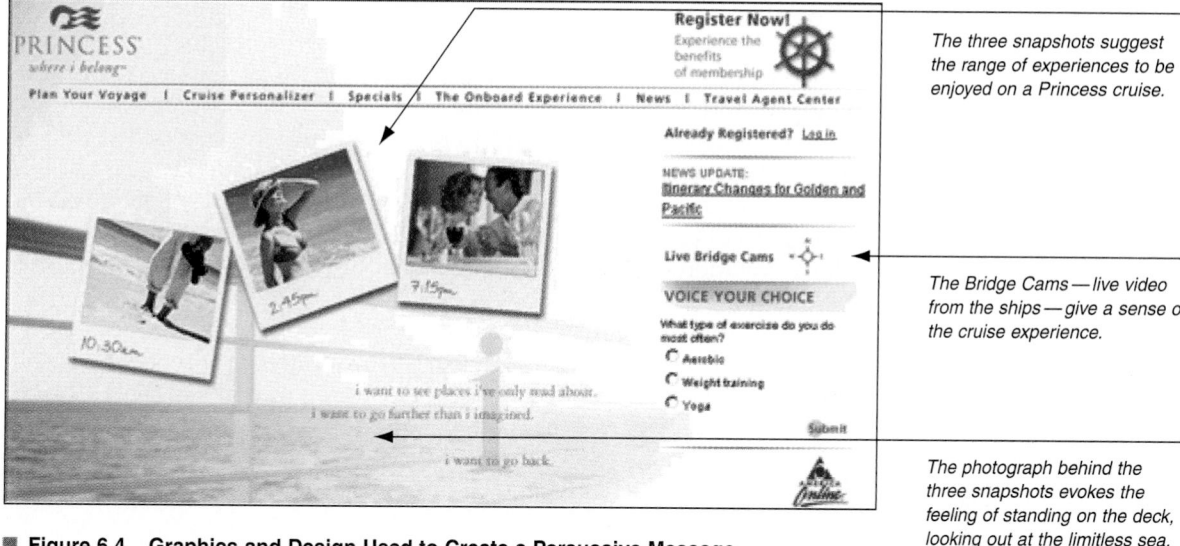

The three snapshots suggest the range of experiences to be enjoyed on a Princess cruise.

The Bridge Cams—live video from the ships—give a sense of the cruise experience.

The photograph behind the three snapshots evokes the feeling of standing on the deck, looking out at the limitless sea.

■ **Figure 6.4   Graphics and Design Used to Create a Persuasive Message**

In this sample, the words do some of the persuasive work, but the graphics and design do much more of it. The photographs and the live videos are the main persuasive elements.

Graphics and design can also be used to present evidence in a less technical way. Figure 6.4 (Princess Cruises, 2001) shows a portion of the main page from a cruise line's Web site.

LINKING PERSUASION AND ETHICS

The young actor asks the old actor, "What's the key to great acting?" The old actor replies, "Sincerity. Once you learn how to fake sincerity. . . ." Any discussion of image and persuasion has to address the question at the heart of this ancient joke. Does a writer have to be cooperative to appear cooperative? Does there have to be any reality behind the image, or is the image enough?

This is too big a question for our discussion here. It belongs in a full-length book on ethics or politics or marriage. For our purposes, the comment by Mayberry at the start of the chapter is useful: writers "project an image of intelligence . . . and trustworthiness. There is nothing false or superficial about this kind of image"). You cannot *pretend* to be honest and cooperative and moderate. Or, to be more precise, you probably cannot pretend forever. Although some people succeed for a long while, most are revealed in the end.

The easiest way to appear honest and cooperative is to be honest and cooperative. As suggested in Chapter 2, you need to tell the truth and not mislead your readers. As suggested in Chapter 4, you need to be cooperative, diplomatic, and constructive. And as suggested in this chapter, you need to remember people's broader goals: to protect their own security, to reinforce their own egos, and to learn and grow in their professional lives.

When you hear an ad that says "Quality is Job One," or "Quality is our most important product," you don't automatically believe it. The most generous interpretation of claims like these is that the company would like you to think that it is serious about quality. But of course you wait for the specifics. You want to know exactly what the company has done to earn your business. In the same way, you as a writer have to earn your image. You persuade readers by being a credible and trustworthy person who presents valid arguments.

PERSUADING MULTICULTURAL AUDIENCES

The discussion in Chapter 5 about addressing multicultural audiences made the point that cultures differ significantly not only in such matters as business customs but also in their most fundamental values. In western cultures, for example, individualism is a primary value: you are responsible for your own success or failure. In the east, the group is much more important than the individual.

It follows, therefore, that what makes an argument persuasive for one set of readers will not necessarily make it so for another group. Statistics and experimental data are fundamental kinds of evidence in the west, but testimony from respected authority figures is much more persuasive in the east.

How to structure an argument — whether to put the claim up front or delay it — is also culturally determined. In a western culture, the claim is usually presented up front. In an eastern culture, it is likely to be delayed or to remain unstated but implied.

The advice here is the same as the advice in Chapter 5. When you write for an audience from another culture, you need to study that culture and adjust the content, structure, and style of your arguments. And because nobody can ever become a true expert in another culture, you should be sure to set aside the resources to have your important documents reviewed and edited by a person from the target culture.

A LOOK AT SEVERAL PERSUASIVE ARGUMENTS

The following examples of technical communication show how the persuasive elements of an argument differ depending on a writer's purpose. Figure 6.5 presents two paragraphs from a student's job-application letter.

Figure 6.6, from the General Electric Web site (2001), illustrates an effective use of tone and appeals to an audience's broader goals.

At Western State University, I have earned 87 credits toward a degree in Technical Communication. I have been a full-time student (no fewer than 12 credit hours per semester) while working full time for the Northwest Watershed Research Center. The four upper-division courses I am taking this semester, including Advanced Technical Communication and Technical Editing, are required for the B.A. in Technical Communication.

In addition to my formal education, I have completed 34 training courses on the job. These courses have included diverse topics such as financial management, the Fair Labor Standards Act, the Americans with Disabilities Act, career-development opportunities in public affairs, and software applications such as MS Office, Quark Xpress, and RoboHelp.

Without making her claim explicit, the writer presents evidence that she is hard working and lets the prospective employer draw his or her own conclusions.

In listing some of the training courses she has taken, the writer supports an earlier claim that her broad background might be of use to her next employer.

■ **Figure 6.5   Persuading a Prospective Employer**

A student writer uses specific examples in an effort to persuade a prospective employer.

GE in the Community

Every day, GE people and the GE Fund are helping to improve the communities the Company calls home around the world. GE Elfun, the global volunteer organization of employees and retirees, undertakes more than 1,000 community service projects every year. From mentoring in schools to building playgrounds to serving meals for the elderly to renovating parks and cleaning rivers, GE's people volunteer more than one million hours each year.

Education is a primary focus of the GE Fund. Fund programs help improve literacy and math skills in the early grades; prepare high school students to attend college; and support women and minority students seeking to earn engineering and business degrees and move into the faculty ranks at universities.

In all, GE and its employees, the GE Fund and GE Elfun contributed nearly $100 million in 2000 to support nonprofit community organizations, schools, colleges and universities around the world.

In Pittsfield, Mass., Elfuns renovated Burbank Park, used by residents and visitors to Pittsfield for over 112 years. The project included a complete renovation of a pavilion, installation of a children's play structure, painting and restoring a beach house, and cleaning over 2,000 feet of shoreline and play areas.

The photograph of GE employees doing volunteer work gives the statistics a human face.

The argument begins with a statement about the scope of the community activities undertaken by GE employees. Notice the use of specific evidence about the number of hours devoted to philanthropy each year.

The second paragraph focuses on GE's contribution to education.

The final paragraph provides more evidence and links community support and education initiatives.

■ **Figure 6.6   Persuading Employees and Customers**

Published on General Electric's Web site, this description of the company's community programs aims to persuade readers that GE is a socially responsible company.

Figure 6.7 ("Death," 1999) on page 136, an article about the security risks involved when employees take company laptops with them on business trips, uses researched examples and commonsense logic as its main evidence.

The specific, memorable images make the scenarios and the argument seem plausible.

Theft is the most prevalent risk while the notebook is out of the office. Your users will undoubtedly want to leave their PCs behind when they have lunch or dinner with their clients. Even the most disciplined will put the notebook-bearing briefcase down while they browse through the duty free at the airport. No one we spoke to took their notebook out of the car when they went into a store; most didn't even lock it in the trunk out of sight of the opportunist thief.

Conclusions based on common-sense logic enhance the writers' credibility.

Data loss or data theft is a serious risk — though most users dismiss this. The vast majority of people do not believe that the contents of their drive are of serious interest to individuals outside their own organization. In fact, they are wrong. Our research team has spoken to several organizations who, while they insist on remaining anonymous, testify to incidents in which data has been stolen from company laptops. Recent stories have come to light of government notebooks being stolen — and while these attract more media attention they are probably no more common than the theft of PCs from commercial organizations.

Here an informal tone suggests that the writers are experts, well-acquainted with the culture of business travelers.

Viruses and trojans are also a problem for the out-of-office worker. The most effective anti-virus solutions tend to be those that require regular updating. Unfortunately, infrequent connection to the corporate network means that road warriors do not always have the latest update to their virus defenses. This could mean catastrophic results in the case of viruses like Melissa or CIH (the so-called Chernobyl virus). The effects might also not be limited to the damage done to the single notebook user. The payload might have corporate-wide consequences, and the road warrior might be a source of infection for the rest of the organization.

Arranging the examples from most to least prevalent avoids ending with a pronouncement of doom, making the argument seem more fair-minded and reasonable.

Connection to the corporate network is at once a great boon for the road warrior and a great problem for system administrators. It is essential that you properly identify and authenticate the remote use, and it is also important that you choose a system that is easy for your remote users to employ. This system must itself be protected since if a user leaves his or her machine in a hotel room, even though the machine may be physically protected preventing its removal, it may not be adequately protected against someone using the machine to access the corporate network. This latter possibility, while remote, is nonetheless real.

■ **Figure 6.7   Persuading an Expert Audience**

This passage from a trade magazine about computer security systems aims to warn system administrators that their companies take serious risks when their employees travel with laptops. The writers enhance their credibility in the eyes of experts by laying out their argument logically, focusing on one example of a security risk in each paragraph.

✔ **Revision Checklist**

1. In analyzing your audience, did you consider the broader goals of
    - ❏ maintaining security?
    - ❏ reinforcing ego?
    - ❏ achieving self-actualization?

2. In planning, did you consider the following constraints:
    - ❏ ethical?
    - ❏ legal?
    - ❏ political?
    - ❏ informational?
    - ❏ personnel?
    - ❏ financial?
    - ❏ time?
    - ❏ format and tone?

3. In crafting a persuasive argument, did you
    - ❏ use the three-part structure of claim, evidence, and reasoning?
    - ❏ choose the appropriate kinds of evidence?
    - ❏ consider opposing viewpoints?
    - ❏ decide where to present the claim?

4. In writing the argument, did you avoid the following logical fallacies:
    - ❏ *ad hominem* argument?
    - ❏ argument from ignorance?
    - ❏ appeal to pity?
    - ❏ argument from authority?
    - ❏ circular argument?
    - ❏ either-or argument?
    - ❏ *ad populum* argument?
    - ❏ hasty generalization?
    - ❏ *post-hoc* reasoning?
    - ❏ oversimplifying?

5. In drafting your argument, did you create a persona that is
    - ❏ cooperative?
    - ❏ moderate?
    - ❏ fair-minded?
    - ❏ modest?

6. In addressing a multicultural audience, did you consider what types of evidence and what argument structures would be most effective?

Exercises

Some of the following exercises ask you to write a memo. See Chapter 15, page 430, for a discussion of memos.

1. Visit the Web site of a car manufacturer, such as Ford (www.fordvehicles.com) or Mercedes Benz (www.mbusa.com). Identify the major techniques of persuasion used in the words and graphics on the site. For example, what claims are made? What types of evidence are used? Is the reasoning sound?

2. The manager of an expensive hotel sent the following letter to guests who had experienced the problems he refers to in the first paragraph. In a brief memo to your instructor, evaluate the letter. Which elements of the letter are persuasive? Which are not?

June 18, 2000

Dear Valued Guest:

Once again, please accept our sincerest apologies for the inconvenience that you experienced on Sunday morning when our Hotel's emergency alarm system was activated and again on Monday when there was a shortage of hot water. We regret that it took an extended time for you to be able to return to your rooms, but we and the Boston Fire Department had to be completely sure that there was no immediate danger.

As a token of our gratitude for your patience and understanding, we would like to offer you the opportunity to visit our Hotel once again at a discounted rate. Please accept this letter as a certificate to stay overnight at The Eastin Boston at 50% off our best available rate. Please contact our reservations department locally at 617.555.9600 to coordinate your one-night stay prior to December 28, 2000.

We assure you that the safety and comfort of our guests is always in the forefront of our minds. We appreciate your confidence and sincerely hope that you will take advantage of our offer, as it would be our pleasure to welcome you back. We promise to do everything in our power to ensure a pleasant and peaceful night's sleep on your return visit.

Sincerely,
Arthur Plant
General Manager

3. The following text, called "SEC to Investors: Don't Chase Performance," appeared on The Vanguard Group Web site (Vanguard, 2000). The underscored items in the text are links to other documents. What techniques does the writer use to make a persuasive case? How persuasive is the text? Present your findings in a brief memo to your instructor.

Last year, according to the Securities and Exchange Commission (SEC), a record number of mutual funds posted returns of 100% or more. Such results have prompted the SEC to urge investors to look further than short-term performance when it comes to evaluating funds.

The SEC recently issued tips for the 83 million Americans invested in mutual funds, warning that high-performing funds often fail to repeat their gains. Instead of chasing performance, investors would be better off shopping for funds that best match their long-term financial goals and risk tolerance.

Vanguard Chairman and CEO John J. Brennan has similar concerns. "Buying last year's or last quarter's top-performing investment is a dangerous strategy," Mr. Brennan said. He offers additional insights in a recently published "In The Vanguard" interview, Brennan Urges Investors to Tune Out Noise, Focus on Basics.

Historically, investors who follow fads and jump in and out of funds typically fare worse than those who buy and hold. In addition, buying into a well-performing fund after the fact could result in a considerable tax liability if the fund distributes its gains to shareholders—gains that recent investors didn't benefit from. Also, investors who hop from fund to fund may incur their own taxable gains.

The SEC suggests that investors thoroughly read the prospectus and annual report for any fund they're considering. Assessing a fund's cost is important—a difference of one percent in a fund's annual fee could reduce the fund's ending balance by 18% after 20 years.

The SEC offers the following guidelines for fund evaluation:

- Scrutinize other costs, such as sales charges and expenses.
- Know how a fund will impact taxes.
- Consider the age and size of the fund.
- Think about the volatility of the fund.
- Factor in the risks the fund takes to achieve its returns.
- Ask about recent changes in the fund's operations.
- Check the types of services offered by the fund.
- Assess how the fund will impact portfolio allocation.

For more information, read the SEC's Mutual Funds Tips.

4. Study the following excerpt from "Statement of the Director of Central Intelligence George J. Tenet On Diversity" (Tenet, 1999). Which elements of persuasion are used effectively? Which elements of persuasion are used ineffectively? Present your findings in a brief memo to your instructor.

Our country is home to gifted people of virtually every national origin, creed, and culture. In our diversity there is tremendous strength. We must learn to recognize this diversity as the valuable asset that it is. If we fail to do so, we will waste an enormous amount of talent and resources. That is a waste which our country cannot afford and which I will not tolerate.

I regard our diversity as a powerful tool that can help us meet the intelligence challenges of the coming century. That is why I have made advancing diversity within our Agency and Community an important part of my Strategic Direction planning. I am determined to increase the diversity of our workforce and to use the many talents of the men and women who are already with us to optimum advantage.

The demographic trends are unmistakable. Projections indicate that over the next twenty years, women and people of color will constitute a growing majority of new entrants into the American labor market. The Millennium Generation, today's 15-to-25-year-olds, is the most racially mixed generation in our history.

Diversity is already an imperative for the business community. Corporate America was among the first to recognize that diversity pays dividends. CEOs know that organizations which value diversity — and know how to use it — will have the competitive edge, not only in recruiting and retaining the best employees, but in operating successfully worldwide.

Fortune 500 companies are out every day aggressively targeting women and minorities for recruitment. Their sophisticated recruiting ads prominently feature women, people with disabilities, and a range of races, ethnicities, and ages. The corporate slogans say it all. I will cite just a few: "We view the world through many different lenses" (Kodak). "Working together to change the world" (Raytheon). "It only works if we work together" (Citibank). "Equal Opportunity Empowerer" (First USA). The ads aren't merely slick public relations tools, they are genuinely representative of a corporate population that is more diverse than ever before and growing more so every day.

Our Intelligence Community is competing with the private sector for the best and the brightest of the rising generation. CIA is now engaged in the biggest recruiting drive since the establishment of our Agency and I have made it a priority to strengthen our recruitment capabilities.

5. Write one paragraph identifying the logical flaws in each of the following items:

   a. The election couldn't have been fair — I don't know anyone who voted for the winner.

   b. It would be wrong to prosecute Allied for age discrimination; Allied has always been a great corporate neighbor.

   c. Increased restrictions on smoking in public are responsible for the decrease in smoking.

   d. Bill Jensen's proposal to create an on-site day-care center is just the latest of his hare brained ideas.

   e. Since the introduction of cola drinks at the start of this century, cancer has become the second greatest killer in the United States. Cola drinks should be outlawed.

   f. If mutual-fund guru Peter Lynch recommends this investment, I think we ought to buy it.

   g. We should not go into the DRAM market; we have always been a leading manufacturer of integrated processors.

   h. The other two hospitals in the city have implemented computerized patient recordkeeping; I think we need to do so, too.

   i. Our Model X500 didn't succeed because we failed to sell a sufficient number of units.

   j. No research has ever established that Internet businesses can earn money; they will never succeed.

Research Project

6. Form small groups to study the Web site of a large organization such as the American Red Cross (www.redcross.org), the White House (www.whitehouse.gov), Fidelity Investments (www.fidelity.com), or Microsoft Corporation (www.microsoft.com). Have each person in the group study the site separately and consider the following questions:

- What are the various audiences for the site? (Briefly describe the major pages it contains.)

- What are the various purposes of the site?
- What are the various techniques of persuasion used in the words and the graphics?
- How effective is the site as a persuasive document? Which techniques are effective and which ineffective? How might the site be made more persuasive?

Come together as a group and share your perceptions about the site. Draft a memo to your instructor presenting your findings.

For additional cases, click on
Case of the Month and
Archive on TechComm Web
(www.bedfordstmartins.com/
techcomm).

C A S E 1
Writing an Apology

Every once in a while, something goes wrong with a company's product or service, prompting the company to apologize to its customers. A problem that occurred at an online brokerage service recently led the company to publish an apology on its Web site. Use the following steps to analyze the company's apology.

1. Study the apology presented below. Identify the function of each paragraph and how it contributes to the argument. Consider the audience. In what ways does the writer address the needs and tastes of the audience? How successful are the writer's attempts?

2. Think about the ways in which the writer projects a professional image throughout the article. Does the writer appear to be a professional according to the criteria presented in this chapter? What might the writer have done differently?

3. Write a 500-word memo to your instructor presenting your findings.

Responding to the Recent Denial-of-Service Attack

Since our founding in 1999, RapidTrade has taken pride in its position of leadership in the online brokerage industry. We offer the lowest rates in the industry, along with highly secure, near-instantaneous transactions, and a wealth of other investment information services that enable you to stay current with up-to-the-minute financial news. For these reasons, RapidTrade has received the highest evaluations by independent consumer groups for the quality of its overall services to its valued customers.

As you are no doubt aware, last week RapidTrade was victimized by an as-yet unknown hacker or hackers, who caused a denial-of-service event that lasted for over fourteen hours. "Denial of service" is a term that refers to a computer problem that prevents an online organization, such as an Internet service provider or a business, from offering its customary services to its customers.

This denial of service occurred at an especially unfortunate time, when market volatility was extremely high due to provocative statements made earlier in the day by the chairman of the Federal Reserve Board and the publication the previous day of discouraging inflation reports.

An additional hacker-related problem occurred. The hacker or hackers who caused the denial-of-service event also managed to breach the security of our customer database. While we are confident that no significant harm was done, it appears that unauthorized people somehow gained access to our customer information and the log of all transactions made by the customers.

A number of our valued customers have expressed their concern that they were unable to carry out transactions during the period of the denial of service. Of course, we sincerely regret that this problem occurred, but, unfortunately, RightTrade can take no legal responsibility for the denial-of-service event. As stated in our Terms of Agreement, section 4, paragraph 19c, "RightTrade shall under no circumstances be held liable for consequential damages resulting from denial of service caused by technical malfunctions." Clearly, the denial of service falls under this category.

RapidTrade has now completed a thorough investigation of the situation. We want to point out, initially, that this investigation has been undertaken voluntarily and carried out according to the highest standards of professionalism. RightTrade has not been required to perform this investigation by any federal, state, or industry oversight agency.

Our principal conclusion is that while there was no apparent negligence by the vendor of our security systems, the denial-of-service event and the related compromising of our customer information was clearly unsatisfactory. As a result, RightTrade has taken swift action to resolve the situation.

We have met with the vendors of our security system in intensive sessions to determine whether we can determine what factors contributed to the denial-of-service event and the related situation with the customer database. At the time of this writing, it is unclear how the two events are related. We will of course take all necessary steps to prevent future events of this sort.

Once again, we regret and apologize for any inconvenience caused by the denial-of-service event and the related situation with the customer database. We remain deeply commited to offering the highest quality service, and we are deeply committed to the privacy concerns of our valued customers. We encourage you to contact us with your comments and suggestions at feedback@rapidtrade.com.

This case is on TechComm Web (www.bedfordstmartins .com/techcomm).

C A S E   2
Analyzing an Audience

You work in the Customer Service department at Greenlawn, a company that applies chemical fertilizers, herbicides, and pesticides to lawns. Recently Greenlawn received a letter from Gwen Smith, a customer with a five-year-old child and an infant. Ms. Smith is very concerned about the safety of the chemical you spray on her lawn; recently she read an article in the newspaper about a man who died suddenly after playing a round of golf on a course that had been treated by a similar company. The article stated that although commercial lawn treatments are generally safe, every year, two or three people die because of an unusual vulnerability to the chemicals.

Your supervisor, Helen Lewis, has drafted the following letter to Ms. Smith and has asked you to review it. "Do you think this responds to Ms. Smith's concerns?" Helen asks you. Write a memo to Helen Lewis evaluating how persuasive the letter is and recommending any revisions you think would improve it.

April 5, 20XX

Dear Ms. Smith:

Thank you for inquiring about the safety of the Greenlawn program. The materials purchased and used by professional landscape companies are effective, nonpersistent products that have been extensively researched by the Environmental Protection Agency. Scientific tests have shown that diluted tank-mix solutions sprayed on customers' lawns are rated "practically nontoxic," which means that they have a toxicity rating equal to or lower than such common household products as cooking oils, modeling clays, and some baby creams. Greenlawn applications present little health risk to children and pets. A child would have to ingest almost 10 cupfuls of treated lawn clippings to equal the toxicity of one baby aspirin. The child's stomach could not possibly hold enough lawn clippings to prove dangerous!

Research published in the *American Journal of Veterinary Research* in February 1984 demonstrated that a dog could not consume enough grass treated at the normal rate of application to ingest the amount of spray material required to produce toxic symptoms. The dog's stomach simply is not large enough.

A check at your local hardware or garden store will show that numerous lawn, ornamental, and tree-care pesticides are available for purchase by homeowners either as a concentrate or combined with fertilizers as part of a weed-and-feed mix. Label information shows that these products contain generally the same pesticides as those programmed for use by professional lawn-care companies but at higher concentrations than found in the diluted tank-mix solutions applied to lawns and shrubs. By using a professional service, homeowners can eliminate the need to store pesticide concentrates and avoid the problems of improper overapplication and illegal disposal of leftover products in sewers or household trash containers.

On the basis of these facts, I am sure that you will be pleased to know that the Greenlawn program is a safe and effective way to protect your valued home landscape. I have also enclosed some additional safety information. I encourage you to contact me directly should you have any questions.

Sincerely,

Helen Lewis

Helen Lewis
Branch Manager

7

Researching Your Subject

The science writer Laurence A. Moore (qtd. in Harris, 1999) on using the Internet and traditional research tools:

Searching for data on the Internet can be frustrating but what you find often can't be found in a library — the same is true in reverse. I didn't stop using the library when I started using the Internet.

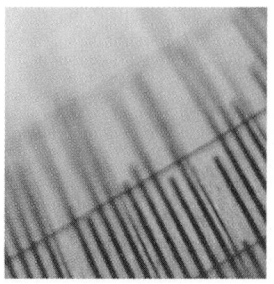

One of the main points in Chapter 6 was that you need to present clear and compelling evidence to make a persuasive argument. This chapter focuses on finding information to use as evidence in your documents.

 See TechComm Web (www .bedfordstmartins.com /techcomm) for guidelines boxes, additional examples, and links related to topics in this chapter.

The discussion in this chapter covers *primary research* and *secondary research*. Primary research involves creating technical information yourself. Secondary research is the process of collecting information that other people have discovered or created. The most common ways to perform secondary research are to read books and journals and search the Internet, but researchers also talk with colleagues, consult databases, and attend conferences.

This chapter presents secondary research first. Only rarely would you conduct primary research before doing secondary research, because to design the experiments or the field research that goes into primary research, you need a thorough understanding of the information that already exists about your subject.

First, the chapter explains how to perform secondary research: understanding the different research media, using basic research tools, researching government information, and understanding and evaluating the information you have gathered. It then discusses techniques of primary research, including how to conduct inspections, experiments, field research, and interviews; how to write letters of inquiry; and how to administer questionnaires.

PLANNING A RESEARCH STRATEGY

Once you have clearly defined your writing topic, you will need to plan a strategy for researching your topic, taking into account the kind of information you need as well as the scope of your research. To determine a research strategy, consider three factors:

See Ch. 5 for more on audience and purpose.

- *Your audience.* Are your most important readers experts, technicians, managers, or general readers? What are their personal characteristics, their attitudes toward your subject, their motivations for reading? If you are writing to an expert audience that might have good reasons to be skeptical about your message, you will need to do a lot of research to gather the evidence to make a convincing argument.

- *Your purpose.* What are you trying to accomplish by writing? Understanding your purpose often helps you understand the types of information readers expect in the document. For example, if you are proposing that your company investigate purchasing new equipment, you will need to gather the kinds of information usually presented in proposals: expert opinions on your subject, management plans for carrying out the project, budget information, and so forth.

See Ch. 3, p. 45, for more on ways to generate ideas about your topic.

- *Your subject.* What do you already know about your subject? What do you still need to find out? Using techniques such as freewriting and brainstorming, you can determine those aspects of the subject you need to investigate.

Your ideas about your audience, purpose, and subject will develop and change as you get further into the project. Students often think that, once they do their research, they're done with that phase of the project. In fact, the process of researching a subject and writing a document is anything but linear. During the drafting or even the revision stage, you might realize that you need to revise your assessment of your audience and purpose and to do more research. Experienced writers know that they're done researching only when they see the final draft of the document slide out of the printer.

CONDUCTING SECONDARY RESEARCH

For students, the best places to find information are the Internet and a college or university library. Most college libraries have substantial reference collections and receive the major professional journals. Large universities have more comprehensive collections. Many large universities have specialized libraries that complement selected graduate programs, such as those in zoology or architecture. Large cities often have special scientific or business libraries that are open to the public.

For professionals, the best place to begin a search might be the organization's information center. An *information center* is the organization's library, a resource that collects different kinds of information critical to the organization's operations. At some information centers, the staff does all the searching for the employee. For instance, you can email a question such as "How many cell phones were sold in the United States last year?" and the staff will find an answer and get back to you. At other information centers, the staff provides guidance, but you carry out the search.

At both public and private libraries, the most important information sources are the reference librarians. They are always willing to suggest new ways to find what you need — specialized directories, bibliographies, or collections that you didn't know existed. They can also assist you in using online databases. And they will tell you if the library doesn't have the information you need and suggest other libraries to try.

Reference librarians can save you much time, effort, and frustration. Don't be afraid to ask them questions when you run into problems.

Understanding the Research Media

Today, most technical information is being distributed not only in print but also through one or several digital media. For instance, the federal government is likely to publish census information in printed reports and books, on CDs, and on Web sites.

You will probably use information published in five major media.

Media used to publish information:
Print
Online databases
Digital disks, such as CD-ROM
Web sites
Online discussion groups

Of these media, digital information is generally easier and more convenient to use than printed information for three reasons:

- Digital information is usually more current than printed information because it is easier to revise and distribute.

- Digital information is often cumulative. That is, for indexes and other information published periodically, new information is integrated with the old; therefore, you do not have to search through back issues of the periodical; there are no back issues.

- Digital information can be saved to a disk, so you don't have to make photocopies or take notes. You don't introduce spelling errors into your document as you take notes and then type up the information.

Printed Information

Books, journals, reports, and other documents will continue to be produced in print because printed information is portable, and you can write on it. For documents that do not need to be updated periodically, print remains a useful and popular medium. And to find it you will continue to use online catalogs, as you do now.

Online Databases

Most libraries — even many public libraries — have facilities for online searching. Libraries today lease access to different database services, with which they communicate by computer. The largest database service is DIALOG Information Services, which offers electronic access to more than six billion pages of articles, conference proceedings, news, and statistics in scientific, technical, and medical literature, as well as business, trade, and academic studies. DIALOG also provides access to more than 100 full-text newspapers and thousands of magazines and journals in more than nine hundred databases. The big disadvantage of commercial online databases is their expense. A simple search can cost more than fifty dollars.

The *Gale Directory* is available at turboguide.com /cdprod1/cdhrec/006/756 .shtml.

For information on database services, see the journal *Online*. For information on how to use databases, see the *Gale Directory of Databases and Information Companies* (Detroit: Gale, 2002). Or search the Web using the "reference" category of a directory search engine.

Digital Disks

Digital disks, including DVD and CD-ROM, are likely to become the dominant form of information storage in all research libraries because of their low cost and small size. Currently, digital disks are used primarily to store research tools such as indexes, abstract services, and reference texts. However, full-text disks — holding the full texts of journal articles and books — are fast coming on the market. *Fulltext Sources Online* (Medford, NJ: Information Today), a semiannual directory, lists fifteen thousand journals, newspapers, newsletters, and television and radio transcripts available in full text online.

Fulltext Sources Online is available at www .infotoday.com.

Searching a bibliography on disk, like using a search engine to find Web sites, calls for the use of keywords and the Boolean operators.

GUIDELINES

Using Boolean Operators in Keywords

Boolean operators let you modify your search to include more or fewer hits. Because each system uses slightly different syntax, you need to read the "search tips" or "syntax" section to learn how to use that system's Boolean operators.

To get more hits:
- Use *or* (such as *ethics or morals*): finds any site with either of the words.

	• Use a wild card (usually an asterisk, such as *moral**): finds any site with *morals, morality,* or any other word that begins with the five letters *moral.*
To get fewer hits:	• Use *and* (*ethics and morals*): finds any site with both of the words.
	• Use *not* (*ethics not morals*): finds any site with *ethics* but excludes any site with *morals.*
	• Use a date specifier (*ethics – 1995*): finds any site published in 1995 or more recently.

Because the same reference source is often available both online and on disk, online directories often include the disk sources. For more information on how to use disks, see the *Gale Directory of Databases and Information Sources.*

Web Sites

By mid-2001, there were some thirty million Web sites (Zakon, 2001). Private companies, organizations, governments, and individuals all have Web sites. Searching for information on the Web, however, can be a challenge. As one anonymous writer put it, the Web "is an enormous library in which someone has turned out the lights and tipped the index cards all over the floor." It takes practice to learn to use the Web effectively.

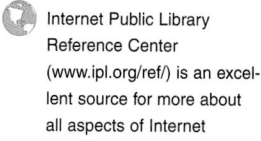

Internet Public Library Reference Center (www.ipl.org/ref/) is an excellent source for more about all aspects of Internet research.

There are three basic ways to locate Web sites:

• *Enter the address of the site you wish to visit.* Every site has a unique address, called a *Uniform Resource Locator* (URL). If you know the URL of the site you want, you enter it in the *address* or *location* portion near the top of your browser screen.

• *Go to a search engine, then enter a keyword or phrase.* For example, if you want to find the site for the Centers for Disease Control, you would go to a search engine such as AltaVista, then enter the key phrase "centers disease control." In this case, you are using the search engine like a book index.

For more information on search engines, click on Links Library for Ch. 7 on TechComm Web (www.bedfordstmartins.com/techcomm).

• *Use a directory search engine to do a subject search.* A directory search engine classifies sites by subject. For instance, you want to find companies that produce palmtop computers. You select the category "business," then the subcategory "computers," then the subsubcategory "palmtops." When you have drilled down to a sufficiently specific category, you ask the search engine to display the results: a list of links to companies that make palmtops.

Figures 7.1 through 7.6 on pages 150–152 show how to use Web pages.

If you know the address of the site you want to visit, type it here.

Or type your keywords here.

Here are a few of the directory categories you can use to search. Notice the selected subcategories that let you narrow your search.

Using an advanced search can save you time and frustration.

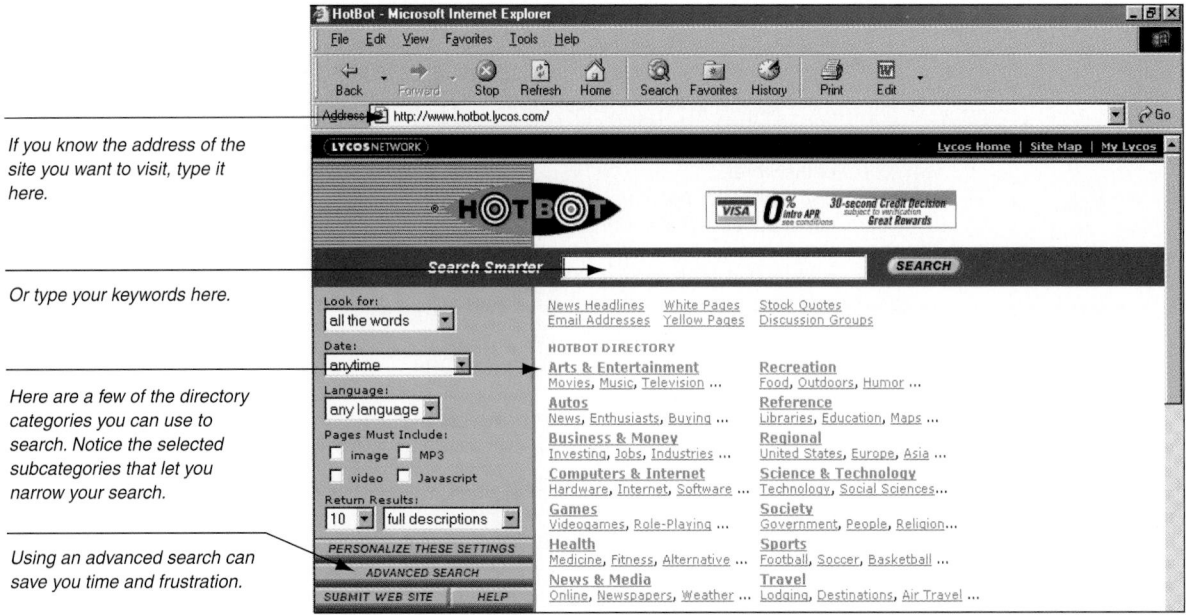

■ **Figure 7.1   Using a Directory Search Engine**

The pull-down menus let you customize your search according to a number of parameters, including language, date, and kind of information contained in the site. Perhaps the most important parameter is the first one: the phrasing of the keywords. You can search for "all the words," "any of the words," "exact phrase," and several other options.

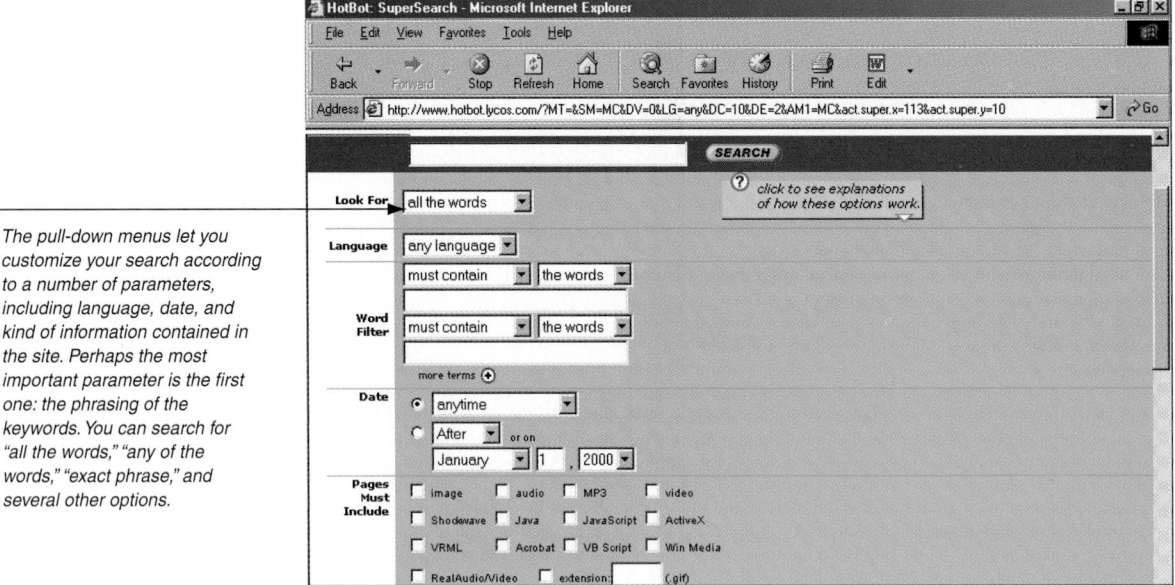

■ **Figure 7.2   Using the Advanced Search Function of a Search Engine**

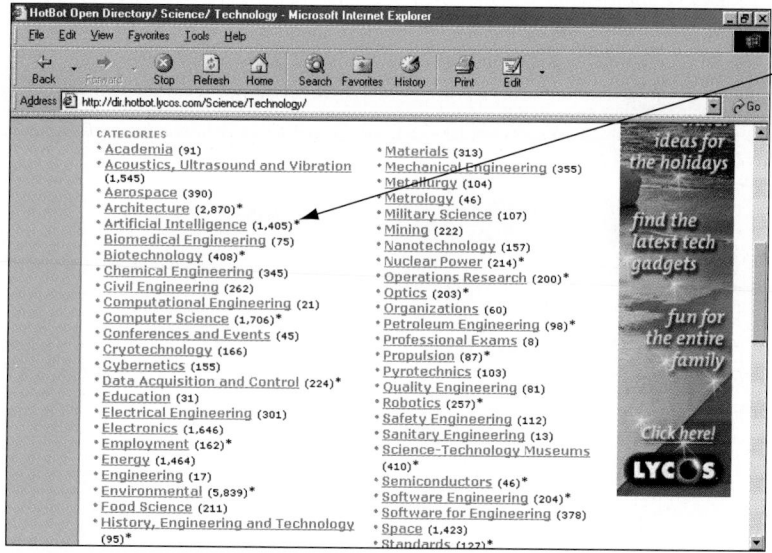

The number after the category shows the number of items in the category. There are 1,405 items in the category Artificial Intelligence.

▨ Figure 7.3   Directory Listings

Use the listings to drill down deeper into the Web. This screen shows HotBot's listings for the technology category.

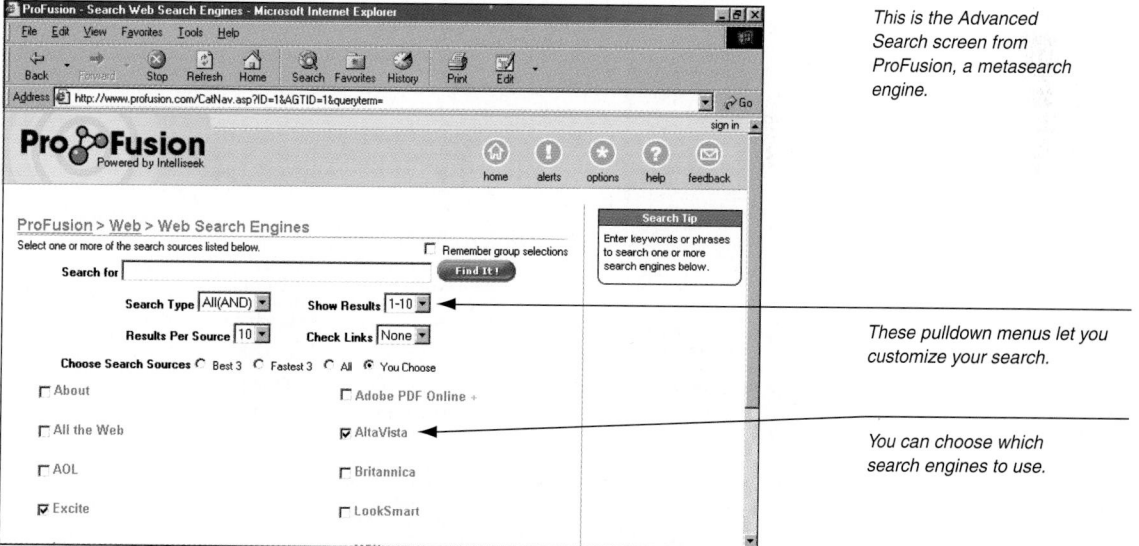

This is the Advanced Search screen from ProFusion, a metasearch engine.

These pulldown menus let you customize your search.

You can choose which search engines to use.

▨ Figure 7.4   Using a Metasearch Engine

Metasearch engines, such as ProFusion, submit your keywords to a number of different search engines at once.

If you get a message indicating that the site you want is not at the address where you think it should be, first check your spelling of the URL. Also try checking back later: the server could be overloaded or down for maintenance.

Then try pruning the URL by removing "ezeditor.htm," and trying again. If that doesn't work, remove "products" too. Eventually, you may end up at the organization's home page, where you might be able to use a site map or index to find the pages you want.

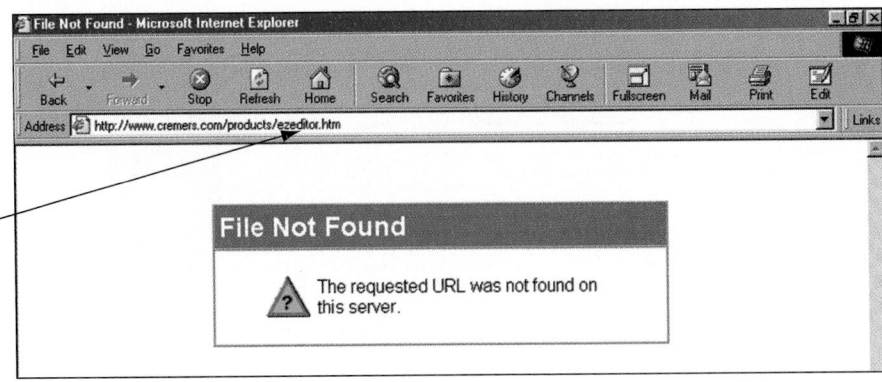

■ **Figure 7.5   Pruning the URL**

You can save a file (with or without the HTML codes) to your computer using the "save as" command in the "file" pull-down menu.

If you think you might want to visit the site again, bookmark it.

You can print the page by using the print command or the print icon.

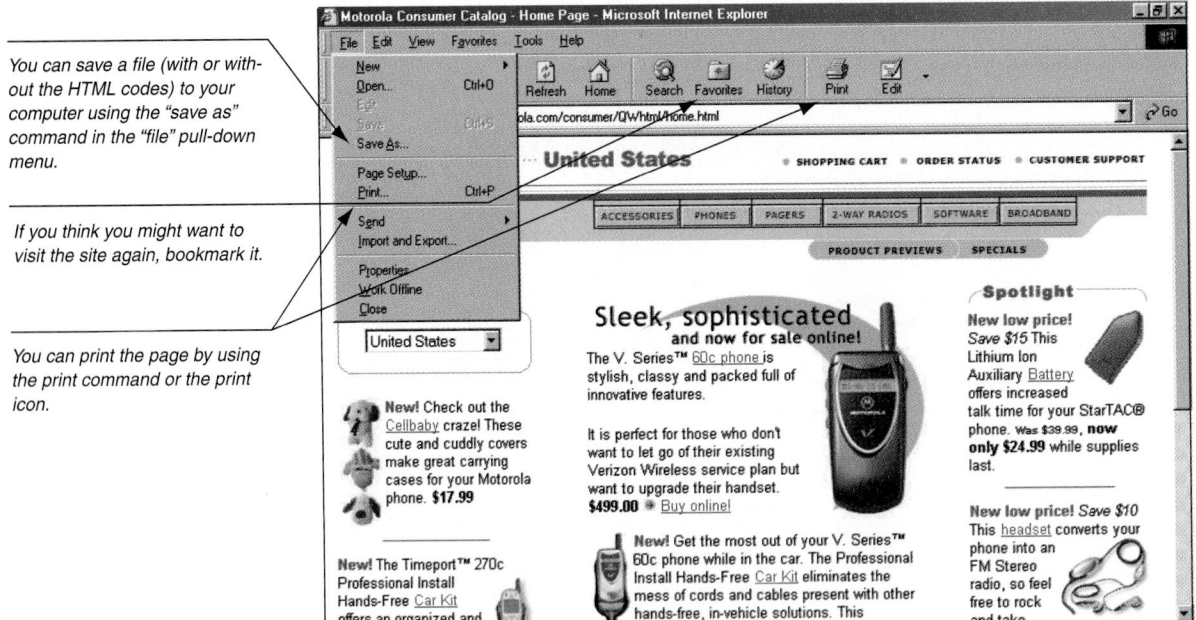

■ **Figure 7.6   Saving and Printing a Web Page**

Information on the Web is covered by copyright, just like printed information. When you save material from the Web to your computer, be sure to record the source's bibliographic information.

Online Discussion Groups

There are two major forums for online discussions: *Usenet newsgroups* and *electronic mailing lists.*

Usenet newsgroups, sometimes called *bulletin boards*, publish email messages sent by members of the group. Newsgroups give participants an opportunity to discuss issues, ask questions, and get answers. Usenet consists of thousands of newsgroups organized according to ten basic categories, including computer science, science, recreation, and business. In a Usenet newsgroup, mail is not sent to individual computers but stored on databases, which you then access.

Electronic mailing lists are like newsgroups in that they publish email messages sent by people who subscribe to the service. The basic difference is that mailing lists send email messages to every person who subscribes. The mail comes to you; you don't go to it, as you do with a newsgroup.

Discussions on newsgroups and electronic mailing lists vary greatly in quality. Discussions on topics such as astronomy tend to contain high-level talk by experts from universities, research institutes, and the government. Discussions on topics such as UFOs or *Star Trek* tend not to.

For sites that list newsgroups and mailing lists, click on Links Library for Ch. 7 on TechComm Web (www.bedfordstmartins.com /techcomm).

Using Basic Research Tools

There is a tremendous amount of information in the different media. The trick is to learn how to find what you want. This section discusses five basic research tools.

Basic research tools:

Online catalogs

Reference works

Periodical indexes

Newspaper indexes

Abstract services

Online Catalogs

In most libraries, the card catalog has been replaced by an online catalog of almost all the library's holdings: its books, microforms, films, compact discs and phonograph records, tapes, and other materials. Most online catalogs work essentially the same way, although their screens and commands differ. Study the instructions for the system you will be using, and read the on-screen commands.

The online catalog lists and describes the holdings at one particular library or a group of libraries. Using the Internet, you can search the catalogs of hundreds of major libraries around the world. For more on how Telnet (remote login) works, see Chapter 4, p. 81.

Figures 7.7 through 7.9 show how to search one online catalog.

Using the pull-down menus, you can search for keywords in subjects, titles, etc.

You can also limit the search by publication date, publication type, and language.

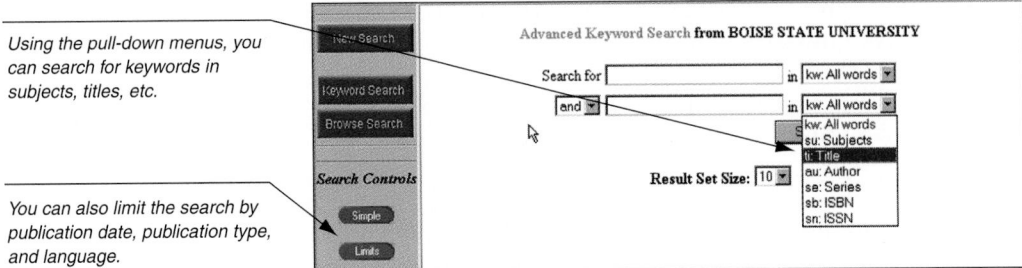

■ **Figure 7.7   Searching for an Item on a Web-based Online Catalog**

The search for "technical writing" at this library resulted in 203 items.

You can click to see the full record (see Figure 7.9) or "mark" the record, so that you can download it when you finish searching the catalog.

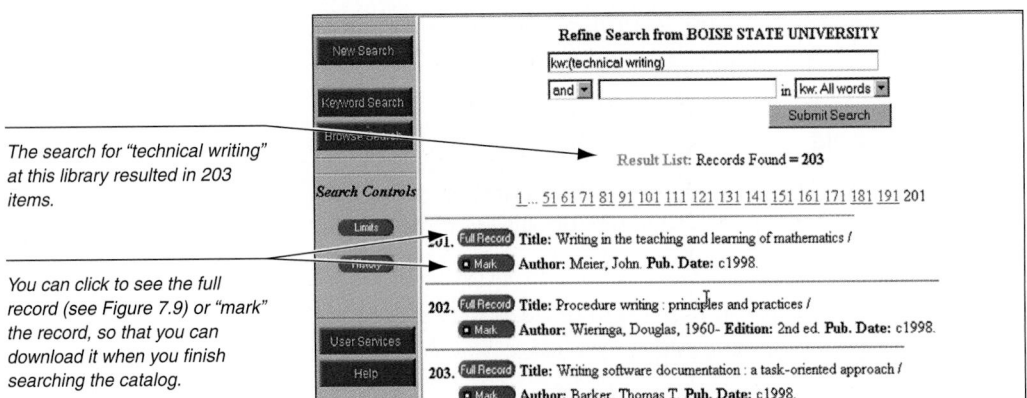

■ **Figure 7.8   List of Items Resulting from a Keyword Search on an Online Catalog**

This screen shows the brief records of the last three items in the list.

The record is hyperlinked to related information. This link will take you to other titles about software documentation.

The record also indicates whether the item is checked out of the library.

■ **Figure 7.9   Detailed Record of a Title from the Keyword Search**

The detailed record provides all the bibliographic information you will need to document your source.

Reference Works

Some books listed in the catalog have call numbers preceded by the abbreviation "Ref." These books are part of the reference collection, a separate grouping of books that normally may not be checked out of the library. Many reference works in the library also exist on the Web.

The reference collection in print and online includes general dictionaries and encyclopedias, biographical dictionaries (*International Who's Who*), almanacs (*Facts on File*), atlases (*Rand McNally Commercial Atlas and Marketing Guide*), and dozens of other general research tools. The reference collection also contains encyclopedias (*Encyclopedia of Banking and Finance*), dictionaries (*Psychiatric Dictionary*), and handbooks (*Biology Data Book*). These specialized books are especially useful when you begin a writing project because they provide an overview of the subject and often list the major works in the field.

How do you know if there is a dictionary of the terms used in a given field, such as nutrition? Searching the Web is probably the fastest and most accurate way to find out. Check, too, those reference books that list the many others available. Among such guides-to-the-guides are the following:

Harris, S. (Ed.). (1994). *The New York Public Library book of how and where to look it up*. New York: Macmillan.

Kieft, R. (Ed.). (forthcoming). *Guide to reference books* (12th ed.). Chicago: American Library Association.

Mullay, M., & Schlicke, P. (Eds.). (1998–2000). *Walford's guide to reference material* (8th ed.). 3 vols. London: Library Association.

The most comprehensive guide-to-the-guides is Sheehy's *Guide to Reference Books* (also see its recent supplements), a valuable resource that lists bibliographies, indexes, abstract services, dictionaries, directories, handbooks, encyclopedias, and many other sources. However, Sheehy is updated infrequently, and obviously, these guides do not provide current information about reference works on the Web. For that information, go to the Web itself.

To find information on the Web, use a search engine and go to its "reference" section. There you will find numerous sites that contain links to excellent collections of reference works online. A few examples:

- The Best Information on the Net (http://www.sau.edu/bestinfo/index .htm)
- CyberStacks(sm) (www.public.iastate.edu/~CYBERSTACKS/)
- The Internet Public Library (www.ipl.org/ref/)

Another strategy is to use a directory search engine to find a category closely related to your topic. For instance, if you are writing about helicopters, you could search the categories in Yahoo! by clicking on "science," then "aviation and aeronautics," then "organizations," then "American Helicopter Society." There you will see a set of links to other sites about helicopters. Many of these sites list only online information, but others include valuable bibliographies of printed information as well.

Periodical Indexes

Periodicals are an excellent source of information for most research projects because they offer recent, authoritative discussions of limited subjects. The biggest challenge in using periodicals is identifying and locating the dozens of relevant articles that are published each month. Although only half a dozen major journals may concentrate on your field, a useful article might appear in one of hundreds of other publications. A periodical index, which is simply a list of articles classified according to title, subject, and author, can help you determine which journals you want to locate.

There are periodical indexes in all fields. The following brief list gives you a sense of the diversity of titles:

- *Applied Science & Technology Index*
- *Business Periodicals Index*
- *Readers' Guide to Periodical Literature*
- *Engineering Index*

You can also use a directory search engine. Many directory categories include a subcategory called "journals" or "periodicals" listing online and printed sources.

Once you have created a bibliography of printed articles you want to study, you have to find them. Check your library's online catalog or *serials holding catalog*, the book that lists all the journals your library receives. If your library does not have an article you want, you can use one of two techniques for securing it:

- *Interlibrary loan.* Your library uses an online directory to learn which nearby library has the article. That library makes a photocopy of the article and sends it to your library. The advantage of an interlibrary loan is that it is a free (or very inexpensive) service to you. The disadvantage is that it can take up to two weeks or longer for the article to arrive at your library.

- *Document-delivery services.* If you are in a hurry, you can log on to a document-delivery service, such as ingenta (www.ingenta.com), which searches a database of 11 million articles in 26,000 periodicals. If the service has the article, it faxes it to you or makes an electronic copy available.

Newspaper Indexes

Many major newspapers around the world are indexed by subject. The three most important indexed U.S. newspapers are the following:

- *The New York Times*
- *The Christian Science Monitor*
- *The Wall Street Journal*

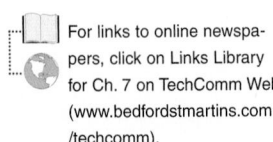
For links to online newspapers, click on Links Library for Ch. 7 on TechComm Web (www.bedfordstmartins.com /techcomm).

The first two are highly reputable general newspapers; the third is the authoritative source on business, finance, and the economy. Many newspapers are now available for free on the Web and can be searched electronically. Keep in mind, however, that the print version and the electronic version of a newspaper can vary greatly. If you wish to cite a quotation from an article in a newspaper, the print version is the preferred one.

Abstract Services

Abstract services are like indexes but also provide abstracts: brief summaries of an article's important findings. In most cases, reading the abstract will enable you to decide whether to search out the full article. The title of an article, alone, is often a misleading indicator of its contents.

See Ch. 12, p. 309, for more about abstracts.

Some abstract services, such as *Chemical Abstracts*, cover a broad field, but many are specialized rather than general. *Adverse Reaction Titles*, for instance, covers research on the subject of adverse reactions to drugs. Figure 7.10 (Chemical, 2001), a screen from the *Chemical Abstracts* CD-ROM, shows the basic components of an abstract.

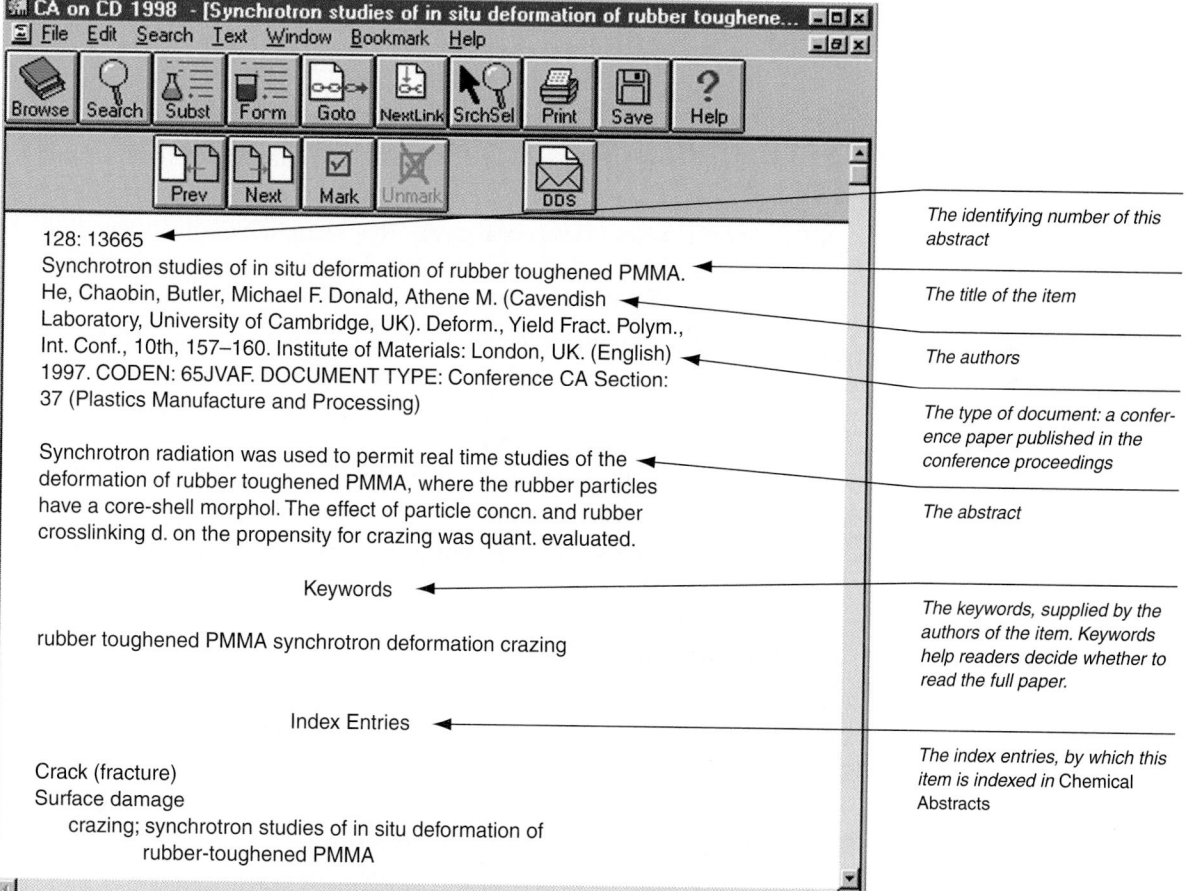

The identifying number of this abstract

The title of the item

The authors

The type of document: a conference paper published in the conference proceedings

The abstract

The keywords, supplied by the authors of the item. Keywords help readers decide whether to read the full paper.

The index entries, by which this item is indexed in Chemical Abstracts

■ **Figure 7.10   An Abstract**

Researching Government Information

The U.S. government is the world's biggest publisher. In researching any field of science, engineering, or business, you are likely to find that a government agency or department has produced a relevant brochure, report, or book.

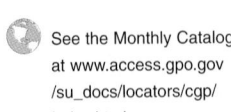 See the Monthly Catalog at www.access.gpo.gov /su_docs/locators/cgp/ index.html.

Government publications are not usually listed in indexes and abstract journals. The *Monthly Catalog of United States Government Publications*, available on paper, on CD, and on the Web, provides extensive access to these materials. The digital versions of the *Monthly Catalog* are fully searchable.

Printed government publications are usually cataloged and shelved separately from other kinds of materials. They are classified according to the Superintendent of Documents system, not the Library of Congress system. See the reference librarian or the government documents specialist for information about finding government publications in your library.

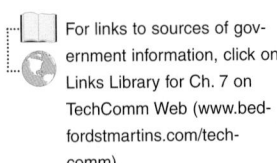 For links to sources of government information, click on Links Library for Ch. 7 on TechComm Web (www.bedfordstmartins.com/techcomm).

You can also access many government sites and databases on the Internet. Two Web sites are particularly useful:

- Fedworld Information Network (www.fedworld.gov). This searchable site contains more than 15,000 files on numerous fields in business, health, safety, and the environment. The site also links to the U.S. Government Information Locator Service, a database of federal government information.
- GPO Access (www.access.gpo.gov). This searchable site offers links to hundreds of sites for all three branches of the federal government, including dozens of departments and agencies.

Patricia Cruse and Sherry DeDecker have created an excellent site, "How to Effectively Locate Federal Government Information on the Web" (gort.ucsd.edu/pcruse/universe/intro.html) that clearly explains how to research government information.

For additional information on government publications, consult these two printed guides:

Hoffman, F. W. (1998). *Guide to popular U.S. government publications* (5th ed.). Englewood, CO: Libraries Unlimited.
Morehead, J. (1999). *Introduction to United States government information sources* (6th ed.). Englewood, CO: Libraries Unlimited.

Keep in mind, however, that books cannot keep up with the rapid transition from paper to screen. Check the two Web sites listed above and consult the reference librarian or government document librarian.

Skimming Your Sources and Taking Notes

To record the information that will eventually go into your document, you will need to skim your potential sources and take notes.

Don't try to read every potential source. A careful reading of a work that looks useful might prove disappointing. You might also get halfway through a book and realize that you must start writing immediately to submit your document on time.

GUIDELINES

Skimming Books and Articles

To skim effectively, read the following parts of books and articles.

In a book, skim:	*In an article, skim:*
• *the preface and introduction:* to understand the writer's approach and methods • *the acknowledgments section:* to learn about assistance the author received from other experts in the field, or about the author's use of primary research or other resources • *the table of contents:* to understand the scope and organization of the book • *the notes at the ends of chapters or at the end of the book:* to understand the nature and extent of the author's research • *the index:* to determine theextent of the coverage of the information you need • *a few paragraphs from different portions of the text:* to gauge the quality and relevance of the information	• *the abstract:* to get an overview of the article's content • *the introduction:* to understand the article's purpose, main ideas, and organization • *the notes and references:* to understand the nature and extent of the author's research • *the headings and several of the paragraphs:* to understand the article's organization and the quality and relevance of the information • *the acknowledgments section (if there is one):* to learn about assistance the author received from other experts in the field, or about the author's use of primary research or other resources

---

Skimming will not always tell you whether a book or article is going to be useful, but it can tell you if a work is *not* going to be useful: because it doesn't cover your subject, for example, or because it is too superficial or too advanced. Eliminating the sources you don't need will give you more time to spend on the ones you do.

Note taking is often the first step in writing the document. Because you will refer to your notes over and over, it is smart to take them logically and systematically.

The best way to take notes is electronically. If you can download files from the Internet, download bibliographic references from a CD-ROM database, and take notes on a laptop computer, you will save a lot of time and prevent many errors.

If you do not have access to these electronic tools, get two packs of note cards: one large (4" × 6" or 5" × 8"), the other small (3" × 5"). There is nothing special about commercial note cards; you can make your own from scrap paper. The advantage of using cards is that they are easy to rearrange later, when you want to draft an outline of the document.

GUIDELINES

Recording Bibliographic Information

Record the bibliographic information for each source from which you take notes. If you print information from a Web site, in most cases the bibliographic information is at the top or bottom of the page.

Information to Record for a Book	*Information to Record for an Article*
• author • title • publisher • place of publication • year of publication • call number	• author • title of the article • title of the periodical • volume • number • date of publication • pages on which the article appears • call number of the periodical

---

On the large cards, write your notes. Write on one side only and limit each card to a narrow subject so that you can easily reorder the information to suit the needs of your document.

Most note taking involves three kinds of activities: paraphrasing, quoting, and summarizing. Knowing how to paraphrase, quote, and summarize is important for two reasons:

- The work you do at this point will to a large extent determine the quality of your finished product. You want to record the information accurately and clearly. Mistakes made at this point can be hard to catch later, and they can ruin your document.

- You want to use your sources responsibly. You don't want to plagiarize unintentionally.

See Appendix, Part A, p. 660, for a discussion of plagiarism.

Paraphrasing

A paraphrase is a restatement, in your own words, of someone else's words. *In your own words* is crucial: if you simply copy someone else's words — even a mere two or three in a row — you must use quotation marks.

In taking notes, what kind of material should you paraphrase? Any information that you think might be useful in writing the document: background data, descriptions of mechanisms or processes, test results, and so forth.

GUIDELINES

Paraphrasing Accurately

▶ *Study the original until you understand it thoroughly.*

▶ *Rewrite the relevant portions of the original.* Use complete sentences, fragments, or lists, but don't compress the material so much that you'll have trouble understanding it later.

▶ *Title the card so that you'll be able to identify its subject at a glance.* The title should include the general subject and the author's attitude or approach to it, such as "Criticism of open-sea pollution-control devices."

▶ *Include the author's last name, a short title of the article or book, and the page number of the original.* You will need this information later in citing your source.

Figure 7.11 on page 162 shows two paraphrased note cards based on the following discussion (adapted from Lovgren, 1994, pp. 87–88). The author is describing his business: creating metaphors to be used in graphical user interfaces on computers. The student has paraphrased each paragraph on a separate note card.

How do you find good metaphors? To be usable, a metaphor must be in the user's sphere of knowledge. The lack of understanding of this requirement frequently causes problems in interface design. Many of the interface metaphors I have seen are metaphors within the knowledge of the developer, but not within the knowledge of the user.

You have to talk to the users, not just to other developers. In my consultancy, we use a technique called contextual interviews to gather this information. I'll explain it by walking you through a session with a hypothetical potential user I'll call Sally.

The first thing we do is phone Sally and schedule a two-hour meeting at her place of employment. We always try to meet the person where he or she works. When we interview users outside their work environment, they give us an overly organized picture of their work, apparently forgetting that real work tends to be less structured and is often interrupted. Also, we have found that two hours is just about the right time. Longer sessions tend to distort results because the session is exhausting for both interviewer and interviewee.

Before beginning the interview, we ask Sally if she has any comments, questions, or issues of a general nature. Our intent is to put her at ease. As part of the preinterview, we inform Sally that the interview will be audiotaped.

We then ask Sally what she would be doing if we weren't here and ask her to start the task. While she is doing it, we ask her to describe what she is doing, why, and for whom. We also ask what kinds of difficulties she might run into that we don't see.

As follow-up, we prepare a complete report of the interview, including an annotated transcript, sketches, and copies of any forms, calendars, and other supplemental material. In annotating the transcript, I mark the nouns Sally used, the actions she mentioned, and any relationships among the nouns.

After repeating this process for several other potential users, we generally get a sense of the metaphors that might be appropriate. For example, if several people in addition to Sally refer to a "red book," and the system is to manage information in this book, then the red book becomes a likely metaphor. It works because when anybody in the organization mentions this term, everyone else knows what they are talking about.

Lovgren

Why some metaphors don't work:

they're understandable to the developer, not the user;
not from the user's sphere of understanding.

"How to Choose Good Metaphors" p. 87

Notice that a heading provides a focus for each card. The student has omitted the information he does not need, but he has recorded the necessary bibliographic information so that he can document his source easily or return to it.

■ **Figure 7.11**
Paraphrased Notes

There is no one way to paraphrase: You have to decide what to paraphrase — and how to do it — on the basis of your own analysis of the audience and the purpose of your document.

Lovgren

How to carry out "contextual interviews."
2-hour meeting at job site; put person at ease, audiotape interview; have person do normal tasks and talk aloud during them; write up results — annotate transcript; repeat with others; see if a metaphor emerges.

"How to Choose Good Metaphors" p. 87

For more on formatting quotations, see "Quotation Marks," "Ellipses," and "Square Brackets" in Appendix, Part B. For a discussion of how to document quotations, see Appendix, Part A.

Quoting

Sometimes you will want to quote a source, either to preserve the author's particularly well-expressed or emphatic phrasing or to lend authority to your discussion. In general, avoid quoting passages of more than two or three sentences; otherwise, your document will look like a mere compilation. Your job is to integrate an author's words and ideas into your own thinking, not merely to introduce a series of quotations.

Although you probably won't be quoting long passages in your document, recording a complete quotation in your notes will help you recall its meaning and context more accurately when you are ready to integrate it into your own work.

The simplest form of quotation is an author's exact statement:

As Jones states, "Solar energy won't make much of a difference in this century."

To add an explanatory word or phrase to a quotation, use brackets:

As Nelson states, "It [the oil glut] will disappear before we understand it."

Use ellipses (three spaced dots) to show that you are omitting part of an author's statement:

original statement: "The generator, which we purchased in May, has turned out to be one of our wisest investments."

elliptical quotation: "The generator . . . has turned out to be one of our wisest investments."

If you are using the documentation style recommended by the Modern Language Association (MLA), you should also add brackets around the ellipses that you introduce:

elliptical quotation: "The generator [. . .] has turned out to be one of our wisest investments."

Summarizing

Summarizing, a way of taking comprehensive notes, is the process of rewriting a passage in your own words to make it shorter while still retaining its essential message. The two main reasons to summarize are the following:

- *To learn a body of information.* When you find a useful source, you might want to summarize it so that you can integrate it with other information in creating your document.

- *To create a draft of one or more of the summaries that will go into your document.* Most long technical documents contain several kinds of summaries:

  – a letter of transmittal (see page 308) that provides an overview of the document
  – an abstract (see page 309), a brief technical summary
  – an executive summary (see page 316), a brief nontechnical summary directed to the manager
  – a conclusion (see page 209) that draws together a complicated discussion

The guidelines and examples in this chapter describe the process of summarizing the printed information you uncover in your research.

Summarizing

Although the following advice will prove useful as you prepare abstracts and parts of other kinds of documents that include summaries, these guidelines focus on extracting the essence of a passage by summarizing it.

1. *Read the passage carefully several times.*
2. *Underline the key ideas.* Most writers put their main ideas in a few obvious places: titles, headings, topic sentences, transitional paragraphs, concluding paragraphs.
3. *Combine the key ideas.* Take a break, then study what you have underlined. Paraphrase the underlined ideas. Don't worry about your grammar, punctuation, or style; you just want to see if you can reproduce the essence of the original.
4. *Check your draft against the original for accuracy and emphasis.* Check that you record statistics and names correctly and that your version of a complicated concept faithfully represents the original. Check that you get the proportions right; if the original devotes 20 percent of its space to a particular point, your draft should not devote 5 percent or 50 percent to that point.
5. *Record the bibliographic information carefully.* Even though a summary might contain all your own words, you still must cite it, because the main ideas are someone else's. If you don't have the bibliographic information in an electronic form, put it on note cards.

Figure 7.12 (based on McComb, 1991, pp. 19–21) is a narrative history of television technology addressed to the general reader. Figure 7.13 on page 166 is a summary that includes the key terms. This summary is 10 percent of the length of the original.

Evaluating the Information

You have done all kinds of secondary research. With more information than you can possibly use, you try to figure out what it all means. You realize that you still have some questions, that some of the information is incomplete, some contradictory, and some just unclear. There is no shortage of information; the challenge is to find good information. Look for information that is

- *accurate.* If you are researching whether your company should consider flextime scheduling, start by determining the number of employees who might be interested in flextime. If you estimate that number to be 500 but

A BRIEF HISTORY OF TELEVISION

Although it seems as if television has been around for a long time, it's a relatively new science, younger than rocketry, internal medicine, and nuclear physics. In fact, some of the people that helped develop the first commercial TV sets and erect the first TV broadcast antennas are still living today.

The Early Years
The first electronic transmission of a picture was believed to have been made by a Scotsman, John Logie Baird, in the cold month of February 1924. His subject was a Maltese Cross, transmitted through the air by the magic of television (also called "Televisor" or "Radiovision" in those days) the entire distance of ten feet.

To say that Baird's contraption was crude is an understatement. His Televisor was made from a cardboard scanning disk, some darning needles, a few discarded electric motors, piano wire, glue, and other assorted odds and ends. The picture reproduced by the original Baird Televisor was extremely difficult to see — a shadow, at best.

Until about 1928, other amateur radiovision enthusiasts toyed around with Baird's basic design, whiling away long hours in the basement transmitting Maltese Crosses, model airplanes, flags, and anything else that would stay still long enough under the intense light required to produce an image. (As an interesting aside, the lighting for Baird's 1924 Maltese Cross transmission required 2,000 volts of power, produced by a roomful of batteries. So much heat was generated by the lighting equipment that Baird eventually burned his laboratory down.)

Baird's electromechanical approach to television led the way to future developments in transmitting and receiving pictures. The nature of the Baird Televisor, however, limited the clarity and stability of images. Most of the sets made and sold in those days required the viewer to peer through a glass lens to watch the screen, which was seldom over seven by ten inches in size. What's more, the majority of screens had an annoying orange glow that often marred reception and irritated the eyes.

Modern Television Technology
In the early 1930s, Vladimir Zworykin developed a device known as the iconoscope camera. About the same time, Philo T. Farnsworth was putting the finishing touches on the image dissector tube, a gizmo that proved to be the forerunner of the modern cathode ray tube or CRT — the everyday picture tube. These two devices paved the way for the TV sets we know and cherish today.

The first commercially available modern-day cathode ray tube televisions were available in about 1936. Tens of thousands of these sets were sold throughout the United States and Great Britain, even though there were no regular television broadcasts until 1939, when RCA started what was to become the first American television network, NBC. Incidentally, the first true network transmission was in early 1940, between NBC's sister stations WNBT in New York City (now WNBC-TV) and WRGB in Schenectady.

■ **Figure 7.12   Original Passage**

Postwar Growth

World War II greatly hampered the development of television, and during 1941–1945, no television sets were commercially produced (engineers were too busy perfecting radar, which, interestingly enough, contributed significantly to the development of conventional TV). But after the war, the television industry boomed. Television sets were selling like hot-cakes, even though they cost an average of $650 (based on average wage earnings, that's equivalent to about $4,000 today).

Progress took a giant step in 1948 and 1949 when the four American networks, NBC, CBS, ABC, and Dumont, introduced quality, "class-act" programming, which at the time included *Kraft Television Theatre, Howdy Doody*, and *The Texaco Star Theatre* with Milton Berle. These famous stars of the stage and radio made people want to own a television set.

Color and Beyond

Since the late 1940s, television technology has continued to improve and mature. Color came on December 17, 1953, when the FCC approved RCA's all-electronic system, thus ending a bitter, four-year bout between CBS and RCA over color transmission standards. Television images beamed via space satellite caught the public's fancy in July of 1962, when Telstar 1 relayed images of AT&T chairman Frederick R. Kappell from the U.S. to Great Britain. Pay-TV came and went several times in the 1950s, 1960s, and 1970s; modern-day professional commercial videotape machines were demonstrated in 1956 by Ampex; and home video recorders had appeared on retail shelves by early 1976.

■ **Figure 7.12**   *(Continued)*

Summary: A Brief History of Television

In 1924, Baird made the first electronic transmission of a picture. The primitive equipment produced only a shadow. Although Baird's design was modified by others in the 1920s, the viewer had to look through a glass lens at a small screen that gave off an orange glow.

Zworykin's iconoscopic camera and Farnsworth's image dissector tube — similar to the modern CRT — led in 1936 to the development of modern TV. Regular broadcasts began in 1939 on the first network, NBC. Research stopped during WWII, but after that, sales grew, even though sets cost approximately $650, the equivalent of $4,000 today.

Color broadcasts began in 1953; satellite broadcasting began in 1962; and home VCRs were introduced in 1976.

Key terms: television, history of television, NBC, color television, satellite broadcasting, video cassette recorders, Baird, Zworykin, Farnsworth.

■ **Figure 7.13   Summary of the Original Passage**

it is in fact closer to 50, you might end up wasting a lot of time doing an unnecessary study.

- *unbiased.* You want sources that have no financial stake in the project. A private company that transports workers in vans might be a biased source of information on flextime because it might be interested in contracting with your company.

- *comprehensive.* You want to hear from different kinds of people — in terms of gender, cultural characteristics, and age — and from people representing all views of the topic. The last person you interview might be the first one to point out something you have never considered.

- *appropriately technical.* Good information is sufficiently detailed to respond to the needs of your readers but not so detailed that they cannot understand it. For instance, for the study of flextime, you need to find out whether opening your building an hour early and closing it an hour late will have a significant effect on your utility costs. You will want to interview people in operations and in security; you will not need to do a detailed analysis of all the utility records of the company.

- *current.* If your information about other companies that have tried flextime is 10 years old, it might not accurately reflect the needs of today's workers.

- *clear.* You want information that is easy to understand; otherwise, you'll waste time figuring it out, and you might misinterpret it.

The most difficult kind of material to evaluate is information from the Internet. The Internet's greatest strength — that it is open to anyone who is connected, and that one person's voice is heard as loudly as the next person's — also poses a great challenge for researchers. In most cases, information appears on the Internet without passing through the formal review procedure that is characteristic of books and professional journals. Therefore, you have to be particularly careful in evaluating any information you find on the Internet.

GUIDELINES

Evaluating Print and Online Sources

The five criteria in the chart below may be used to evaluate sources found either in print or on the Web.

Criteria	For Printed Sources	For Online Sources
Authorship	Do you recognize the name of the author? Can you learn about the author's credentials and current position from a biographical note? If this	If you do not recognize the author's name, did you find the site by linking from another reputable site? Does the site contain links to other

Criteria	For Printed Sources	For Online Sources
	information is not included in the document itself, can you find it in a who's who or by searching for other books or other journal articles by the author?	reputable sites? Does the site contain biographical information — the author's current position and credentials? Can you use a search engine to find other references to the author's credentials or other documents by the author?
Publishing body	What is the publisher's reputation? To be reliable, a book should be published by a reputable trade, academic, or scholarly publisher; a journal should be sponsored by a professional association or university. Are the editorial board members well-known names in the field? Trade publications — magazines about a particular industry or group — often promote the interests of that industry or group. For example, don't automatically assume the accuracy of information in trade publications for loggers or environmentalists. If you doubt the authority of a book or journal, ask the reference librarian or a professor.	Can you determine the publishing body's identity from headers or footers? Is the publishing body reputable in the field? If the site comes from a personal account on an Internet service provider, the author might be writing outside his or her field of expertise. Many Internet sites exist largely for public relations or advertising. For instance, the home page for the White House is not going to provide information critical of the administration. Likewise, the Web sites of corporations and other organizations are unlikely to contain information critical of those corporations or organizations.
Knowledge of the literature	Does the author appear to be knowledgeable about the major literature in the field? Is there a bibliography? Are there notes throughout the document?	Analyze the Internet source as you would any other source. Often, references to other sources will take the form of links.
Accuracy and verifiability of the information	Does the author clearly describe the methods and theories used in producing the information? Are these methods and theories appropriate to the subject?	Is the site well constructed? Is the information well written? Are the claims supported by appropriate evidence? Are sources cited?
Timeliness	Does the document rely on recent data? Was the document published recently?	Was the document created recently? Was it updated recently? If a site is not yet complete, be wary.

 Evaluating online sources is easier if you start searching from a reputable list of links, such as that of the WWW Virtual Library (vlib .org/Home.html), sponsored by the World Wide Web Consortium.

CONDUCTING PRIMARY RESEARCH

Although the library and the Internet offer a wealth of secondary sources, you will often need to conduct primary research to acquire new information. There are six major types of primary research.

Types of primary research:
Inspections
Experiments
Field research
Interviews
Letters of inquiry
Questionnaires

Inspections

Regardless of your field, you are likely to encounter many sentences that begin "An inspection was conducted to determine. . . ." A civil engineer can often determine what caused the crack in a foundation by inspecting the site; an accountant can learn a lot about the financial health of an organization by inspecting the company's financial records.

These professionals are looking at a site, an object, or a document (for example, the financial records) and applying their knowledge and professional judgment to what they see. Sometimes the inspection techniques are more complicated. A civil engineer inspecting foundation cracking might want to test hunches by studying a soil sample. An accountant checking the books might need to perform some computerized analyses on the information.

When you carry out an inspection, take good notes. Try to answer the appropriate journalistic questions — *who, what, when, where, why,* and *how* — as you go, or as soon as possible after you finish. Where appropriate, photograph or sketch the site or print the output from computer-assisted inspections. You will probably need the data later for your document.

Experiments

Learning to conduct the many kinds of experiments used in a particular field can take months or even years. This discussion can serve only as a brief introduction.

In many cases, conducting an experiment involves the following four phases:

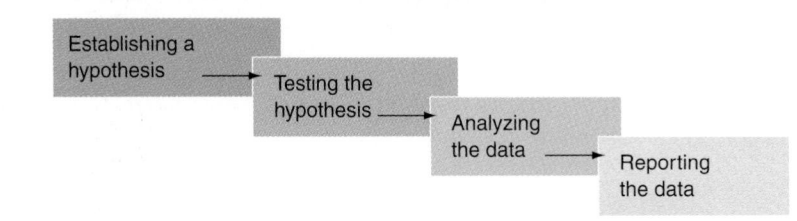

- *Establishing a hypothesis.* A hypothesis is an informed guess about the relationship between two factors. To determine the relationship between gasoline octane and miles per gallon, you could test the following hypothesis: "A car will get better mileage with 89-octane gas than with 87-octane gas."

- *Testing the hypothesis.* Usually, you need an experimental group and a control group. These two groups are identical except for the condition you are studying, in the above example, the gasoline. The control group would be a car running on 87 octane. The experimental group would be an identical car running on 89 octane. The experiment would consist of driving the two cars over an identical course at the same speed — preferably in some sort of controlled environment, such as a laboratory — over a given distance, such as 1,000 miles. At the end of the 1,000 miles you would calculate the miles per gallon. The results would either support or refute your original hypothesis.

  An experiment as neat as this one is often difficult to set up. You probably wouldn't be able to get two identical cars and run them for 1,000 miles in a lab. So you do the best you can: you run your own car with two or three tanks of 87 octane and calculate your miles per gallon; then you run it on the same amount of 89 octane and do the same calculations. Of course, you try to make sure that you are subjecting the car to the same driving conditions to control other variables.

- *Analyzing the data.* You must try to understand whether your data show merely a correlation — one factor changing along with another — or a causal relationship. For example, we know that sports cars are involved in more fatal accidents than sedans, but we don't know whether the car has much to do with that fact, or whether driving habits are the important factor. Analyzing the data objectively is a challenge because it is tempting to see what you want to see.

For more about reports, see Chs. 18 and 19.

- *Reporting the data.* When researchers report their findings, they try to explain as clearly as they can what they did, why they did it, what they saw, what it means, and what ought to be done next.

Field Research

Whereas an experiment is quantitative (it yields statistical data that can be measured), field research is usually qualitative: it yields data that cannot be measured or, at least, not as precisely as experimental data. Often in field research you seek to understand the quality of an experience. For instance, you might want to understand how a new seating arrangement would affect group dynamics in a classroom. You could design a study in which you observed and recorded the classes and perhaps interviewed the students and the instructor about their reactions. Then you could do the same in a traditional classroom and compare the results.

Some kinds of studies have both quantitative and qualitative elements. In the case of the classroom seating arrangements, you could include some quantitative measures, such as the number of times students talked with each other or the length of the discussions. In addition, you could distribute questionnaires to elicit the opinions of the students and the instructor. If you used these same quantitative measures on enough classrooms, you could gather valid quantitative information.

When you are doing quantitative or qualitative studies on the behavior of animals — from rats to monkeys to people — try to minimize two common problems:

- *The effect of the experiment on the behavior you are studying.* When a television camera crew covers a protest demonstration, the protest becomes more animated. In studying the effects of the classroom seating arrangement, try to minimize the effects of your own presence. Make sure that the camera is placed unobtrusively and that it is set up before the students arrive, so they don't see the process. Still, any time you bring in a camera, you can never be sure that what you witness is typical. Even an outsider who sits quietly can disrupt typical behavior.

- *Bias in the recording and analysis of the data.* Bias can occur because researchers want to confirm their hypotheses. In an experiment to determine whether word processors help students write better, the researcher is likely to see improvement where other people don't. For this reason, the experiment should be designed so that it is *double blind*. That is, the students doing the writing the researcher is studying shouldn't know what the experiment is about so they won't change their behavior to support or negate the hypothesis. And the data being analyzed should be disguised so that researchers don't know whether they are examining the results from the control group or the experimental group. If the control group wrote in ink and the experimental group used word processors, for example, the control group's papers should be formatted on a word processor, so that all the papers look identical.

Conducting an experiment or field research is relatively simple; the hard part is designing your study so that it accurately measures what you want it to measure.

Interviews

Interviews are extremely useful when you need information on subjects that are too new to have been discussed in the professional literature or inappropriate for widespread publication (such as local political questions). Most students are inexperienced at interviewing and hence are reluctant to do it. Interviewing, like any other communication skill, requires practice. The following discussion explains how to make interviewing less intimidating and more productive.

In choosing a respondent, a person to interview, determine three things:

- *What you want to find out.* Only then can you begin to search for a person who can provide the information.
- *Who could provide this kind of information.* The ideal respondent is an expert willing to talk. Unless the respondent is an obvious choice, such as the professor carrying out the research you are studying, use directories, such as local industrial guides, to locate the names and addresses of potential respondents.
- *Whether the person is willing to be interviewed.* On the phone or in writing, state what you want to ask about. The person might not be able to help you but might be willing to refer you to someone who can. And explain to the respondent why you have decided to ask him or her. A compliment works better than admitting that the person you really wanted to interview is out of town. Explain what you plan to do with the information, such as write a report or give a talk. Then, if the person is willing to be interviewed, set up an appointment at his or her convenience.

GUIDELINES

Conducting an Interview

Preparing for the interview

- *Do your homework.* Never give the impression that you are conducting the interview to avoid doing other kinds of research. If you ask questions that are already answered in the professional literature, the respondent might become annoyed and uncooperative.
- *Prepare good questions.* Good questions are clear, focused, and open.
  - Be clear. The respondent should be able to understand what you are asking.

UNCLEAR	Why do you sell Trane products?
CLEAR	What are the characteristics of Trane products that led you to include them in your product line?

The unclear question can be answered in a number of unhelpful ways: "Because they're too expensive to give away" or "Because I'm a Trane dealer."
  - Be focused. The question must be narrow enough to be answered briefly. If you want more information, you can ask a follow-up question.

Preparing for the interview (continued)	UNFOCUSED	What is the future of the computer industry?
	FOCUSED	What will the American chip industry look like in 10 years?

– Be open. Your purpose is to get the respondent to talk. Don't ask a lot of questions that have yes or no answers.

	CLOSED	Do you think the federal government should create industrial partnerships?
	OPEN	What are the advantages and disadvantages of the federal government's creating industrial partnerships?

- *Check your equipment.* If you will be taping the interview, test your tape recorder or video camera to make sure it is operating properly.

Beginning the interview	• Arrive on time. • Thank the respondent for taking the time to talk with you. • Repeat the subject and purpose of the interview and what you plan to do with the information. • If you wish to tape the interview, ask permission.
Conducting the interview	• *Take notes.* Write down important concepts, facts, and numbers, but don't take such copious notes that you are still writing when the respondent finishes an answer. • *Start with prepared questions.* Because you are likely to be nervous at the start, you might forget important questions. It is wise to begin with prepared questions. • *Be prepared to ask follow-up questions.* Listen carefully to the respondent's answer and be ready to ask a follow-up question or request a clarification. Have your other prepared questions ready, but be willing to deviate from them. The respondent probably will lead you in directions you had not anticipated. • *Be prepared to get the interview back on track.* Gently return to the point if the respondent begins straying unproductively, but don't interrupt rudely or show annoyance.
Concluding the interview	• *Thank the respondent.* • *Ask for a follow-up interview.* If a second meeting would be useful — and you think the person would be willing to talk with you further — ask to arrange a second meeting now.

 See Business Communication: Managing Information and Relationships (spider.hcob.wmich .edu/bis/faculty/bowman /dyads.html) for an excellent discussion of interview questions.

Concluding the interview (continued)	• *Ask for permission to quote the respondent.* If you think you might want to quote the respondent by name, ask permission now.
After the interview	• *Write down the important information while the interview is fresh in your mind.* (This step is unnecessary, of course, if you have recorded the interview.) If you will be printing a transcript of the interview, make the transcript now.
	• *Send a brief thank-you note.* Within a day or two, send a note that shows you appreciate the courtesy and that you value what you have learned. In the letter, confirm any previous offers you have made, such as sending the respondent a copy of your final document.

Figure 7.14 is from a transcript of an interview with an attorney specializing in information technology. The interviewer is a student writing about legal aspects of software ownership.

Letters of Inquiry

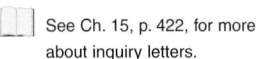

See Ch. 15, p. 422, for more about inquiry letters.

A letter of inquiry is often a useful alternative to a personal interview. If you are lucky, your respondent will provide detailed and helpful answers. However, the person might not clearly understand what it is you want to know or might choose not to help you. In addition, you can't ask follow-up questions in a letter, as you can in an interview. Although the strategy of the inquiry letter is essentially that of a personal interview, inquiry letters can be less successful, because the recipient has not already agreed to provide information and sometimes does not respond. Also, an inquiry letter, unlike an interview, gives you little opportunity to follow up by asking for a clarification.

Questionnaires

Questionnaires enable you to solicit information from a large group of people. However, questionnaires rarely yield completely satisfactory results, for three reasons:

- *Some of the questions will misfire.* No matter how careful you are in drafting your questions, respondents will misinterpret some of them or supply useless answers.

- *You won't obtain as many responses as you want.* The response rate will almost never exceed 50 percent. In most cases, it will be closer to 10–20 percent.

- *You cannot be sure the respondents are representative.* In general, people who feel strongly about an issue are much more likely to respond than are

Interview Transcript, Page 1

Q. Why is copyright ownership important in marketing software?

A. If you own the copyright, you can license and market the product and keep other people from doing so. It could be a matter of millions of dollars if the software is popular.

Q. Shouldn't the programmer automatically own the copyright?

A. If the programmer wrote the program on personal time, he or she should and does own the copyright.

Notice how the student prompts the attorney to expand her answers.

Q. So "personal time" is the critical concept?

A. That's right. We're talking about the "work-for-hire" doctrine of copyright law. If I am working for you, anything I make under the terms of my employment is owned by you.

Q. What is the complication, then? If I make the software on my machine at home, I own it; if I'm working for someone, my employer owns it.

A. Well, the devil is in the details. Often the terms of employment are casual, or there is no written job description or contract for the particular piece of software.

Also notice how the student responds to the attorney's answers, making the interview more of a discussion.

Q. Can you give me an example of that?

A. Sure. There was a 1992 case, *Aymes* v. *Bonelli*. Bonelli owned a swimming pool and hired Aymes to write software to handle recordkeeping on the pool. This was not part of Bonelli's regular business; he just wanted a piece of software written. The terms of the employment were casual. Bonelli paid no health benefits, Aymes worked irregular hours, usually unsupervised — Bonelli wasn't a programmer. When the case was heard, the court ruled that even though Bonelli was paying Aymes, Aymes owned the copyright because of the lack of involvement and participation by Bonelli. The court found that the degree of skill required by Aymes to do the job was so great that, in effect, he was creating the software by himself, even though he was receiving compensation for it.

Q. How can such disagreements be prevented? By working out the details ahead of time?

A. Exactly. The employer should have the employee sign a statement that the project is being carried out as work-for-hire, and should register the copyright with the U.S. Copyright Office in Washington. Conversely, employees should try to have the employer sign a statement that the project is not work-for-hire, and should try to register the copyright themselves.

Q. And if agreement can't be reached ahead of time?

A. Then stop right there. Don't do any work.

■ **Figure 7.14**
Excerpt from an Interview

those who do not. For this reason, you need to be careful in drawing conclusions based on a small number of responses to a questionnaire.

When you send a questionnaire, you are asking the recipient to do you a
favor. Of course, if the questionnaire requires only two or three minutes to
complete, you are more likely to receive a response than if it requires an hour.
Your goal, then, should be to construct questions that will elicit the information you need as simply and efficiently as possible.

Asking Effective Questions

To ask effective questions, keep two points in mind:

- *Use unbiased language.* Don't ask "Should U.S. clothing manufacturers protect themselves from unfair foreign competition?" Instead, ask "Are you
  in favor of imposing tariffs on imported clothing?"
- *Be specific.* If you ask "Do you favor improving the safety of automobiles?"
  only an eccentric would answer no. Instead, ask "Do you favor requiring
  automobile manufacturers to equip new cars with side-impact air bags,
  which would raise the price by an average of $300 per car?"

Table 7.1 explains common types of questions used in questionnaires.

After you have decided on the questions, write a letter or memo to accompany the questionnaire. For someone outside your organization, this is basically an inquiry letter (sometimes with the questions themselves on a separate
sheet); therefore, it must clearly indicate who you are, why you are writing,
what you plan to do with the information, and when you will need it. For
people within your organization, provide the same information in a memo or
email accompanying a questionnaire.

Testing the Questionnaire

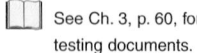 See Ch. 3, p. 60, for more on
testing documents.

Before you send out *any* questionnaire, show it and the accompanying letter
or memo to a few people who can help you identify any problems. After you
have revised the materials, test them on people whose backgrounds are similar to those of your real respondents. Revise the materials a second time, and,
if possible, test them again. Remember, once you have sent out the questionnaire, you cannot revise it and resend it to the same people.

Administering the Questionnaire

Having drafted the questions and tested them, administer the questionnaire.
Determining who the recipients should be can be simple or difficult. If you
want to know what the residents of a particular street think about a proposed
construction project, your job is easy. But if you want to know what mechan-
ical-engineering students in colleges across the country think about their

▉ **Table 7.1   Common Types of Questions Used in Questionnaires**

Type of Question	Example	Comments
Multiple choice	Would you consider joining a company-sponsored sports team? Yes_____ No_____	The respondent selects one of the alternatives.
Likert scale	The flextime program has been a success in its first year. strongly strongly disagree _ _ _ _ _ _ agree	The respondent ranks the degree to which he or she agrees or disagrees with the statement. Most statisticians recommend using an even number of possible responses (six, in this case); with an odd number, too many respondents choose the middle response, which does not provide useful data.
Semantic differentials	simple _ _ _ _ _ _ difficult interesting _ _ _ _ _ _ boring	The respondent registers a response along a continuum between a pair of opposing adjectives. Usually, these questions are used to measure a person's feelings about a task, an experience, or an object. As with Likert scales, most statisticians recommend using an even number of possible responses (six, in this case); with an odd number, too many respondents choose the middle response, which does not provide useful data.
Ranking	Please rank the following work schedules in order of preference. Put a 1 next to the schedule you would most like to have, a 2 next to your second choice, and so on. 8:00–4:30 _____ 9:00–5:30 _____ 8:30–5:00 _____ flexible _____	The respondent indicates a priority among a number of alternatives.
Short answer	What do you feel are the major advantages of the new parts-requisitioning policy? 1. _____ 2. _____ 3. _____	The respondent writes a brief answer using phrases or sentences.
Short essay	The new parts-requisitioning policy has been in effect for a year. How well do you think it is working? _____ _____ _____ _____	Although an essay question can elicit information you never would have found using closed-end questions, you will receive fewer responses, simply because they require more effort from the respondent. Also, essays cannot be quantified precisely, as the data from the other types of questions can.

curricula, you will need background in sampling techniques to isolate a representative sample.

Include a self-addressed, stamped envelope with the questionnaires sent to people outside your organization. Send a memo or email to people within your organization.

Figure 7.15 on page 178 shows a sample questionnaire.

September 6, 20XX

To: All employees
From: William Bonoff, Vice President of Operations
Subject: Evaluation of the Lunches Unlimited food service

As you may know, every two years we evaluate the quality and cost of the food service that caters our lunchroom. We would like you to help in our evaluation by sharing your opinions about the food service. Please note that your responses will remain anonymous. Please drop the completed questionnaires in the marked boxes near the main entrance to the lunchroom.

1. Approximately how many days per week do you eat lunch in the lunchroom?
    0 _____   1 _____   2 _____   3 _____   4 _____   5 _____

2. At approximately what time do you eat in the lunchroom?
    11:30–12:30 _____   12:00–1:00 _____   12:30–1:30 _____   varies _____

3. A clean table is usually available.
    Strongly Disagree _____ _____ _____ _____ _____ _____ Strongly Agree

4. The Lunches Unlimited personnel are polite and helpful.
    Strongly Disagree _____ _____ _____ _____ _____ _____ Strongly Agree

5. Please comment on the quality of the different kinds of food you have eaten in the lunch-room.
    a. Daily specials
        excellent _____ good _____ satisfactory _____ poor _____
    b. Hot dogs and hamburgers
        excellent _____ good _____ satisfactory _____ poor _____
    c. etc.

6. What *foods* would you like to see served that are not served now?
    _____

7. What *beverages* would you like to see served that are not served now?
    _____

8. Please comment on the prices of the foods and beverages served.
    a. Hot meals (daily specials)
        too high _____ fair _____ a bargain _____
    b. Hot dogs and hamburgers
        too high _____ fair _____ a bargain _____
    c. etc.

9. Would you be willing to spend more money for a better-quality lunch if you thought the price was reasonable?
    yes, often _____   sometimes _____   not likely _____

10. On the other side of this sheet, please provide whatever comments you think will help us evaluate the catering service.

Thank you for your assistance.

Likert-scale questions 3 and 4 make it easy for the writer to quantify data about subjective impressions.

Short-answer questions 6 and 7 are best for soliciting ideas from respondents.

■ **Figure 7.15**
Questionnaire

The process of researching a topic is cyclical: as you learn more about the topic, you do more research, which makes you see the topic differently. You will probably find that you continue to do research through the entire process, refining your topic as you proceed.

Once you have performed your basic primary and secondary research, you can turn your attention to shaping the document itself. Chapters 9–12 discuss techniques for drafting and revising the text. Chapters 13 and 14 discuss document design, layout, and graphics. Because you will probably do further research, it is smart to keep all your materials in order. Don't toss out any information you have already gathered, even if you think you will never need it again.

✔ **Revision Checklist**

Did you
1. analyze your audience and purpose?
2. choose a topic?
3. determine what you know — and don't know — about the topic?
4. determine how to carry out the research?
5. consult the appropriate reference books, including periodical indexes, newspaper indexes, abstract services, government publications, guides to business and industry, online databases, CD-ROMs, and the Web?
6. in evaluating information, carefully assess
   ❑ the author's credentials?
   ❑ the publishing body?
   ❑ the author's knowledge of literature in the field?
   ❑ the accuracy and verifiability of the information?
   ❑ the timeliness of the information?
7. study the information by skimming, taking notes, and summarizing?
8. if appropriate,
   ❑ conduct inspections?
   ❑ conduct experiments?
   ❑ perform field research?
   ❑ conduct interviews?
   ❑ send letters of inquiry?
   ❑ administer questionnaires?

Exercises

1. Use a search engine to find at least 10 sites about some key term or concept in your field, such as "genetic engineering," "hospice care," or "fuzzy logic." For each site, write a brief paragraph explaining why it would or would not be a credible source of information for a research report.

2. Now use a metasearch engine (such as www .dogpile.com or www.go2net.com) to search for sites about a term, as in exercise 1. Compare the hits you get from three different search engines. Is there a pattern? For example, does one search engine find more sites, or find more scholarly sites? Does one

search engine return some or all of the sites returned by the other two?

3. Using a search engine, locate two different manufacturers of software used by professionals in your field. Compare how each company's Web site distinguishes its product from those of its competitors.

4. Find a Web site for a company that makes consumer products, such as a soft-drink manufacturer, a movie studio, or an auto maker. Describe some of the most interesting and creative features of the site. You might consider addressing such questions as the following:

   • What is the ratio of information to entertainment on the site?
   • Overall, how effective is the site?

5. You are planning to do a research report on whether local employers think your school's recent graduates in your major are competent in the kinds of computer skills used in the workplace. Which research techniques would enable you to gather the most appropriate data for your study? For each technique, write a brief paragraph explaining the kind of information it will yield.

6. Using a search engine, answer the following questions. Provide the URL of each site you mention. If your instructor requests it, submit your answers as an email to him or her.

   a. What are the three largest or most important professional organizations in your field (for example, if you are a construction management major, your field is construction management or civil engineering or industrial engineering)?
   b. What are three important journals read by people in your field?
   c. What are the three most important listservs or bulletin boards read by people in your field?
   d. What are the date and location of an upcoming national or international professional meeting for people in your field?
   e. Name and describe, in one paragraph for each, three major issues being discussed by practitioners or academics in your field. Nurses, for instance, might be discussing the effect of managed care on the quality of medical care delivered to patients.

7. Revise the following interview questions to make them more effective. In a brief paragraph for each, explain why you have revised it as you have.

   a. What is the role of communication in your daily job?
   b. Do you think it is better to relocate your warehouse or to go to just-in-time manufacturing?
   c. Isn't it true that it's almost impossible to train an engineer to write well?
   d. Where are your company's headquarters?
   e. Is there anything else you think I should know?

8. Revise the following questionnaire questions to make them more effective. In a brief paragraph for each, explain why you have revised it as you have.

   a. Does your company provide tuition reimbursement for its employees? Yes_____ No_____
   b. What do you see as the future of bioengineering?
   c. How satisfied are you with the computer support you receive?
   d. How many employees work at your company? 5–10_____ 10–15_____ 15 or more_____
   e. What kinds of documents do you write most often? memos_____ letters_____ reports_____

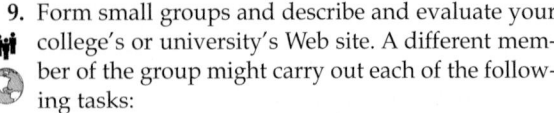

Research Projects

9. Form small groups and describe and evaluate your college's or university's Web site. A different member of the group might carry out each of the following tasks:

   • In an email to the site's Webmaster, ask questions about the process of creating the site. For example, how involved with the content and design of the site was the Webmaster? What is the Webmaster's role in maintaining the site?
   • Analyze the kinds of information the site contains and determine whether the site is intended primarily for faculty, students, alumni, legislators, or prospective students.
   • Determine the overlap between information on the site and information in printed documents

published by the school. In those cases in which they overlap, is the information on the site merely a duplication of the printed information, or has it been revised to take advantage of the unique capabilities of the Web?

In a memo to your instructor, present your findings and recommend how the site might be improved. See Chapter 15, page 430, for a discussion of memos.

10. Form groups according to major and analyze the Web site of a major professional organization in your field. Each member of the group might investigate one of the following questions:

   • Who is responsible for creating and maintaining the site?
   • What are the major kinds of information provided on the site?
   • What is the major function of the organization? Whom does it serve? What services does it provide? What activities does it sponsor?
   • What is the role of students in the organization? Are there special activities or opportunities for students? Can you suggest ways to increase the role of students in the organization?

Present your findings to the appropriate officer at the organization and to your instructor.

11. Check with your instructor about whether this assignment should be done individually or in a small group. Choose a topic on which to write a report for this course (See Chapters 18 and 19 for a discussion of reports.). Make sure the topic is sufficiently focused so that you will be able to cover it in some detail. Subjects that can be focused into topics include the following:

   – *A campus problem, such as the need to revise the requirements for a particular major.* What is the evidence that the major needs to be revised? What is the process for revising a major? What are the challenges that need to be met in order to carry out the process successfully? Conduct your own analysis of the situation, draw conclusions, and recommend what should be done next.
   – *Alternative fuels and energy sources for automobiles.* What alternatives are being developed now? What are their strengths and weaknesses? What factors encourage their development? What are

the social or political barriers to their widespread introduction? Which fuels are most likely to succeed in the marketplace?

   – *Job prospects in a particular field.* What data exist on job prospects, both locally and nationally? What are the trends? Does one aspect of the field appear to have better prospects than others? What steps can a student take to ensure the best preparation for the job market? How well does the curriculum meet the needs of industry?

   – *The evolving role of communication skills for the technical professional.* Determine the different skills and backgrounds of technical professionals in your field. What kinds of documents do they produce? What kinds of equipment do they use? Where did they learn their skills? How well equipped is your college to prepare tomorrow's technical professionals to communicate in the workplace? What steps could be taken to improve their preparation?

a. Using the Web and one of the guides-to-the-guides, plan a strategy for researching your topic.
   – Which guides, handbooks, dictionaries, and encyclopedias contain the background information you should read first?
   – Which basic reference books discuss your topic?
   – Which major indexes and abstract journals cover your topic?
   – Which of these basic tools are available online in your library? In which media (for instance, online bibliographies, CD-ROM, the Web) are they available?

b. Write down the call numbers or Internet addresses of the three indexes and abstract journals most relevant to your topic.

c. Make a preliminary bibliography of two books, five articles, and five Web sites that relate to your topic.

d. Find one of the works listed in your bibliography and write a brief assessment of its value.

e. Paraphrase any three paragraphs, each on a separate note card, from the first two pages of an article listed in your bibliography. Also note at least two quotations, each on a separate card: one should be a complete sentence, and one an excerpt from a sentence. Include a photocopy of the original when you submit the assignment, with the sentences you have quoted highlighted or underlined.

f. Write a 500-word summary of an article in your bibliography. Include a photocopy of the article when you submit the assignment.

g. Using a local industrial guide, list five people who might have firsthand knowledge of your topic. Arrange and carry out an interview with one of the five.

12. Form small groups for this project about search engines. As a group, choose a topic in the news, such as an advance in computer technology, and compile a list of keywords for researching the topic. Then have each member choose a different search engine and keep a log of the kinds of information the search engine uncovers. As a group, write a memo to your instructor explaining your findings. Does one search engine work better than another for this topic? Can you draw any inferences about the strengths and weaknesses of the different search engines you have investigated?

For additional cases, click on Case of the Month and Archive on TechComm Web (www.bedfordstmartins .com/techcomm).

C A S E
Compiling a Research Guidebook for Your Major

Information sources for students in your major are changing so rapidly that new students need a current guidebook describing how to carry out research. The chair of your department has asked your group to assemble a guidebook that can be distributed at the library and the department office. Each member of the group will want to concentrate on one of the following tasks:

- interviewing professors for their suggestions about the best sources of information for undergraduate majors

- using the printed information in the library: dictionaries, directories, handbooks, specialized encyclopedias, indexes, and so forth

- using the online resources in the library: online databases and CD-ROMs

- using the Web

Compile your findings into a guidebook. Be sure to submit the electronic file to your instructor so that it can be updated, because this version will be out of date by next semester.

Organizing Your Information

8

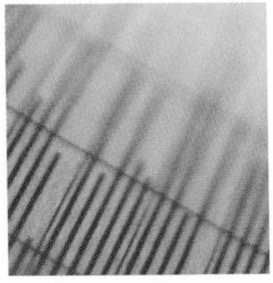

The technical communicators and scholars Charles T. Brusaw, Gerald J. Alred, and Walter E. Oliu (1997, p. 411) on the role of organization in technical communication:

An appropriate method of development is the writer's tool for keeping things under control and the reader's means of following the writer's development of a theme.

See TechComm Web (www .bedfordstmartins.com /techcomm) for guidelines boxes, additional examples, and links related to topics in this chapter.

At this point, you know for whom you are writing, and why, and you have done most of your research. Now it is time to start organizing the information that will make up the body of your document.

This chapter begins with a discussion of basic principles of organizing technical information. Then it examines eight patterns that can serve as templates to modify and combine as you organize the information for your readers. Finally, it discusses ways to introduce and conclude the body of your document.

BASIC PRINCIPLES OF ORGANIZING TECHNICAL INFORMATION

You should keep three principles in mind as you organize your information:

- analyze your audience and purpose
- use conventional patterns of organization
- display your organizational pattern prominently in the document

Analyzing Your Audience and Purpose

See Ch. 5 for more about audience and purpose.

Although you have thought about your audience and purpose as you planned your document and researched your subject, your initial analyses of audience and purpose are likely to change as you continue. For example, you might have begun your research with one purpose: to explain the advantages of purchasing several portable computer projection devices for your company. However, in the process of reading and gathering information, you realize that portable devices are not yet small enough for employees to carry on plane trips, especially if they are already carrying laptop computers. Your purpose, then, might change to investigating alternatives for company employees who want a reliable way to make computer presentations.

At this point, it is useful to review your assessments of audience and purpose before you proceed.

Does your audience like the message you will present? If they do, you can announce your main point early in the document. If they don't, you need to consider an organizational pattern that presents your important arguments before your main message. Is your audience used to seeing a particular organizational pattern in the kind of document you will be writing? If they do, you will probably want to use that pattern, unless you have a very good reason to use a different one.

What is your purpose in writing the document? Do you want your audience to understand a body of information or to accept a point of view and perhaps act on it? One purpose might call for you to present a brief report without any appendices; the other might require a detailed report, complete with appendices that enable some of your readers to carry out tasks.

Using Conventional Patterns of Arrangement

This chapter presents a number of conventional patterns of arrangement, such as the chronological pattern and the spatial pattern. When you sit down to organize your document, you should start by asking yourself whether a conventional pattern for presenting your information already exists. Using a conventional pattern makes things easier for you as a writer and for your audience.

For you, a conventional pattern serves as a template or checklist, helping you remember which information to include where. If you plan to write a proposal, for example, make sure to include a budget section, which you know your readers will expect. If possible, start with the computer file of an existing proposal, then delete the body text that doesn't belong in the new proposal.

For your audience, a conventional pattern makes your document easier to read and understand. Readers familiar with proposals can find the information they want — your budget, for instance — because you put it where others have put similar information. Your use of conventional patterns conveys the message that you are a professional who knows how technical information is generally presented, and that you want to make locating that information easy for your readers.

Does this mean that technical communication is merely the process of filling in the blanks? Not at all. As every chapter of this book emphasizes, you need to assess the writing situation constantly as you work. If you think you could communicate your ideas more effectively by modifying a conventional pattern or by devising a new pattern, of course you should do so. However, you gain nothing if an existing pattern would work just as well. Your readers don't want to be entertained. They simply want to read what you have to say so they can get on to the next task.

Displaying Your Arrangement Prominently

Whether you place your main point at the start of a passage or at the end, you should make it easy for your readers to understand the overall arrangement of your information. Displaying your arrangement prominently involves three main steps:

For more on tables of contents, see Ch. 12, p. 313. For more on headings and topic sentences, see Ch. 10, pp. 252 and 257.

- *Creating a detailed table of contents.* If your document has a table of contents, include at least two levels of headings. A detailed table of contents helps readers find the information they seek.
- *Using headings liberally.* Headings break up the text, making the page more interesting visually. In addition, headings state the subject of the section, improving the readers' understanding.
- *Using topic sentences at the beginnings of your paragraphs.* The topic sentence announces the main point of a paragraph. Putting the topic sentence at the start helps the reader understand the details that follow.

BASIC PATTERNS OF ORGANIZING INFORMATION

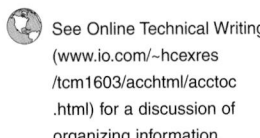

See Online Technical Writing (www.io.com/~hcexres /tcm1603/acchtml/acctoc .html) for a discussion of organizing information.

Every argument is different and calls for its own organizational pattern. Long, complex arguments often require several organizational patterns. For instance, one part of a document might be a causal analysis of the problem you are writing about, whereas another might be a comparison and contrast of two options for solving that problem.

This section discusses eight organizational patterns.

Patterns typically used in organizing information:
Chronological
Spatial
General to specific
More-important to less-important
Comparison and contrast
Classification and partition
Problem-methods-solution
Cause and effect

Chronological

The chronological — or time-line — pattern is commonly used to describe events. Following are some examples of the use of chronology as an organizing pattern:

In an *accident report*, you describe the events in the order in which they occurred.

In the background section of a *report*, you describe the events that led to the present situation.

In a *reference manual*, you explain how to carry out a task by describing the steps in sequence.

In the methods section of a *journal article*, you describe the actions you took in carrying out the experiment.

GUIDELINES

Organizing Information Chronologically

▶ *Provide signposts.* If the passage is more than a few hundred words long, use headings. Choose words such as *step, phase, stage,* and *part,* and consider numbering them. Add descriptive phrases to focus readers' attention on the topic of the section:

Phase One: Determining Our Objectives

Step 3: Installing the Lateral Supports

At the paragraph and sentence levels, transitional words such as *then, next, first,* and *finally* help your reader follow your discussion.

▶ *Consider using graphics to complement the text.* Graphics can clarify and emphasize chronological passages. Flowcharts are a particularly effective way of emphasizing chronological passages for all kinds of readers, from the most expert to the general reader.

▶ *Analyze events where appropriate.* Although chronology is an easy pattern to use, it doesn't explain why or how an event happened, or what it means. For instance, the largest section of an accident report is usually devoted to the chronological discussion, but the report is of little value unless it explains what caused the accident, who bears responsibility, and how such accidents can be prevented.

For more on transitions, see Ch. 10, p. 260.

For more on graphics, see Ch. 14.

Figure 8.1 on page 188 (Boeing, 2001), an excerpt from the Boeing Web site, uses the chronological pattern to present the history of the company.

 To view Figure 8.1 in context on the Web, click on Links Library for Ch. 8 on TechComm Web (www.bedfordstmartins .com/techcomm).

Spatial

The spatial pattern is commonly used to describe objects and physical sites. Here are some examples of the use of spatial organization:

In an *accident report*, you describe the physical scene of the accident.

In a *feasibility study* about building a facility, you describe the property on which it would be built.

In a *proposal* to design a new microchip, you describe the new chip.

In the results section of a *journal article* about the effectiveness of an antitoxin, you describe the tissue sample on which you tested the antitoxin.

GUIDELINES

Organizing Information Spatially

▶ *Provide signposts.* Help your readers follow the argument by using words and phrases that indicate location (*to the left, above, in the center*) in headings, topic sentences, and support sentences.

▶ *Consider using graphics to complement the text.* Diagrams, drawings, photographs, and maps help readers understand the argument.

▶ *Analyze events where appropriate.* A spatial arrangement doesn't explain itself; you still have to do the analysis: a diagram of a floor plan cannot explain why the floor plan is effective or ineffective.

Links for each year covered in this page.

Links to other pages for more information on particular subjects.

Links to other pages in the history section of the site.

a brief history

building for the future: 1983-1998

1983 | 1984 | 1985 | 1986 | 1987 | 1988 | 1989 | 1990 | 1991 | 1992 | 1993 | 1994 | 1995 | 1996 | 1997 | 1998

The Beginnings
The War Years
Post-War Developments
Jets and Rockets Take Off
New Markets
Building for the Future
Chronology
Biographies
Indexes

1995 May 17: Delivery of the first 777 to United Airlines.
June 21: The 767 Freighter makes its first flight.
June 26: Board of directors authorizes production of 777-300.
Dec. 18: Sea Launch gets first order for 10 commercial space satellite launches from Hughes Space and Communication Co.

1996 Jan. 4: RAH-66 Comanche makes first flight.
Feb. 15: The 777 wins Robert J. Collier Trophy as top aeronautical achievement for 1995.
March 29: First flight of the DarkStar, an unmanned aerial vehicle designed and built by Boeing and Lockheed Martin.
April 29: Phil Condit named chief executive officer.
Nov. 16: Boeing wins Joint Strike Fighter concept demonstration contract

■ **Figure 8.1 Information Organized According to the Chronological Pattern**

The clear, easy-to-navigate layout highlights the chronological organization of this section of the site.

Figure 8.2 (University, 2001), a Web-based map of the University of Texas campus, illustrates a spatial arrangement of information. A spatial arrangement could also be used in the text. For instance, each of the eleven regions of the campus could be discussed in accompanying paragraphs.

To view Figure 8.2 in context on the Web, click on Links Library for Ch. 8 on TechComm Web (www.bedfordstmartins .com/techcomm).

General to Specific

The general-to-specific pattern is based on the idea that readers need a general understanding of a subject before they can understand and remember the details. The general-to-specific pattern is used in many kinds of technical documents:

In a *process description*, you explain the overall process before you describe each step in detail. (In the detailed descriptions of each step, you will probably use other organizational patterns.)

In a *report*, you include an executive summary — an overview for managers — before the body of the report.

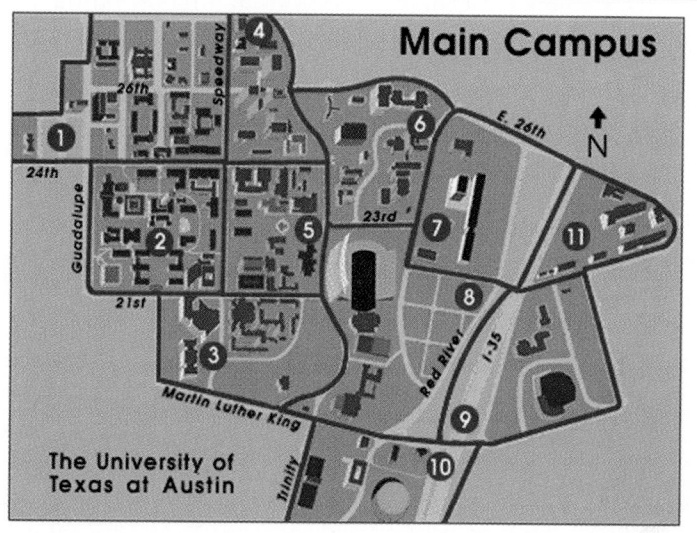

This map is divided into eleven regions, each marked with a numeral. The user clicks on a numbered region beneath the map for a more detailed map of that region.

1 Communication   2 Tower   3 Perry-Castañeda Library 4 Engineering
5 East Mall   6 Law School   7 LBJ School & Museum 8 Stadium
9 Disch-Falk   10 Erwin Center   11 Physical Plant

■ **Figure 8.2   Information Organized Spatially**

The spatial pattern helps the reader by breaking a very large area into smaller portions, making the campus easier to navigate.

In a set of *instructions*, you provide general information about the necessary tools and materials and about safety measures before providing the step-by-step instructions.

In a *memo*, you present background information before going into the details.

GUIDELINES

Organizing Information from General to Specific

▶ *Provide signposts.* In the introduction, explain that you will address general issues first and then move on to specific concerns. If appropriate, incorporate the words *general* and *specific* or other relevant terms in the major headings or at the start of the text for each item you are describing.

▶ *Consider using graphics to complement the text.* Diagrams, drawings, photographs, and maps help your reader understand the organization of the information.

Figure 8.3 (U.S. Agency, 1999), an excerpt from a report on the federal government's international food aid program, is an example of how to organize information from general to specific.

This excerpt begins with general information: a paragraph listing the three main U.S international food assistance programs.

Note the outline on the right that presents general-to-specific information in graphic form. (The paragraph and the outline, however, should present programs 2 and 3 in the same order.)

Here the writer begins a specific discussion of one of the three programs, Public Law 480, followed by a more specific discussion of one of its four programs, Title I.

B. U.S. International Food Assistance Programs

In responding to the challenge posed by global food insecurity, the U.S. government provides international food assistance through three channels: Public Law 480 (the Agricultural Trade Development and Assistance Act of 1954), Section 416(b) of the Agricultural Act of 1949 and the Food for Progress Act of 1985. All of these programs were re-authorized by the 1996 Federal Agriculture Improvement and Reform (FAIR) Act, also known as the 1996 Farm Bill.

U.S. INTERNATIONAL FOOD ASSISTANCE PROGRAMS

1. P.L.480:
   A. Title I
   B. Title II:
      i. Emergency
      ii. Development
   C. Title III
   D. Farmer-to-Farmer
2. Food For Progress
3. Section 416 (b)

PUBLIC LAW 480

The Agricultural Trade Development and Assistance Act of 1954, Public Law 480, is the preeminent mechanism for US international food assistance. P.L. 480, also known as Food for Peace, has three food aid titles. Each title has different objectives and provides commodity assistance to countries at different levels of economic development.

P.L. 480 TITLE I: TRADE AND DEVELOPMENT ASSISTANCE

■ **Figure 8.3   Information Organized from General to Specific**

More Important to Less Important

The more-important-to-less-important pattern recognizes that readers of technical communication often want the bottom line — the most important information — first.

Here are examples of the use of the more-important-to-less-important pattern:

> In an *accident report*, you describe the three most important factors that led to the accident before describing the less-important factors.

> In a *feasibility study* about building a facility, you present the major reasons that the site is appropriate, then the minor reasons.

> In a *proposal* to design a new microchip, you describe the major applications for the new chip, then the minor applications.

> In the conclusion of a *journal article* about the effectiveness of an antitoxin, you first describe the major conclusions of the experiment, then the less-important conclusions.

GUIDELINES

Organizing Information from More Important to Less Important

▶ *Provide signposts.* Tell your readers how you are organizing the passage. For instance, in the introduction of a proposal to design a new microchip, you might write, "The three applications for the new chip, each of which is discussed below, are arranged from most important to least important."

   Be straightforward. If you have two very important points and three less-important points, present them that way: group the two important points and label them, as in "Major Reasons to Retain Our Current Management Structure." Then present the less-important factors as "Other Reasons to Retain Our Current Management Structure." Being straightforward makes the material easier to follow and enhances your credibility.

▶ *Explain why one point is more important than another.* Don't just say that you will be arranging the items from more important to less important. Explain why the more important point is, in fact, more important.

▶ *Consider using graphics to complement the text.* Diagrams and numbered lists often help to suggest levels of importance.

   Figure 8.4 on page 192, from a memo written by an executive at a company that sells equipment for manufacturing semiconductors, shows the more-important-to-less-important organizational structure.

A THREE-POINT PROGRAM TO IMPROVE SERVICE

As you know, our most significant goal for this year has been to improve our customer service. Over the past six weeks, we have attempted to learn what our customers expect — and demand — in the service they receive. Toward that end, we have attended numerous conferences and conducted many focus groups.

What we have learned from recent conferences of semiconductor purchasers and from the focus groups is that customers expect and demand better service than the industry currently provides. By better service I don't mean merely returning phone calls. I mean something much more ambitious and difficult to attain: helping our customers do their jobs by anticipating and addressing their total needs. For this reason, I have formed a Customer Satisfaction Panel, chaired by Maureen Bedrich, whose job will be to develop policies that will enable us to improve the quality of the service we offer our customers.

I have asked the panel, under Maureen Bedrich's direction, to consider three major areas:

- improving the ease of use of our equipment
- improving preventive and corrective maintenance
- improving our compatibility with other vendors' products

Improving the Ease of Use of Our Equipment
User friendliness is the most important area we need to improve, because it affects our customers during the entire lifetime of the product. When we deliver a new product, we have to sit down with customers and explain how to integrate it into their manufacturing processes. This session is time-consuming and costly for us and for them. Therefore, we must explore the option of automating it....

Improving Preventive and Corrective Maintenance
The second most important area for study is improving preventive and corrective maintenance. Our customers will no longer tolerate down times approaching 10 percent; they will accept no more than 2 percent to 3 percent. Preventive maintenance is critical in our industry because gases used in vapor-deposition systems periodically have to be removed from the inside of the equipment. Customers want to be able to plan for these stoppages to reduce costs. Currently, we have no means of helping them do so....

Improving Our Compatibility with Other Vendors' Products
Finally, we have to accept the fact that because no one in our industry is likely to control the market, we have to make our products more compatible with those of other manufacturers. This means that we must be willing to put our people on-site to see what the customers' setup is and help them determine how to modify our product to fit in efficiently. We can no longer offer a "take-it-or-leave-it" product.

I hope you will extend every effort to work constructively with Maureen and her committee over the coming months to ensure that we improve the overall service we offer our customers.

The writer states his organizational pattern in the topic sentence and explains why this first area is the most important one he will discuss.

The writer again indicates his organizational pattern.

■ **Figure 8.4**
Information Organized from More Important to Less Important

Comparison and Contrast

Typically, the comparison-and-contrast organizational pattern is used to describe and evaluate two or more options. This pattern lies at the heart of the feasibility study, a document whose purpose is to compare two or more options under consideration. Here are examples of the use of the comparison-and-contrast pattern:

See Ch. 19, p. 549, for more on feasibility reports.

> In a *memo*, you compare and contrast the credentials of three finalists for a job.
>
> In a *proposal* to design a new microchip, you compare and contrast two different strategies for designing the chip.
>
> In a *report* describing a legal challenge that your company faces, you compare and contrast several options for responding.

The first step in comparing and contrasting two or more items is to determine the criteria: the standards or needs you will use in studying the items. For example, a comparison and contrast between two music keyboards might use the number of keys as one criterion. One keyboard might have 48 keys, whereas the other might have 64. Another criterion might be weight. One might weigh 25 pounds; the other, 46 pounds.

Often you compare and contrast several items as part of a decision-making process. That is, you need to decide which item best suits your needs. For instance, if you need to choose an elective course to take next semester, your only criterion might be the time it is offered: it must meet at 10 o'clock on Mondays, Wednesdays, and Fridays. For this criterion, MWF 10 is a *required characteristic*, sometimes called a *minimum specification*. However, you probably have other criteria: you would like the course to be interesting, to look good on your transcript, and so on. These other criteria are examples of *desired characteristics*, sometimes called *evaluative criteria*.

Almost always, you will need to consider several criteria in writing a document using comparison and contrast as an organizational scheme. For a recommendation report on which computer to buy, for example, you would probably include some required characteristics and some desired characteristics. A required characteristic might be that you be able to connect the computer to your company's network. If a particular computer cannot be connected, you will not consider it. But most of the criteria might call for desired characteristics, such as ease of operation, reliability, and ease of maintenance. You will evaluate each option — each computer — first by eliminating those that fail to meet the required characteristics and then by comparing and contrasting the remaining options according to the desired characteristics.

Two typical patterns for organizing a comparison-and-contrast discussion are *whole by whole* and *part by part*. The following table illustrates the difference between them. In this table, two printers — Model 5L and Model 6L — are being compared and contrasted according to three criteria: price, resolution, and print speed.

The whole-by-whole pattern provides a coherent picture of each option: the 5L and the 6L. This pattern works best if your readers need an overall assessment of each option, or if each option is roughly equivalent according to the criteria.

Whole by Whole	Part by Part
Model 5L	Price
• price	• Model 5L
• resolution	• Model 6L
• print speed	
	Resolution
Model 6L	• Model 5L
• price	• Model 6L
• resolution	
• print speed	Print Speed
	• Model 5L
	• Model 6L

The part-by-part pattern lets you focus your attention on the criteria. If, for instance, Model 5L produces much better resolution than Model 6L, the part-by-part pattern reveals this difference more effectively than the whole-by-whole pattern does. The part-by-part pattern is best for detailed comparisons and contrasts.

You can have it both ways. If you want to use a part-by-part pattern to emphasize particular aspects, you can begin the discussion with a general description of the various items.

Once you have chosen the overall pattern — whole-by-whole or part-by-part — you can decide how to organize the second-level items. That is, in a whole-by-whole passage, you have to sequence the "aspects"; in a part-by-part passage, you have to sequence the "options." For most documents, a more-important-to-less-important pattern will work well because readers want to get to the bottom line as soon as possible.

For some documents, however, other patterns might work better. People who write for readers outside their own company often reverse the more-important-to-less-important pattern because they want to make sure their audience reads the whole discussion. This pattern is also popular with writers who are delivering bad news. If, for instance, you want to justify recommending that your organization not go ahead with a popular plan, the reverse sequence lets you explain the problems with the popular plan before you present the plan you recommend. Otherwise, readers might start to formulate objections before you have had a chance to explain your position.

GUIDELINES

Organizing Information by Comparison and Contrast

▶ *Establish criteria for the comparison and contrast.* Choose criteria that are consistent with the needs of your audience.

▶ *If appropriate, determine whether each criterion calls for a required characteristic or a desired characteristic.* Follow this step only if you will be using the comparison and contrast pattern as part of a decision-making process.

▶ *Evaluate each item according to the criteria you have established.* Draw your conclusions.

▶ *Organize the discussion.* Choose either the *whole-by-whole* or *part-by-part* pattern, or some combination of the two. Then organize the second-level items.

▶ *Consider using graphics to complement the text.* Graphics can clarify and emphasize comparison-and-contrast passages. Diagrams, drawings, and tables are common ways to provide such clarification and emphasis.

Figure 8.5 on page 196 (Hsiao, 1998), excerpts from a technical article about the Linux computer operating system, shows an interesting variation on the two comparison-and-contrast organizational patterns described here.

Classification and Partition

Classification is the process of assigning items to categories. For instance, all the students at a university could be classified by sex (males and females), age (18 years old, 19, and so forth), major (nursing, forestry), and any number of other characteristics. You can also create categories within categories. For instance, within the category of students majoring in business at your college or university, you can create subcategories: male business majors and female business majors.

Here are examples of the use of classification as an organizing pattern:

In a *feasibility study* about building a facility, you classify sites into two categories: domestic and foreign.

In a *journal article* about ways to treat a medical condition, you classify the treatments as surgical and nonsurgical.

In a description of a major in a *college catalog*, you classify courses as required or elective.

Partition is the process of breaking a unit into its components. For example, a stereo system could be partitioned into the following components: cassette deck, CD player, tuner, amplifier, and speakers. Each component is separate, but together they form a whole stereo system. Each component can, of course, be partitioned further.

Partition is used in descriptions of objects, mechanisms, and processes (see Chapter 9). Here are examples of the use of partition:

In an *equipment catalog*, you use partition to describe one of your products.

In a *proposal*, you use partition to describe an instrument being proposed for development.

In a *brochure*, you describe how to operate a product by describing its features.

The writer first presents the tasks that Linux does better than the other systems, then the tasks the other systems do better than Linux.

These two paragraphs are based on the criteria that make Linux a superior system.

The criteria that favor Windows and Mac are presented second. This excerpt, therefore, is organized part-by-part, with the parts (or criteria) arranged from most favorable to least favorable.

Why should I use Linux instead of Windows or Mac-OS?

There are a number of situations in which Linux will outperform or outmaneuver other operating systems such as Windows or the Mac-OS. You might want to try Linux if:

You handle large amounts of information.
With an incredibly fast native filesystem, powerful database engines available at no cost with source code included, the ability to reduce thousands of mouse-clicks to single command lines, support for high-performance configurations (see below), and a networked-distributed core designed for data processing automation across the network, Linux is the master of data storage, retrieval, manipulation, and sharing.

You can't afford the software you need.
Linux itself is a free operating system — no licensing fees, whether per user, per copy, or per use. No fees whatsoever. That's right. Free. But even more importantly, nearly every piece of Linux software is available for free. Powerful webservers such as Apache (the most widely used server on the Internet), SQL engines, typesetting engines, scientific data acquisition systems, and even research-oriented clustering tools are all free for Linux users, even for commercial use. No program for Linux has ever started free and then later become a proprietary or pay-per-license product; all products which began free have remained available at zero cost.

Where does Linux fall short?

Now we'll cover the few major things which Linux doesn't do as well as Windows or the Mac-OS. If you use your computer almost exclusively for one of the following, Linux probably won't be such a revelation to you.

Proprietary solutions. Linux = costly.
If a major part of your operation is based on a proprietary solution created for another operating system, don't assume that the company which provided the solution will be willing to support Linux. Chances are that it won't. In situations like this, installing Linux generally requires a complete switch in the software upon which your operation is based. Depending on how badly your company needs the benefits which Linux can offer, installing Linux may or may not make financial sense.

General layman-style home use. Linux = overkill.
The multiuser, high-performance, networked distributed nature of Linux makes it overkill for most home applications. Yes, home Linux users can browse the Web, write a letter, draw a picture, and maintain the family budget in a spreadsheet, but running Linux also requires some amount of Unix-style systems administration. If you are a computer geek and father of seven, you can install Linux and give each of your children their own account on the system. If, however, you are a barber, weekend golfer, and father of seven, you probably don't have any time to administer a Linux system, and don't need high performance or two-year-uptime capable stability.

■ **Figure 8.5 Information Organized by Comparison and Contrast**

In this passage, addressed to expert readers, the writer compares and contrasts the Linux operating system with Windows and Mac operating systems.

Organizing Information by Classification or Partition

▶ *Choose a basis of classification or partition that fits your audience and purpose.* If you are writing a warning about snakes for hikers in a particular state park, your basis of classification will probably be whether the snakes are poisonous. You will describe all the poisonous snakes, then all the nonpoisonous ones.

▶ *Use only one basis of classification or partition at a time.* If you are classifying graphics programs according to their technology — paint programs and draw programs — do not include another basis of classification, such as cost.

▶ *Avoid overlap.* In classifying, make sure that no single item could logically be placed in more than one category. In partitioning, make sure that no listed component includes another listed component. Overlapping generally occurs when you change the basis of classification or the level at which you are partitioning a unit. In the following classification of bicycles, for instance, the writer introduces a new basis of classification that results in overlapping categories:

– mountain bikes
– racing bikes
– touring bikes
– ten-speed bikes

The first three items share a basis of classification: the type of bicycle. The fourth item has a different basis of classification: number of speeds. Adding the fourth item is illogical because a particular ten-speed bike could be a mountain bike, a touring bike, or a racing bike.

▶ *Be inclusive.* Include all the categories necessary to complete your basis of classification. For example, a partition of an automobile by major systems would be incomplete if it included the electrical, fuel, and drive systems but not the cooling system. If your purpose or audience requires that you omit a category, tell your readers that you are doing so.

▶ *Arrange the categories in a logical sequence.* Use a reasonable plan: chronology (first to last), spatial development (top to bottom), importance (most important to least important), and so on.

▶ *Consider using graphics to complement the text.* Block diagrams are commonly used to illustrate classification passages; drawings and diagrams are often used to illustrate partition passages.

In Figure 8.6, a discussion of nondestructive testing techniques, the writer uses classification effectively in introducing nondestructive testing to a technical audience. Notice that the writer could have used another basis for classification: sensitivity. The four techniques range from very sensitive to less sensitive.

TYPES OF NONDESTRUCTIVE TESTING

Nondestructive testing of structures permits early detection of stresses that can cause fatigue and ultimately structural damage. The least sensitive tests isolate macrocracks. More sensitive tests identify microcracks. The most sensitive tests identify slight stresses. All sensitivities of testing are useful because some structures can tolerate large amounts of stress — or even cracks — before their structural integrity is threatened.

Currently there are four techniques for nondestructive testing, as shown in Figure 1. These techniques are presented from least sensitive to most sensitive.

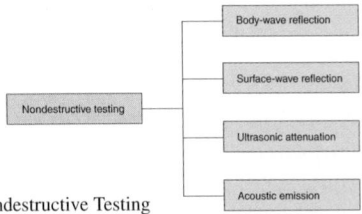

Figure 1. Types of Nondestructive Testing

Body-Wave Reflection
In this technique, a transducer sends an ultrasonic pulse through the test material. When the pulse strikes a crack, part of the pulse's energy is reflected back to the transducer. Body-wave reflection cannot isolate stresses: the pulse is sensitive only to relatively large cracks.

Surface-Wave Reflection
The transducer generates an ultrasonic pulse that travels along the surface of the test material. Cracks reflect a portion of the pulse's energy back to the transducer. Like body-wave reflection, surface-wave reflection picks up only macrocracks. Because cracks often begin on interior surfaces of materials, surface-wave reflection is a poor predictor of serious failures.

Ultrasonic Attenuation
The transducer generates an ultrasonic pulse either through or along the surface of the test material. When the pulse strikes cracks or the slight plastic deformations associated with stress, part of the pulse's energy is scattered. Thus, the amount of the pulse's energy decreases. Ultrasonic attenuation is a highly sensitive method of nondestructive acoustic testing.

There are two methods of ultrasonic attenuation. One technique reflects the pulse back to the transducer. The other uses a second transducer to receive the pulses sent through or along the surface of the material.

Acoustic Emission
When a test specimen is subjected to a great amount of stress, it begins to emit waves; some are in the ultrasonic range. A transducer attached to the surface of the test specimen records these waves. Current technologies make it possible to interpret these waves accurately for impending fatigue and cracks.

Notice that the writer clearly explains the sequence of the document's organization.

This simple block diagram helps the readers get an overview of the subject.

Here the writer introduces a second level of classification.

■ **Figure 8.6**
Information Organized by Classification and Subclassification

The writer classifies nondestructive testing into four categories.

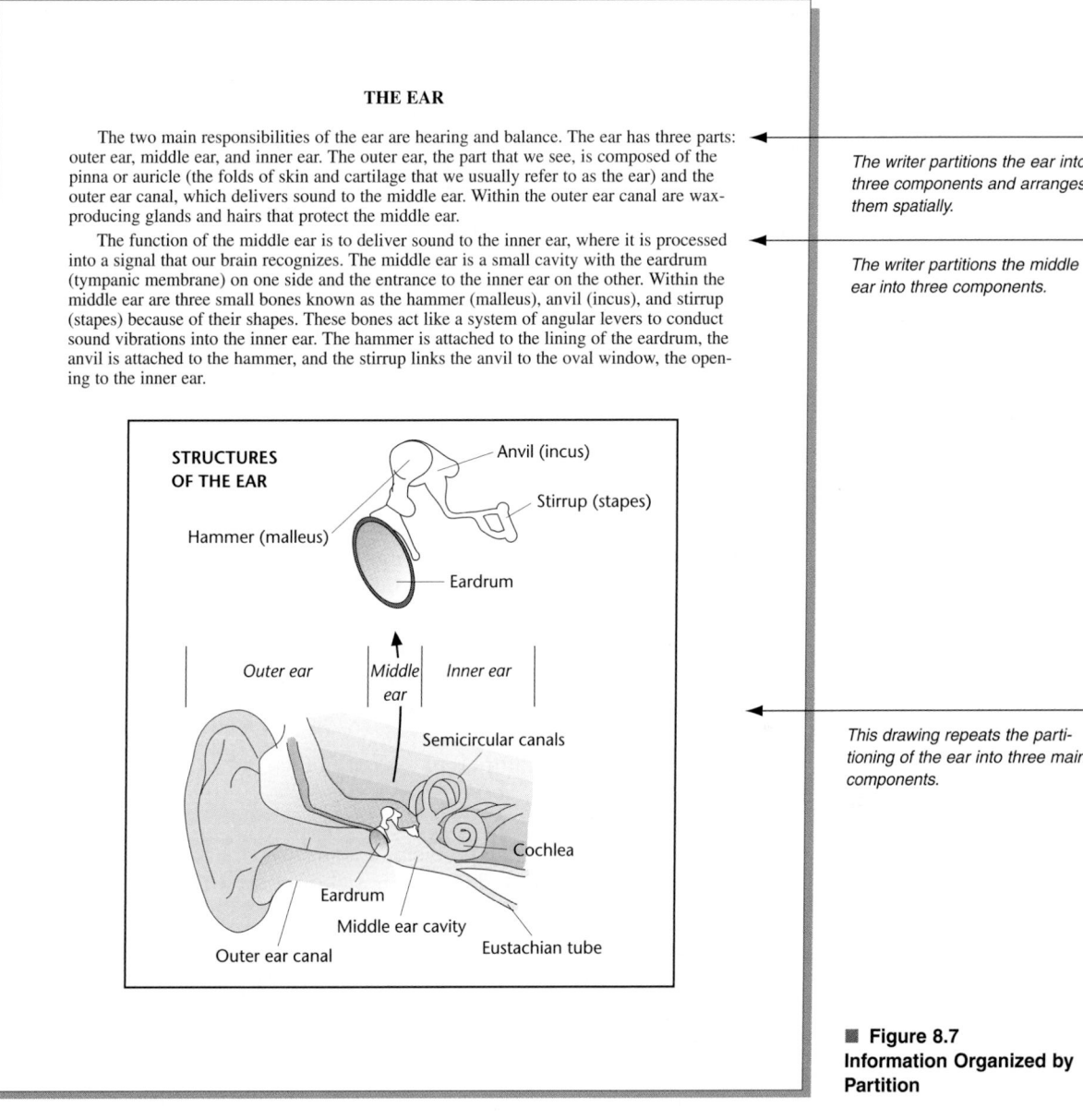

THE EAR

The two main responsibilities of the ear are hearing and balance. The ear has three parts: outer ear, middle ear, and inner ear. The outer ear, the part that we see, is composed of the pinna or auricle (the folds of skin and cartilage that we usually refer to as the ear) and the outer ear canal, which delivers sound to the middle ear. Within the outer ear canal are wax-producing glands and hairs that protect the middle ear.

The function of the middle ear is to deliver sound to the inner ear, where it is processed into a signal that our brain recognizes. The middle ear is a small cavity with the eardrum (tympanic membrane) on one side and the entrance to the inner ear on the other. Within the middle ear are three small bones known as the hammer (malleus), anvil (incus), and stirrup (stapes) because of their shapes. These bones act like a system of angular levers to conduct sound vibrations into the inner ear. The hammer is attached to the lining of the eardrum, the anvil is attached to the hammer, and the stirrup links the anvil to the oval window, the opening to the inner ear.

The writer partitions the ear into three components and arranges them spatially.

The writer partitions the middle ear into three components.

STRUCTURES OF THE EAR

Anvil (incus)

Stirrup (stapes)

Hammer (malleus)

Eardrum

Outer ear   Middle ear   Inner ear

Semicircular canals

Cochlea

Eardrum

Middle ear cavity

Outer ear canal

Eustachian tube

This drawing repeats the partitioning of the ear into three main components.

■ **Figure 8.7**
Information Organized by Partition

Figure 8.7 (Larson, 1990, p. 674) is an example of partition. For more examples of partition, see Chapter 9 (p. 232), which includes descriptions of objects, mechanisms, and processes.

Problem-Methods-Solution

The problem-methods-solution pattern is an excellent way to organize information about most kinds of problem-solving projects. It is easy to write — and also easy to read — because it reflects the logic used in carrying out a project.

The three components of this pattern are simple to identify:

- *Problem.* A description of what was not working (or not working as effectively as it should), or what opportunity exists for improving current processes.
- *Method.* The procedure performed to confirm the analysis of the problem, solve the problem, or exploit the opportunity.
- *Solution.* The statement of whether the analysis of the problem was correct, or of what was discovered or devised to solve the problem or capitalize on the opportunity.

In most cases, you will present these three components in the order you see here. Sometimes, however, you might vary the sequence. For example, you might want to present the problem first and then go directly to the solution, leaving the methods for last. This sequence deemphasizes the methods, a strategy appropriate for readers who already know them well or don't need to understand them. When readers want to focus on the solution, you can begin with the solution and then discuss the problem and methods.

Following are some examples of the use of the problem-methods-solution pattern:

In a *proposal*, you describe a problem in your business, how you plan to carry out your research, and how your end product (an item or a report) can help solve the problem.

In a *completion report* about a project to improve a manufacturing process, you describe the problem that motivated the project, the methods you used to carry out the project, and the findings: the results, conclusions, and recommendations.

In a *journal article* about ways to treat a medical condition, you describe the problem (the disadvantages of the current techniques), your methods, and your results and conclusions.

The example of the problem-methods-solution argument in Figure 8.8 is an excerpt from a newsletter article written by the president of a company that manufactures personal computers.

THE PROBLEM

Earlier this year, we were proud to offer the industry's largest array of add-on multimedia products for both our own computers and those of other manufacturers. Our offerings in cards, DVD drives, speakers, and other peripherals were unrivaled in both quantity and quality. And the response was terrific: in our first three months we sold more than 12,000 multimedia kits and 58,000 other peripheral units.

But growing pains soon became apparent: we logged more than 9,000 multimedia-related customer-support calls in that same period. What was the cause of this unprecedented customer-support problem? After considerable analysis of our customer-support data, we concluded that two factors were at work:

The writer describes the problem in detail.

- Add-on multimedia kits, even those meant for our own computers, were not necessarily compatible with the hardware or software our customers were using. We heard too many horror stories with the same theme: the kits were installed properly, but when the customer tried to reboot, the operating system was gone.

The writer speculates on the causes of the problem.

- Some 70 percent of the customers were novices, as opposed to a rate of less than 40 percent for our other product lines, and our documentation was simply inadequate to the task.

Meeting the Challenge

We recognized that being a pioneer in the industry had its costs: we were the first to encounter the problems that are now pervasive in the industry and well publicized in the literature. And because we were first, we were also the first to receive negative coverage from the trade journals for the resulting problems with customer satisfaction.

We instituted a four-point plan to meet the challenge:

The writer describes the methods — the steps the company took to solve the problems.

- We instituted a new quality-control program. Now every product is taken out of the box, plugged in, and turned on. We make sure that the printer setup is accurate and that the hardware and the bundled software are compatible. At our weekly audit meetings, we

■ **Figure 8.8**
Information Organized by the Problem-Methods-Solution Pattern

review that week's quality-control data; each team leader can now stop production to investigate a recurrent or unexplained problem.

- We expanded our use of novices in our preproduction focus groups and in the quality-control program. We are concentrating on learning how the novice uses our products; in our expansion into the family market we expect to find that an increasing percentage of our customers are first-time computer owners.
- We instituted a Process-Improvement Team, a group of 12 veteran employees committed to improving customer support and customer satisfaction. Among the first innovations of the Process-Improvement Team was the creation of more than 200 documents designed to assist users with the common problems they encounter when installing our kits and using common software. These documents are on our Web site and can be faxed to customers at no charge when they call a special toll-free number.
- We instituted a Quality Team of 15 employees charged with seeking Manufacturing's ideas about quality and efficiency standards.

The Results

These measures have been in place for only two months, but the preliminary data are encouraging. Customer-support calls on our multimedia kits are down more than 15 percent the last two months. Catastrophic problems — such as destruction of the operating system — are down more than 30 percent. Finally, the increased use of novices in design and focus groups has led to three interface improvements that were noted in a *PC Week* article earlier this month. The work of the Quality Team has resulted in a 7 percent decrease in rejection rates of our multimedia kits.

In short, I think we are on the right track. But quality improvement is a frame of mind and an ongoing commitment, not a goal that can ever be reached. I pledge to you that we shall continue to strive to make RST the best place to buy PCs and PC-related products.

The writer describes the results — the data on the reduction in the problem.

▥ **Figure 8.8**
(Continued)

GUIDELINES

Organizing Information by Problem-Methods-Solution

▶ *In describing the problem, be clear and specific.* Don't say that "our energy expenditures are getting out of hand." Instead, say that "the energy usage has increased 7 percent in the last year" and that "the utility costs have risen 11 percent." Then calculate the total increase in energy costs.

▶ *In describing your methods, help your readers understand not only exactly what you did but also why you did it that way.* Because most technical problems can be approached using several methods, you might have to justify your choices. Why, for example, did you use a *t*-test in calculating the statistics in an experiment? If you can't defend your choice, you lose credibility.

▶ *In describing the solution, don't overstate.* Avoid claims such as "this project will increase our market share from 7 percent to 10 percent within 12 months." Instead, be cautious: "This project promises to increase our market share from 7 percent to 10 percent or even 11 percent." This way, your document won't come back to haunt you if things don't turn out as well as you had hoped.

▶ *Choose a logical sequence.* The most common sequence is to start with the problem and conclude with the solution. However, different sequences work equally well as long as you provide some kind of preliminary summary to give readers an overview and headings or some other design elements (see Chapter 11) to help readers find the information they want.

▶ *Consider using graphics to complement the text.* Graphics, such as flowcharts, diagrams, and drawings, can clarify and emphasize problems-methods-solution passages.

Cause and Effect

Technical communication often involves cause-and-effect discussions. Sometimes you will reason forward, from cause to effect. If we raise the price of a particular product we manufacture (cause), what will happen to our sales (effect)? Or, the government forbids us to use a particular chemical in our production process (cause); what will we have to do to keep the production process running smoothly (effect)?

Sometimes you will reason backward, from effect to cause. Productivity went down by 6 percent in the last quarter (effect); what factors led to this decrease (causes)? Or, the federal government has decided that used-car dealers are not required to tell potential customers about the cars' defects (effect); why did the federal government reach this decision (causes)?

Cause-and-effect reasoning, therefore, provides a way to answer the following two questions:

- What will be the effect(s) of X?
- What caused X?

Here are examples of the use of cause and effect as an organizing pattern:

In an *environmental impact statement,* you argue that a proposed construction project would have three important effects on the ecosystem.

In the recommendation section of a *report,* you argue that a recommended solution would improve operations in two major ways.

In the introduction to a *journal article,* you argue that your topic is worthy of study because it helps readers understand an important but overlooked aspect of the subject you are addressing.

Cause-effect relationships are difficult to describe because there is no scientific way to determine causes or effects. You draw on your common sense and your knowledge of your subject. When you try to determine, for example, why the product you introduced last year did not succeed, you start with the obvious possibilities: the market for this kind of product was saturated, the product was of low quality, the product was poorly marketed, and so forth. The more you know about your subject, the more precise and more insightful your assessment of the possible reasons will be.

A causal discussion can never be certain. Nobody can determine, with complete certainty, why a product failed in the marketplace. But you can explain why you think the causes or effects you are identifying are the most plausible ones. For instance, to make a plausible case that the main reason your product did not sell is that it was poorly marketed, you can show that in the past, your company's other unsuccessful products were marketed in similar ways. This argument does not confirm that poor marketing was the problem, but it does make the claim plausible. Similarly, if you can show that your best-selling products have been marketed very differently from the unsuccessful one, that reasoning helps make the claim plausible.

Figure 8.9, an excerpt from an accident report about the explosion of a propane tank (U.S. Chemical, 1999), illustrates an effective cause-and-effect argument.

To view Figure 8.9 in context on the Web, follow the links in Chapter 8 on TechComm Web (www.bedfordstmartins .com/techcomm).

GUIDELINES

Organizing Information by Cause and Effect

▶ *Explain your reasoning.* If appropriate, explain your reasoning. If your point is that the marketing for the unsuccessful product was inadequate, use specific facts and figures — the low marketing budget, delays in beginning the marketing campaign, and so forth — that support your claim.

4.3 DAMAGE TO LIQUID LINE

CSB investigators based the conclusion that the impact of the ATV broke off the liquid pipe at its connection to the shut-off valve on a number of factors. The piece of 3/4-inch liquid pipe broken off was labeled A1 following the incident (see Figures 29 and 36 in Appendix B). As these figures illustrate, a bend was discovered in the A1 segment between the threads and the union (a coupling used to connect two segments of pipe). This bend indicates that this segment of A1 was subjected to a significant horizontally applied load before fracture occurred. Horizontal stress patterns identified on the edge of the A1 pipe segment that connected to the A20 valve also revealed that the pipe failed in a horizontal plane (see Figure 37, Appendix B). Had the A1 piece been completely connected to the A20 valve at the time of the explosion, these stress patterns would almost certainly have been twisted with a vertical orientation, not a horizontal one. Thus, the force required to produce these stress patterns most likely occurred when the ATV hit the liquid line. The direction of thread deformation on the A1 edge (Appendix B, Figures 36–39) also indicates that the pipe was subjected to a horizontally applied load such as that likely caused by the impact of the ATV.

In addition, the debris map in Appendix C illustrates that the A1 piece was discovered in the immediate area of the original tank location following the blast. The discovery of A1 at this location suggests that it was severed from the shut-off valve prior to the explosion. The shut-off valve, which had been connected to A1, was thrown a significant distance from its original location by the explosion (see Appendix C, item A20).

If the A1 piece were still connected to the shut-off valve at the time of the explosion, there would most likely be dents or other types of damage visible on the pipe surface similar to the damage observed on other pieces of debris. For example, the shut-off valve (A20) exhibited a number of abrasions that likely occurred as a result of the explosion. As Figure 29 in Appendix B shows, the A1 piece was not dented or damaged. Finally, eyewitness reports of fire under the north and west end of the tank following the impact of the ATV were consistent with a propane leak in the vicinity of the A20/A1 connection under the tank.

Notice that in the first sentence, the writers have "concluded" that the crash caused the explosion. They are not stating it as a fact.

Throughout this passage, the writers carefully refer to physical evidence suggesting that their conclusion is correct and that alternative conclusions would be incorrect.

Figure 8.9 A Discussion Organized by the Cause-and-Effect Pattern

The writers are describing what they think caused the explosion of a large propane tank: an all-terrain vehicle (ATV) crashed into the tank, severing one of its pipes.

▶ *Avoid overstating your argument.* You cannot prove a causal relationship. The best you can do is to make a persuasive case. Therefore, avoid overstating your argument. If you describe a causal relationship using a word that is more appropriate for a factual statement, you can undermine your argument and alienate your readers. For instance, if you write that Steve Jobs, the founder of Apple, "created the computer revolution," you are claiming too much. It is better to write that Steve Jobs "was one of the central players in creating the computer revolution."

▶ *Avoid logical fallacies.* Logical fallacies, such as hasty generalizations or *post-hoc* reasoning, can also undermine your discussion.

▶ *Consider using graphics to complement the text.* Graphics, such as flowcharts, organizations charts, diagrams, and drawings, can clarify and emphasize cause-and-effect passages.

For more on logical fallacies, see Ch. 6, p. 129.

INTRODUCING AND CONCLUDING THE BODY

Drafting the passages in the body of a document involves using and modifying the basic organizational patterns described in this chapter. Two more elements of the body — the introduction and conclusion — are also fundamental to the success of your document.

Introducing the Body

An introduction has one main goal: to help readers understand your discussion by explaining *what* information you are going to present, *how* you are going to present it, and *why* you choose to present it that way. If you communicate these points clearly, your readers will be more willing to read the document and better able to understand and remember it.

Your document can have one introduction (at the beginning) or several introductions (one at the beginning and one at the start of each major section).

Every document calls for a different kind of introduction. A brochure on backyard pool safety obviously needs little introduction, and then perhaps only a brief review of the statistics on injuries and deaths. But the introduction to a scholarly article, which needs to conform to the practices in a particular academic field, might require many elements. To draft the introduction to the body of your document, use your common sense and study similar documents to note the conventions in your field and the expectations of your readers.

Readers rely on an introduction to prepare them for the information that follows.

An introduction should answer these questions:
What is the subject?
What are the key terms that will be used in the argument?
What is the purpose of the argument?
What is the background of the subject?
What is the relevant literature, and what are its limitations?
What is the scope of the argument?
What is the organization of the argument?

- *What is the subject?* Answer this question explicitly, even though your readers might already know the answer.

  This report describes the relationship between the courts and the scientific community on the subject of phantom risk. We live in an age of tremendous technological advances, and some new products and practices might pose health risks. The conflict explored in this report is that, whereas the courts require certainty (they need to know whether a plaintiff was or was not injured by a technology), the scientific community often cannot supply a definitive answer.

- *What are the key terms that will be used in the argument?* Define all key terms you intend to discuss in the document.

Phantom risk refers to alleged but uncertain risks associated with a scientific or technological practice. For instance, weak magnetic fields represent a phantom risk in that the scientific community is uncertain whether they represent any health risk at all, or, if they do, what the threshold for such risks might be.

For more about definitions, see Ch. 9, p. 220.

If the reader needs to know a key term to understand the introduction, present the definition early in the introduction.

- *What is the purpose of the argument?* Explain what you hope to achieve in the discussion.

This report has two main purposes: to summarize the conflict between the legal and scientific communities and to propose three principles to guide future debate on the legal ramifications of phantom risk.

- *What is the background of the subject?* Present the information that readers need to understand the discussion.

It is inevitable that some people will be injured by technological advances, and one of the foundations of our legal system is that victims be empowered to seek redress of their injuries through the legal system. However, science does not offer the kind of certainty the legal system requires. Are breast implants dangerous? Cautious scientists would have to admit that they don't know, but different constituencies answer yes and no with self-assurance and conviction. When courts render false-positive judgments — by finding that harmless products and services have caused injuries — they put companies out of business and deprive people of their products and services. When courts render false negatives — by finding that harmful products and services have not injured anyone — they endanger the public. Currently, there is no way to reconcile the different cultures and practices of the legal and scientific communities.

- *What is the relevant literature, and what are its limitations?* Show your readers that you have done the necessary research on your topic.

A number of researchers have studied the topic of phanton risk, but most of the studies have examined individual cases from either the legal or scientific perspective. For instance, the asbestos issue has been discussed from the scientific perspective by Wilkins (1998), Thomas (1996), and Rivera (1999), and the legal perspective by Halloran (1997), Bradford (1998), and De Moss (1999). The edited collection by Foster et al. (1999) is the first to attempt a detailed look at the conflict; however, it offers piecemeal opinions about the different phantom-risk issues but does not provide a theoretical framework for reconciling the two cultures.

Introduction

Epidemiologic studies have defined neck MSDs in one of two ways: (a) by symptoms occurring in the neck (usually with regard to a specific duration, frequency, or intensity), or (b) by using both symptoms and physical examination findings.

The prevalence of reported MSDs is generally lower when they are defined using both symptoms and physical examination results than when defined using symptoms alone. For example, the prevalence rate of tension neck syndrome (TNS) among male industrial workers in the United States was reported to be 4.9% from interview data and 1.4% when case definitions included physical exam findings [Hagberg and Wegman 1987]. The percent of work-related MSD cases defined by physical examination findings to those defined solely by symptoms has ranged from approximately 50% (Silverstein et al. [1987]; Blider et al. [1991]; Bernard et al. [1993]; Hales et al. 1994]) to about 85% (Andersen and Gaardboe [1993b]). Forty-seven of the listed studies referenced included physical examination findings in their health outcome assessment criteria.

Many of the neck and neck/shoulder MSD studies referenced in the tables were part of larger studies that inquired about musculoskeletal symptoms and physical findings in multiple body sites. In most of these studies, there were no separate ergonomic exposure observations or measurements made that pertained to the neck region (e.g., there were no neck posture observations, neck angle measurements, neck work-load assessment, trapezius electromyographic testing, etc.). In these studies, the primary interest and measurement strategies focused on the hand and wrist region (e.g., Kuorinka and Koskinen [1979]; Ohlsson et al. [1989]; Hales et al. [1989]; Kiken et al. [1990]; Baron et al. [1991]). In the studies, workers were categorized only by hand/wrist exposures. Hand/wrist categorization will not reflect exposures of the neck region (or other musculoskeletal sites). For example, workers who may have frequent and rapid awkward postures of the neck but less frequent or extreme postures of the hand and wrist region may be misclassified as low risk if classification depends only on hand/wrist exposure. In general, we have given these studies less weight because of a significant potential for misclassification.

The text of this section on neck and neck/shoulder MSDs is organized by work-related exposure factor. The discussion within each factor is organized according to the criteria for evaluating evidence for work-relatedness in epidemiologic studies using the strength of association, the consistency of association, temporal relationships, exposure-response relationship, and coherence of evidence. Conclusions are presented with respect to neck and neck/shoulder MSDs as a single disorder for each exposure factor. Summary information relevant to the criteria used to evaluate study quality is presented in Tables 2-1 through 2-6. A more extensive summary, which includes information on health outcome, covariates, and exposure measures, is presented at the end of this chapter.

*The authors discuss the difficulty of **defining their key term**: neck MSDs. If neck MSDs are defined only by reported symptoms, the incidence of the condition is much higher than if physical exams are required. Notice that the authors **review the literature** carefully.*

*A related problem is that the technical studies of MSDs often do not distinguish neck MSDs from all other MSDs. This discussion of the literature helps readers understand the **complexity of the subject**, and it helps establish the authors' credibility.*

*The writers explain **the organization of the report**.*

■ **Figure 8.10   A Sample Introduction**

*The title of the report explains both the **subject** and the **purpose** of the study: "Musculoskeletal Disorders (MSDs) and Workplace Factors: A Critical Review of Epidemiologic Evidence for Work-Related Musculoskeletal Disorders of the Neck, Upper Extremity, and Low Back."*

- *What is the scope of the argument?* Explain what you are including or excluding.

  This report focuses on four phantom-risk issues: electrical and magnetic field risks, spermicide cancer risks, asbestos risks, and secondhand smoke risks. It reviews the scientific evidence and the legal opinions for each issue. Where appropriate, it discusses the social policy resulting from the legal actions.

- *What is the organization of the argument?* Help readers concentrate on the information without worrying about what will come next.

  In the sections that follow, I treat each phantom-risk issue separately, as a case study. After a background discussion, I fill in the scientific consensus, followed by the legal precedents and, where appropriate, the resultant social policy. I conclude the article with a recommendation that we discuss three principles for reconciling the legal and scientific approaches to phantom-risk issues.

Figure 8.10 is an excerpt from the introduction to a report about musculo-skeletal disorders (MSDs) in the workplace (Bernard, 1997).

Concluding the Body

In discussions of writing, the word *conclusion* has two meanings. One refers to the inferences drawn from technical data. For instance, while investigating whether your company should switch from paper-based to online documentation, you might conclude that the change would be unwise at this time; that conclusion, along with others, would appear in the "Conclusions" section of your feasibility report. This kind of conclusion is discussed in Chapter 19.

Conclusion also refers to the final part of a document or a section of a document. The following discussion concerns this second sense of the word.

Although some kinds of documents, such as parts catalogs, do not generally have a conclusion, most do.

A conclusion should answer these questions:
What are the main ideas communicated in the argument?
What should be done next?
How can the reader find more information?
How can we help you in the future?

- *What are the main ideas communicated in the argument?* Readers can forget material, especially material from the beginning of a long document. Therefore, it is a good idea to summarize the important ideas. The following examples are from a report on reuse-adoption programs: company-wide programs to increase the reuse of systems and software.

  Our analysis yielded two main conclusions. First, reuse adoption is good engineering and good business; it can reduce expenses and increase productivity and quality. Second, reuse adoption is a complex procedure that requires a substantial amount of planning and supervision. If the program is carried out casually or thoughtlessly, it can backfire and cause more problems than it solves.

• *What should be done next?* Offer recommendations.

I recommend that we convene an ad hoc committee to study the feasibility of converting to a reuse-adoption plan. Specifically, we should carry out a Phase I analysis to determine the answers to the following questions:

1. Is demand in our market sufficient for the kinds of products that are appropriate for reuse adoption?
2. What effect will reuse adoption have on . . .

---

*The authors begin by summarizing their **major finding:** there is considerable evidence that workplace factors can cause MSDs.*

The authors also point out the shortcomings of the existing research.

*The authors speculate on the possible **scope** of the problem.*

*The authors discuss the **implications of their study**.*

Conclusion

A substantial body of credible epidemiologic research provides strong evidence of an association between MSDs and certain work-related physical factors when there are high levels of exposure and especially in combination with exposure to more than one physical factor (e.g., repetitive lifting of heavy objects in extreme or awkward postures) [Table 1].

The consistently positive findings from a large number of cross-sectional studies, strengthened by the limited number of prospective studies, provides *strong evidence* (+++) for increased risk of work-related MSDs for some body parts. This evidence can be seen from the strength of the associations, lack of ambiguity in temporal relationships from the prospective studies, the consistency of the results in these studies, and adequate control or adjustment for likely confounders. . . .

In general, there is limited detailed quantitative information about exposure-disorder relationships between risk factors and MSDs. The risk of each exposure depends on a variety of factors such as the frequency, duration, and intensity of physical workplace exposures. Most of the specific exposures associated with the *strong evidence* (+++) involved daily whole shift exposure to the factors under investigation.

The number of jobs in which workers routinely lift heavy objects, are exposed on a daily basis to whole body vibration, routinely perform overhead work, work with their necks in chronic flexion position, or perform repetitive forceful tasks is unknown. While these exposures do not occur in most jobs, a large number of workers may indeed work under these conditions. The BLS data indicate that the total employment is over three million in the industries with the highest incidence rates of cases involving days away from work from overexertion in lifting and repetitive motion. . . .

This critical review of the epidemiologic literature identified a number of specific physical exposures strongly associated with specific MSDs when exposures are intense, prolonged, and particularly when workers are exposed to several risk factors simultaneously. This scientific knowledge is being applied in preventive programs in a number of diverse work settings. While this review has summarized an impressive body of epidemiologic research, it is recognized that additional research would be quite valuable. The MSD components of the National Occupational Research Agenda efforts are principally directed toward stimulation of greater research on MSDs and occupational factors, both physical and psychosocial. Research efforts can be guided by the existing literature, reviewed here, as well as by data on the magnitude of various MSDs among U.S. workers.

■ **Figure 8.11   A Sample Conclusion**

*The Web site on which this report is presented **invites readers to inquire for more information** on how the organization can provide further assistance.*

- *How can the reader find more information?* Sometimes this portion of the conclusion is a sales message; sometimes it is not.

    I have asked Corporate Information to get the six articles listed in the references section. These articles will be gathered and routed next week. Please read them and be prepared to offer any questions or concerns at our meeting on August 15, at which time we will discuss the recommendations.

- *How can we help you in the future?* Offer to provide services if you can.

    I think reuse adoption offers the promise to improve our business in a number of ways. Please feel free to get in touch with me, either before or after you read the articles, to discuss this initiative.

Figure 8.11 is an excerpt from the conclusion of the report on musculoskeletal disorders in Figure 8.10 (Bernard, 1997).

 To view Figure 8.11 in context on the Web, click on Links Library for Ch. 8 on TechComm Web (www .bedfordstmartins.com /techcomm).

✔   **Revision Checklist**

Use the following checklist to ensure that the information in your document is organized and that your introduction and conclusion are complete.

Did you
  1. analyze your audience and purpose?
  2. consider using a conventional pattern of arrangement?
  3. display your organization prominently by
      ❑ creating a detailed table of contents?
      ❑ using headings liberally?
      ❑ using topic sentences at the start of your paragraphs?

The following checklists cover the eight organizational patterns discussed in this chapter.

Chronological and Spatial

Did you
      ❑ provide signposts, such as headings or transitional words or phrases?
      ❑ consider using graphics to complement the text?
      ❑ analyze events where appropriate?

General to Specific

Did you
      ❑ provide signposts, such as headings or transitional words or phrases?
      ❑ consider using graphics to complement the text?

More-Important to Less-Important

Did you
      ❑ provide signposts, explaining clearly that you are using this organizational pattern?

❑ make clear why the first point is the most important, the second is the second most important, and so forth?

❑ consider using graphics to complement the text?

Comparison and Contrast

Did you

❑ establish criteria for the comparison and contrast?

❑ choose a structure — whole-by-whole or part-by-part — that is most appropriate for your audience and purpose?

❑ choose appropriate organizational patterns for your second-level items?

❑ consider using graphics to complement the text?

Classification and Partition

Did you

❑ choose a basis consistent with the audience and purpose of the document?

❑ use only one basis at a time?

❑ avoid overlap?

❑ include all the appropriate categories?

❑ arrange the categories in a logical sequence?

❑ consider using graphics to complement the text?

Problem-Methods-Solution

Did you

❑ describe the problem clearly and specifically?

❑ if appropriate, justify your methods?

❑ avoid overstating your solution?

❑ arrange the discussion in a sequence consistent with the audience and purpose of the document?

❑ consider using graphics to complement the text?

Cause and Effect

Did you

❑ explain your reasoning?

❑ avoid overstating your argument?

❑ avoid logical fallacies?

❑ consider using graphics to complement the text?

The following questions cover introductions and conclusions.

Introductions

Did you

❑ explain the subject of the document?

❑ define all key terms that will be used in the discussion?

❑ explain the purpose of the discussion?

❑ explain the background of the subject?

❑ review the relevant literature, explaining its limitations?

❑ explain the scope of the discussion?

❑ explain the organization of the discussion?

Conclusions

Did you

- ☐ summarize the main ideas in the document?
- ☐ recommend what should be done next?
- ☐ explain how the reader can find more information?
- ☐ describe how you can help?

Exercises

Some of the following exercises ask you to write a memo. See Chapter 15, page 430, for a discussion of memos.

1. Using a search engine, find the Web site of a company that makes a product used by professionals in your field (personal computers are a safe choice). Locate three discussions on the site. For example, there will probably be the following: a passage devoted to ordering a product from the site (using a chronological pattern), a description of a product (using a partition argument), a passage describing why the company's products are superior to those of the competitors (using a cause-and-effect argument). How effective is each of the three discussions? What in particular makes each passage effective or ineffective? How might each one be improved?

2. Write a memo to your instructor evaluating the effectiveness of the following introduction to a student report about how audits are conducted. What works well in the introduction? How might the introduction be improved?

Introduction

An audit is a formal examination and verification of financial accounts. According to the tenth edition of *Montgomery's Auditing* textbook, historians believe that audits were conducted as early as 4000 B.C. in Babylonia. Governments were concerned with establishing controls, including audits, to reduce errors and fraud in the tax collection system. Auditing has been seen in history ever since. The Bible (generally believed to cover the period of time between 1800 B.C. and A.D. 95) refers to internal controls and surprise audits. The earliest records in English-speaking countries are from England and Scotland. They have accounting records and references to audits dating back to A.D. 1130.

Auditing was fairly slow to progress in the United States. In the late 1800s the U.S. railroad companies employed audi-

tors, and in 1887 the American Institute of Accountants, now the American Institute of Certified Public Accountants (AICPA), was established. In 1935 audits became more common for two reasons:

- The Securities Act of 1933 and the Securities Exchange Act of 1934, which required listed companies to file audited financial statements, were enacted.
- The AICPA collaborated with the New York Stock Exchange to improve reporting standards.

Today all major corporations conduct audits every year. Audits are used to analyze the financial condition of a company. Audits are usually conducted by certified public accountants who work for independent accounting firms such as Arthur Anderson. The purpose of performing an audit is to identify any problematic areas, such as embezzlement. An audit consists of six steps:

1. understanding the client's business
2. planning the audit
3. evaluating internal accounting controls
4. performing compliance tests
5. performing substantive tests
6. preparing the financial statement and issuing an opinion

3. The following conclusion is from the same student report as the introduction in exercise 2. Write a memo to your instructor evaluating the conclusion's effectiveness. Which aspects are effective? How would you improve it?

All major corporations and most minor corporations require an audit to be conducted every year. Audits are the primary tool used to ensure that financial irregularities such as embezzlement, conflict of interest, and inaccurate pricing are kept to a minimum.

The next nine exercises call for you to write an argument using a different organizational pattern. For each of these nine exercises, write a 500–1,000 word discussion on one of the topics specified or on a topic of your choice. For each exercise, attach a statement indicating the type of organizational pattern the discussion uses, your audience and purpose, and the type of document in which it might be included. For example, "This argument uses the chronological pattern and is addressed to first-semester college students. Its purpose is to help them understand the policies they are to follow in using the university's mainframe. The argument would be part of the *Student Handbook*."

4. Using chronological organization:
   a. how to register for courses at your college or university
   b. how to learn to use a software package
   c. how to buy the right car for your needs
   d. how to prepare for a job interview
   e. how to determine job prospects in your field

5. Using spatial organization:
   a. your bicycle
   b. your car's dashboard
   c. the room in which you are sitting
   d. the space shuttle
   e. the remote-control device from a television or stereo set

6. Using general-to-specific organization:
   a. advances in manufacturing technology
   b. alternatives to incarceration for nonviolent criminals
   c. pay scales for general practitioners and medical specialists
   d. cooperative education and internships for college students
   e. energy efficiency in computers and printers

7. Using more-important-to-less-important organization:
   a. the reasons you chose your college or major
   b. the effects of acid rain on a particular area
   c. the reasons you should (or should not) be required to study a foreign language
   d. the three most important changes you would like to see at your school
   e. the reasons that residential recycling should (or should not) be mandatory

8. Using comparison-and-contrast organization:
   a. lecture classes and discussion classes
   b. the tutorials that come with two different software packages
   c. manual transmission and automatic transmission automobiles
   d. black-and-white and color photography
   e. two different word-processing programs

9. Using problem-methods-solution organization:
   a. how you solved a recent problem related to your education
   b. how you went about deciding on a recent major purchase, such as a car, a personal computer, or a bicycle
   c. how you would propose reducing the time required to register for classes or to change your schedule
   d. how you would propose increasing the ties between your college or university and local business and industry

10. Using classification:
    a. foreign cars
    b. smoke alarms
    c. college courses
    d. personal computers
    e. cameras

11. Using partition:
    a. a student organization on your campus
    b. an audiocassette tape
    c. a portable radio
    d. a bicycle
    e. a guitar

12. Using causal reasoning — either forward or backward:
    a. women serving in combat in the military
    b. the price of gasoline
    c. the emphasis on achieving high grades in college
    d. the prospects for employment in your field
    e. computer literacy and your job

Research Project

13. Form small groups for this project about organizing information. Review the background section in the government regulation (U.S. Department of Labor, 1997) presented in the case that follows. Write a memo to your instructor describing and analyzing the organization of this passage. Which techniques discussed in this chapter do the authors of this excerpt use? How effectively do they use them? What would you do differently? To see the full document, go to www.osha-slc.gov/Publications /Osha3144.pdf.

C A S E
Introducing a Document

You are an intern at the U.S. Department of Labor. Your supervisor, Allen Young, has asked you to study the document on methylene chloride that is excerpted below. The full document, which can be found at www.osha-slc .gov/Publications/Osha3144.pdf, is intended for a managerial and general audience. "It's got a disclaimer statement at the front of the document, and a background statement," Allen tells you, "but I think it needs an introduction. Would you mind taking a look at it and drafting an introduction?" Write an introduction for this document.

For additional cases, click on Case of the Month and Archive on TechComm Web (www.bedfordstmartins.com /techcomm).

BACKGROUND

When established under the authority of the *Occupational Safety and Health Act of 1970,* the Occupational Safety and Health Administration (OSHA) had 2 years to adopt existing federal standards or national consensus standards[1] so it would have standards in place to enforce. OSHA chose to adopt existing federal standards issued under the *Walsh-Healey Public Contracts Act*, which were derived from threshold limit values of the American Conference of Governmental Industrial Hygienists and consensus standards from standards-developing organizations such as the American National Standards Institute (ANSI).

For methylene chloride, OSHA adopted an ANSI standard under Subpart Z of *Title 29 Code of Federal Regulations* (CFR), Part 1910.1000 to ensure that employee exposure did not exceed 500 parts per million parts of air (500 ppm) as an 8-hour time-weighted average (TWA) — i.e., the average exposure during an 8-hour period.

Since 1971, however, industrial experience, new developments in technology, and emerging scientific data clearly indicate that this limit did not adequately protect worker health. The agency realized the need to better control worker exposure to methylene chloride due to its harmful health effects.

Methylene chloride, also called dichloromethane, is a volatile, colorless liquid with a chloroformlike odor. Inhalation and skin exposure are the predominant means of exposure to methylene chloride. Inhaling the vapor causes mental confusion, lightheadedness, nausea, vomiting, and headache. With acute, or short-term exposure, methylene chloride acts as an anesthetic; continued exposure may cause staggering, unconsciousness, and even death. High concentrations of the vapors may cause irritation of the eyes and respiratory tract and aggravate the symptoms of angina. Skin contact with liquid methylene chloride causes irritation and burns. Splashing methylene chloride into the eyes causes irritation. Studies on laboratory animals indicate that long-term (chronic) exposure causes cancer.

Methylene chloride is used in various industrial processes in many different industries: paint stripping, pharmaceutical manufacturing, paint remover manufacturing, metal cleaning and degreasing, adhesives manufacturing and use, polyurethane foam production, film base manufacturing, polycarbonate resin production, and distribution and formulation of solvents.

The agency adopted the methylene chloride final rule on January 10, 1997, as published in the *Federal Register*. The rule became effective on April 10, 1997.

[1]Consensus standards are developed by private, standards-developing organizations and are discussed and substantially agreed upon through consensus by industry, labor, and other representatives.

PART THREE

DEVELOPING THE TEXTUAL ELEMENTS

Drafting and Revising Definitions and Descriptions

9

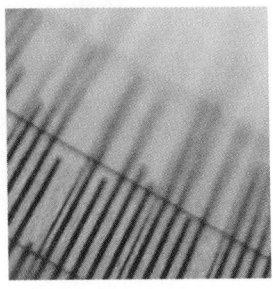

The technical-communication scholar Anne Eisenberg (1992, p. 159) on the role of definitions in technical communication:

Usually your reader knows less *about the subject than you. This is true whether you write for administrators outside your particular area, for marketing people, or for people following a set of instructions. Only rarely will you have peers for an audience. This means you'll need to explain. Definitions are crucial for these explanations, for they are the kernels out of which the reader's understanding will grow.*

See TechComm Web (www .bedfordstmartins.com /techcomm) for guidelines boxes, additional examples, and links related to topics in this chapter.

This chapter discusses two related techniques fundamental to technical communication: definitions and descriptions. Both use words and, in most cases, graphics to help readers understand concepts.

Definitions help readers understand what you mean by a word or phrase. For instance, in a discussion of how inflation is calculated by the government, you might provide a definition of *consumer price index*. Or, in a discussion of how geologists analyze soil strata, you might define *cone penetration test*.

Descriptions usually provide a fuller picture of an object, a mechanism, or a process, often going into detail about the components of the object or the stages of the process. You might, for example, write a description of a *dark object* in a discussion of discoveries in astronomy and explain theories about the dimensions and materials of a dark object. Or you might write a description of *PET scans* in a discussion for clinical nurses, explaining how the process was devised, how it works, and when it is used. As you have probably guessed, definitions are often included in descriptions. The description of how PET scans are done might begin with a definition of a PET scan.

This chapter will discuss definitions first, then descriptions.

UNDERSTANDING THE ROLE OF DEFINITIONS

The world of business and industry depends on clear definitions. Suppose you learn at a job interview that the employer pays tuition and expenses for employees' job-related education. That's good news, of course, if you are planning to continue your education. But until you study the employee-benefits manual, you won't know with any certainty just what the company will pay for.

Who, for instance, is an *employee?* Is it anyone who works for and is paid by the company? Maybe. But you might find that the company defines an *employee* as someone who has worked for the company full time (40 hours per week) for at least six uninterrupted months. What is *tuition?* Does the definition include incidental laboratory or student fees? What is *job-related education?* Does a course about time management qualify under the company's definition? What, in fact, constitutes *education?* All these terms, and many others, must be defined so that employees will understand their rights and responsibilities. Definitions are common in communicating policies and standards "for the record."

Definitions also have many uses outside legal or contractual contexts. Two such uses occur frequently:

- *Definitions clarify a description of a new development or a new technology in a technical field.* For instance, a zoologist who has discovered a new animal species names and defines it. A scientist who has devised a new laboratory procedure publishes an article in a technical journal defining and describing it.

- *Definitions help specialists communicate with less knowledgeable readers.* A manual explaining how to tune up a car includes definitions of parts and tools. A researcher at a manufacturing company uses definitions in describing a new product to the sales staff.

Definitions, then, are crucial in many kinds of technical communication, from brief letters and memos to technical reports, manuals, and journal articles. All readers, from the general reader to the expert, need effective definitions to carry out their jobs.

Writing effective definitions requires thought and planning. Before you attempt to do so, you should carry out three steps.

> **Steps in preparing to write a definition:**
>
> Analyze the writing situation.
>
> Determine the kind of definition needed.
>
> Decide where to place the definition.

ANALYZING THE WRITING SITUATION FOR DEFINITIONS

The first step in writing effective definitions is to analyze the writing situation: the audience and the purpose of your document.

Unless you know who the readers of your document will be and how much they know about the subject, you cannot determine which terms to define or what kind of definition to write. Physicists wouldn't need a definition of *entropy,* but lawyers might. Builders know what a Molly bolt is, but

See Ch. 5 for more on audience and purpose.

many insurance agents don't. If you are aware of your audience's background and knowledge, however, you can easily devise effective informal definitions. For example, if you are describing a DVD player to a group of readers who understand automobiles, you can use a familiar analogy to define the function of the pause button: "The PAUSE button is the clutch pedal of the DVD player."

Think, too, about your purpose. If you want to give your readers only a basic understanding of a concept — say, time-sharing vacation resorts — a brief, informal definition is usually sufficient. However, if you want your readers to understand an object, process, or concept thoroughly and be able to carry out tasks associated with it, a more formal and elaborate definition is required. For example, the definition of a "Class 2 Alert" for operators at a nuclear power plant must be comprehensive, specific, and precise.

DETERMINING THE KIND OF DEFINITION TO WRITE

Definitions can be short or long, informal or formal; it depends on your audience and your purpose. There are three basic types.

Types of definitions:
Parenthetical
Sentence
Extended

Writing Parenthetical Definitions

A *parenthetical definition* is a brief clarification placed unobtrusively in a sentence. Sometimes a parenthetical definition is simply a word or phrase, enclosed in parentheses or commas, or introduced by a colon or a dash. In the following examples, the term being defined is shown in italics, and the definition is underscored:

The crane is located on the *starboard* (right) side of the ship.

Summit Books announced its intention to create a new *colophon* (emblem or trademark).

United Engineering is seeking to purchase the *equity stock*, the common stock, of Minnesota Textiles.

A parenthetical definition can also take the form of a longer explanatory phrase or clause:

Motorboating is permitted in the *Jamesport Estuary*, the portion of the bay that meets the mouth of the Jamesport River.

Before the metal is plated, it is immersed in the *pickle*, <u>an acid bath that removes scales and oxides from the surface</u>.

Parenthetical definitions are not meant to be comprehensive; they serve mainly as quick and convenient ways of introducing terms. When addressing general readers especially, make sure that the definition itself is clear. You have gained nothing if your readers don't understand it:

Next, check for blight on the epicotyl, the stem portion above the cotyledons.

This parenthetical definition would be clear if your readers were botanists but not if they were general readers. The problem is that botanists don't need a definition of *epicotyl*, whereas general readers do.

Writing Sentence Definitions

A sentence definition — a one-sentence clarification — is more formal than a parenthetical definition. A sentence definition usually follows a standard pattern: the item to be defined is placed in a category of similar items and then distinguished from them.

Item	= Category	+ Distinguishing Characteristics
A flip flop	is a circuit	containing active elements that can assume either one of two stable states at any given time.
An electrophorous	is a laboratory instrument	used to generate static electricity.
Hypnoanalysis	is a psychoanalytical technique	in which hypnosis is used to elicit information from a patient's unconscious mind.
An electron microsocope	is a microscope	that uses electrons rather than visible light to produce magnified images.
A Bunsen burner	is a small laboratory heating device	consisting of a vertical metal tube connected to a gas source.

In many cases, a sentence definition also includes a graphic. For example, the definitions of electrophorus, Bunsen burner, and electron microscope would probably be accompanied by photographs, diagrams, or drawings of these items.

Sentence definitions are useful when your readers require a more formal or more informative clarification than brief parenthetical definitions can provide. Writers often use sentence definitions to present a working definition for a particular document: "In this report, the term *electron microscope* is used to refer to any microscope that uses electrons rather than visible light to produce magnified images." Such definitions are sometimes called *stipulative definitions* because the writer is stating how the term will be used in the document.

Writing Effective Sentence Definitions

The following four suggestions can help you write effective sentence definitions:

▶ *Be specific in stating the category and the distinguishing characteristics.* If you write "A Bunsen burner is a burner that consists of a vertical metal tube connected to a gas source," the imprecise category — "a burner" — defeats the purpose of your definition: many types of large-scale burners use vertical metal tubes connected to gas sources.

▶ *Don't describe a specific item if you are defining a general class of items.* If you wish to define *catamaran*, don't describe a particular catamaran. The catamaran you see on the beach in front of you might be made by Hobie and have a white hull and blue sails, but those characteristics are not essential to catamarans in general.

▶ *Avoid writing circular definitions, that is, definitions that merely repeat the key words or the distinguishing characteristics of the item being defined in the category.* The sentence "A required course is a course that is required" is useless: required of whom, by whom? However, in defining electron microscopes, you can repeat *microscope* because *microscope* is not the difficult part of the item. The purpose of defining *electron microscope* is to clarify the word *electron* as it applies to a particular type of microscope.

▶ *Be sure the category contains a noun or a noun phrase rather than a phrase beginning with* when, what, *or* where.

INCORRECT	A brazier is what is used to . . .
CORRECT	A brazier is a metal pan used to . . .
INCORRECT	An electron microscope is when a microscope . . .
CORRECT	An electron microscope is a microscope that . . .
INCORRECT	Hypnoanalysis is where hypnosis is used to . . .
CORRECT	Hypnoanalysis is a psychological technique in which . . .

Writing Extended Definitions

An *extended definition* is a long, detailed clarification — usually consisting of one or more paragraphs — of an object, a process, or an idea. Often an extended definition begins with a sentence definition, which is then elaborated.

For instance, the sentence definition "An electrophorus is a laboratory instrument used to generate static electricity" tells you the basic function of the device, but it doesn't explain how it works, what it is used for, and its strengths and limitations. An extended definition would address these and other topics.

There is no one way to "extend" a definition. Your analysis of the audience and purpose of your communication will help you decide which method to use. In fact, an extended definition sometimes employs several of the following nine techniques.

Techniques used in extended definitions:
Graphics
Examples
Partition
Principle of operation
Comparison and contrast
Analogy
Negation
Etymology
History of the term

Graphics

Perhaps the most common way to present an extended definition in technical communication is to use a graphic — such as a photograph, diagram, schematic, or flowchart — and then to explain the graphic. Graphics are, obviously, useful in defining physical objects, but they are also an effective way to present concepts and ideas. A definition of the meteorological concept of *temperature inversion*, for instance, might include a diagram showing the forces that create temperature inversion. The same approach could be used to define lasers, compound fractures, and the process of buying stocks on margin.

The following excerpt from an extended definition of *parallelogram* shows an effective combination of words and illustrations.

A *parallelogram* is a four-sided plane whose opposite sides are parallel with each other and equal in length. In a parallelogram, the opposite angles are the same, as shown in the sketch below.

Examples

Examples are particularly useful in making an abstract term easier to understand. The following paragraph is an extended definition of the psychological defense mechanism called *conversion:*

This extended definition is effective because the writer has chosen a clear and interesting (and therefore memorable) example of the idea he is describing.

> A third mechanism of psychological defense, "conversion," is found in hysteria. Here the conflict is converted into the symptom of a physical illness. In a case of conversion made famous by Freud, a young woman went out for a long walk with her brother-in-law, with whom she had fallen in love. Later, on learning that her sister lay gravely ill, she hurried to her bedside. She arrived too late and her sister was dead. The young woman's grief was accompanied by sharp pain in her legs. The pain kept recurring without any apparent physical cause. Freud's explanation was that she felt guilty because she desired the husband for herself, and unconsciously converted her repressed feelings into an imaginary physical ailment. The pain struck her in the legs because she unconsciously connected her feelings for the husband with the walk they had taken together. The ailment symbolically represented both the unconscious wish and a penance for the feelings of guilt which it engendered. (Wilson, 1964, p. 84).

Partition

See Ch. 8, p. 197, for more about partitioning.

Partitioning is the process of dividing a thing or an idea into smaller parts so that the reader can understand it more easily. The following example, from a study of technology in nursing (U.S. Congress, 1995a, p. 133), uses partition to define *clinical decision support system.*

> A knowledge-based system designed for clinical use, sometimes called a *clinical decision support system* (CDSS), usually involves three components:
>
> 1. *Data on the patient* being diagnosed or treated are either entered into the system manually, captured automatically from diagnostic or monitoring equipment, or drawn from an electronic patient record.
>
> 2. A *knowledge base* contains rules and decision algorithms that incorporate knowledge and judgment about the health problem at hand and alternative tests and treatments for it, mainly in the form of "if-then" statements, such as "if the patient's potassium is less than 3.0 mEq/dl and the patient is on digoxin, then warn the clinician to consider potassium supplementation."
>
> 3. An *inference engine* combines information from both the patient data and the knowledge base to perform specified tasks, outlined in appendix C.

Principle of Operation

Describing the principle of operation — the way something works — is an effective way to develop an extended definition, especially for an object or a process. The following excerpt from an extended definition of a parabolic dish solar-energy system (U.S. Congress, 1995b, pp. 173, 174) is based on the mechanism's principle of operation.

Parabolic dish systems use a large dish or set of mirrors on a single frame with two-axis tracking to reflect sunlight onto a receiver mounted at the focus. Most commonly, a free piston stirling engine is mounted on the receiver, but hot fluids can also be piped to a central turbine as in the parabolic trough and central receiver systems. Current research is focusing on lowering the cost of the mirror systems through the use of stretched membranes and to improve the reliability and performance of the stirling engine. Stirling engine lifetimes of 50,000 hours (about 10 years) with little or no maintenance are needed and are being developed. In comparison, the typical automobile engine must have minor maintenance every 250 hours or so, and a major overhaul perhaps every 2,500 hours. Parabolic dishes can achieve the highest temperatures (8008C or 1,5008F) and thus the highest efficiencies of concentrating solar thermal systems. Parabolic dish systems currently hold the efficiency record of 31 percent (gross) and 29 percent (net) for converting sunlight into electricity.

The first two sentences describe the principle of operation. This principle is illustrated in the diagram.

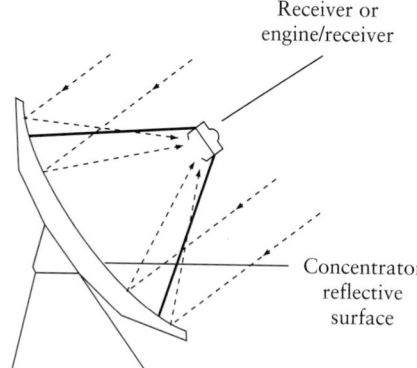

Receiver or
engine/receiver

Concentrator
reflective
surface

Comparison and Contrast

Using comparison and contrast, a writer discusses the similarities or differences between the item being defined and an item with which readers are more familiar. The following definition of *digital communication links* is based on a comparison to analog communication links.

The best way to understand digital communication links is to compare them to analog communication links. In a traditional analog environment such as a telephone system, a person speaks into the phone mouthpiece; the phone transforms the sounds into an analog signal, which is sent along wires to the earpiece in the recipient's phone, which then converts the signals back to sound. In an analog environment, only one signal can be carried on one wire at one time. By contrast, in a digital environment, the caller's signal is digitized — turned into 0's and 1's — almost immediately and *multiplexed* with other signals; that is, the different signals are carried as packets of information on a single set of wires, then separated at their destinations. In addition, in a digital environment, signals do not need to be transmitted along wires; they can be sent as pulses of light on fiberoptic cables or as

First the writer describes the more familiar item — the analog environment. Beginning with the more familiar item helps readers understand the less familiar item.

See Ch. 8, p. 193, for more about comparison and contrast.

microwaves between satellites or land stations. Compared to analog signals, digital signals are much less susceptible to noise and require much less amplification along the way.

Analogy

An *analogy* is a specialized kind of comparison. In a traditional comparison, the writer compares one item to another similar item: an electron microscope to a common microscope, for example, or a bit brace to a power drill. In an analogy, however, the item being defined is compared to an item that is in some ways completely different but shares some essential characteristic.

For instance, the central processing unit of a computer is often compared to a brain. Obviously, these two items are very different, except that the relationship of the central processing unit to the computer is similar to that of the brain to the body. People often compare computer software to phonograph records, cassette tapes, or compact discs. Again, the differences are many, but the similarity is essential: the compact disc contains the code that enables the stereo system to fulfill its function, just as the software contains the code that enables the computer to do its job.

The following example, from an extended definition of *computer literacy*, shows how an analogy can clarify an unfamiliar concept.

Computer literacy is the ability to use computers effectively. If you can operate a personal computer to do word processing or create a database, you are computer literate. If you can operate a digital watch or program a VCR or use an automated teller machine at a bank, you also can be said to be computer literate. To use an analogy, computer literacy is like automotive literacy: if you know how to operate an automobile safely to get from one place to another, you possess automotive literacy. Just as you don't have to understand the principle of the internal-combustion engine to drive a car, you don't have to understand the concepts of RAM and ROM to use a computer.

You do not need to use the term analogy, as this writer does, but there is nothing wrong with doing so.

Negation

A special kind of contrast is sometimes called *negation* or *negative statement*. Negation clarifies a term by distinguishing it from a different term with which the reader might confuse it. The following example uses negation to distinguish the term *ambulatory* from *ambulance*.

An ambulatory patient is not a patient who must be moved by ambulance. On the contrary, an ambulatory patient is one who can walk without assistance from another person.

Negation is rarely the only technique used in an extended definition; in fact, it is used most often in a sentence or two at the start. Once you have stated what the item is not, you still have to state what it is.

Etymology

Etymology, the derivation of a word, is often a useful and interesting way to develop a definition. The following example uses the etymology of *mortgage* to define it.

> The word *mortgage* was originally a compound of *mort* (dead) and *gage* (pledge). The meaning of the word has not changed substantially since its origin in Old French. A mortgage is still a pledge that is "dead" upon either the payment of the loan or the forfeiture of the collateral and payment from the proceeds of its sale.

Etymology is a popular way to begin definitions of *acronyms*, which are abbreviations pronounced as words, as illustrated in the following examples:

> SCUBA stands for self-contained underwater breathing apparatus.

> COBOL, which is the Common Business-Oriented Language, was originally invented to . . .

Etymology, like negation, is rarely used alone in technical communication, but it is an effective way to introduce an extended definition.

History of the Term

A common way to define a term is to explain its history. Often the extended definition explains the original use of the term and then describes how the meaning has changed in response to historical events or technological advances. The following example (Roblee & McKechnie, 1981, pp. 15–16) uses the history of the term *arson* to define it:

> In general, the common-law definition of *arson* was traditionally the willful burning of the house of another, including all outhouses or outbuildings adjoining thereto. The emphasis was on another's habitation, and his life and safety at the place where he or she resided. Then, many legal issues began to arise. Was a school a dwelling? a jail? a church? The common-law courts began to view the crime of arson as being against the habitation or possessions of another.

> Gradually, laws were enacted to plug the loopholes of the common-law definition of *arson*. The first laws brought all buildings or structures into the scope of arson, provided they had human occupancy of any kind on a regular basis. Later, the occupancy requirements were dropped. Today, *arson* is a term applied to the willful and intentional burning of all types of structures, vehicles, forests, fields, and so on.

A Sample Extended Definition

Figure 9.1 on page 230 (National Institutes of Health, 1993) is an example of an extended definition addressed to a general audience.

 To view Figure 9.1 in context on the Web, follow the links in Chapter 9 on TechComm Web (www.bedfordstmartins.com/techcomm).

DON'T LOSE SIGHT OF GLAUCOMA
Information for People at Risk

1. What is glaucoma?
Glaucoma is an eye disease in which the normal fluid pressure inside the eyes slowly rises, leading to vision loss — or even blindness. This brochure is about open-angle glaucoma, the most common form of the disease.

Question 1 is answered with a sentence definition.

2. What causes it?
At the front of the eye, there is a small space called the anterior chamber. Clear fluid flows in and out of the chamber to bathe and nourish nearby tissues. In glaucoma, for still unknown reasons, the fluid drains too slowly out of the eye. As the fluid builds up, the pressure inside the eye rises. Unless this pressure is controlled, it may cause damage to the optic nerve and other parts of the eye and cause loss of vision.

Question 2 is answered with the "principle of operation" strategy accompanied by a graphic.

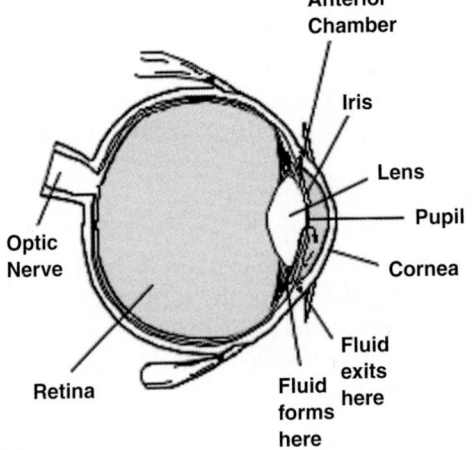

3. Who is most likely to get it?
Nearly 3 million people have glaucoma, a leading cause of blindness in the United States. Although anyone can get glaucoma, some people are at higher risk. They include:

The answer to Question 3 begins with a parenthetical definition.

• Blacks over the age of 40

• anyone over the age of 60

• people with a family history of glaucoma

■ **Figure 9.1   An Extended Definition**

The question-and-answer format seen here uses the journalistic questions (who, what, when, where, why, and how) to present the information of most interest to general readers.

Among Blacks, studies show that glaucoma is:

- five times more likely to occur in Blacks than in Whites

- about four times more likely to cause blindness in Blacks than in Whites

- fifteen times more likely to cause blindness in Blacks between the ages of 45–64 than in Whites of the same age group

4. What are the symptoms?
At first, there are no symptoms. Vision stays normal, and there is no pain. However, as the disease progresses, a person with glaucoma may notice his or her side vision gradually failing. That is, objects in front may still be seen clearly, but objects to the side may be missed. As the disease worsens, the field of vision narrows and blindness results.

5. How is it detected?
Many people may know of the "air puff" test or other tests used to measure eye pressure in an eye examination. But this test alone cannot detect glaucoma. Glaucoma is found most often during an eye examination through dilated pupils. This means drops are put into the eyes during the exam to enlarge the pupils. This allows the eye care professional to see more of the inside of the eye to check for signs of glaucoma.

"Dilated pupils" is followed by a one-sentence principle of operation.

6. How can it be treated?
Although open-angle glaucoma cannot be cured, it can usually be controlled. The most common treatments are . . .

7. What research is being done?
A large amount of research is being done in the U.S. to learn what causes glaucoma and to improve its diagnosis and treatment. For instance, . . .

8. What can you do to protect your vision?
Studies have shown that the early detection and treatment of glaucoma, before it causes major vision loss, is the best way to control the disease. . . .

See Ch. 3, p. 45, for more on the journalistic questions.

■ **Figure 9.1   (Continued)**

DECIDING WHERE TO PLACE THE DEFINITION

In many cases, you might not need to consider where to place a definition, because most of your readers will not be familiar with a term you want to use and you can easily include a parenthetical definition that will satisfy your readers' needs.

In assessing the writing situation before beginning your draft, however, you might conclude that you will need to introduce one or more terms and that these terms will require more detailed and comprehensive clarifications, perhaps sentence definitions and extended definitions. In these cases, you should plan, at least tentatively, where you are going to place the definitions.

Definitions are typically placed in one or more of these six locations:

- *In the text.* The text is an appropriate location for parenthetical and sentence definitions that many or most of your readers will need, and for extended definitions of terms that are central to the discussion. For instance, a discussion of recent changes in workers' compensation insurance would probably begin with an extended definition of that kind of insurance.

- *In a marginal gloss.* Parenthetical and sentence definitions can be placed in the margin of the document.

- *In a hyperlink.* In a hypertext document such as a Web page, you can put definitions in a separate file, which allows the reader to click on highlighted or underlined words to view their definitions.

- *In footnotes.* Footnotes are a logical place for an occasional sentence definition or extended definition. The reader who doesn't need it will ignore it. However, footnotes can slow readers down by interrupting the flow of the discussion. If you think you will need more than one footnote for a definition on every two to three pages, consider including a glossary.

- *In a glossary.* A glossary — an alphabetized list of definitions — can accommodate sentence definitions and extended definitions of fewer than three or four paragraphs in one convenient location. A glossary can be placed at the beginning of a document (for example, after the executive summary in a report) or at the end, preceding the appendices.

See Ch. 12, pp. 320 and 321, for more about glossaries and appendices.

- *In an appendix.* An appendix is an appropriate place for an extended definition of one page or longer, which would be cumbersome in a glossary or in a footnote and, unless it explains a crucial term, too distracting in the text.

UNDERSTANDING THE ROLE OF DESCRIPTIONS

 See Online Technical Writing (www.io.com/~hcexres /tcm1603/acchtml/acctoc .html) for sample descriptions.

Whereas technical communication requires a shared understanding of terms, it is also filled with descriptions — verbal and visual representations of objects, mechanisms, and processes.

- *Objects.* An object is anything ranging from a physical site, such as a volcano or some other kind of natural phenomena, to a synthetic artifact, such as a hammer. A tomato plant is an object, as is an automobile tire or a book.

- *Mechanisms.* A mechanism is a synthetic object consisting of a number of identifiable parts that work together as a system. A compact-disc player is a mechanism, as is a voltmeter, a lawnmower, a submarine, or a steel mill.

- *Processes.* A process is an activity that takes place over time: the earth was formed; steel is made; animals evolve; plants perform photosynthesis. *Descriptions of processes,* which explain how something happens, differ from *instructions,* which tell how to do something. Readers of a process description want to *understand* the process; readers of instructions want a

See Ch. 20, p. 588, for more about instructions.

step-by-step guide to help them *perform* the process. A process description answers the question "How is wine made?" A set of instructions answers the question "How do I make wine?"

Descriptions of objects, mechanisms, and processes appear in virtually every kind of technical communication. Here are a few examples:

A company studying the feasibility of renovating an old factory as opposed to building a new one produces a report that includes a description of the old factory. The old factory is a complex mechanism. The company's managers can make a rational decision only if they understand it thoroughly.

An employee who wants to persuade management to authorize the purchase of some equipment includes a mechanism description in the proposal to buy the equipment.

An engineer trying to explain to the sales staff how a product works, so they can advertise it effectively, includes a mechanism description in the form of product specifications.

A company manufacturing a consumer product provides a description and a graphic on its Web site to attract buyers.

A developer who wants to build a housing project includes in his environmental impact statement descriptions of the geographical area and of the process he will use in developing that area.

Notice that a description is usually part of a larger document. A maintenance manual for a boiler system, for example, might begin with a mechanism description of the system to help the reader understand how it operates.

ANALYZING THE WRITING SITUATION FOR DESCRIPTIONS

Before you begin to write a description, consider carefully how the audience and the purpose of the document will affect what you write.

What does the audience already know about the general subject? If, for example, you are going to describe an electron microscope, you first have to know whether your readers understand what a microscope is. If you want to describe how the next generation of industrial robots will affect car manufacturing, you first have to know whether your readers understand the current process and whether they understand robotics.

See Ch. 5 for more on audience and purpose.

Your sense of your audience will determine not only whether you use technical vocabulary but also your sentence and paragraph structure and length. Another audience-related factor is your use of graphics. Less knowledgeable readers need simple graphics; they might have trouble understanding sophisticated schematics or decision charts.

Consider, too, your purpose. What are you trying to accomplish with this description? If you want your readers to understand how a personal computer works, you will write a *general description* that applies to several varieties of computers. If you want your readers to understand how a specific computer works, you will write a *particular description*. A general description of personal computers might classify them by size, then go on to describe palmtops, laptops, and desktops in general terms. A particular description, however, will describe only one, such as the Millennia 200. Your purpose will determine every aspect of the description, including its length, the amount of detail, and the number and type of graphics.

The following introduction to a description of job-related stress was written by a sociologist studying law enforcement. The writer builds upon her sense of her audience (managers in law-enforcement) and of her purpose (to explain the need to combat stress actively) in developing her description.

> All police officers are subjected to stress. For a police officer to perform his or her duties effectively — and for the police unit to perform effectively — stress must be recognized and managed.
>
> No officer should be forced to recognize and deal with the symptoms of stress in isolation. It is the responsibility of all police officers, but especially those in managerial positions, to recognize and identify the signs of stress and take an active role in reducing the stress. Otherwise, the stress can become debilitating for both the officer and the unit.
>
> The following description of stress . . .

WRITING THE DESCRIPTION

There is no single structure or format to use in writing descriptions. Because they are written for different audiences and different purposes, they can take many shapes and forms. There are, however, four principles for writing descriptions that apply to most situations.

> **Principles for writing descriptions:**
>
> Clearly indicate the nature and scope of the description.
>
> Introduce the description clearly.
>
> Provide appropriate detail.
>
> Conclude the description.

Clearly Indicate the Nature and Scope of the Description

An obvious point: if the description is to be a separate document, give it a title. If the description is to be part of a longer document, give it a section heading. In either case, clearly state the subject and indicate whether the description is general or particular. For instance, a general description of an object might be entitled "Description of a Minivan," and a particular descrip-

See Chs. 10 and 12, pp. 251 and 309, for more about titles. See Ch. 10, p. 252, for more about headings.

tion, "Description of the 2001 Honda Odyssey." A general description of a process might be called "Description of the Process of Designing a New Production Car," and a particular description, "Description of the Process of Designing the General Motors Saturn."

Introduce the Description Clearly

Give readers the information they need to understand the detailed information to follow. Introductions to descriptions are usually general: you want to give your readers a broad understanding of the object, mechanism, or process. You might also provide a graphic that introduces readers to the overall concept. For example, in describing a process, you might include a flowchart summarizing the steps in the body of the description; in describing an object, such as a bicycle, you might include a photograph or a drawing showing the major components you will detail in the body.

Table 9.1 shows some of the basic kinds of questions you might want to answer in introducing object, mechanism, and process descriptions. If the answer is obvious, simply move on to the next question.

■ **Table 9.1   Questions to Answer in Introducing a Description**

For Mechanism and Object Descriptions	*For Process Descriptions*
• *What is the item?* You might start with a sentence definition.	• *What is the process?* You might start with a sentence definition.
• *What is the function of the item?* If the function is not implicit in the sentence definition, state it clearly: "Electron microscopes magnify objects that are smaller than the wavelengths of visible light."	• *What is the function of the process?* Unless the function is obvious, state it: "The central purpose of performing a census is to obtain up-to-date population figures, which legislators and government agencies use to revise legislative districts and determine revenue-sharing."
• *What does the item look like?* Include a photograph or drawing if possible (see Chapter 14 for more information about incorporating graphics with text). If not, use an analogy or compare it to a familiar item: "The cassette that encloses the tape is a plastic shell, about the size of a deck of cards." Mention the material, texture, color, and the like, if relevant. Sometimes an object is best pictured with both graphics and words.	• *Where and when does the process take place?* "Each year the stream is stocked with hatchery fish in the first week of March." Omit these facts only if you are certain your readers already know them.
• *How does the item work?* In a few sentences, define the operating principle of the item. Sometimes objects do not "work"; they merely exist. For instance, a ship model has no operating principle.	• *Who or what performs the process?* If there is any doubt about who or what performs the process, make this information explicit; confusion at this early stage can ruin the document.
• *What are the principal parts of the item?* Limit your description to the item's principal parts. A description of a bicycle, for instance, would not mention the dozens of nuts and bolts that hold the mechanism together; it would focus on the chain, gears, pedals, wheels, and frame.	• *How does the process work?* "The four-treatment lawn-spray plan is based on the theory that the most effective way to promote a healthy lawn is to apply different treatments at crucial times during the growing season. The first two treatments — in spring and early summer — consist of . . ."
	• *What are the principal steps of the process?* Name the steps in the order in which you will later describe them. The principal steps in changing an automobile tire, for instance, include jacking up the car, replacing the old tire with the new one, and lowering the car back to the ground. Changing a tire also includes secondary steps, such as placing chocks against the tires to prevent the car from moving once it is jacked up. Explain or refer to these secondary steps at the appropriate points in the description.

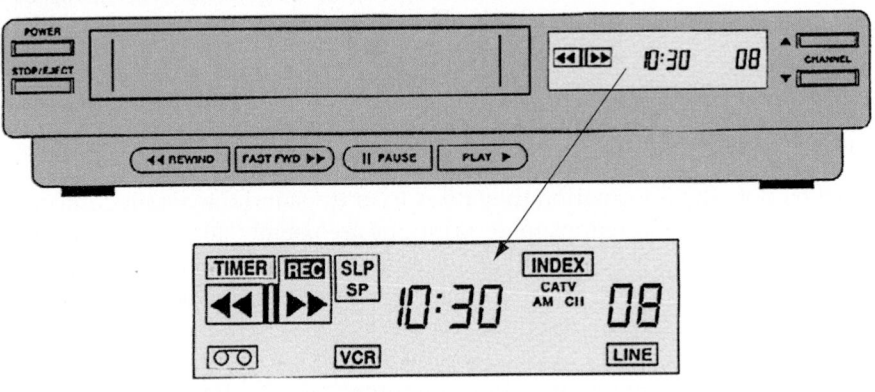

■ **Figure 9.2   Introductory Graphic and Detailed Graphic**

Figure 9.2 shows the introductory graphic accompanying a description of a VCR and the graphic for one of the parts discussed in the detailed description.

Provide Appropriate Detail

The body of a description — the part-by-part or step-by-step section — treats each major part or step as a separate item. In describing an object or a mechanism, you define each part and then, if applicable, describe its function, operating principle, and appearance. Your discussion of the appearance should include shape, dimensions, material, and physical details such as texture and color (if essential). In some descriptions, other qualities, such as weight or hardness, might also be appropriate. If a part has important subparts, describe them in the same way.

In describing a process, treat each major step as if it were a separate process. Do not repeat your answer to the question about who or what performs the action unless a new agent performs a particular step, but do answer the other principal questions — what the step is, what its function is, and when, where, and how it occurs. If a step has important substeps, explain them clearly.

A description closely resembles a map with a series of detailed insets. A description of a computer system would include a keyboard as one of its parts. The description of the keyboard, in turn, would include the numeric keypad as one of its parts. And the description of the numeric keypad would include the arrow keys as one of its parts. This ever-increasing specificity continues as required by the complexity of the item and the needs of the readers. The same principle applies in describing processes: a step might have substeps. For each substep, you would probably need to describe who or what performs the substep (if it is not obvious), what the substep is, what its function is, and when, where, and how it occurs.

Providing Appropriate Detail in Descriptions

Use the following techniques to flesh out your descriptions.

For Mechanism and Object Descriptions	For Process Descriptions

Choose an appropriate organizing principle. Two organizational principles are common:

- Functional: the way the item works or is used. In a radio, the sound begins at the receiver, travels into the amplifier, and then flows out through the speakers.
- Spatial: based on the physical structure of the item: from top to bottom, east to west, outside to inside, and so forth.

Most descriptions can be organized in various ways. For instance, the description of a house could be organized functionally (the different electrical and mechanical systems) or spatially (top to bottom, inside to outside, east to west, and so on). A complex description can use a combination of patterns at different levels in the description.

- *Use graphics.* Try to supply a graphic for each major part. Use photographs to show external surfaces, drawings to emphasize particular items on the surface, and cutaways and exploded diagrams to show details beneath the surface. Other kinds of graphics, such as graphs and charts, are often useful supplements (see Chapter 14).

- *Structure the step-by-step description chronologically.* If the process is a closed system, such as the cycle of evaporation and condensation, and thus has no first step, explain to your readers that the process is cyclical and simply begin with any principal step.

- *Explain causal relationships among steps.* Don't present the steps as if they have nothing to do with one another. In many cases, one step causes another. In the operation of a four-stroke gasoline engine, for instance, each step sets up the conditions under which the next step can occur.

- *Use the present tense.* Discuss steps in the present tense unless you are writing about a process that occurred in the historical past. For example, a description of how the Snake River aquifer was formed would be written in the past tense: "The molten material condensed. . . ." The past tense helps readers understand that the process was a one-time occurrence that has since concluded. However, a description of how steel is made would be written in the present tense: "The molten material is then poured into. . . ." Usually the present tense helps readers to understand that steel is always made in this fashion.

- *Use graphics.* Whenever possible, use graphics to clarify each point. Additional flowcharts are useful, but other kinds of graphics, such as photographs, drawings, and graphs, are also helpful.

↓

For Mechanism and Object Descriptions	For Process Descriptions
	For example, in a description of how a four-stroke gasoline engine operates, you could use diagrams to illustrate the position of the valves and the activity occurring during each step.

Conclude the Description

Descriptions generally do not require an elaborate conclusion. A brief conclusion is necessary, however, if only to summarize the description and prevent readers from overemphasizing the part or step discussed last.

A common technique for concluding descriptions of mechanisms and of some objects is to state briefly how the parts function together. At the end of a description of a telephone, for example, the conclusion might include the following paragraph:

> When you make a phone call, everything that happens depends on the flow of current through the phone lines, and by what your phone and the other person's phone do with that current. When the phone is taken off the hook, a current flows through the carbon granules. The intensity of the speaker's voice causes a greater or lesser movement of the phone's diaphragm and thus a greater or lesser intensity in the current flowing through the carbon granules. The phone receiving the call converts the electrical waves back into sound waves by means of an electromagnet and a diaphragm. The varying intensity of the current transmitted by the phone line alters the strength of the current in the electromagnet, which in turn changes the position of the diaphragm. The movement of the diaphragm reproduces the speaker's sound waves.

Like object and mechanism descriptions, process descriptions do not usually require long conclusions. If the description itself is brief — less than a few pages — a short paragraph summarizing the principal steps is all you need. Here, for example, is the concluding section of a description of how a four-stroke gasoline engine operates:

> In the intake stroke, the piston moves down, drawing the air-fuel mixture into the cylinder from the carburetor. As the piston moves up, it compresses this mixture in the compression stroke, creating the conditions necessary for combustion. In the power stroke, a spark from the spark plug ignites the mixture, which burns rapidly, forcing the piston down. In the exhaust stroke, the piston moves up, expelling the burned gases.

For longer descriptions, a discussion of the implications of the process might be appropriate. For instance, a description of the Big Bang, one theory of how the universe began, might conclude with a discussion of how the theory has been supported and challenged by recent astronomical discoveries and theories.

In technical communication, a description often concludes with a sales message that summarizes the benefits of the product or process and motivates the audience to take further action, such as phoning or writing to request more information or to set up a visit from a sales representative.

ANALYZING SAMPLE DESCRIPTIONS

A look at sample descriptions will give you an idea of how different writers adapt basic approaches for a particular audience and purpose.

Figure 9.3 (U.S. Congress, 1995b) shows the extent to which a process description can be based on graphics. The topic is drivetrain efficiencies for vehicles powered by internal-combustion engines and for vehicles powered by electricity. The audience is the general reader.

Figure 9.4 (U.S. Environmental Protection Agency, 1991, pp. 45–47) is an excerpt from a discussion of the process of diagnosing indoor air-quality (IAQ) problems. The audience consists of building owners and facility managers. Following this excerpt, the authors of this report provide detailed instructions for each of the steps, starting with the initial walkthrough.

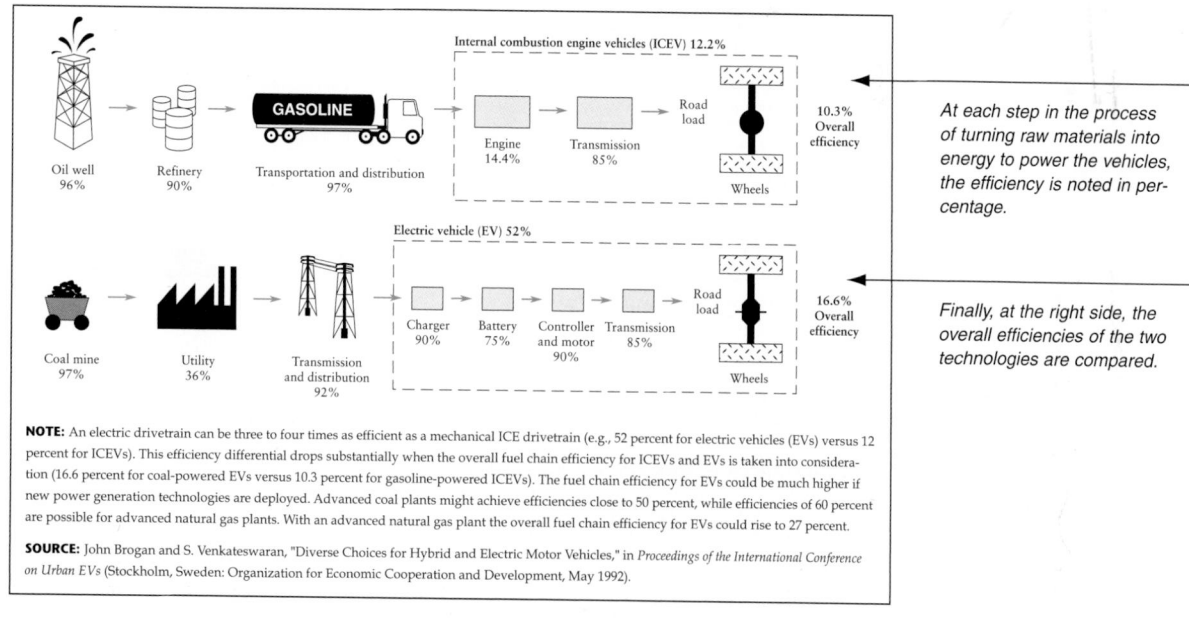

At each step in the process of turning raw materials into energy to power the vehicles, the efficiency is noted in percentage.

Finally, at the right side, the overall efficiencies of the two technologies are compared.

NOTE: An electric drivetrain can be three to four times as efficient as a mechanical ICE drivetrain (e.g., 52 percent for electric vehicles (EVs) versus 12 percent for ICEVs). This efficiency differential drops substantially when the overall fuel chain efficiency for ICEVs and EVs is taken into consideration (16.6 percent for coal-powered EVs versus 10.3 percent for gasoline-powered ICEVs). The fuel chain efficiency for EVs could be much higher if new power generation technologies are deployed. Advanced coal plants might achieve efficiencies close to 50 percent, while efficiencies of 60 percent are possible for advanced natural gas plants. With an advanced natural gas plant the overall fuel chain efficiency for EVs could rise to 27 percent.

SOURCE: John Brogan and S. Venkateswaran, "Diverse Choices for Hybrid and Electric Motor Vehicles," in *Proceedings of the International Conference on Urban EVs* (Stockholm, Sweden: Organization for Economic Cooperation and Development, May 1992).

■ **Figure 9.3   A Process Description Based on a Graphic**

Notice how effectively graphics show the relative efficiencies of an internal combustion engine vehicle (top row) and an electric vehicle (bottom row). The graphics clarify the process and make it interesting.

This introduction begins by explaining the purpose of performing a diagnostic building investigation. The introduction explains the strategy for conducting the investigation and states who should perform it.

Notice the use of the flowchart to summarize the process.

Diagnosing IAQ Problems

6

The goal of the diagnostic building investigation is to identify and solve the indoor air quality complaint in a way that prevents it from recurring and that does not create other problems. This section describes a method for discovering the cause of the complaint and presents a "toolbox" of diagnostic activities to assist you in collecting information.

Just as a carpenter uses only the tools that are needed for any given job, an IAQ investigator should use only the investigative techniques that are needed. Many indoor air quality complaints can be resolved without using all of the diagnostic tools described in this chapter. For example, it may be easy to identify the source of cooking odors that are annoying nearby office workers and solve the problem by controlling pressure relationships (e.g., installing exhaust fans) in the food preparation area. Similarly, most mechanical or carpentry problems probably require only a few of the many tools you have available and are easily accomplished with in-house expertise.

The use of in-house personnel builds skills that will be helpful in minimizing and resolving future problems. On the other hand, some jobs may be best handled by contractors who have specialized knowledge and experience. In the same way, diagnosing some indoor air quality problems may require equipment and skills that are complex and unfamiliar. Your knowledge of your organization and building operations will help in selecting the right tools and deciding whether in-house personnel or outside professionals should be used in responding to the specific IAQ problem.

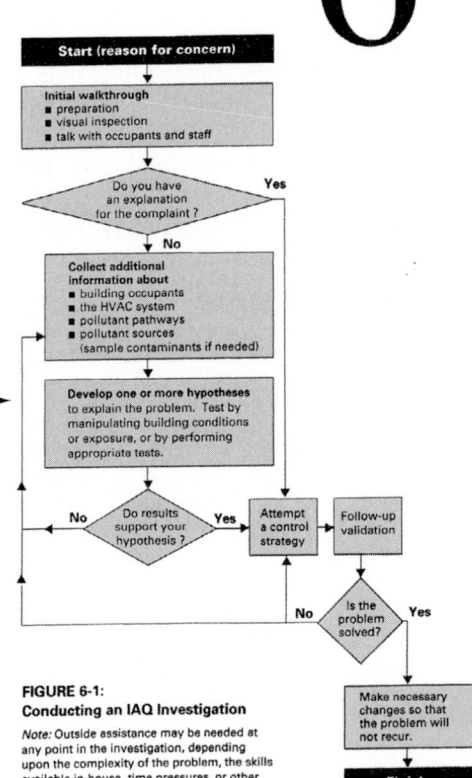

FIGURE 6-1:
Conducting an IAQ Investigation

Note: Outside assistance may be needed at any point in the investigation, depending upon the complexity of the problem, the skills available in-house, time pressures, or other factors.

▦ **Figure 9.4**
A Process Description

The IAQ investigation is often a repetitive cycle of information-gathering, hypothesis formation, and hypothesis testing.

OVERVIEW: CONDUCTING AN IAQ INVESTIGATION

An IAQ investigation begins with one or more reasons for concern, such as occupant complaints. Some complaints can be resolved very simply (e.g., by asking a few common sense questions of occupants and facility staff during the walkthrough). At the other extreme, some problems could require detailed testing by an experienced IAQ professional. In this section "the investigator" refers to in-house staff responsible for conducting the IAQ investigation.

The flowchart on page 45 shows that the IAQ investigation is a cycle of information-gathering, hypothesis formation, and hypothesis testing. The goal of the investigation is to understand the IAQ problem well enough so that you can solve it. Many IAQ problems have more than one cause and may respond to (or require) several corrective actions.

Initial Walkthrough

An initial walkthrough of the problem area provides information about all four of the basic factors influencing indoor air quality (occupants, HVAC system, pollutant pathways, and contaminant sources). The initial walkthrough may provide enough information to resolve the problem. At the least, it will direct further investigation. For example, if the complaint concerns an odor from an easily identified source (e.g., cooking odors from a kitchen), you may want to study pollutant pathways as a next step, rather than interviewing occupants about their patterns of discomfort.

Developing and Testing Hypotheses

As you develop an understanding of how the building functions, where pollutant sources are located, and how pollutants move within the building, you may think of many "hypotheses," potential explana-

tions of the IAQ complaint. Building occupants and operating staff are often a good source of ideas about the causes of the problem. For example, they can describe changes in the building that may have occurred shortly before the IAQ problem was noticed (e.g., relocated partitions, new furniture or equipment).

Hypothesis development is a process of identifying and narrowing down possibilities by comparing them with your observations. **Whenever a hypothesis suggests itself, it is reasonable to pause and consider it.** Is the hypothesis consistent with the facts collected so far?

You may be able to test your hypothesis by modifying the HVAC system or attempting to control the potential source or pollutant pathway to see whether you can relieve the symptoms or other conditions in the building. If your hypothesis successfully predicts the results of your manipulations, then you may be ready to take corrective action. Sometimes it is difficult or impossible to manipulate the factors you think are causing the IAQ problem; in that case, you may be able to test the hypothesis by trying to predict how building conditions will change over time (e.g., in response to extreme outdoor temperatures).

Collecting Additional Information

If your hypothesis does not seem to be a good predictor of what is happening in the building, you probably need to collect more information about the occupants, HVAC system, pollutant pathways, or contaminant sources. Under some circumstances, detailed or sophisticated measurements of pollutant concentrations or ventilation quantities may be required. Outside assistance may be needed if repeated efforts fail to produce a successful hypothesis or if the information required calls for instruments and procedures that are not available in-house.

The detailed section of the process description begins with this overview section that provides general information about the process and refers to the flowchart on the previous page.

The overview contains a number of subsections, each of which has its own heading. Note that although the writer uses the second-person pronoun — "you" — the document is not a set of instructions but a process description. A set of instructions would include detailed information about tools and materials as well as step-by-step instructions.

Figure 9.4
(Continued)

Results of the Investigation

Analysis of the information collected during your IAQ investigation could produce any of the following results:

The apparent cause(s) of the complaint(s) is (are) identified.

Remedial action and follow-up evaluation will confirm whether the hypothesis is correct.

Other IAQ problems are identified that are not related to the original complaints.

These problems (e.g., HVAC malfunctions, strong pollutant sources) should be corrected when appropriate.

A better understanding of potential IAQ problems is needed in order to develop a plan for corrective action.

It may be necessary to collect more detailed information and/or to expand the scope of the investigation to include building areas that were previously overlooked. Outside assistance may be needed.

The cause of the original complaint cannot be identified.

A thorough investigation has found no deficiencies in HVAC design or operation or in the control of pollutant sources, and there have been no further complaints. In the absence of new complaints, the original complaint may have been due to a single, unrepeated event or to causes not directly related to IAQ.

Using Outside Assistance

Some indoor air quality problems may be difficult or impossible for in-house investigators to resolve. Special skills or instruments may be needed. Other factors can also be important, such as the benefit of having an impartial outside opinion or the need to reduce potential liability from a serious IAQ problem. You are best able to make the judgment of when to bring in an outside consultant. See *Section 8* for a discussion of hiring professional assistance to solve an IAQ problem.

■ **Figure 9.4**
(Continued)

Figures 9.5 and 9.6 are excerpts from a description of a piece of atmospheric-testing hardware used by the Naval Research Laboratory in its Defense Meteorological Satellite Program (Praxis, 1997). Figure 9.5, taken from the Web site, is the main page for the description.

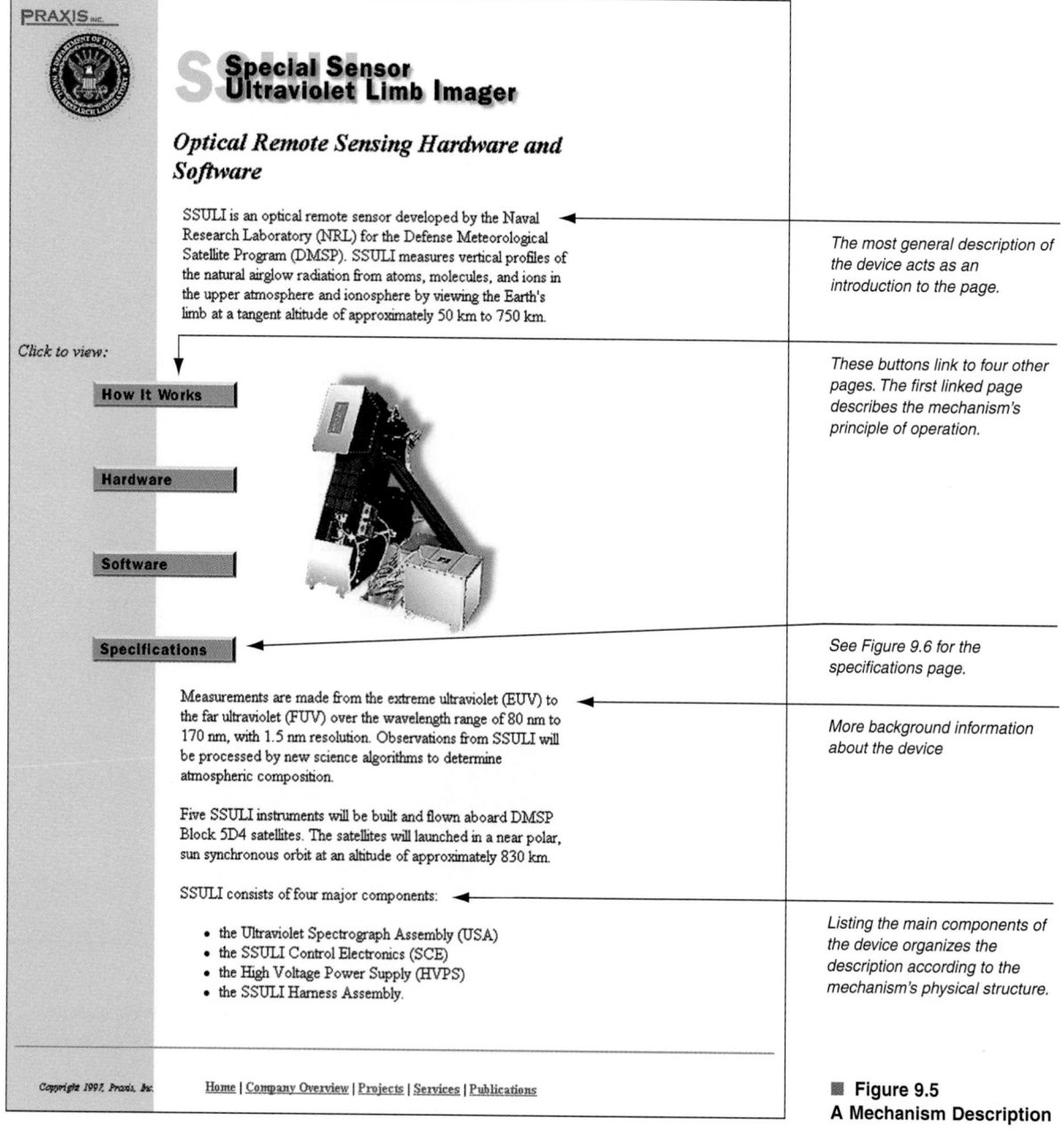

PRAXIS INC.

Special Sensor Ultraviolet Limb Imager

Optical Remote Sensing Hardware and Software

SSULI is an optical remote sensor developed by the Naval Research Laboratory (NRL) for the Defense Meteorological Satellite Program (DMSP). SSULI measures vertical profiles of the natural airglow radiation from atoms, molecules, and ions in the upper atmosphere and ionosphere by viewing the Earth's limb at a tangent altitude of approximately 50 km to 750 km.

The most general description of the device acts as an introduction to the page.

Click to view:

How It Works

These buttons link to four other pages. The first linked page describes the mechanism's principle of operation.

Hardware

Software

Specifications

See Figure 9.6 for the specifications page.

Measurements are made from the extreme ultraviolet (EUV) to the far ultraviolet (FUV) over the wavelength range of 80 nm to 170 nm, with 1.5 nm resolution. Observations from SSULI will be processed by new science algorithms to determine atmospheric composition.

More background information about the device

Five SSULI instruments will be built and flown aboard DMSP Block 5D4 satellites. The satellites will launched in a near polar, sun synchronous orbit at an altitude of approximately 830 km.

SSULI consists of four major components:

- the Ultraviolet Spectrograph Assembly (USA)
- the SSULI Control Electronics (SCE)
- the High Voltage Power Supply (HVPS)
- the SSULI Harness Assembly.

Listing the main components of the device organizes the description according to the mechanism's physical structure.

Copyright 1997, Praxis, Inc.   Home | Company Overview | Projects | Services | Publications

■ **Figure 9.5**
A Mechanism Description

Special Sensor Ultraviolet Limb Imager

SSULI Specifications

SSULI Specifications

Field of View:	0.1 degrees vertical and 2.4 degrees horizontal
Field of Regard:	30 degrees by 2.4 degrees
Scanning Range:	10 to 40 degrees below the host spacecraft y-direction
Sensitivity:	0.5 count per second per Rayleigh at 83.4 nm
Spectral Range:	80 to 170 nm with a resolution of 1.5 nm or less
Scanning Rate:	Up to 6 degrees per second

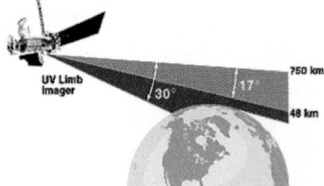

Geometry of SSULI Limb Scans

SSULI Observables

Parameter	Species	Wavelength
Dayside Ionosphere	O+	834 Å
Dayside Ionosphere	O	911 Å, 1304 Å, 1356 Å
Dayside Neutral Density	O	989 Å, 1304 Å, 1356 Å, 1641 Å
Dayside Neutral Density	N₂	1085 Å (N+),N2 LBH Bands
Dayside Neutral Density	O₂	Schumann-Runge, Absorption of N2 LBH
Nightside Neutral Density	O	1356 Å
Temperatures	---	Emission Scale Heights

Return to Introduction | Hardware | Software | How It Works

■ **Figure 9.6 Specifications**

An important kind of description is called a specification. *A typical specification consists of a graphic and a set of statistics about the device and its performance characteristics. You will see specifications on devices as small as transistors and as large as aircraft carriers.*

Revision Checklist

This checklist covers parenthetical, sentence, and extended definitions:

1. Are all necessary terms defined?
2. Are the parenthetical definitions
    ❏ appropriate for the audience?
    ❏ clear?
    ❏ smoothly integrated into the sentences?
3. Does each sentence definition
    ❏ contain a sufficiently specific category and distinguishing characteristics?
    ❏ avoid describing one particular item when a general class of items is intended?
    ❏ avoid circular definition?
    ❏ contain a noun or a noun phrase in the category?
4. Are the extended definitions developed logically and clearly?
5. Are the definitions placed in the location most useful to readers?

The following questions cover **descriptions of objects and mechanisms:**

1. Did you clearly indicate the nature and scope of the description?
2. In introducing the description, did you answer, if appropriate, the following questions:
    ❏ What is the item?
    ❏ What is its function?
    ❏ What does it look like?
    ❏ What is its principle of operation?
    ❏ What are its principal parts?
3. Did you include a graphic identifying all the principal parts?
4. In providing detailed information, did you
    ❏ answer, for each of the major components, the questions in item 2 above?
    ❏ choose an appropriate organizing principle?
    ❏ include graphics for each of the components?
5. In concluding the description, did you
    ❏ summarize the major points in the part-by-part description?
    ❏ include (where appropriate) a description of the item performing its function or an attempt to motivate the reader to take action?

The following questions cover **process descriptions:**

1. Did you clearly indicate the nature and scope of the description?
2. In introducing the description, did you answer, if appropriate, the following questions:
    ❏ What is the process?
    ❏ What is its function?
    ❏ Where and when does the process take place?
    ❏ Who or what performs it?
    ❏ How does the process work?
    ❏ What are its principal steps?

3. Did you include a graphic identifying all the principal steps?

4. In providing detailed information, did you
   - ❏ answer, for each of the major steps, the questions in item 2?
   - ❏ discuss the steps in chronological order or other logical sequence?
   - ❏ make clear the causal relationships among the steps?
   - ❏ include graphics for each of the principal steps?

5. In concluding the description, did you
   - ❏ summarize the major points in the step-by-step description?
   - ❏ discuss, if appropriate, the importance or implications of the process?
   - ❏ attempt (if appropriate) to motivate the reader to take action?

Exercises

1. In the following sentences, add a parenthetical definition for each italicized term:
   a. Reluctantly, he decided to *drop* the physics course.
   b. Last week the computer was *down*.
   c. The department is using *shareware* in its drafting course.
   d. The tire plant's managers hope they do not have to *lay off* any more employees.
   e. Please submit your assignments *electronically*.

2. Write a sentence definition for each of the following terms:
   a. catalyst
   b. DVD player
   c. job interview
   d. Web site
   e. automated teller machine
   f. fax machine
   g. intranet

3. Revise any of the following sentence definitions that need revision:
   a. Dropping a course is when you leave the class.
   b. A thermometer measures temperature.
   c. The spark plugs are the things that ignite the air-gas mixture in a cylinder.
   d. Double-parking is where you park next to another car.
   e. A strike is when the employees stop working.
   f. Multitasking is when you do two things at once while you're on the computer.

4. Identify the techniques used in writing the following extended definition:

   Holography, from the Greek *holos* (entire) and *gram* (message), is a method of photography that produces images that appear to be three-dimensional. A holographic image seems to change as the viewer moves in relation to it. For example, as the viewer moves, one object on the image appears to move in front of another object. In addition, the distances between objects in the image seem to change.

   Holographs are produced by coherent light, that is, light of the same wavelength, with the waves in phase and of the same amplitude. This light is produced by laser. Stereoscopic images are created by incoherent light — random wavelengths and amplitudes, out of phase. The incoherent light, which is natural light, is focused by a lens and records the pattern of brightness and color differences of the object being imaged.

   How are holographic images created? The laser-produced light is divided as it passes through a beam splitter. One portion of the light, called the *reference beam,* is directed to the emulsion — the "film." The other portion, the object beam, is directed to the subject and then reflected back to the emulsion. The reference beam is coherent light, whereas the object beam becomes incoherent because it is reflected off the irregular surface of the subject. The resulting dissonance between the reference beam and the object beam is encoded; it records not only the brightness of the different parts of the subject but also the different distances from the laser. This encoding creates the three-dimensional effect of holography.

5. Evaluate the effectiveness of the following description of a digital camera (based on Farkas, 2000). Which aspects of the description are successful? Which parts of the description would you change? Why?

Digital Cameras

A digital camera is a camera that captures images electronically, storing them in memory instead of on film. Digital cameras capture images on a sensor array instead of on film, then store them in flash memory.

Digital and Analog Cameras

When you click the button to take a picture—digitally or with old-fashioned film—light passes through a lens. In traditional cameras, the light contacts light-sensitive film, changing the film's chemistry and capturing a negative of an image. This image is revealed when the film is developed. Digital cameras capture an image with a light-sensitive sensor array instead of film. The most common sensor array is the charge-coupled device (CCD). Essentially a silicon chip about the size of a fingernail, a CCD has light-sensitive diodes arranged in a grid across its face.

Registering an Image

A CCD array can have millions of sensors, each of which registers the brightness of a color (red, green, or blue) at its position. An array with a million sensors is said to capture a "megapixel" of data. This brightness data is read, one row at a time, by an analog-to-digital converter (ADC) that changes the brightness readings (captured by the CCD as differences in voltage) to digital data. By comparing data from adjacent red, green, and blue pixels, the ADC determines and records the color of each pixel. For example, if red, green, and blue are each at maximum brightness, the pixel is white. If all three are at minimum brightness, it's black. Millions of colors can be described by the differences in the three brightness readings at each pixel. The data is then color-corrected, compressed, and stored in memory.

In a film camera, the film acts as both image capturer and storage medium. Digital cameras store the images as files, just as you store files on your computer. Older digital cameras and some entry-level cameras save images on memory chips in the camera itself. When the memory is full, you must delete unwanted images or upload the images to a computer to clear the memory. Most newer digital cameras store files in flash memory. Flash memory is RAM, similar to the RAM used by your computer.

Uploading the Image

Once you have stored the image, you can upload the file to your PC for editing or printing. Most digital cameras come with serial or Universal Serial Bus cables and software that lets you upload images to your computer. Alternatively, you can use a flash card reader that lets your computer read a memory card as if it were a removable disk. Some printers let you print directly from flash memory.

Megapixels and Prices

Cameras with one-megapixel captures are fine for snapshots and Web images, and cost about $300 from vendors such as Olympus, Epson, and Kodak. Midrange cameras, which cost from $500 to $900, capture approximately two megapixels of data, resulting in image sizes of about 1600 by 1200 pixels. Such cameras let you print a letter-size photo on an ink jet printer, or a high-quality 5-by-7-inch image.

The latest cameras capture more than three megapixels, for images of about 2048 by 1536 pixels—large enough for very high quality 8-by-10-inch prints. They cost about $1000. Professional-level cameras, with features comparable to those in high-end film cameras, start at about $4000. Although most new cameras come with flash memory, you can buy additional cards to expand your storage capacity. CompactFlash and SmartMedia are the two most popular forms of this memory. CompactFlash cards come in denominations up to 128MB; an 8MB card costs about $40. SmartMedia cards can store up to 64MB, and an 8MB card costs about $30.

Editing an Image

After you've saved an image (or several dozen) on your hard drive, you'll want to tweak the image. Most cameras come with basic image editing software (such as Adobe PhotoDeluxe) that lets you edit and enhance your pictures. If you want more options, you can buy an advanced package such as Adobe's Photoshop or Corel Photo-Paint.

Printing an Image

Several companies have special ink jet printers that make photo prints that are as good as or better than those developed from film. Such printers print at up to 1440-dot-per-inch resolution and cost between $300 and $900. They often require special paper to produce the best results.

Research Projects

Some of the following projects ask you to write a memo. See Chapter 15, page 430, for a discussion of memos.

6. Write a 500–1,000-word extended definition of one of the following terms, or of a term used in your field of study. If you do secondary research, cite your sources clearly and accurately. In addition, check that the graphics are appropriate for your audience and purpose. In a brief note at the start, indicate the audience and purpose for your definition.

   a. flextime
   b. binding arbitration
   c. robotics
   d. an academic major (don't focus on any particular major; instead, define what a major is)
   e. quality control
   f. bioengineering
   g. fetal-tissue research
   h. community policing
   i. software

7. Locate an extended definition in one of your textbooks, in a journal article, or on a Web site. In a memo to your instructor, describe the techniques the author uses to define the term. Then evaluate the effectiveness of these techniques. Submit a photocopy of the definition along with your assignment.

8. Write a 500–1,000-word description of one of the following items or of a piece of equipment used in your field. Include appropriate graphics. In a note preceding the description, specify your audience and indicate the type of description (general or particular) you are writing.

   a. carburetor
   b. locking bicycle rack
   c. deadbolt lock
   d. folding card table
   e. lawn mower
   f. photocopy machine
   g. cooling tower at a nuclear power plant
   h. jet engine
   i. telescope
   j. ammeter
   k. television set
   l. automobile jack
   m. stereo speaker
   n. refrigerator
   o. personal computer

9. Write a 500–1,000-word description of one of the following processes or a similar process with which you are familiar. Include appropriate graphics. In a note preceding the description, specify your audience and indicate the type of description (general or particular) you are writing. If you use secondary sources, cite them properly (see Appendix, Part A, p. 660 for documentation systems).

   a. how steel is made
   b. how a nuclear power plant works
   c. how a food co-op works
   d. how a suspension bridge is constructed
   e. how we see
   f. how a dry battery operates
   g. how a baseball player becomes a free agent

10. The following topics are appropriate for both process descriptions and instructions. Write a 500–1,000-word process description. Later in the semester, when you are studying Chapter 20, write a set of instructions on the same topic.

    a. how an angler cleans a fish
    b. how a fax machine operates
    c. how a student should study for a test
    d. how an audit is conducted
    e. how the owner of a recreational vehicle prepares the vehicle for winter storage

11. Form small groups according to major for this project on definitions and descriptions on the Web. Find a description of an item used in your field and, in a brief memo to your instructor, respond to the following questions:

    • Does the description contain definitions? If so, what kind? How clear are they? How would you revise them?
    • What is the purpose of the description, and how does that purpose determine its nature, scope, and organization?

- In what ways has the author used the hypertext medium in this description? In other words, how is this version of the description different from one that would appear as a traditional printed document?

- How effective is the description? Which aspects are well done? Which aspects would you do differently?

C A S E

Describing a New Fighter Jet

You are a student intern working for the Seattle, Washington, Chamber of Commerce. To publicize the business environment in Seattle, the chamber is assembling a directory of leading high-technology manufacturers in the Seattle area. One of the major corporations to be featured in this directory is Boeing. You have been asked to write a brief description — no more than three or four double-spaced pages — of one of Boeing's newest fighter jets: the F-22 Raptor. When you phone the public-affairs office at Boeing, you are told that all the information you will need about the F-22 is on their Web site (www.boeing.com) and that you are free to refer to the text and graphics in doing your project as long as you cite Boeing in quoting any text or reproducing any graphics. Write the description of the F-22 for a general audience.

 For additional cases, click on Case of the Month and Archive on TechComm Web (www.bedfordstmartins.com/ techcomm).

10 Drafting and Revising Coherent Documents

The scholar and former corporate affairs officer Charles Darling (1999) on coherence:

The most convincing ideas in the world, expressed in the most beautiful sentences, will move no one unless those ideas are properly connected. Unless readers can move easily from one thought to another, they will surely turn to something else or turn on the television.

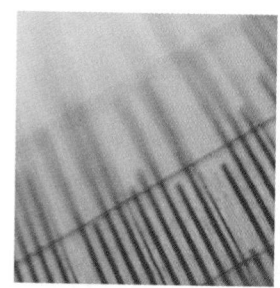

Coherence is a measure of how effectively a document hangs together. A coherent document seems to flow smoothly from one part to the next, allowing the reader to concentrate on understanding the technical information it contains. An incoherent document is harder to read; the reader can easily misunderstand the information or become confused, unable to determine how a particular point relates to one that preceded it. This chapter focuses on how to draft and revise your documents to make them coherent.

Although a document might begin to take shape during the planning stage, as you develop an outline, that shape will probably change, or become more refined, as you draft and revise. This chapter describes four elements that help to produce coherence in a document: titles, headings, lists, and paragraphs. Paying attention to these elements as you draft and revise will ensure that your document is coherent. The chapter also explains how to revise the whole document for coherence.

See TechComm Web (www .bedfordstmartins.com /techcomm) for guidelines, exercises, and links related to topics in this chapter.

For more about planning, drafting, and revising a document, see Ch. 3.

WRITING COHERENT TITLES

You might want to put off giving a final title to your document. Until you have completed it, you cannot be sure that the subject and purpose you established during the planning stages will not change. However, you should jot down a working title before you start drafting to give you a sense of direction, then come back to revise it at the end.

The title is a crucial element because it is your first opportunity to define the subject and purpose of the document for your readers. Everything else that follows should relate clearly to the title.

Precision is the key to a good title. For example, if you are writing a feasibility study on the subject of offering free cholesterol screening at your company, the title should contain the key terms *cholesterol screening* and *feasibility*. The following title would be effective:

Offering Free Cholesterol Screening at Thrall Associates: A Feasibility Study

If your document is an internal report discussing company business, you might not need to identify the organization. In that case, the following would be clear:

Offering Free Cholesterol Screening: A Feasibility Study

Or you could present the purpose before the subject:

A Feasibility Study of Offering Free Cholesterol Screening

Avoid general terms, such as *health screening* for *cholesterol screening*, because key terms from your title could be used in various kinds of indexes; the more precise your terms, the more useful your readers will find the title. Your readers should be able to paraphrase it in a clear, meaningful sentence. For instance, "A Feasibility Study of Offering Free Cholesterol Screening" could be paraphrased as "This document reports on whether it is feasible to offer free cholesterol screening for our employees." Notice what happens, however, when the title is incomplete: "Free Cholesterol Screening." The reader knows that the document has something to do with free cholesterol screening, but is the writer recommending that free cholesterol screening be instituted — or discontinued? Or is the writer reporting on how well an existing program is working?

Here are a few more examples of effective titles:

Choosing a Laptop: A Recommendation

An Analysis of the Kelly 1013 Packager

Open Sea Pollution-Control Devices: A Summary

A Forecast of Smoking Habits in the United States in the Coming Decade

WRITING COHERENT HEADINGS

A heading is a lower-level title inside a document. A clear and informative heading is vital in technical communication because it announces the subject and purpose of the discussion that follows it. This information helps readers understand what they will be reading or, in some cases, helps them decide whether they need to read further. For the writer, a heading eliminates the need to announce the subject in a sentence such as "Let us now turn to the advantages of the mandatory enrollment process."

Because headings are used to introduce text, you should avoid back-to-back headings. In other words, avoid following one heading directly with another heading:

3. Approaches to Neighborhood Policing

　3.1 Community Policing

Instead, separate the headings with text, as in this example:

3. Approaches to Neighborhood Policing

Over the past decade, the scholarly community has reached a consensus that community policing offers significant advantages over the traditional patrol car-based approach. However, the traditional approach has some distinct strengths. In the following discussion, we define each approach and then explain its advantages and disadvantages. Finally, we profile three departments that have successfully made the transition to community policing while preserving the major strengths of the traditional approach.

　3.1 Community Policing

The text after the heading "3. Approaches to Neighborhood Policing," called an *advance organizer*, introduces the material in section 3. It indicates the background, purpose, scope, and organization of the discussion that follows it. By using an advance organizer, you improve coherence by giving readers an overview of the discussion before they encounter the details.

GUIDELINES

Writing Effective Headings

▶ *Avoid long noun strings.* The following example is hard to understand:

　　Production Enhancement Proposal Analysis Techniques

Instead, add prepositions to make the title clearer:

　　Techniques for Analyzing the Proposal for Enhancing Production

This version says more clearly that the writer is going to describe the techniques.

▶ *Be informative.* In the preceding example, you could add information about how many techniques will be described:

　　Three Techniques for Analyzing the Proposal for Enhancing Production

And you can go one step further by indicating what you wish to say about the three techniques:

　　Advantages and Disadvantages of the Three Techniques for Analyzing the Proposal for Enhancing Production

For more about noun strings, see Ch. 11, p. 290.

Don't worry if the heading seems too long; clarity is more important than brevity.

▶ *Use a grammatical form appropriate to your audience.* The question form works well for less knowledgeable readers (Benson, 1985):

> What Are the Three Techniques for Analyzing the Proposal for Enhancing Production?

The "how to" form is best for instructional material, such as manuals:

> How to Analyze the Proposal for Enhancing Production

The verbal form (*-ing*) works well for processes:

> Analyzing the Proposal for Enhancing Production

▶ *Avoid back-to-back headings.* Use advance organizers to separate the headings.

For more about how to format headings, see Ch. 13, p. 351.

WRITING COHERENT LISTS

Lists are fundamental to technical communication because they add a visual dimension to the text that makes it easier for readers to understand the discussion. Lists work especially well for any kind of information that can be enumerated or expressed in a sequence. For instance, the following sentences introduce material that would be well suited to a list:

> Contractors typically use one of four methods to shore the walls of an excavation:

> Solar-energy research stalled in the 1990s for three major reasons:

> When you arrive at the site, please make the following arrangements:

For readers, the chief advantage of a list is that it makes the information easier to read and remember: the logic of the discussion is evident right on the page. The key terms in the list are set off with bullets or numbers, which helps readers see them easily. This arrangement enhances coherence: readers see the overall structure before they read the detailed discussion.

The following discussion covers paragraphs that can be turned into lists. (For a discussion of using lists in individual sentences, see Chapter 11, page 273.)

For the writer, turning paragraphs into lists has four advantages:

- *It forces you to look at the big picture.* As you start to draft your document, it is easy to lose sight of the information outside the paragraph you are working on. By looking for opportunities to create lists as you revise, you force yourself to focus on the key idea in each paragraph. This practice increases your chances of noticing that an important item is missing or that an item needs to be clarified.

- *It forces you to examine the sequence.* As you turn some of your paragraphs into bulleted or numbered lists with key phrases, you get a clearer look at the sequence of the information.

- *It forces you to create a clear lead-in.* In the lead-in, you can add a number designation that further forecasts the content and organization of the material that follows:

> Auto imports declined last year because of four major factors:

You can also add the same kind of number signal in a traditional paragraph, but you are less likely to be thinking in these terms if you are not focusing on the bulleted list of items.

- *It forces you to tighten and clarify your prose.* When you spot an opportunity to make a list, look for a word, phrase, or sentence that identifies each item. Your focus shifts from weaving sentences together in a paragraph to highlighting key ideas. And once you have formatted the list, you can look at it critically and revise it until it is clear and concise.

Figure 10.1 (based on Cohen & Grace, 1994, p. 15) shows a passage displayed in a paragraph form and in a list form. The authors are discussing the idea that engineers have a special social responsibility.

Paragraph Format	*List Format*	
Currently, there are three conceptions of the relation between engineering as a profession and society as a whole.	Currently, there are three conceptions of the relation between engineering as a profession and society as a whole.	*Turning the paragraph into a list forces the writer to create headings that sharply focus each bulleted entry.*
The first conception is that there is no relation. Engineering's proper regard is properly instrumental, with no constraints at all. Its task is to provide purely technical solutions to problems.	• *There is no relation.* Engineering's proper regard is properly instrumental, with no constraints at all. Its task is to provide purely technical solutions to problems.	*By deleting the wordy topic sentences from the paragraph version, the writer saves space. The list version of the passage is not significantly longer than the paragraph version, despite the indentations and extra white space.*
The second conception is that engineering's role is to protect. It must be concerned, as a profession, with minimizing the risk to the public. The profession is to operate on projects as presented to it, as an instrument; but the profession is to operate in accordance with important safety constraints, which are integral to its performing as a profession.	• *The engineer's role is to protect society.* Engineering is concerned, as a profession, with minimizing the risk to the public. The profession is to operate on projects as presented to it, as an instrument; but the profession is to operate in accordance with important safety constraints, which are integral to its performing as a profession.	
The third conception is that engineering has a positive social responsibility to try to promote the public good, not merely to perform the tasks that are set for it, and not merely to perform those tasks such that risk is minimized or avoided in performing them. Rather, engineering's purpose as a profession is to promote the social good.	• *The engineer's role is to promote social responsibility.* Engineering has a positive social responsibility to try to promote the public good, not merely to perform the tasks that are set for it, and not merely to perform those tasks such that risk is minimized or avoided in performing them. Rather, engineering's purpose as a profession is to promote the social good.	

■ **Figure 10.1   Paragraph Format and List Format**

WRITING COHERENT PARAGRAPHS

There are two kinds of paragraphs: body paragraphs and transitional paragraphs.

A *body paragraph* is the basic structural unit for communicating technical information. A body paragraph could be defined as a group of sentences (or sometimes a single sentence) that is complete and self-sufficient but that also contributes to a larger discussion. The challenge of creating an effective paragraph is to make sure, first, that all the sentences clearly and directly substantiate one main point, and second, that the whole paragraph follows logically from the material that precedes it.

 See the Guide to Grammar and Writing (http://webster .commnet.edu/grammar /index.htm) for more about paragraphing.

Readers tend to pause between paragraphs (not between sentences) to digest the information given in one paragraph and link it with that given in previous paragraphs. For this reason, the paragraph is the key unit of composition. Readers might forgive or at least overlook a slightly fuzzy sentence. But if they can't figure out what a paragraph says or why it appears where it does, communication is likely to break down.

A *transitional paragraph* helps readers move from one major point to another. Usually it summarizes the previous point, introduces the next point, and helps readers understand how the two are related.

The following example of a transitional paragraph is taken from a manual explaining how to write television scripts. The writer has already described six principles of writing for an episodic program, including introducing characters, pursuing the plot, and resolving the action at the end of the episode.

The first sentence contains the word "then" to signal that it is introducing a summary.

The six basic principles of writing for episodic television, then, are the following:

- Reintroduce the characters.
- Make the extra characters episode specific.
- Present that week's plot swiftly.
- Make the characters react according to their personalities.
- Resolve the plot neatly.
- Provide a denouement that hints at further developments.

The final sentence clearly indicates the relationship between what precedes it and what follows it.

But how do you put these six principles into action? The following section provides specific how-to instructions.

Structure Paragraphs Clearly

If you draft quickly — as you should — your paragraphs will need a lot of revision before they are easy to read and understand. A hastily written paragraph often starts off with a number of details: who worked on the problem before and what equipment or procedure they used; the ups and downs of the project; the specifications, dimensions, and computations. The paragraph may wind its way down the page until, finally, it concludes: "No problems were found."

This structure — moving from the particular to a general statement — reflects the way we think: we accumulate details and draw conclusions from them. But the paragraph will be easier to read if you present the main point first, in a topic sentence, and then support it in the remainder of the paragraph.

The Topic Sentence

Put the point — the topic sentence — up front. The topic sentence in technical communication functions just as it does in any other kind of writing: it summarizes or forecasts the main point of the paragraph. Technical communication should be clear and easy to read, not full of suspense. If a paragraph describes a test you performed on a piece of equipment, include the result in your first sentence:

> The point-to-point continuity test on Cabinet 3 revealed no problems.

Then go on to explain the details. If the paragraph describes a complicated idea, start with an overview. In other words, put the "bottom line" on top:

> Mitosis occurs in four stages: (1) prophase, (2) metaphase, (3) anaphase, and (4) telophase.

Notice how difficult the following paragraph is to read. The writer has structured the discussion in the same order in which she performed her calculations:

> DRAFT   Our estimates are based on our generating power during eight months of the year and purchasing it the other four. Based on the 1999 purchased power rate of $0.034/KW (January through April cost data) inflating at 8 percent annually, and a constant coal cost of $45–$50, the projected 2000 savings resulting from a conversion to coal would be $225,000.

Putting the bottom line on top makes the paragraph much easier to read. In the revision, notice that the writer has added a numbered list after the topic sentence:

> REVISION   The projected 2000 savings resulting from a conversion to coal are $225,000. This estimate is based on three assumptions: (1) that we will be generating power during eight months of the year and purchasing it the other four, (2) that power rates inflate at 8 percent from the 1999 figure of $0.034/KW (January through April cost data), and (3) that coal costs remain constant at $45–$50.

Make sure each of your topic sentences relates clearly to the organizational pattern you are using. In a discussion of why water consumption in the United States is declining even though the population is increasing, for example, you might be using a more-important-to-less-important format and start a paragraph with the following topic sentence:

> The most important reason for the decline in water usage is the increasing use of water-saving devices such as low-flow shower heads and toilets.

Your next paragraph should begin with a topic sentence that continues the more-important-to-less-important organization:

Also important in the decline of water usage is the role of increasing utility rates.

The phrase "also important" suggests that increasing utility rates are important, but less important than the water-saving devices discussed in the previous paragraph. Similarly, if your first topic sentence is "First, we need to . . . ," your next topic sentence should explicitly refer to the chronological pattern: "Second, we should. . . ."

The Support

The supporting information, which follows the topic sentence, makes the topic sentence clear and convincing. Sometimes a few explanatory details provide all the support needed. At other times, however, this part of the paragraph must carry a heavier load, that of clarifying a difficult thought or defending a controversial one.

Because every paragraph is unique, it is impossible to generalize about the exact function of the supporting information. Usually, it fulfills one of these five roles:

- It defines a key term or idea included in the topic sentence.
- It provides examples or illustrations of the situation described in the topic sentence.
- It identifies causes: factors that led to the situation.
- It defines effects: implications of the situation.
- It defends the assertion made in the topic sentence.

For more about organizing information, see Ch. 8.

The techniques writers use in developing the supporting information are those used in most nonfiction writing, including definition, comparison and contrast, classification and partition, and causal analysis. An important aspect of all of them is the use of specific, concrete details.

Paragraph Length

In technical communication, how long should a paragraph be? In general, 75 to 125 words are enough for a topic sentence and four or five supporting sentences. Long paragraphs are more difficult to read than short paragraphs simply because they require more focused concentration. Long, unbroken stretches of type can actually intimidate some readers, who then skip over them.

But don't let arbitrary guidelines about length take precedence over your own analysis of audience and purpose. You might need only one or two sentences to introduce a graphic, for example, and transitional paragraphs are also likely to be quite short. If a brief paragraph fulfills its function, let it be. Do not combine two ideas in one paragraph simply to achieve a minimum word count.

You may need to break up your discussion of one idea into two or more paragraphs. An idea that requires 200 or 300 words to develop should probably not be squeezed into one paragraph.

Dividing Long Paragraphs

Here are three techniques for dividing long paragraphs.

Technique	Example
Break the discussion at a logical place.	High-tech companies have been moving their operations to the suburbs for two main reasons: cheaper, more modern space and a better labor pool. A new office complex in the suburbs will charge anywhere from half to two-thirds of the rent charged for the same square footage in the city. And that money goes a lot further, too. The new office complexes are bright and airy, with picture windows looking out on lush landscaping. New office space is already wired for the computers; and exercise clubs, shopping centers, and even libraries are often on-site.
The most logical place to divide this paragraph is at the introduction of the second factor. Because the paragraphs are still relatively long, this strategy works best for skilled readers.	The second major factor attracting high-tech companies to the suburbs is the availability of experienced labor. Office workers and middle managers are abundant; many suburbanites, especially women returning to the labor force after their children start school, are highly trained and willing to make the short trip to the office complex. In addition, the engineers and executives, who tend to live in the suburbs anyway, are happy to forego the commuting, the city wage taxes, and the noise and stress of city life.
Make the topic sentence a separate paragraph and break up the support.	High-tech companies have been moving their operations to the suburbs for two main reasons: cheaper, more modern space and a better labor pool.
This revision is easier for all readers to understand because the brief paragraph at the start clearly introduces the information. In addition, each of the two main paragraphs now has a clear topic sentence.	First, office space is a bargain in the suburbs. A new office complex will charge anywhere from half to two-thirds of the rent charged for the same square footage in the city. And that money goes a lot further, too. The new office complexes are bright and airy, with picture windows looking out on lush landscaping. New office space is already wired for the computers; and exercise clubs, shopping centers, and even libraries are often on-site.
	Second, experienced labor is plentiful. Office workers and middle managers are abundant; many suburbanites, especially women returning to the labor force after their children start school, are highly trained and willing to make the short trip to the office complex. In addition, the engineers and executives, who tend to live in the suburbs anyway, are happy to forego the commuting, the city wage taxes, and the noise and stress of city life.

Technique	Example
Use a list. *This is the easiest of the three versions for all readers because of the extra visual cues provided by the list format.*	High-tech companies have been moving their operations to the suburbs for two main reasons: • *Cheaper, more modern space.* Office space is a bargain in the suburbs. A new office complex will charge anywhere from half to two-thirds of the rent charged for the same square footage in the city. And that money goes a lot further, too. The new office complexes are bright and airy, with picture windows looking out on lush landscaping. New office space is already wired for the computers; and exercise clubs, shopping centers, and even libraries are often on-site. • *A better labor pool.* Office workers and middle managers are abundant; many suburbanites, especially women returning to the labor force after their children start school, are highly trained and willing to make the short trip to the office complex. In addition, the engineers and executives, who tend to live in the suburbs anyway, are happy to forego the commuting, the city wage taxes, and the noise and stress of city life.

Use Coherence Devices within and between Paragraphs

The thoughts in a coherent paragraph are linked together clearly and logically. Parallel ideas are expressed in parallel grammatical constructions. Even if the paragraph already moves smoothly from sentence to sentence, however, you can emphasize coherence in three ways.

Emphasizing coherence:

Add transitional words and phrases.

Repeat key words.

Use demonstrative pronouns followed by nouns.

Transitional Words and Phrases

Transitional words and phrases help the reader understand a discussion by explicitly signaling the logical relationship between one idea and another. Table 10.1 lists the most common logical relationships between two thoughts and some of the common transitions that express those relationships.

■ **Table 10.1 Transitional Words and Phrases.**

Relationship	Transition
addition	also, and, finally, first (second, etc.), furthermore, in addition, likewise, moreover, similarly
comparison	in the same way, likewise, similarly
contrast	although, but, however, in contrast, nevertheless, on the other hand, yet
illustration	for example, for instance, in other words, to illustrate
cause-effect	as a result, because, consequently, hence, so, therefore, thus
time or space	above, around, earlier, later, next, to the right (left, west, etc.), soon, then
summary or conclusion	at last, finally, in conclusion, to conclude, to summarize

In the following examples, the first version contains no transitional words or phrases. Notice how much clearer the second version is.

WEAK   Neurons are not the only kind of cell in the brain. Blood cells supply the brain with oxygen and nutrients.

IMPROVED Neurons are not the only kind of cell in the brain. *For instance,* blood cells supply the brain with oxygen and nutrients.

WEAK   The project was originally expected to cost $300,000. The final cost was $450,000.

IMPROVED The project was originally expected to cost $300,000. *However,* the final cost was $450,000.

WEAK   The manatee population of Florida has been stricken by an unknown disease. Marine biologists from across the nation have come to Florida to assist in manatee-disease research.

IMPROVED The manatee population of Florida has been stricken by an unknown disease. *As a result,* marine biologists from across the nation have come to Florida to assist in manatee-disease research.

Place transitions as close as possible to the beginning of the second element. For example, the link between two sentences should be near the start of the second sentence:

The new embossing machine was found to be defective. *However,* the warranty on the machine will cover replacement costs.

Use transitional words to maintain coherence *between* paragraphs just as you use them to maintain coherence *within* paragraphs. The link between two paragraphs should be near the start of the second paragraph:

> . . . The complete system, then, would be too expensive for us to purchase now.

> *In addition*, a more advanced system is expected on the market within six months. . . .

Key Words

Repeating key words — usually nouns — helps readers follow the discussion. In the following example, the first version could be confusing:

UNCLEAR   For months the project leaders carefully planned their research. The cost of the work was estimated to be over $200,000.

What is the work: the planning or the research?

CLEAR   For months the project leaders carefully planned their research. The cost of the research was estimated to be over $200,000.

From a misguided desire to be interesting, some writers keep changing their important terms. *Plankton* becomes *miniature seaweed*, then the *ocean's fast food*. Avoid this kind of word game; technical communication must be clear and precise.

Of course, too much repetition can be boring. You can vary nonessential terms as long as you don't sacrifice clarity.

SLUGGISH   The purpose of the new plan is to reduce the problems we are seeing in our accounting operations. We hope to see a reduction in the problems by early next quarter.

BETTER   The purpose of the new plan is to reduce the problems we are seeing in our accounting operations. We hope to see an improvement by early next quarter.

Demonstrative Pronouns Followed by Nouns

In addition to using transitional words and phrases and repeating key phrases, carefully using demonstrative pronouns — *this, that, these*, and *those* — can help you maintain the coherence of a discussion by linking ideas securely together. In almost all cases, demonstrative pronouns should serve as adjectives rather than as pronouns; that is, they should be followed by nouns. In the following examples, notice that a demonstrative pronoun by itself can be vague and confusing.

UNCLEAR   New screening techniques are being developed to combat viral infections. *These* are the subject of a new research effort in California.

What is being studied in California: new screening techniques or viral infections?

CLEAR   New screening techniques are being developed to combat viral infections. *These techniques* are the subject of a new research effort in California.

UNCLEAR   The task force could not complete its study of the mine accident. *This* was the subject of a scathing editorial in the union newsletter.

What was the subject of the editorial: the mine accident or the task force's inability to complete its study of the accident?

CLEAR   The task force failed to complete its study of the mine accident. *This failure* was the subject of a scathing editorial in the union newsletter.

Even when the context is clear, a demonstrative pronoun used without a noun might interrupt readers' progress by referring them back to an earlier idea.

INTERRUPTIVE   The law firm advised that the company initiate proceedings. *This* caused the company to search for a second legal opinion.

FLUID   The law firm advised that the company initiate proceedings. *This advice* caused the company to search for a second legal opinion.

Transitional words and phrases, the repetition of key words, and demonstratives followed by nouns cannot *give* your writing coherence: they can only help readers appreciate the coherence that already exists. Your job is, first, to make your writing coherent and, second, to highlight that coherence.

Improving the Coherence of a Sample Paragraph

The following discussion shows how these techniques can improve the coherence of a weak paragraph. The paragraph is taken from a status report written by a branch manager of a utility company. In it, the writer explains how he decided on a method for increasing the company's business within his particular branch. (The sentences are numbered to connect them with the related marginal comment.)

(1) There were two principal alternatives considered for improving the Montana Branch. (2) The first alternative was to drill and equip additional sources of supply with sufficient capacity to provide for the present and projected system deficiencies. (3) The second alternative was to provide for said deficiencies through a combination of additional sources of supply and a storage facility. (4) Unfortunately, groundwater studies, which were conducted in the Southeast Montana area by the consulting firm of Smith and Jones, indicated that although groundwater is available within this general area of our system, it is limited as to quantity, and there is considerable separation between said sources. (5) This being the case, it becomes necessary to utilize the sources that are available or that can be developed in the most efficient manner,

Sentence 1 focuses on the two alternatives, not on the final decision the writer made. Throughout the paragraph, the focus is on the process of the study, not on the results.

By listing two alternatives in sentences 2 and 3, and then addressing the first alternative in sentence 4, the writer makes a leap that may confuse some readers.

which means operating them in conjunction with a storage facility. (6) In this way, the sources only have to be capable of providing for the average demand on a maximum day, and the storage facility can be utilized to provide for the peaking requirements plus fire protection. (7) Consequently, the second alternative as mentioned above was determined to be the more desirable alternative.

First, let's be fair. This paragraph has been taken out of context (a 17-page report); it was never meant to stand alone on a page. Also, it was written not for the general reader but for an executive of the water company — someone who, in this case, is technically knowledgeable in the writer's field. Still, an outsider's analysis of an essentially private communication can at least isolate the weaknesses.

Following is a revision of this sample paragraph. (The sentences are numbered to connect them with the related marginal comment.)

Sentence 1 is an effective topic sentence clearly stating the main point of the paragraph. Sentence 1 is justified in sentence 2.

Sentences 3 and 4 focus on the other alternative considered, explaining why it was rejected.

(1) We found that the best way to improve the Montana branch would be to add a storage facility to our existing supply sources. (2) Currently, we can handle the average demand on a maximum day; the storage facility will enable us to meet peaking requirements and fire-protection needs. (3) In conducting our investigation, we considered developing new supply sources with sufficient capacity to meet current and future needs. (4) This alternative was rejected, however, when our consultants (Smith and Jones) did groundwater studies that revealed that insufficient groundwater is available and that the new wells would have to be located too far apart if they were not to interfere with each other.

The revision is shorter and more direct, and therefore easier to read and understand. Notice the use of transitional words ("currently" in sentence 2 and "however" in sentence 4); the repetition of key words ("storage facility," "sources," and "needs"); and the use of demonstrative pronouns followed by nouns ("this alternative"). These devices provide a sense of coherence that was lacking in the original.

The only possible objection to the streamlined version is that it is *too* clear, that it leaves the writer vulnerable if his decision turns out to have been wrong. But the writer will be responsible anyway, if the decision doesn't work out, and poor writing will not endear him to his supervisor. Good writing is the best bet under any circumstances.

REVISING THE WHOLE DOCUMENT FOR COHERENCE

In looking for problems that need fixing, most writers prefer a top-down approach. They look for the largest and most important problems first and then proceed to the smaller, less important ones. This way, they don't waste time on awkward paragraphs they might eventually decide to throw out. They begin revising by considering the document as a whole.

GUIDELINES

Revising the Whole Document

In revising your document for coherence, answer the following eight questions:

▶ *Have you left out anything in turning your outline into a draft?* As you write a draft, it is easy to leave out a topic. Check the document against the outline to see that all the topics in the outline are presented in the document itself.

▶ *Have you included all the elements your readers expect to see?* If, for instance, the readers of a report are expecting a transmittal letter, a table of contents, and an abstract, they might be distracted and unable to understand the document easily if any of these elements is missing.

▶ *Is the organization logical?* Readers should be able to understand the logical progression from one topic to the next. Check the opening passages of each section to be sure they clearly and logically connect that section to the one that preceded it.

▶ *Is the content strong?* Have you provided sufficient — and appropriate — evidence to support your claims? Is your reasoning valid and persuasive?

▶ *Do you come across as reliable, honest, and helpful?* Check to see that your persona is fully professional.

▶ *Are all the elements presented consistently?* Check to see that all parallel items are presented consistently. For example, are all your headings on the same level structured the same way: as noun phrases or as *-ing* verb phrases? And check for grammatical parallelism, particularly in lists.

▶ *Is the emphasis appropriate throughout the document?* If a relatively minor topic seems to be treated at great length, with numerous subheadings, check the text itself. Maybe you introduced more headings and subheadings at that point than you did in treating some of the other topics. But if your treatment of the minor topic is in fact excessive, mark passages for possible condensing.

▶ *Are the cross-references accurate?* In a technical document you often refer to other elements in the document. Check to see that items you refer to are where you say they are.

See Online Technical Writing: (www.io.com/~hcexres /tcm1603/acchtml/acctoc .html) for more advice on revising the whole document.

See Ch. 6, pp. 125 and 131, for more on evidence and persona.

✔   **Revision Checklist**

1. Did you check the whole document to make sure that
   - ❏ you didn't leave out anything in turning your outline into a draft?
   - ❏ you include all the elements your readers expect to see?
   - ❏ the organization is logical?
   - ❏ the content is strong?
   - ❏ you come across as reliable, honest, and helpful?
   - ❏ all the elements are presented consistently?
   - ❏ the emphasis is appropriate throughout the document?
   - ❏ all the cross-references are accurate?

2. Did you revise the title so that it
   - ❏ clearly refers to your audience and the purpose of your document?
   - ❏ is sufficiently precise and informative?

3. Did you revise the headings to
   - ❏ avoid long noun strings?
   - ❏ be informative?
   - ❏ use the question form for less knowledgeable readers?
   - ❏ use the "how to" form in instructional materials, such as manuals?
   - ❏ use the verbal form (-*ing*) to suggest a process?
   - ❏ avoid back-to-back headings by including an advance organizer?

4. Did you look for opportunities to turn traditional paragraphs into lists?

5. Did you revise your paragraphs so that each one
   - ❏ begins with a clear topic sentence?
   - ❏ has adequate and appropriate support?
   - ❏ is not too long for readers?
   - ❏ uses coherence devices such as transitional words and phrases, repetition of key words, and demonstratives followed by nouns?

Exercises

1. Write a one-paragraph evaluation of each of the following titles. How clearly does the title indicate the subject and purpose of the document? On the basis of your analysis, rewrite each title.
   a. Recommended Forecasting Techniques for Haldane Company
   b. Robotics in Japanese Manufacturing
   c. A Study of Disc Cameras
   d. Agriculture in the West: A 10-Year View
   e. Synfuels: Fact or Hoax?

2. Write a one-paragraph evaluation of each of the following headings. How clearly does the heading indicate the subject of the text that will follow it? On the basis of your analysis, rewrite each heading to make it clearer and more informative. Invent any necessary details.
   a. Multigroup Processing Technique Review Board Report Findings
   b. The Great Depression of 1929
   c. Low-Level Radiation and Animals
   d. Minimize Down Time
   e. Intensive Care Nursing

3. Revise the following passage (based on Snyder, 1993) using a list format. The subject is bioremediation, the process of using microorganisms to restore natural environmental conditions.

Scientists are now working on several new research areas. One area involves using microorganisms to make some compounds less dangerous to the environment. Although coal may be our most plentiful fossil fuel, most of the nation's vast Eastern reserve cannot meet air-pollution standards because it emits too much sulfur when it is burned. The problem is that the aromatic compound dibenzothiothene (DBT) attaches itself to hydrocarbon molecules, producing sulfur dioxide. But the Chicago-based Institute of Gas Technology last year patented a bacterial strain that consumes the DBT (at least 90 percent, in recent lab trials) while leaving the hydrocarbon molecules intact.

A second research area is the genetic engineering of microbes in an attempt to reduce the need for toxic chemicals. In 1991, the EPA approved the first genetically engineered pesticide. Called Cellcap, it incorporates a gene from one microbe that produces a toxin deadly to potato beetles and corn borers into a thick-skinned microbe that is hardier. Even then, the engineered bacteria are dead when applied to the crops.

A third research area is the use of microorganisms to attack stubborn metals and radioactive waste. Microbes have been used for decades to concentrate copper and nickel in low-grade ores. Now researchers are exploiting the fact that if certain bacteria are given special foods, they excrete enzymes that break down metals and minerals. For example, researchers at the U.S. Geological Survey found that two types of bacteria turn uranium from its usual form — one that easily dissolves in water — into another one that turns to a solid that can be easily removed from water. They are now working on doing the same for other radioactive waste.

4. Provide a topic sentence for each of the following paragraphs:

a. _____. The goal of the Web Privacy Project is to make it simple for users to learn the privacy practices of a Web site and thereby decide whether to visit the site. Site owners will electronically "define" their privacy practices according to a set of specifications. Users will enter their own preferences through settings on their browsers. When a user attempts to visit a site, the browser will read the site's practices. If those practices match the user's preferences, the user will seamlessly enter the site. However, if the site's practices do not match the user's preferences, the user will be asked whether he or she wishes to visit the site.

b. _____. The reason for this difference is that a larger percentage of engineers working in small firms may be expected to hold high-level positions. In firms with fewer than 20 engineers, for example, the median income was $52,200. In firms of 20 to 200 engineers, the median income was $50,345. For the largest firms, the median was $48,600.

5. Develop the following topic sentences into full paragraphs:

a. Job candidates should not automatically choose the company that offers the highest salary.

b. Every college student should learn at least the fundamentals of computer science.

c. The one college course I most regret not having taken is _____.

d. Sometimes two instructors offer contradictory advice about how to solve the same kind of problem.

6. The following paragraph was written by the contractor for a nuclear power plant. The audience is a regulator at the Nuclear Regulatory Commission (NRC) and its purpose, to convince the regulator to waive one of the regulations. In this paragraph, transitional words and phrases have been removed. Add an appropriate transition in each blank space. Where necessary, add punctuation.

As you know, the current regulation requires the use of conduit for all cable extending more than 18 inches from the cable tray to the piece of equipment. _____ conduit is becoming increasingly expensive: up 17 percent in the last year alone. _____ we would like to determine whether the NRC would grant us any flexibility in its conduit regulations. Could we _____ run cable without conduit for lengths up to 3 feet in low-risk situations such as wall-mounted cable or low-traffic areas? We realize _____ that conduit will always remain necessary in high-risk situations. The cable specifications for the Unit Two report to the NRC are due in less than two months; _____ we would appreciate a quick reply to our request, because this matter will seriously affect our materials budget.

7. In each of the following exercises, the second sentence begins with a demonstrative pronoun. Add a noun after the demonstrative to enhance coherence.

a. The Zoning Commission has scheduled an open hearing for March 14. This _____ will enable concerned citizens to voice their opinions on the proposed construction.

b. The university has increased the number of parking spaces, instituted a shuttle system, and increased parking fees. These _____ are expected to ease the parking problems.

c. Congress's decision to withdraw support for the supercollider in 1994 was a shock to the U.S. particle-physics community. This _____ is seen as instrumental in the revival of the European research community.

8. The three paragraphs that follow are taken from a brochure explaining to the general reader some of the responsibilities involved in owning a horse. The three topic sentences do not clearly indicate the organizational pattern the writer is using. Rewrite them to provide better coherence between and within paragraphs.

The first investment needs to be in education. Any horse owner needs to have some knowledge of horse care, riding, and horse psychology. Without education, the owner can unintentionally cause tragic consequences. A horse that gets into the grain bin can die a very painful death. An owner who is rough or inconsistent can turn a well-behaved horse into a rebel. An owner who is not paying attention can get an unexpected kick when the horse becomes frightened. . . .

These consequences can be minimized in direct proportion to the time spent with the horse. Horses are social animals that need contact with others. If there are no other horses available, the owner will be its focus and will need to spend more time with it. The horse should be handled and worked daily so that any physical or mental problems that occur can be spotted quickly. This will also build the communication that is necessary for a happy relationship. . . .

Horse expenses can be divided into the initial outlay and maintenance costs. You can spend any amount for a horse. A good price for an average horse is $1,200 to $3,500. You should always get an expert's evaluation before you buy. The saddle and tack will probably cost from $350 to $2,000. The cost of keeping a horse can range from $500 to $3,500 a year, depending on the services you need. . . .

Research Projects

Some of the following exercises ask you to write a memo. See Chapter 15, page 430, for a discussion of memos.

9. Form small groups. Have each person in the group contribute a multipage document he or she has written recently, either in this class or in another. Make copies of this document for each group member. Have each member annotate the document according to the principles of coherence discussed in this chapter and then write a summary statement at the end of the document highlighting those techniques that are done well and those that could be improved. Meet as a group to study these annotated documents. Write a memo to your instructor describing those aspects of the annotations on your document cited by more than one group member and those aspects cited by only one group member. Overall, what basic differences do you see among the annotations and the summary statement? Do you think that, as a general practice, it would be worthwhile to have a draft reviewed and annotated by more than one person? What have you learned about the usefulness of peer review?

10. Find a report on the Web for this project. The report should be at least several pages long and contain some of the elements discussed in this chapter, such as titles, headings, and lists. You might use a search engine to find sites of companies and professional organizations in your field (see Chapter 7, page 149), which are likely to contain reports. In a memo to your instructor, discuss the report's coherence. Have the authors presented information in a different way than in printed text? For instance, do headings appear as they would in a printed document, or are they links to other files or sites? Does the paragraphing style seem tailored to the Web, or would a printed version of the information look essentially the same? Would you make any changes in the writing to improve its coherence? Present your findings in a memo to your instructor. Include copies of representative pages from the site.

CASE
Writing Guidelines about Coherence

You are a public-information officer recently hired by the Agency for Health Care Policy and Research. One of your responsibilities is to make sure that your agency's public information on the Web is clear and accurate. Your supervisor, José Martinez, has asked you to write a set of guidelines for physicians and other researchers who contribute the articles you put on your site. You ask him why he thinks they need the guidelines. "Their writing is factually correct," José replies, "but because they are taking excerpts from longer, more scientific studies, their documents can be choppy. They need to be smoothed out." You ask your supervisor to point you in the right direction by identifying a sample that shows the qualities he wants you to describe. He directs you to the following brief report (U.S. Department of Health and Human Services, 1997) about therapies for treating pneumonia. "This is a good sample of how to write to the general reader," José tells you. Study this report, noting the different techniques the writer has used to achieve coherence. Focus on the title, the headings, and the paragraphs. Write a brief set of guidelines using excerpts from this sample to illustrate your advice.

For additional cases, click on Case of the Month and Archive on TechComm Web (www.bedfordstmartins.com/ techcomm).

Pneumonia: More Patients May Be Treated at Home
Research Findings for Consumers

Overview

A promising new model for doctors developed by U.S. and Canadian researchers could lead to more patients being treated for pneumonia in the comfort of their own homes as safely and as effectively as if they were hospitalized, according to new research supported by the federal government's Agency for Health Care Policy and Research (AHCPR). The model — a clinical algorithm — can help doctors quickly and easily determine a patient's risk level, which is essential for deciding where treatment should take place and the type of therapy to be used.

Although pneumonia — a disease characterized by a bacterial or viral infection of the lungs — can be deadly, research has shown that the majority of patients are low-risk, meaning that they are in little danger of dying from the disease or of suffering serious consequences because of it.

Some patients already being treated in the hospital develop pneumonia. However, this research addresses the vast majority of pneumonia cases, patients not already hospitalized. About 4 million Americans a year develop this "community-acquired" pneumonia and 600,000 of these, or 15 percent, are hospitalized.

Although about 85 percent of pneumonia cases currently are treated outside the hospital, medical experts believe that an even higher percentage are eligible for outpatient care, and others could be hospitalized just for short periods. Treating more pneumonia patients at home also could help lower the cost of care. Inpatient treatment of pneumonia costs an estimated 10 to 15 times as much as outpatient care.

The problem is that doctors do not have science-based criteria to guide their decisions for admitting pneumonia patients. Part of the solution, researchers believe, may lie in this new model, which helps physicians identify pneumonia patients who do not need intensive treatment by accurately estimating their progress, or prognosis, from basic medical information.

To ensure the accuracy of the algorithm, the researchers tested it using data on thousands of pneumonia patients, including roughly 2,300 individuals in Pittsburgh, Boston, and Halifax, Nova Scotia, who were treated at home or in the hospital. If the prediction model had been available to doctors in those three cities, roughly a quarter to nearly one-third of the hospitalized patients could have been assigned outpatient care, and slightly over a tenth to almost one-fifth could have been kept only briefly for observation instead of having a longer stay.

What Patients Prefer

Most of the low-risk patients in the study who were surveyed, including those hospitalized for initial treatment, said they generally preferred home-based care. But the researchers found that patients usually are not asked where and how they would like their pneumonia to be treated. Of the doctors from the three-city study who were surveyed, 83 percent said that they alone made the decisions for outpatients, and 72 percent said they did so for the inpatients. A number of factors weigh against home care, such as the lack of a family caregiver, limited availability of home nursing services, inability to drink fluids and take medication by mouth, and certain severe medical conditions.

About the Study

These findings are from a recently completed, five-year study of variations and outcomes in pneumonia care. The research is part of a series of studies on the quality, effectiveness, and cost-effectiveness of current therapies for treating some of the most common and costly medical conditions in the United States. The project is supported by AHCPR — the U.S. Department of Health and Human Services agency spearheading federal efforts to improve the quality of American medical care. The study was directed by Wishwa N. Kapoor, M.D., M.P.H., from the University of Pittsburgh School of Medicine.

Drafting and Revising Effective Sentences

11

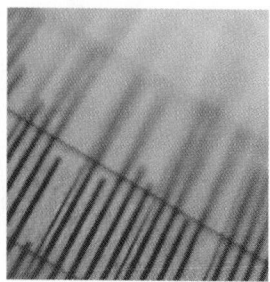

**Writing teacher William
Strunk (1918) on concise-
ness:**

*Vigorous writing is concise. A sentence should contain
no unnecessary words . . . for the same reason that a
drawing should have no unnecessary lines and a machine
no unnecessary parts.*

 See TechComm Web (www
.bedfordstmartins.com
/techcomm) for guidelines,
exercises, and links related
to topics in this chapter.

The previous chapter discussed techniques for improving the coherence of
the whole document, of long passages, and of paragraphs. This chapter concentrates on sentences and the clauses, phrases, and words they contain.

Technical communication is meant to get a job done, not to show off the
writer's skill. Therefore, the advice offered in this chapter is based on a simple
idea: sentences and words should be clear, concise, and easy to understand.

Your readers should not be aware of your presence. They should not notice that you have a wonderful vocabulary or that your sentences flow beauti-
fully, even if these things are true. In the best kind of technical communication, the writer stays in the background.

This is as it should be. Few people read technical communication for
pleasure. They read it to learn how to carry out a task, to keep abreast of developments, or to gather information. For this reason, experienced writers do
not try to be fancy. The old saying has never been more appropriate: write to
express, not to impress.

STRUCTURING EFFECTIVE SENTENCES

Good technical communication is characterized by clear, correct, and graceful
sentences that convey information without calling attention to themselves.
This section consists of seven principles for structuring effective sentences.

To structure effective sentences:
Use lists.
Place new and important information at the end of the sentence.

Choose an appropriate sentence length.

Focus on the "real" subject.

Focus on the "real" verb.

Express parallel elements in parallel structures.

Use modifiers effectively.

Use Lists

Chapter 10 (p. 259) contains a discussion of lists as a technique for organizing paragraphs. This section discusses lists as a technique in constructing sentences.

 See Online Technical Writing (www.io.com/~hcexres /tcm1603/acchtml/acctoc .html) for more about using lists.

Many sentences in technical communication are long and complicated:

> We recommend that more work on heat-exchanger performance be done with a larger variety of different fuels at the same temperature, with similar fuels at different temperatures, and with special fuels such as diesel fuel and shale-oil-derived fuels.

Here readers cannot concentrate fully on the information because they are trying to remember all the "with" phrases following "done." If they could "see" how many phrases they had to remember, their job would be easier.

Revised as a list, the sentence is easier to follow:

> We recommend that more work on heat-exchanger performance be done:
>
> - with a larger variety of different fuels at the same temperature
> - with similar fuels at different temperatures
> - with special fuels such as diesel fuels and shale-oil-derived fuels

In this version, the arrangement of the words on the page reinforces the meaning. The bullets direct the reader's eyes to three items in a series, and the fact that each item begins at the same left margin helps, too.

If you don't have enough space to list the items vertically, or if you are not permitted to do so, number the items within the sentence:

> We recommend that more work on heat-exchanger performance be done (1) with a larger variety of different fuels at the same temperature, (2) with similar fuels at different temperatures, and (3) with special fuels such as diesel fuels and shale-oil-derived fuels.

GUIDELINES

Creating Effective Lists

▶ *Indent the items in the list.* The amount of indentation depends on the length of the items. Single words or short phrases might be indented so that the list appears centered; longer items might be indented only two or three characters.

▶ *Set off each listed item with a number, a letter, or a symbol (usually a bullet).*

– Use numbered lists to suggest sequence (as in the numbered steps in a set of instructions) or priority (the first item is the most important). Sometimes using numbers helps to emphasize the total number of items in a list (as in the "Seven Warning Signals of Cancer" from the American Cancer Society). For sublists, use lowercase letters:

1. Item
    a. subitem
    b. subitem

2. Item
    a. subitem
    b. subitem

– Use bullets when you do not wish to suggest sequence or priority. Avoid using numbers for lists of people; everyone except number 1 gets offended. For sublists, use hyphens:

• Item
    – subitem
    – subitem

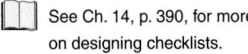
See Ch. 14, p. 390, for more on designing checklists.

– Use an open (unshaded) box (❑) for checklists.

▶ *Break up long lists.* Most people can remember only 5 to 9 items easily; break up lists of 10 or more items.

Original List	Revised List
Tool kit:	*Tool kit:*
• hand saw	• Saws
• coping saw	– hand saw
• hacksaw	– coping saw
• compass saw	– hacksaw
• adjustable wrench	– compass saw
• box wrench	• Wrenches
• Stillson wrench	– adjustable wrench
• socket wrench	– box wrench
• open-end wrench	– Stillson wrench
• Allen wrench	– socket wrench
	– open-end wrench
	– Allen wrench

Carliner (1987) recommends breaking up a bulleted list of more than 5 items and a numbered list of more than 10 (the numbers give readers extra cues to help them remember the information).

▶ *Present the items using a parallel structure.*

Nonparallel	Parallel
Here is the sequence we plan to follow:	Here is the sequence we plan to follow:
1. construction of the preliminary proposal	1. write the preliminary proposal
2. do library research	2. do library research
3. interview with the Bemco vice president	3. interview the Bemco vice president
4. first draft	4. write the first draft
5. revision of the first draft	5. revise the first draft
6. after we get your approval, typing of the final draft	6. type the first draft, after we get your approval

The nonparallel list is sloppy, a mixture of noun phrases (items 1, 3, 4, and 5), a verb phrase (item 2), and a participial phrase preceded by a dependent clause (item 6). The revision uses parallel verb phrases and deemphasizes the dependent clause in item 6 by placing it after the verb phrase.

See page 281 for more on parallelism.

▶ *Structure and punctuate the lead-in correctly.* Although standards vary from organization to organization, the most common lead-in consists of a grammatically complete sentence followed by a colon:

Following are the three main assets:

The three main assets are as follows:

The three main assets are the following:

If you cannot use a grammatically complete lead-in, use a colon, a dash, or no punctuation at all:

The committee found that the employee

- did not cause the accident
- acted properly immediately after the accident
- reported the accident properly

▶ *Punctuate the list correctly.* Because rules for punctuating lists vary, you should find out whether people in your organization have a preference. If not, you can generally punctuate lists as follows:

– If the items are sentence fragments, use a lowercase letter at the start and do not use a period or a comma at the end.

The new facility will offer three advantages:

- lower leasing costs
- easier commuting distance
- a greater pool of potential workers

The last item in a list of fragments is generally not followed by a period. The white space to the right of the last item and the white space that separates the list from the following line clearly indicate that this is the end of the list. Some writers, however, prefer to add a period after the last item.

– If the items are complete sentences, use an uppercase letter at the start and a period at the end.

The new facility will offer three advantages:

- The leasing costs will be lower.
- The commuting distance for most employees will be shorter.
- The pool of potential workers will be larger.

– If the items are fragments followed by complete sentences, use an uppercase letter in the first word of the fragment and end with a period. Then begin the complete sentences with uppercase letters and end them with periods.

The new facility will offer three advantages:

- *Lower leasing costs.* The lease will cost $1,800 per month; currently we pay $2,300.
- *Easier commuting distance.* According to a recent questionnaire, our workers now spend an average of 18 minutes traveling to work. At the new location, the average would drop to 14 minutes.
- *A greater pool of potential workers.* In the last decade, the population has begun to shift westward to the area near the new facility. We would be able to increase our potential workforce, especially in the semiskilled and managerial categories.

– If the list consists of two kinds of items — fragments and fragments followed by complete sentences — punctuate both in the same way, with uppercase letters and periods.

The new facility will offer three advantages:

- Lower leasing costs.
- Easier commuting distance. According to a recent questionnaire, our workers now spend an average of 18 minutes traveling to work. At the new location, the average would drop to 14 minutes.
- A greater pool of potential workers. In the last decade, the population has begun to shift westward to the area near the new facility. We would be able to increase our potential workforce, especially in the semiskilled and managerial categories.

Note that in the usual design specifications for lists, the second and subsequent lines, called *turnovers,* are indented to align under the first letter of the first line, highlighting the bullet or the number to the left of the text. This form, called *hanging indentation,* helps the reader to see and follow the organization of the passage.

Place New and Important Information at the End of the Sentence

In general, sentences are easiest to understand and more emphatic if new information appears at the end.

For instance, if your company is experiencing labor problems, and you want to describe the possible results, you would structure the sentence like this:

> Because of the labor problems, we anticipate a three-week delay.

In this case, the *three-week delay* is the new and important information.

If your readers already know there will be a three-week delay but don't know why, you would structure the sentence this way:

> We anticipate the three-week delay in the production schedule because of labor problems.

Here, *labor problems* is the new and important information. The articles *the* and *a* signal which is the old and which is the new information. In the second sentence, the definite article *the* is used with old information; readers already know about the delay. In contrast, in the first sentence, the indefinite article *a* is used with new information; readers don't know about the delay.

Try not to end the sentence with qualifying information that blunts the impact of the new information:

WEAK	The joint could fail under special circumstances.
IMPROVED	Under special circumstances, the joint could fail.

Put new or difficult terms at the end of the sentence:

WEAK	You use a wired glove to point to objects.
IMPROVED	To point to objects, you use a wired glove.

Choose an Appropriate Sentence Length

Sometimes sentence length affects the quality of the writing. In revising a draft, you might want to compute the average sentence length of a representative passage. (Many software programs do this for you.)

There are no firm guidelines for appropriate sentence length, but in general, an average of 15 to 20 words is effective for most technical communication. A series of 10-word sentences would be choppy. A series of 35-word sentences would probably be too demanding. And a succession of sentences of approximately the same length would be monotonous.

See Online Technical Writing (www.io.com/~hcexres /tcm1603/acchtml/acctoc .html) for more about varying sentence length.

Avoid Overly Long Sentences

How long is too long? There is no simple answer, because ease of reading depends on the vocabulary and structure of the sentence as well as its length; the reader's knowledge of the topic; and the purpose of the communication.

Often a draft will include sentences such as the following:

> The construction of the new facility is scheduled to begin in March, but it might be delayed by one or even two months by winter weather conditions, which can make it impossible or nearly impossible to begin excavating the foundation.

This 40-word sentence is confusing to read. To make it more readable, divide it into two sentences:

> The construction of the new facility is scheduled to begin in March. However, construction might be delayed until April or even May by winter weather conditions, which can make it impossible or nearly impossible to begin excavating the foundation.

Sometimes an overly long sentence can be fixed by creating a list:

WEAK   To connect the DVD player to the amplifier, first be sure that the power is off on both units, then insert the plugs firmly into the jack (the red plug into the right-channel jack and the black plug into the left-channel jack), making sure that you leave a little slack in the connecting cord to prevent shock or vibration.

IMPROVED   To connect the DVD player to the amplifier, follow these steps:

1. Be sure that the power is off on both units.
2. Insert the plugs firmly into the jacks. The red plug goes into the right-channel jack, and the black plug into the left-channel jack.

Make sure that you leave a little slack in the connecting cord to prevent shock or vibration.

As this revision demonstrates, sometimes the best way to communicate technical information is to switch to a more visually oriented format using lists or graphics.

Avoid Overly Short Sentences

Just as sentences can be too long, they can also be too short:

> The fan does not oscillate. It is stationary. The blade is made of plastic. This is done to increase safety. Safety is especially important because this design does not include a guard around the blade. A person could be seriously injured by putting his or her hand into the turning blade.

The problem here is not that the word count of these sentences is too low, but rather that the sentences are choppy and contain too little information. In cases like this, the best way to revise is to combine sentences:

> The fan is stationary, not oscillating, with a plastic blade. Because the fan does not have a blade guard, the blade is made of plastic to prevent injury if a person accidentally touches it when it is moving.

Another problem with excessively short sentences is that they needlessly repeat key terms. Again, consider combining sentences:

SLUGGISH	Computronics, a medium-sized consulting firm, consists of many diverse groups. Each group handles and develops its own contracts.
BETTER	Computronics, a medium-sized consulting firm, consists of many diverse groups, each of which handles and develops its own contracts.
SLUGGISH	I have experience working with various microprocessor-based systems. Some of these systems include the T90, RCA 9600, and the AIM 7600.
BETTER	I have experience working with various microprocessor-based systems, including the T90, RCA 9600, and the AIM 7600.

Focus on the "Real" Subject

The conceptual or "real" subject of the sentence should also be the grammatical subject, and it should be prominent. Don't bury the real subject in a prepositional phrase following a useless or "phantom" grammatical subject. In the following examples, the limp subjects disguise the real subjects (the grammatical subjects are italicized).

See Online Technical Writing (www.io.com/~hcexres /tcm1603/acchtml/acctoc .html) for more about using "real" subjects.

WEAK	The *use* of this method would eliminate the problem of motor damage.
STRONG	This *method* would eliminate the problem of motor damage.
WEAK	The *presence* of a six-membered lactone ring was detected.
STRONG	A six-membered lactone *ring* was detected.

Another way to make the subject of the sentence prominent is to reduce the number of grammatical expletives: *it is, there is,* and *there are.* In most cases, these constructions just waste space.

WEAK	There is no alternative for us except to withdraw the product.
STRONG	We have no alternative except to withdraw the product.
WEAK	It is hoped that testing the evaluation copies of the software will help us make this decision.
STRONG	I hope that testing the evaluation copies of the software will help us make this decision.

This second example uses the expletive *it is* with the passive voice. (See page 286 for more about using the passive voice.)

Expletives are not errors. Rather, they are conversational expressions that can clarify the meaning of a sentence by emphasizing the information that follows them.

WITH THE EXPLETIVE   It is hard to say whether the recession will last more than a few months.

WITHOUT THE EXPLETIVE   Whether the recession will last more than a few months is hard to say.

The second version, without the expletive, is a little more difficult to understand because the reader has to remember a long subject — "Whether the recession will last more than a few months" — before getting to the verb — "is." However, the sentence could also be rewritten in other ways to make it easier to understand and to eliminate the expletive.

I don't know whether the recession will last for more than a few months.

Nobody really knows whether the recession will last more than a few months.

Use the search function of your word processor to locate most weak subjects (usually they precede the word *of*) and expletives.

Focus on the "Real" Verb

A "real" verb, like a "real" subject, should stand out in every sentence. Few stylistic problems weaken a sentence more than *nominalizing* verbs. Writers nominalize a verb by changing it into a noun and then adding another verb, usually a weaker one, to clarify the meaning. *To install* becomes *to effect an installation; to analyze* becomes *to conduct an analysis*. Notice how nominalizing the verbs (the nominalized verbs are italicized) makes the following sentences both awkward and unnecessarily long.

WEAK   Each *preparation* of the solution is done twice.

STRONG   Each solution is prepared twice.

WEAK   An *investigation* of all possible alternatives was undertaken.

STRONG   All possible alternatives were investigated.

WEAK   *Consideration* should be given to an acquisition of the properties.

STRONG   We should consider acquiring the properties.

Like expletives, nominalizations are not errors. In fact, many common nouns are nominalizations: *maintenance, requirement,* and *analysis,* for example. In addition, nominalizations are often an effective way to summarize an idea from a previous sentence (the nominalizations in the following example are italicized).

The new *legislation* could delay our *entry* into the HDTV market. This *delay* could cost us millions.

Some software programs search for the most common nominalizations. With any word processor, you can identify most of the nominalizations by searching for character strings such as *tion, ment, sis, ence, ing,* and *ance.* If you search for the word *of,* you will also find many nominalizations.

Express Parallel Elements in Parallel Structures

A sentence is parallel if its coordinate elements follow the same grammatical form. For example, the clauses are either passive or active, the verbs are either infinitives or participles, and so on. A recognizable pattern makes a sentence easier for the reader to follow.

Click on the link to Exercise Central on TechComm Web (www.bedfordstmartins.com /techcomm) for interactive exercises on parallelism and other topics discussed in this chapter.

Notice how faulty parallelism weakens the following sentences:

NONPARALLEL   Our present system is costing us profits and reduces our productivity.

The verbs in this sentence are unparallel because they are not the same tense.

PARALLEL   Our present system costs us profits and reduces our productivity.

NONPARALLEL   The dignitaries watched the launch, and the crew was applauded.

The first clause is written in active voice; the second, in passive voice. Voice of verbs in the same sentence should be parallel. See the discussion of voice on page 286.

PARALLEL   The dignitaries watched the launch and applauded the crew.

NONPARALLEL   The typist should follow the printed directions; do not change the originator's work.

The first clause is subjunctive; the second, imperative. Mood in a sentence should be parallel.

PARALLEL   The typist should follow the printed directions and not change the originator's work.

When using parallel constructions, make sure that parallel items in a series do not overlap, which could change or confuse the meaning of the sentence:

CONFUSING   The speakers will include partners of law firms, businesspeople, and civic leaders.

CLEAR   The speakers will include businesspeople, civic leaders, and partners of law firms.

In the original sentence, "partners" appears to apply to "businesspeople" and "civic leaders," as well as to law firms. The revision solves the problem by rearranging the items so that "partners" can apply only to "law firms."

CONFUSING   We need to buy more lumber, hardware, tools, and hire the subcontractors.

CLEAR   We need to buy more lumber, hardware, and tools, and we need to hire the subcontractors.

In the confusing sentence, the writer has linked two ideas inappropriately. The first idea is that we need to buy three things: lumber, hardware, and tools; the second, that we need to hire the subcontractor. Hiring is not in the same category as the items we need to buy. In the revision, each idea now has its own subject and verb.

Use Modifiers Effectively

Technical communication frequently employs modifiers — words, phrases, and clauses that describe other elements in the sentence. To make your meaning clear, you must communicate to your readers whether a modifier provides necessary information about the word or phrase it refers to (its *referent*) or whether it simply provides additional information. You must also make sure that the referent itself is clearly identified.

Distinguish between Restrictive and Nonrestrictive Modifiers

See the Appendix "Reference Handbook," p. 697, for more about restrictive modifiers.

A *restrictive modifier*, as the term implies, restricts the meaning of its referent; it provides the information necessary in order to identify the referent. In the following examples, the restrictive modifiers are italicized:

> The aircraft *used in the exhibitions* are slightly modified.
>
> The phrase *used in the exhibitions* identifies which aircraft.

> Please disregard the notice *you recently received from us.*
>
> The phrase *you recently received from us* identifies which notice.

In most cases, the restrictive modifier doesn't require a relative pronoun, such as *that* or *which*. If you choose to use a pronoun, however, use *that*:

> The aircraft *that* are used in the exhibits are slightly modified.

(If the pronoun refers to a person or persons, use *who*.) Notice that restrictive modifiers are not set off by commas.

A *nonrestrictive modifier* does not restrict the meaning of its referent: the information it provides is not necessary in order to identify the referent. In the following examples, the nonrestrictive modifiers are italicized:

> The Hubble telescope, *intended to answer fundamental questions about the origin of the universe*, was repaired in 1997.

> When you arrive, go to the Registration Area, *which is located on the second floor.*

Like the restrictive modifier, the nonrestrictive modifier usually does not require a relative pronoun. If you use one, however, choose *which* (*who* or *whom* when referring to a person). Note that nonrestrictive modifiers are separated from the rest of the sentence by commas.

Avoid Misplaced Modifiers

The placement of the modifier often determines the meaning of the sentence. Notice, for instance, how the placement of *only* changes the meaning of the following sentences.

Only Turner received a cost-of-living increase last year.

Meaning: Nobody else received one.

Turner received *only* a cost-of-living increase last year.

Meaning: He didn't receive a merit increase.

Turner received a cost-of-living increase *only* last year.

Meaning: He received a cost-of-living increase as recently as last year.

Turner received a cost-of-living increase last year *only*.

Meaning: He received a cost-of-living increase in no other year.

Misplaced modifiers — those that appear to modify the wrong referent — are a common problem in technical communication. The best solution is usually to place the modifier as close as possible to its intended referent. The misplaced modifier is often a phrase or a clause:

MISPLACED	The subject of the meeting is the future of geothermal energy *in the downtown Webster Hotel.*
CORRECT	The subject of the meeting *in the downtown Webster Hotel* is the future of geothermal energy.
MISPLACED	*Jumping around nervously in their cages*, the researchers speculated on the health of the mice.
CORRECT	The researchers speculated on the health of the mice *jumping around nervously in their cages*.

A *squinting modifier* is a special kind of misplaced modifier that falls ambiguously between two possible referents, so the reader cannot tell which one is being modified:

UNCLEAR	We decided *immediately* to purchase the new system.

Did we decide immediately, or did we decide to make the purchase immediately?

CLEAR	We immediately decided to purchase the new system.
CLEAR	We decided to purchase the new system *immediately.*
UNCLEAR	The people who worked on the Eagle assembly line *reluctantly* picked up their last paychecks.

Did they work reluctantly, or did they pick up their last checks reluctantly?

CLEAR	The people who worked *reluctantly* on the Eagle assembly line picked up their last paychecks.

CLEAR   The people who worked on the Eagle assembly line picked up their last paychecks *reluctantly*.

A subtle form of misplaced modification can also occur with correlative constructions, such as *either . . . or, neither . . . nor,* and *not only . . . but also:*

NONPARALLEL The new refrigerant not only decreases energy costs but also spoilage losses.

PARALLEL   The new refrigerant decreases not only energy costs but also spoilage losses.

In this example, "decreases" applies to both "energy costs" and "spoilage losses." Therefore, the first half of the correlative construction should follow "decreases." Note that if the sentence contains two different verbs, the first half of the correlative construction should precede the verb:

The new refrigerant not only decreases energy costs but also reduces spoilage losses.

Avoid Dangling Modifiers

A dangling modifier has no referent in the sentence:

DANGLING   Trying to solve the problem, the instructions seemed unclear.

In this sentence, the writer has not identified who is doing the trying; as written, it would appear that the instructions are trying to solve the problem. To correct the ambiguity, rewrite the sentence, adding the clarifying information either within the modifier or next to it:

CORRECT   As I was trying to solve the problem, the instructions seemed unclear.

CORRECT   Trying to solve the problem, I thought the instructions seemed unclear.

Sometimes you can correct a dangling modifier by switching from the *indicative mood* (a statement of fact) to the *imperative mood* (a request or command):

DANGLING   To initiate the procedure, the BEGIN button should be pushed.

CORRECT   To initiate the procedure, push the BEGIN button.

In the imperative, the referent — *you* — is understood.

CHOOSING AN APPROPRIATE LEVEL OF FORMALITY

Although no standard definition of levels of formality exists, most experts would agree on three:

INFORMAL   The Acorn 560 is a real screamer. With 1.7GHz of pure computing power, it slashes through even the thickest spreadsheets before you can say 2 + 2 = 4.

MODERATELY FORMAL	With its 1.7GHz microprocessor, the Acorn 560 can handle even the most complicated spreadsheet problems quickly.
HIGHLY FORMAL	With a 1.7GHz microprocessor, the Acorn 560 is a high-speed personal computer designed for computation-intensive applications such as large spreadsheets.

In general, technical communication requires either a moderately formal or a formal style.

To achieve the appropriate tone, think about your audience, your subject, and your purpose:

- *Audience.* You would use a more formal style in writing to a group of retired executives than to a group of college students. You would likewise use a more formal style in writing to the company vice president than to your subordinates.

- *Subject.* You would use a more formal style in writing about a serious subject — safety regulations or important projects — than about preparations for the office Christmas party.

- *Purpose.* You would use a more formal style in a report to shareholders than in a company newsletter. Instructions, however, tend to be relatively informal, often using the second person and the imperative mood. Sometimes, in fact, they are quite informal, using contractions, the second person, and (occasionally) humor.

In general, it is better to err on the side of formality. Avoid an informal style in any writing you do at the office, for two reasons:

- *Informal writing tends to be imprecise.* In the example "The Acorn 560 is a real screamer," what exactly is a *screamer?*

- *Informal writing can be embarrassing.* If your boss unexpectedly sees an email you have written to a colleague, you might wish that it didn't begin, "How ya doing, loser?"

CHOOSING THE RIGHT WORDS AND PHRASES

Effective technical communication employs the right words and phrases in the right places. The following section discusses three basic principles that will help you to make good choices.

> **Choosing the right words and phrases:**
>
> Be clear and specific.
>
> Be concise.
>
> Use inoffensive language.

Be Clear and Specific

Follow these seven guidelines to make your writing clear and specific:

- Use the active voice and the passive voice appropriately.
- Be specific.
- Avoid unnecessary jargon.
- Use positive constructions.
- Avoid long noun strings.
- Avoid clichés.
- Avoid euphemisms.

Use the Active Voice and the Passive Voice Appropriately

 For more on choosing an appropriate voice, see "The Passive Engineer" by Helen Moody (www.protrainco.com /info/essays/passive.htm).

In a sentence using the active voice, the subject performs the action expressed by the verb: the "doer" of the action is the grammatical subject. In a sentence using the passive voice, by contrast, the subject receives the action. Compare the following examples (the subjects are italicized):

ACTIVE	*Dave Brushaw* drove the launch vehicle.
PASSIVE	The launch *vehicle* was driven by Dave Brushaw.
ACTIVE	Many *physicists* support the big bang theory.
PASSIVE	The big bang *theory* is supported by many physicists.

In most cases, the active voice works better than the passive voice because it emphasizes the agent. In addition, an active-voice sentence is shorter because it does not require a form of the *to be* verb and the past participle as the passive-voice sentence does. In the active version of the second example, for instance, the verb is "support" rather than "is supported," and "by" is unnecessary.

The passive voice, however, is generally better in these four cases:

- When the agent is clear from the context:

  Students are required to take both writing courses.

  Here, the context makes it clear that the college requires students to take both writing courses.

- When the agent is unknown:

  The comet was first referred to in an ancient Egyptian text.

  We don't know who produced this text.

- When the agent is less important than the action:

  The documents were hand delivered this morning.

  It doesn't matter who the messenger was.

- When a reference to the agent is embarrassing, dangerous, or in some other way inappropriate:

  Incorrect data were recorded for the flow rate.

  Here, it might be unwise to specify who recorded the incorrect data, or more tactful to avoid pointing a finger. It is always, however, unethical to use the passive voice to avoid responsibility for an action.

The passive voice can also help you maintain the focus of your paragraph.

  LANs have three major advantages. First, they are inexpensive to run. Second, they can be expanded easily. . . .

Some people believe that the active voice is inappropriate because it emphasizes the person who does the work rather than the work itself, so that the writing seems less objective. In many cases this objection is valid. Why write "I analyzed the sample for traces of iodine" if there is no ambiguity about who did the analysis or if it is unnecessary to identify who did the analysis? Focus on the action rather than the actor: "The samples were analyzed for traces of iodine," but be conservative; if in doubt, use the active voice.

Supporters of the active voice argue that the passive voice produces a double ambiguity. In the sentence "The sample was analyzed for traces of iodine," the reader is not quite sure who did the analysis (you or someone else) or when it was done (as part of the project being described or some time previously). Although a passive-voice sentence can indicate who the actor is, the writer often fails to mention it.

The best approach is to recognize that the two voices differ and to use each of them where they are most effective. The following examples mix active and passive voice for no good reason.

AWKWARD	He lifted the cage door, and a white mouse was seen.
BETTER	He lifted the cage door and saw a white mouse.
AWKWARD	The new catalyst produced good-quality foam, and a flatter mold was caused by the new chute-opening size.
BETTER	The new catalyst produced good-quality foam, and the new chute-opening size resulted in a flatter mold.

Many grammar checkers can help you locate occurrences of the passive voice in your writing. Some will advise you that the passive voice is undesirable, almost an error, but this advice is wrongheaded. Use the passive voice when it works better than the active voice for your purposes.

Any word processor allows you to search for *is, are, was,* and *were,* the forms of the verb *to be* that are most commonly used in passive-voice expressions. In addition, you can also search for *-ed* to isolate past participles, which appear in most passive-voice constructions.

Be Specific

Being specific involves using precise words, providing adequate detail, and avoiding ambiguity.

- *Use precise words.* A Ford Taurus is an automobile, but it is also a vehicle, a machine, and a thing. In describing the Ford Taurus, *automobile* is better than *vehicle*, because the less specific *vehicle* can also refer to pickup trucks, trains, hot-air balloons, and other means of transport. As words become more abstract — from *machine* to *thing*, for instance — chances for misunderstanding increase.

- *Provide adequate detail.* Readers probably know less about your subject than you do. What might be perfectly clear to you might be too vague for them.

  VAGUE   An engine on the plane experienced some difficulties.

  Which engine? What plane? What kinds of difficulties?

  CLEAR   The left engine on the Jetson 411 unaccountably lost power during flight.

- *Avoid ambiguity.* Don't let readers wonder which of two meanings you are trying to convey.

  AMBIGUOUS   After stirring by hand for 10 seconds, add three drops of the iodine mixture to the solution.

  Stir the iodine mixture or the solution?

  CLEAR   Stir the iodine mixture by hand for 10 seconds. Then add three drops to the solution.

  CLEAR   Stir the solution by hand for 10 seconds. Then add three drops of the iodine mixture.

  What should you do if you don't have the specific data? You have two options: to approximate — and clearly tell readers you are doing so — or to explain why the specific data are unavailable and indicate when they will become available.

  VAGUE   The leakage in the fuel system is much greater than we had anticipated.

  CLEAR   The leakage in the fuel system is much greater than we had anticipated; we estimate it to be at least 5 gallons per minute, not 2.

  CLEAR   The leakage in the fuel system is much greater than we had anticipated; we expect to have specific data by 4 P.M. today.

Avoid Unnecessary Jargon

Jargon is shoptalk. To a banker, *CD* means certificate of deposit; to an audiophile, it means compact disc. To the general reader, *ATM* means automated teller machine; to an electrical engineer, it means asynchronous transfer mode. Although jargon is often ridiculed, it is a useful and natural kind of communication in its proper sphere. Two software designers would find it hard to converse about their craft without terms such as *SCSI drive* and *WYSIWYG*.

However, using unnecessary jargon is inadvisable for four reasons:

- *It can be imprecise.* The best current example is the degree to which electronics terminology has crept into everyday English. In offices, we ask employees to provide *feedback* after considering a proposal. Are we asking for a facial expression, body language, a phone call, or a written evaluation?

- *It can be confusing.* If we ask a computer novice to *boot* the system, he or she might have no idea what we're talking about.

- *It is often seen as condescending.* Many people react as if the writer is showing off — displaying a level of expertise that excludes them. But if readers are concentrating on how much they dislike the writer, they miss the message.

- *It is often intimidating.* People feel inadequate or stupid because they do not know what the writer is talking about. Obviously, this reaction undermines communication.

If you are addressing a technically knowledgeable audience, use the jargon recognized in that field. (Remember, however, that a technical term can have very different meanings even to technical professionals with similar backgrounds.) If your audience includes managers or the general public, avoid jargon. If your document has separate sections for different audiences, as in the case of a technical report with an executive summary, use jargon accordingly. A glossary, or list of definitions, is a useful addition if you suspect that managers will read the technical sections.

See Ch. 12, p. 320, for more about creating glossaries.

Use Positive Constructions

The term *positive construction* has nothing to do with a cheerful or optimistic outlook on life. Rather, it indicates that the writer is describing what something is instead of what it is not. In the sentence "I was sad to see this project completed," "sad" is a positive construction. The negative construction would be "not happy."

 See the Security and Exchange Commission's *A Plain English Handbook* (www.sec.gov/consumer /plaine.htm) for advice on positive constructions.

Here are a few more examples of positive and negative constructions:

Positive	Negative	Positive	Negative
most	not all	inefficient	not efficient
few	not many	reject	cannot accept
on time	not late, not delayed	impossible	not possible
positive	not negative		

Why use positive rather than negative constructions? Because readers understand positive constructions more quickly and more easily. When a writer uses several negative constructions in the same sentence, readers must work much harder to untangle the meaning. Consider the following examples:

DIFFICULT	Because the team did not have sufficient time to complete the project, it was not surprising that it was unable to prepare a satisfactory report.
SIMPLER	Because the team had too little time to complete the project, it produced an unsatisfactory report.
DIFFICULT	Without an adequate population of krill, the entire food chain of the Antarctic region would be unable to sustain itself.
SIMPLER	If the krill population were too low, the entire food chain of the Antarctic region would be destroyed.

Avoid Long Noun Strings

A noun string is a phrase consisting of a series of nouns (or nouns and adjectives and adverbs), all of which modify the last noun. For example, in the phrase *parking-garage regulations*, the first two words modify *regulations*. Noun strings save time, and if your readers understand them, they are fine. It is easier to write *passive-restraint system* than *restraint system that is passive*, if it won't confuse readers.

For more on using hyphens, see Appendix, Part B, p. 706.

Hyphens can clarify noun strings by linking words that go together. For example, in the phrase *flat-screen monitor*, the hyphen links *flat* and *screen*. Together they modify *monitor*. In other words, it is not a *flat monitor*, or a *screen monitor*, but a *flat-screen monitor*.

However, noun strings are sometimes so long or so complex that hyphens can't ensure clarity. Consider untangling the phrases and restoring prepositions, as in the following examples:

UNCLEAR	preregistration procedures instruction sheet update
CLEAR	an update of the instruction sheet for preregistration procedures
UNCLEAR	operator-initiated default-prevention technique
CLEAR	a technique for preventing defaults that are initiated by the operator

An additional danger is that noun strings can sometimes sound pompous. If you are writing about a simple smoke detector, there is no reason to call it a *smoke-detection device* — or worse, a *smoke-detection system*.

Avoid Clichés

Good writing is original and fresh. Rather than use a cliché, say what you want to say in plain English. Instead of "It's a whole new ball game," write "The situation has changed completely." Instead of "I am sure the new manager can cut the mustard," write "I am sure the new manager can do the job effectively." Newer clichés have joined these hallowed examples — *movers and shakers; couch potato; fast lane; paradigm shift*; and *been there, done that* — but the advice is the same: if you're used to hearing or reading a phrase, avoid it.

Sometimes, writers embarrass themselves further by getting a cliché wrong: expressions become so timeworn that users forget what the words actually mean. The phrase "I could care less" is often used when the writer means "I couldn't care less." The best solution is not to use clichés.

Compare the following cliché-filled sentence and its translation into plain English:

TRITE   Afraid that we were between a rock and a hard place, we decided to throw caution to the winds with a grandstand play that would catch our competition with its pants down.

PLAIN   Afraid that we were in a difficult position, we decided on a risky, aggressive move that would surprise our competition.

Avoid Euphemisms

A euphemism is a polite way of saying something that makes people uncomfortable. The more uncomfortable the subject, the more often people resort to euphemisms. Dozens of euphemisms deal with drinking, bathrooms, sex, and death. David Lord (as quoted in Fuchsberg, 1990) lists 48 euphemisms for firing someone, including:

personnel surplus reduction	dehiring
workforce imbalance correction	decruiting
degrowing	redundancy elimination
indefinite idling	career-change-opportunity creation
corporate downsizing	

It's fun to think of euphemisms like these, but don't write them. People suffer when they get fired. Don't use language to cloud reality. It's an ethical issue.

Be Concise

The following six principles promote concise technical communication:

- Avoid obvious statements.
- Avoid meaningless modifiers.
- Avoid unnecessary prepositional phrases.
- Avoid wordy phrases.
- Avoid redundant expressions.
- Avoid pompous words.

Avoid Obvious Statements

Writing can become sluggish if it overexplains. The italicized words in the following example are sluggish:

SLUGGISH   The market for *the sale of* flash memory chips is dominated by *two chip manufacturers*: Intel and Advanced Micro Systems. These two *chip manufacturers* are responsible for 76 percent of the $1.3 billion market *in flash memory chips* last year.

IMPROVED   The market for flash memory chips is dominated by Intel and Advanced Micro Systems, two companies that claimed 76 percent of the $1.3 billion industry last year.

The writer doesn't need to state that the chips were made by chip manufacturers or that the $1.3 billion market is in flash memory chips. Only if it were in something else would the writer have to mention it in this context.

Avoid Meaningless Modifiers

Sometimes we use modifiers in our writing that are more suited to speech:

basically	essentially	kind of
certain	sort of	various

Such words are useful when you have to think on your feet, but meaningless filler in written communication.

BLOATED   *I think that, basically,* the board felt *sort of* betrayed, *in a sense,* by the *kind of* behavior the president displayed.

BETTER   The board felt betrayed by the president's behavior.

Modifiers are not always meaningless. For instance, it might be wise to use *I think* or *it seems to me* to suggest your awareness that not everyone shares your view.

BLUNT   Next year we will face unprecedented challenges to our dominance of the market.

LESS BLUNT   In my view, next year we will face unprecedented challenges to our dominance of the market.

Of course, a sentence that sounds blunt to one reader can sound self-confident to another. As you write, keep your audience's preferences in mind.

Avoid Unnecessary Prepositional Phrases

A prepositional phrase consists of a preposition followed by a noun or a noun equivalent. It almost always functions as an adjective or as an adverb. Here are some examples of prepositional phrases:

in the summary

on the engine

under the heading

Unnecessary prepositional phrases, often used along with abstract nouns and nominalizations, can make your writing long and boring. In the following examples, the prepositions are italicized:

LONG The increase *in* the number *of* students enrolled *in* the materials engineering program at Lehigh University is a testament to the regard *in* which that program is held *by* the university's new students.

SHORTER The growth of Lehigh University's materials engineering program suggests that the university's new students consider it a good program.

Avoid Wordy Phrases

Wordy phrases also make writing long and boring. For example, some people will write *on a weekly basis* rather than *weekly*. The long phrase rolls off the tongue easily and appears to carry the weight of scientific truth. But the humble *weekly* says the same thing more concisely.

Table 11.1 lists common wordy phrases and their more concise equivalents. Notice that almost all the wordy phrases contain prepositions and many contain nominalizations.

■ Table 11.1   Wordy Phrases and Their Concise Equivalents

Wordy Phrase	Concise Phrase	Wordy Phrase	Concise Phrase
a majority of	most	in view of the fact that	because
a number of	some, many	it is often the case that	often
at an early date	soon	it is our opinion that	we think that
at the conclusion of	after, following	it is our recommendation that	we recommend that
at the present time	now	it is our understanding that	we understand that
at this point in time	now	make reference to	refer to
based on the fact that	because	of the opinion that	think that
despite the fact that	although	on a daily basis	daily
due to the fact that	because	on the grounds that	because
during the course of	during	prior to	before
during the time that	during, while	relative to	regarding, about
have the capability to	can	so as to	to
in connection with	about, concerning	subsequent to	after
in order to	to	take into consideration	consider
in regard to	regarding, about	until such time as	until
in the event that	if		

Compare the following wordy sentence and its concise translation:

WORDY   I am of the opinion that, in regard to profit achievement, the statistics pertaining to this month will appear to indicate an upward tendency.

CONCISE   I think this month's statistics will show an increase in profits.

Avoid Redundant Expressions

Avoid redundant expressions, such as *collaborate together, past history, end result, any and all, each and every, still remain, completely eliminate,* and *very unique.* Be content to say something once.

REDUNDANT   We initially began our investigative analysis with a sample that was spherical in shape and heavy in weight.

BETTER   We began our analysis with a heavy, spherical sample.

Avoid Pompous Words

 See Pacific Northwest National Laboratory's list of fancy words and redundant expressions (www.pnl.gov /ag/usage/deadwood.html).

Writers sometimes think they will impress readers by using pompous words — *initiate* for *begin, perform* for *do,* and *prioritize* for *rank.* When asked why, some writers say they want to demonstrate that they have a strong vocabulary or prove that they are well educated. In technical communication, however, plain talk is best. Big words won't impress anyone for more than a few seconds.

Compare the following pompous sentences with their plain English versions.

POMPOUS   The purchase of a database program will enhance our record-maintenance capabilities.

PLAIN   Buying a database program will help us maintain our records.

POMPOUS   It is the belief of the Accounting Department that the predicament was precipitated by a computational inaccuracy.

PLAIN   The Accounting Department thinks a math error caused the problem.

Table 11.2 lists commonly used fancy words and their plain equivalents.

Several grammar checkers are able to isolate pompous words and expressions. Of course, any word-processing program will allow you to search for terms you tend to misuse.

Use Inoffensive Language

Writing to avoid offensive is not merely a matter of politeness, it is a matter of perception. Language may reflect people's attitudes, but it also helps to form attitudes. Writing inoffensively is one way to break down stereotypes so that we see people as individuals.

■ **Table 11.2   Fancy Words and Their Plain-Word Equivalents**

Fancy Word	Plain Word	Fancy Word	Plain Word
advise	tell	impact (verb)	affect
ascertain	learn, find out	initiate	begin
attempt (verb)	try	manifest (verb)	show
commence	start, begin	parameters	variables, conditions
demonstrate	show	perform	do
employ (verb)	use	prioritize	rank
endeavor (verb)	try	procure	get, buy
eventuate (verb)	happen	quantify	measure
evidence (verb)	show	terminate	end, stop
finalize	end, settle, agree, finish	utilize	use
furnish	provide, give		

Use Nonsexist Language

Sexist language favors one sex at the expense of the other. Although sexist language can shortchange men — as some writing about nursing and similar female-dominated professions does — in most cases it shortchanges women. Common examples are nouns such as *workman* and *chairman*. In addition, when writers use male pronouns to represent both males and females — "Each worker is responsible for his work area" — they are using sexist language. Some writers sidestep the problem by claiming innocence: "The use of the pronoun *he* does not in any way suggest a male bias," but many readers find this argument unpersuasive and even insincere.

 See Jenny R. Redfern's essay on sexist writing on the Rensselaer Polytechnic Institute Writing Center site (www.rpi.edu/dept/llc/writecenter/ascii/gender.txt).

GUIDELINES

Avoiding Sexist Language

▶ *Replace male-gender words with non–gender-specific words. Chairman,* for instance, can become *chairperson* or *chair. Firemen* are *firefighters, policemen* are *police officers.*

▶ *Switch to a different form of the verb.*

SEXIST   The operator must pass a rigorous series of tests before he is promoted to supervisor.

NONSEXIST   The operator must pass a rigorous series of tests before being promoted to supervisor.

▶ *Switch to the plural.*

SEXIST   The operator must pass a rigorous series of tests before he is promoted to supervisor.

NONSEXIST   Operators must pass a rigorous series of tests before they are promoted to supervisor.

Sometimes, switching to the plural makes the sentence unclear:

UNCLEAR   Operators are responsible for their operating manuals.

Does each operator have one operating manual or more than one?

CLEAR   Each operator is responsible for his or her operating manual.

Some organizations accept the use of plural pronouns with singular nouns, particularly in memos and other informal documents:

If an employee wishes to apply for tuition reimbursement, they should consult Section 14.5 of the Employee Manual.

Careful writers and readers, however, still consider this construction a grammar error.

▶ *Switch to* he or she, he/she, s/he, *or* his or her. *He or she, his or her,* and related constructions are awkward, especially if overused, but at least they are clear and inoffensive.

▶ *Address the reader directly.* Using *you* and *your,* or the understood *you,* avoids the problem.

▶ *Alternate* he *and* she. The language scholar Joseph Williams (1997) and other language authorities recommend alternating *he* and *she* from one paragraph or section to the next.

If you use a word processor, search for *he, man,* and *men,* the words and parts of words most often associated with sexist writing. Some grammar checkers even search out the most common sexist terms and suggest nonsexist alternatives. But use your common sense. You don't want to produce a sentence like this one, which appeared in a benefits manual: "Every employee is responsible for the cost of his or her gynecological examination."

Use Inoffensive Language When Referring to People with Disabilities

According to the National Organization on Disability, one in five Americans — more than 54 million people — has a physical, sensory, emotional, or mental impairment that interferes with daily life (National, 2001).

In writing about people with disabilities, use the "people first" approach: treat the person as someone with a disability, not as someone defined by that disability. The disability is a condition the person has, not what the person is.

GUIDELINES

Using the People-First Approach

When writing about people with disabilities, follow these guidelines, which are based on National (2001).

▶ *Refer to the person first, the disability second.* Write "people with mental retardation," not "the mentally retarded."

▶ *Don't confuse* handicap *with* disability. *Disability* refers to the impairment or condition; *handicap* refers to the interaction between the person and his or her environment. A person can have a disability without being handicapped.

▶ *Don't refer to victimization.* Write "a person with AIDS," not "an AIDS victim" or "an AIDS sufferer."

▶ *Don't refer to a person as "wheelchair bound" or "confined to a wheelchair."* People who use wheelchairs to get around are not confined.

▶ *Don't refer to people with disabilities as abnormal.* They are atypical, not abnormal.

UNDERSTANDING SIMPLIFIED ENGLISH FOR NONNATIVE SPEAKERS

Because English is the language of more than half of the world's scientific and technical communication, millions of nonnative speakers of English read technical communication in English (Peterson, 1990).

In response, many companies and professional associations have created versions of Simplified English. Each version consists of a basic set of grammar rules and a vocabulary of about 1,000 words, each of which has only one meaning: *right* is the opposite of *left*; it does not mean *correct*. Each version of Simplified English is geared toward a specific discipline. For example, "AECMA Simplified English" is intended for aerospace workers.

Here is a sample of text and its Simplified English version.

 For more about Simplified English, see Boeing Simplified English Checker (www .boeing.com/assocproducts/ sechecker/index.html) and Userlab Inc. (http://www .userlab.com/SE.html).

ORIGINAL TEXT	Before filling the gas tank, it is necessary to turn off the propane line to the refrigerator. Failure to do so significantly increases the risk of explosion.
SIMPLIFIED- ENGLISH VERSION	Before you pump gasoline into the gas tank, turn off the propane line to the refrigerator. If you do not turn off the propane tank, it could explode.

For background information on Simplified English, see Peterson (1990). For information on a project to create a software program that assists in turning standard English into Simplified English, see Thomas, Jaffe, Kincaid, and Stees (1992).

UNDERSTANDING STRUCTURED WRITING

For more on proposals, see Ch. 17.

Structured writing is a general term referring to different strategies for imposing a visible structure on text. For example, one structure used by many companies that write proposals displays graphics on one page and the accompanying text on the facing page.

Like Simplified English, structured writing restricts language in order to

TRANSFERS

There are more and more requests for transfers as the company expands and the key workforce adopts a more flexible lifestyle. The company supervisor is a key person in facilitating such transfers and in determining whether they would be in the best interest of the company and the employees. This memorandum covers company policy that has been in effect for the past year and continues to be our policy. It outlines each supervisor's responsibilities when an employee under his or her supervision requests a transfer.

First, it is in the company's interest to retain employees who are performing satisfactorily; therefore, we will try to help employees to move to an area or job that they find more desirable. This is what you should do. Provide an employee who comes to talk about a transfer or to request one with Form 742, Application for Transfer, and tell him or her to fill it out as soon as possible.

If the employee is applying for a new job and not just a new location and if there are any parts of the new job that you as the supervisor consider may disqualify the employee, then you should discuss with the employee those areas immediately. Remember, it is the company policy that if an employee wishes to be transferred, the company will make every effort to find a job acceptable to the employee. So you should not discourage any request for transfer, even if it would disturb the completion of projects or goals in your department.

At the bottom of the form, you should fill out the supervisor's comment. Be brief and to the point. When you have finished that, you should make a photocopy of the employee's latest Performance Evaluation and attach it to the form.

If the employee's current performance rating is unsatisfactory, then your signature and your immediate supervisor's signature on Part C of the form are required. If the current performance rating is outstanding, then attach a copy of any letters of commendation. If the current performance rating is satisfactory, you do not have to attach anything.

Send a copy of the blue copy of the form to the company Placement Bureau and a pink copy to your Department File. The yellow copy should be given to the employee.

■ **Figure 11.1**
Information Mapping®

The first page is a memo without standard headings or design elements. The second page shows the same information drafted in a visual format.

simplify it and standardize it. Structured writing is not intended exclusively for nonnative speakers of English, but to the extent that it simplifies language, it assists them as well as native speakers.

A leading example of structured writing is Information Mapping®, a technique developed by Robert Horn in the mid-1960s. Figure 11.1, from an article by Horn (1985), displays the same information, first in traditional paragraphs and then using his technique.

How to Handle Transfer Requests

Introduction   This procedure outlines each supervisor's responsibilities when an employee under your supervision requests a transfer.

Procedure table

STEP	ACTION	
1	When employees request transfers, provide them with form 742, Application for Transfer.	
2	Discuss with the employee any areas in the new job that you consider may disqualify the employee. Remember, it is our policy that if an employee wishes to be transferred, the company will make every effort to find a job acceptable to the employee. So you may in *no* way discourage a request for transfer.	
3	At the bottom of the form, fill out the supervisor's comments.	
4	Attach a copy of the latest Performance Evaluation.	
5	IF the current performance rating is . . .	THEN
	Unsatisfactory	Your signature and your immediate supervisor's signature on Part C of the form are required.
	Outstanding	Attach a copy of the letter of recommendation.
6	Send this copy of the form	TO
	Blue Copy	Placement Bureau
	Pink Copy	Departmental File
	Yellow Copy	Employee

■ **Figure 11.1**
(Continued)

Revision Checklist

Lists

1. Is each list of the appropriate kind: numbered, lettered, bulleted, or checklist?
2. Does each list contain an appropriate number of items?
3. Are all the items in each list grammatically parallel?
4. Is the lead-in to each list structured and punctuated properly?
5. Are the items in each list punctuated properly?

Sentences

6. Are the sentences structured so that the new or important information comes near the end?
7. Are the sentences the appropriate length: neither long and difficult to understand nor short and choppy?
8. Does each sentence focus on the "real" subject?
9. Have you reduced the number of expletives used as sentence openers?
10. Does each sentence focus on the "real" verb, without unnecessary nominalizations?
11. Did you eliminate nonparallelism from your sentences?
12. Did you use restrictive and nonrestrictive modifiers appropriately?
13. Did you eliminate misplaced modifiers, squinting modifiers, and dangling modifiers?

Level of Formality

14. Did you choose an appropriate level of formality for your audience, subject, and purpose?

Words and Phrases

15. Did you
    - ❑ use active and passive voice appropriately?
    - ❑ use precise words?
    - ❑ provide adequate detail?
    - ❑ avoid ambiguity?
    - ❑ avoid unnecessary jargon?
    - ❑ use positive rather than negative constructions?
    - ❑ avoid long noun strings?
    - ❑ avoid clichés?
    - ❑ avoid euphemisms?
    - ❑ avoid stating the obvious?
    - ❑ avoid meaningless modifiers?
    - ❑ avoid unnecessary prepositional phrases?
    - ❑ use the most concise phrases?
    - ❑ avoid redundancy?
    - ❑ avoid pompous words?
    - ❑ use nonsexist language?
    - ❑ use the people-first approach in referring to people with disabilities?

Exercises

The following exercises are available with this chapter's materials on TechComm Web (www.bedfordstmartins.com/techcomm). Click on Revision Exercises.

1. The information contained in the following sentences could be better conveyed in a list. Rewrite each sentence in the form of a list.
   a. The causes of burnout can be studied from three perspectives: physiological — the roles of sleep, diet, and physical fatigue; psychological — the roles of guilt, fear, jealousy, and frustration; environmental — the role of the physical surroundings at home and at work.
   b. There are many problems with the online registration system used at Dickerson University. First, lists of closed sections cannot be updated as often as necessary. Second, students who want to register in a closed section must be assigned to a special terminal. Third, the computer staff is not trained to handle student problems. Fourth, the Computer Center's own terminals cannot be used on the system; therefore, the university has to rent 15 extra terminals to handle registration.

2. In the following sentences, determine whether the new or important information is emphasized appropriately. If it is not, revise the sentence.
   a. Asynchronous transmission mode is likely to become the dominant communication mode over the next few years.
   b. Multimedia interfaces are most successful when they are designed by engineers, computer scientists, psychologists, and applications experts, all working together.
   c. The program must be carefully designed if it is to succeed.

3. The following sentences might be too long for some readers. Break each sentence into two or more sentences.
   a. If we get the contract, we must be ready by June 1 with the necessary personnel and equipment to get the job done, so with this in mind a staff meeting, which all group managers are expected to attend, is scheduled for February 12.
   b. Once we get the results of the stress tests on the 125-Z fiberglass mix, we will have a better idea of where we stand in terms of our time constraints, because if the mix isn't suitable we will really have to hurry to find and test a replacement by the Phase 1 deadline.
   c. Although we had a frank discussion with Backer's legal staff, we were unable to get them to discuss specifics on what they would be looking for in an out-of-court settlement, but they gave us a strong impression that they would rather settle out of court.

4. The following examples contain choppy, abrupt sentences. Combine sentences to create a smoother prose style.
   a. I need a figure on the surrender value of a policy. The policy number is A4399827. Can you get me this figure by tomorrow?
   b. The program obviously contains an error. We didn't get the results we anticipated. Please ask Paul Davis to test the program.
   c. The supervisor is responsible for processing the outgoing mail. He is also responsible for maintaining and operating the equipment.

5. In the following sentences, the real subjects are buried in prepositional phrases or obscured by expletives. Revise the sentences so that the real subjects appear prominently.
   a. There has been a decrease in the number of students enrolled in our training sessions.
   b. It is on the basis of recent research that I recommend the new CAD system.
   c. The use of in-store demonstrations has resulted in a dramatic increase in business.

6. In the following sentences, unnecessary nominalization obscures the real verb. Revise the sentences to focus on the real verb.
   a. Pollution constitutes a threat to the Wilson Wildlife Preserve.
   b. Evaluation of the gumming tendency of the four tire types will be accomplished by comparing the amount of rubber that can be scraped from the tires.
   c. Reduction of the size of the tear-gas generator has already been completed.

7. Revise the following sentences to eliminate nonparallelism.
   a. The next two sections of the manual discuss how to analyze the data, the conclusions that can be drawn from your analysis, and how to decide

what further steps are needed before establishing a journal list.

b. With our new product line, you would not only expand your tax practice, but your other accounting areas as well.

c. Sections 1 and 2 will introduce the entire system, while Sections 3 and 4 describe the automatic application and step-by-step instructions.

8. The following sentences contain punctuation or pronoun errors related to the use of modifiers. Revise the sentences to eliminate the errors.

a. You press the Greeting-Record Button to record the greeting which is stored on a microchip inside the machine.

b. This problem that has been traced to manufacturing delays, has resulted in our losing four major contracts.

c. Please get in touch with Tom Harvey who is updating the instructions.

9. Revise the following sentences to eliminate the misplaced modifiers.

a. Over the past three years it has been estimated that an average of eight hours per week are spent on this problem.

b. Information provided by this program is displayed at the close of the business day on the information board.

c. The computer provides a printout for the management team that shows the likely effects of the action.

10. Revise the following sentences to eliminate the dangling modifiers.

a. By following these instructions, your computer should provide good service for many years.

b. To examine the chemical homogeneity of the plaque sample, one plaque was cut into nine sections.

c. The boats in production could be modified in time for the February debut by choosing this method.

11. Revise the following informal sentences to make them moderately formal.

a. The learning modules were put together by a couple of professors in the department.

b. The biggest problem faced by multimedia designers is that users freak if they don't see a button — or, heaven forbid, if they have to make up their own buttons!

c. If the University of Arizona can't figure out where to dump its low-level radioactive waste, Uncle Sam could pull the plug on millions of dollars of research grants.

12. Rewrite the following sentences to remove inappropriate usage of the passive voice.

a. Most of the information you need will be gathered as you document the history of the journals.

b. When choosing multiple programs to record, be sure that the proper tape speed has been chosen.

c. During this time I also co-wrote a manual on the Roadway Management System. Frequent trips were also made to the field.

d. Mistakes were made.

e. Come to the reception desk when you arrive. A packet with your name on it can be picked up there.

13. Revise the following sentences by replacing the vague elements with specific information. Make up any reasonable details.

a. The results won't be available for a while.

b. The fire in the lab caused extensive damage.

c. A soil analysis of the land beneath the new stadium revealed an interesting fact.

14. The following sentences addressed to general readers contain unnecessary jargon. Revise the sentences to remove the jargon.

a. Please submit your research assignment in hard-copy mode.

b. The perpetrator was apprehended and placed under arrest directly adjacent to the scene of the incident.

c. The new computer lab supports both platforms.

15. In the following sentences, convert the negative constructions to positive constructions.

a. Williams was accused by management of making predictions that were not accurate.

b. We must make sure that all our representatives do not act unprofessionally to potential clients.

c. The shipment will not be delayed if Quality Control does not disapprove any of the latest revisions.

16. Rewrite the following sentences to eliminate the long noun strings, which the general reader might find awkward or difficult to understand.

a. The corporate-relations committee meeting has been scheduled for next Thursday.

b. The research team discovered a glycerin-initiated, alkylene-oxide-based, long-chain polyether.

c. We are considering purchasing a digital-imaging capable, diffusion-pump equipped, tungsten-gun SEM.

17. Revise the following sentences to eliminate the clichés.

a. We hope the new program will positively impact all our branches.

b. If we are to survive this difficult period, we are going to have to keep our ears to the ground and our noses to the grindstone.

c. DataRight will be especially useful for those personnel tasked with maintaining the new system.

18. Revise the following sentences to eliminate the euphemisms.

a. Downsizing our workforce will enable our division to achieve a more favorable cash-flow profile.

b. Of course, accident statistics can be expected to show a moderate increase in response to a streamlining of the training schedule.

c. Unfortunately, the patient failed to fulfill his wellness potential.

19. The following sentences are verbose because they state the obvious. Revise the sentences to eliminate the obvious material.

a. To register to take a course offered by the university, you must first determine whether the university will be offering that course that semester.

b. The starting date of the project had to be postponed for a certain period of time due to a delay in obtaining the necessary authorization from the Project Oversight Committee.

c. After you have installed DataQuick, please spend a few minutes responding to the questions about the process, then take the card to a post office and mail it to us.

20. Revise the following sentences to remove the meaningless modifiers.

a. It would seem to me that the indications are that the project has been essentially unsuccessful.

b. For all intents and purposes, our company's long-term success depends to a certain degree on various factors that are in general difficult to foresee.

c. This aspect of the presentation was quite well received overall, despite the fact that the meter readings were rather small.

21. Revise the following sentences to eliminate unnecessary prepositional phrases.

a. Another advantage of the approach used by the Alpha team is that interfaces of different kinds can be combined.

b. The complexity of the module will hamper the ability of the operator in the diagnosis of problems in equipment configuration.

c. The purpose of this test of your aptitudes is to help you with the question of the decision of which major to enroll in.

22. Revise the following sentences to make them more concise.

a. The instruction manual for the new copier is lacking in clarity and completeness.

b. The software packages enable the user to create graphic displays with a minimum of effort.

c. We remain in communication with our sales staff on a daily basis.

23. Revise the following sentences to remove the redundancies.

a. In grateful appreciation of your patronage, we are pleased to offer you this free gift as a token gesture of our gratitude.

b. An anticipated major breakthrough in storage technology will allow us to proceed ahead in the continuing evolution of our products.

c. During the course of the next two hours, you will see a demonstration of our new speech-recognition system that will be introduced for the first time in November.

24. Revise the following sentences to eliminate the pomposity.

a. This state-of-the-art soda-dispensing module is to be utilized by the personnel associated with the Marketing Department.

b. It is indeed a not insupportable inference that we have been unsuccessful in our attempt to forward the proposal to the proper agency in advance of the mandated date by which such proposals must be in receipt.

c. Deposit your newspapers and other debris in the trash receptacles located on the station platform.

25. Revise the following sentences to eliminate the sexism.

   a. Each doctor is asked to make sure he follows the standard procedure for handling Medicare forms.
   b. Policemen are required to live in the city in which they work.
   c. Professor Harry Larson and Ms. Anita Sebastian — two of the university's distinguished professors — have been elected to the editorial board of Modern Chemistry.

26. Revise the following sentences to eliminate the offensive language.

   a. This year, the number of female lung-cancer victims is expected to rise because of increased smoking.
   b. Mentally retarded people are finding greater opportunities in the service sector of the economy.
   c. This bus is specially equipped to accommodate the wheelchair-bound.

Research Projects

The following projects ask you to write a memo. See Chapter 15, page 430, for a discussion of memos.

27. Form small groups. Have one person in the group contribute a multipage document he or she has written recently, either in this class or in another. Have each member annotate a copy of this document according to the principles of sentence effectiveness discussed in this chapter. Then write a summary statement of the document, highlighting those techniques of sentence construction that are well done and those that could be improved. Meet as a group and study these annotated documents. As a group, write a memo to your instructor describing those aspects of the annotations cited by more than one group member as well as those cited by only one group member. Overall, what are the basic differences between the annotations and the summary statement from one group member and those from another? Do you think that, as a general practice, it would be worthwhile to have a draft reviewed and annotated by more than one person? What have you learned about the usefulness of peer reviewing of a document?

28. Study the writing styles on Web sites sponsored by three or four companies that compete for a global audience. Consider such factors as sentence structure, level of formality, jargon, and clichés. Also consider the issue of comprehension by nonnative speakers. Overall, which site is most effective? Write a memo comparing and contrasting the writing styles on the companies' Web sites.

This case is available in the Case of the Month and Archive for Ch. 11. For additional cases, click on Case of the Month and Archive on TechComm Web (www.bedfordstmartins.com/techcomm).

CASE
Revising a Draft for Sentence Effectiveness

For three years you have been employed as a work-study student in your university's advising office. The office wishes to distribute a new pamphlet to incoming students describing the services it provides. A new work-study student, Kim Vavrick, has written the following brief introduction to advising. You have been asked to help her with it. Write her a memo evaluating the writing in her draft according to the material presented in this chapter.

> Academic advising is counseling by a university representative, usually a faculty member, to assist the student achieve their goals for their education. The counseling's character, and the relationship that exists between the advisor and the student, change as the student's career in the academic setting progresses.

In the student's freshman and sophomore years, academic advising assists the student to identify, comprehend, and finalizing the sequence of university core requirements; that is, common classes such as English composition and basic science courses. It is also the case that academic advising may also serve to help the student clarify his academic strengths and interests in order to establish a major.

During these first two years, the interpersonal relationship between the student and the advisor are usually general and impersonal. The academic advisor may very well be someone with whom the student has little or no contact beyond obtaining a signature as a formality on paperwork. Similarly, the student may well never be enrolled in a course taught by the advisor, or otherwise become involved in the advisor's activities or academic interests.

This rarely succeeds in giving the student the optimal possible guidance for progressing in their academic career, however it is very economical and usually suitable. Faculty time and resources are expensive, limited commodities. Except in small, private institutions, there is rarely a large enough faculty to provide close and individual attention to each student who needs it. Student attrition rates are high in these first two years, many students flunk out of school before they have an opportunity to benefit from detailed, personal advice. Even among those who stay there is a high percentage of changes in academic majors. The emphasis on ensuring students understand and complete the core requirements ultimately ensures that those who do remain as students are able to progress along their degree path in a relatively smooth fashion.

In the student's junior and senior years, there is a shift in the emphasis. The goal of academic advising now is more to assist the student finish fulfilling their individual educational needs, and less to help the student meet the needs of the university. Academic advising helps the student make the best choices of the remaining options and requirements.

The relationship between the student and the advisor is closer in the last two years as well. The advisor is more personally acquainted with the student; he (or she) has seen the student periodically over a substantial period of time, and may even have instructed the student in one or more classes. The advisor is also more familiar with the major department, the courses it offers, and the colleagues who teach them, and can offer the student personal recommendations regarding many important and critical issues. Owing to the fact that the student is pursuing academic interests related to the advisor's, there is likely to be more interaction between them in academic projects and programs.

Academic advising also helps the student look beyond their undergraduate years. As the student comes close to concluding a degree program, they may be considering the possibility of entering a professional career, for example, or at extending their education in a graduate program. Academic advising serves to assist students again in making the educational choices, which will be most productive in meeting those goals.

12 Drafting and Revising Front and Back Matter

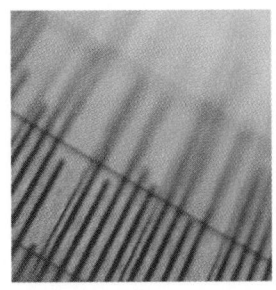

Front matter refers to those elements, such as the title page and the table of contents, that precede the body of a substantial document. Back matter refers to those elements, such as glossaries and indexes, that follow the body. Front and back matter are common in various kinds of documents, including proposals, reports, and manuals. Some elements, such as glossaries, are found in almost any kind of document, including brochures and flyers.

See TechComm Web (www .bedfordstmartins.com /techcomm) for guidelines, additional examples, and links related to topics in this chapter.

Front and back matter elements play various roles:

- *They help readers find the information they seek.* The table of contents and the index guide readers to the information they want.

- *They help readers decide whether to read the document.* The abstract, one kind of summary, helps readers decide whether they are interested enough in the content to read the whole document.

- *They substitute for the whole document.* The executive summary is directed to managers, who often do not read anything else in the document.

- *They help readers understand the document.* The glossary, a list of definitions, clarifies terms for readers who don't know the subject thoroughly.

- *They protect the document.* The cover might not contain any information but serve simply to enclose the document.

Most organizations have established formats for front and back matter. Study the style guide used in your company. If there is no style guide, study examples from the files to see how other writers have assembled their documents.

USING THE WORD PROCESSOR

For more on writing with a word processor, see Ch. 3.

A word processor simplifies the task of assembling the front and back matter in three ways:

- *You can use word-processing software to determine the word count of different elements.* Knowing the word count is useful because many elements, such as abstracts, have length guidelines or restrictions.
- *You can use the copy function in drafting the various summaries and the transmittal letter.* You can make a copy of the body of the document and then revise it to produce the elements you need. Using the copy function rather than drafting from scratch is also more accurate because you are less likely to introduce errors.
- *You can use special features to create tables of contents and indexes.*

WRITING THE FRONT MATTER

The front matter consists of seven elements.

The elements of the front matter are:
Letter of transmittal
Cover
Title page
Abstract
Table of contents
List of illustrations
Executive summary

The Letter of Transmittal

For more about formatting a letter, see Ch. 15, p. 419.

The letter of transmittal introduces the purpose and content of the document to the principal reader. It is attached to the document, bound in with it, or simply placed on top of it. Even though the letter might contain no information that is not included elsewhere in the document, it is important because it is the first thing the reader sees. It establishes a courteous and professional tone. Letters of transmittal are customary even when the writer and the reader both work for the same organization and ordinarily communicate by memo.

In addition to its overview of the main points in the accompanying document, a transmittal letter might state the methods used, acknowledge any assistance the writers have received, and refer to any errors or omissions in the document.

Present the information professionally. The transmittal letter is not a place to apologize to the reader or ask for the reader's patience or compassion. Don't write sentences like these:

> I didn't know a lot about hot-air ballooning before I began this report, so please excuse any errors you might see.

> I realize the figures aren't as neat as they should be, but I ran out of time and didn't have the chance to fix them.

Figure 12.1 on page 310 illustrates a typical transmittal letter.

The Cover

The cover protects the document from normal wear and tear and from harsher environmental conditions such as salt water or grease. The cover usually contains the following information:

For information about the materials used in covers and kinds of bindings, see Ch. 13, pp. 336 and 337.

- the title of the document
- the name and position of the writer
- the date of submission
- the name or logo of the writer's company

Sometimes the cover also includes a security notice or a statement of proprietary information.

The Title Page

Figure 12.2 on page 311 shows a typical title page. A more complex title page might also include a project number, a list of additional personnel who contributed to the document, and a distribution list.

The Abstract

An abstract is a brief technical summary of the document, usually no more than 200 words. It addresses readers already familiar with the technical subject who need to decide whether they want to read the full document. In writing an abstract, you can use technical terminology freely and refer to advanced concepts in the field. Abstracts are sometimes published by abstract services, a useful resource for researchers.

For more about abstract services, see Ch. 7, p. 157.

There are two types of abstracts: the *descriptive abstract* and the *informative abstract*. Descriptive abstracts are most often used when space is at a premium. Some government proposals, for example, call for descriptive abstracts to be placed at the bottom of the title page. The descriptive abstract, sometimes called the *topical* or *indicative* or *table-of-contents abstract*, does only what its name implies: it describes the kinds of information contained in the document. It does not provide the information — important results,

ALTERNATIVE ENERGY, INC.
1399 Soundview Drive
Bar Harbor, ME 00314
555-3267
www.altengy.com

April 3, 20XX
Rivers Power Company
15740 Green Tree Road
Gaithersburg, MD 20760

Attention: Mr. J. R. Hanson, Project Engineering Manager

The title and purpose of the document

Subject: Project #619-103-823

We are pleased to submit "A Proposal for the Riverfront Energy Project" in response to your request of February 6, 20XX.

Who authorized or commissioned the project

The windmill described in the attached proposal uses the most advanced design and materials. Of particular note is the state-of-the-art storage facility described on pp. 14–17. As you know, storage limitations are a crucial factor in the performance of a generator such as this.

Principal findings

If you have any questions, please do not hesitate to call us.

Yours very truly,

Ruth Jeffries

A polite conclusion

Ruth Jeffries
Project Manager

Enclosures 2

■ **Figure 12.1**
Letter of Transmittal

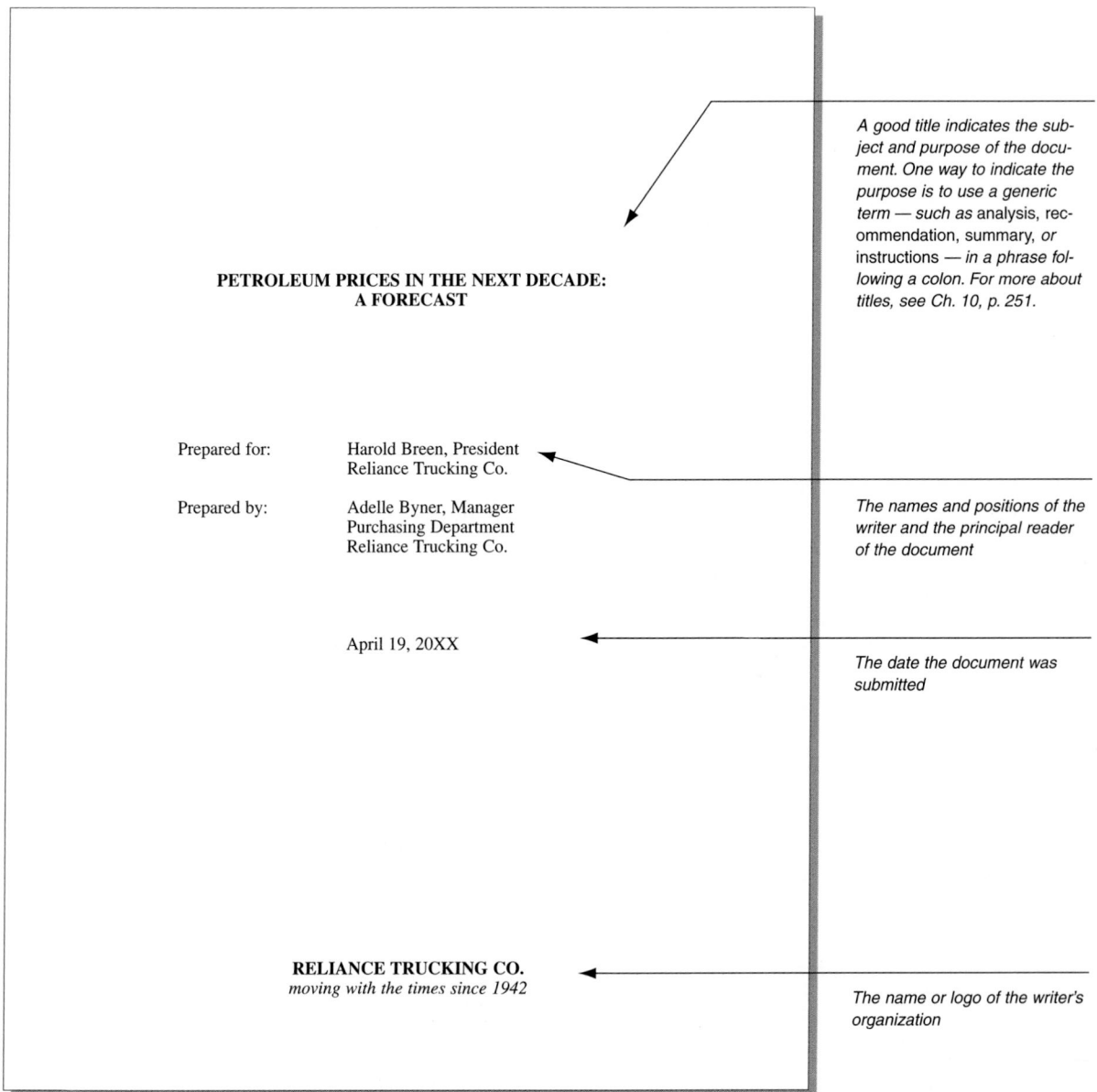

PETROLEUM PRICES IN THE NEXT DECADE:
A FORECAST

Prepared for: Harold Breen, President
Reliance Trucking Co.

Prepared by: Adelle Byner, Manager
Purchasing Department
Reliance Trucking Co.

April 19, 20XX

RELIANCE TRUCKING CO.
moving with the times since 1942

A good title indicates the sub-ject and purpose of the docu-ment. One way to indicate the purpose is to use a generic term — such as analysis, rec-ommendation, summary, *or* instructions — *in a phrase fol-lowing a colon. For more about titles, see Ch. 10, p. 251.*

The names and positions of the writer and the principal reader of the document

The date the document was submitted

The name or logo of the writer's organization

■ **Figure 12.2**
Title Page

conclusions, or recommendations — itself. It simply lists the topics covered, giving equal emphasis to each.

An informative abstract goes a step further by presenting the major findings: results, conclusions, and recommendations. It provides more comprehensive information. If you don't know which kind the reader wants, write an informative one, not a descriptive one.

Abstracts often contain a list of a half-dozen or so keywords, which are entered into electronic databases. As a writer, one of your tasks is to think of the various keywords that will lead people to the information in your document.

Figure 12.3 illustrates the descriptive abstract, Figure 12.4 the informative abstract.

The distinction between descriptive and informative abstracts is not absolute. Sometimes you might have to combine elements of both in a single abstract. For instance, suppose you are writing an informative abstract for a report that includes 15 recommendations, far too many to list. You might identify the major results and conclusions, as you would in any informative abstract, but instead of listing all the recommendations, you might simply add that the report includes numerous recommendations, as you would in a descriptive abstract.

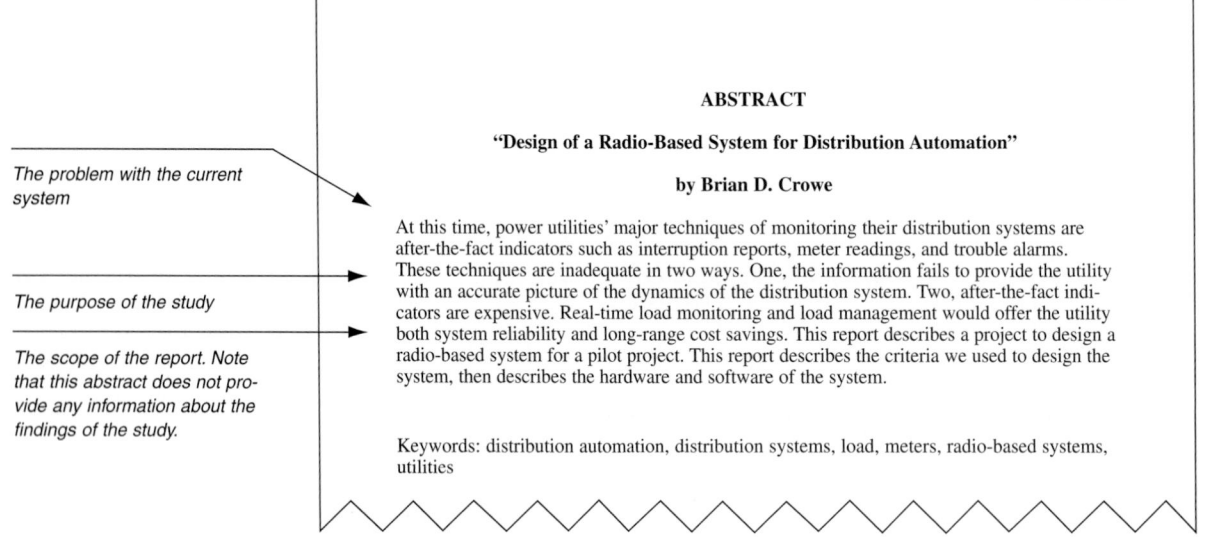

The problem with the current system

The purpose of the study

The scope of the report. Note that this abstract does not provide any information about the findings of the study.

ABSTRACT

"Design of a Radio-Based System for Distribution Automation"

by Brian D. Crowe

At this time, power utilities' major techniques of monitoring their distribution systems are after-the-fact indicators such as interruption reports, meter readings, and trouble alarms. These techniques are inadequate in two ways. One, the information fails to provide the utility with an accurate picture of the dynamics of the distribution system. Two, after-the-fact indicators are expensive. Real-time load monitoring and load management would offer the utility both system reliability and long-range cost savings. This report describes a project to design a radio-based system for a pilot project. This report describes the criteria we used to design the system, then describes the hardware and software of the system.

Keywords: distribution automation, distribution systems, load, meters, radio-based systems, utilities

■ **Figure 12.3   Descriptive Abstract**

The descriptive abstract essentially duplicates the major headings in the table of contents.

ABSTRACT

"Design of a Radio-Based System for Distribution Automation"

by Brian D. Crowe

At this time, power utilities' major techniques of monitoring their distribution systems are after-the-fact indicators such as interruption reports, meter readings, and trouble alarms. This system is inadequate in that it fails to provide the utility an accurate picture of the dynamics of the distribution system, and it is expensive. This report describes a project to design a radio-based system for a pilot project. The basic system, which uses packet-switching technology, consists of a base unit (built around a personal computer), a radio link, and a remote unit. The radio-based distribution monitoring system described in this report is more accurate than the after-the-fact indicators currently used, it is small enough to replace the existing meters, and it is simple to use. We recommend installing the basic system on a trial basis.

Keywords: distribution automation, distribution systems, load, meters, radio-based systems, utilities

An informative abstract, like a descriptive one, begins with the problem and the purpose of the study.

This type of abstract, however, describes some of the technical details of the study, culminating in the major findings: results, conclusions, and recommendations. The writer indicates the system's advantages and recommends that the system be installed on a trial basis.

■ **Figure 12.4   Informative Abstract**

The informative abstract describes the major findings of the study.

The Table of Contents

A table of contents has two main functions:

- *To help readers find the information they want.*
- *To help readers understand the scope and organization of the document.*

The table of contents is the most important guide to navigating the document. It uses the same headings as the document itself. Therefore, to create an effective table of contents, you must first make sure that these headings are effective and that there are enough of them. If the table of contents shows no entry for five or six pages, you probably need to divide the document into additional subsections. In fact, some tables of contents have one entry, or even several, for every document page.

Tables of contents that rely exclusively on generic headings (those that describe an entire class of items) are often too general to be useful. The following table of contents is ineffective because it simply lists generic headings.

Table of Contents

This methods section, which goes from page 4 to page 18, should have subentries to break up the text and to help readers find the information they seek.

For more informative headings, combine the generic and the specific:

Recommendations: Five Ways to Improve Information-Retrieval Materials Used in the Calcification Study

Results of the Commuting-Time Analysis

Then build more subheadings into the report itself. For instance, in the "Recommendations" example above, you could make each of the five recommendations into a subheading. Once you have established a clear system of headings within the document, repeat them on the contents page. Use the same text attributes — capitalization, boldface, italics, and outline style (traditional or decimal) — you have used in the body. If you use the styles in your word processor (see Chapter 3), you can make a table of contents automatically. Unfortunately, most software requires that you modify the text attributes of the table of contents levels to match the text attributes of the styles in the document.

For more about text attributes, see Ch. 13, p. 345.

Figure 12.5 illustrates how to combine generic and specific headings. The report is entitled "Methods of Computing the Effects of Inflation in Corporate Financial Statements: A Recommendation."

In paginating your document, remember two things:

For more on pagination, see Ch. 13, p. 339.

- The table of contents page does not list itself as an entry (see Figure 12.5).
- Front matter is numbered using lowercase roman numerals (*i, ii*, and so forth). The title page of a document is not numbered, although it represents page *i*. The abstract is numbered page *ii*. The table of contents is usually not numbered, although it represents page *iii*. The roman numerals are often centered at the bottom of the page. The body of the document is numbered with arabic numerals (1, 2, and so on), usually in the upper outside corner of the page.

The List of Illustrations

For more on figures and tables, see Ch. 14, p. 375.

A list of illustrations is a "table of contents" for the figures and tables in a document. If the document contains figures but not tables, the list is called a *list of figures*. If the document contains tables but not figures, the list is called a *list of tables*. If the document contains both figures and tables, figures are listed first, then tables, and the two lists together are called a *list of illustrations*.

CONTENTS

Managers can find the executive summary quickly and easily.

Other readers can find the information they need because Part 2 of the report is divided into specific sections.

Using a decimal numbering system to identify the headings helps readers to understand the organization of the document. For more information on numbering systems, see Ch. 3, pp. 51–52.

■ **Figure 12.5
Effective Table of
Contents**

Some writers begin the list of illustrations on the same page as the table of contents; others prefer a separate page. If the list of illustrations begins on a separate page, it is listed in the table of contents.

Figure 12.6 (based on U.S. Congress, 1995, p. 214) shows a list of illustrations.

The Executive Summary

The executive summary (sometimes called the *epitome*, the *executive overview*, the *management summary*, or the *management overview*) is a one- or two-page condensation of the document. It is addressed to managers, who rely on such overviews to cope with the tremendous amount of paperwork they must read every day.

Managers do not need — or want — a detailed understanding of the various projects their organizations undertake; it would be impossible anyway, because of the limits on their time and knowledge. What managers *do* need is a broad understanding of these projects and how they fit together into a coherent whole.

LIST OF ILLUSTRATIONS

Figures

Chapter 1
1-1: Information Technology Applications Currently Being Adopted 13

Chapter 2
2-1: Health Information Pathways to Hospital Information Systems 17
2-2: Health Information on the Internet 18
2-3: Messaging Standards for Clinical Institutions 22

Chapter 3
3-1: Health-Care Industry Trading Relationships 29
3-2: Payment-Related Transactions 34
3-3: Common User Interface 39

Tables

Chapter 2
2-1: Key Information Technologies for Health Care 19
2-2: Storage Requirements for Imaging Techniques 20

Chapter 3
3-1: Cost Savings through the Use of Information Technology 32

■ **Figure 12.6   List of Illustrations**

A one-page (double-spaced) maximum executive summary is the unwritten standard for documents of under 20 pages. For longer documents, the maximum length is often calculated as a percentage of the document, such as 5 percent.

The executive summary presents information to managers in two parts:

- *Background.* Because managers are not necessarily technically competent in the writer's field, this section clearly explains the background of the project. It states the specific problem or opportunity: what was not working — or not working effectively or efficiently — or what potential modification of a procedure or product had to be analyzed.

- *Major findings and implications.* Because managers are not interested in the details of the project, the methods are covered in one or two sentences. The conclusions and recommendations, however, receive a full paragraph.

If the research-and-development division at an automobile manufacturer has devised a composite material that can replace steel in engine components, the technical details of the report might include answers to questions like these:

- How was the composite devised?
- What are its chemical and mechanical structures?
- What are its properties?

Executives are less concerned about chemistry, however, than about how this project can help them make a better automobile for less money. They want a summary that addresses the managerial implications:

- Why is this composite better than steel?
- How much do the raw materials cost? Are they readily available?
- How difficult is it to make the composite?
- Are there physical limitations to the amount we can make?
- Is the composite sufficiently different from similar materials to prevent legal problems?
- Does the composite have other possible uses in cars?

Note how an executive summary differs from an informative abstract: An abstract focuses on the technical subject (can the new radio-based system effectively monitor the energy usage). An executive summary concentrates on whether the system can improve operations *at a particular company.*

Figure 12.7 on page 318 shows the executive summary for the report mentioned in Figure 12.4.

The executive summary describes the symptoms of the problem in financial terms. Notice the use of specific dollar figures.

Notice how brief the discussion of the technology is.

The writer describes the findings in a final paragraph. Notice that this last paragraph clarifies how the pilot program relates to the overall problem described in the first paragraph.

Executive Summary

Currently, we monitor our distribution system using after-the-fact indicators such as interruption reports, meter readings, and trouble alarms. This system is inadequate in two respects:

- It fails to give us an accurate picture of the dynamics of the distribution system. To ensure enough energy for our customers, we must overproduce. Last year we overproduced by 7 percent, for a loss of $273,000.

- It is expensive. Escalating labor costs for meter readers and the increased number of difficult-to-access residences have led to higher costs. Last year we spent $960,000 reading the meters of 12,000 such residences.

This report describes a project to design a radio-based system for a pilot project on these 12,000 homes.

The basic system, which uses packet-switching technology, consists of a base unit (built around a personal computer), a radio link, and a remote unit.

The radio-based distribution monitoring system described in this report is feasible because it is small enough to replace the existing meters and because it is simple to use. It would provide a more accurate picture of our distribution system, and it would pay for itself in 3.9 years. We recommend installing the system on a trial basis. If the trial program proves successful, radio-based distribution-monitoring techniques will provide the best long-term solution to the current problems of inaccurate and expensive data collection.

■ **Figure 12.7**
Executive Summary

GUIDELINES

Writing an Executive Summary

Follow these six suggestions in writing executive summaries:

▶ *Describe the background using specific evidence.* For most managers, the best evidence includes costs and savings. Instead of writing that the equipment you are now using to cut metal foil is ineffective, write that the equipment jams once every 72 hours on average and that every time it jams you lose $400 in materials and $2,000 in productivity. Then add up these figures for a monthly or an annual total.

▶ *In describing research, be specific.* For instance, research suggests that if your company had a computerized energy-management system you could cut your energy costs by 20 to 25 percent. If your energy costs last year were $300,000, you could save $60,000 to $75,000.

▶ *Describe the methods briefly.* In most cases, an executive does not care how you did what you did; he or she assumes you did it competently and professionally. However, if you think your reader is interested, include a brief description — no more than a sentence or two.

▶ *Describe the findings in accordance with your readers' needs.* If your readers want to know your results, provide them. If your readers are not able to understand the technical data or are not interested, go directly to the conclusions and recommendations.

▶ *Ask an outside reader to review your draft.* Give it to someone who has had no connection to the project. That person should be able to read your summary and understand what the project means to the organization.

▶ *Decide how to integrate the executive summary within the body of the document.* The current practice is to place it first, before the body. To further highlight the executive summary, writers commonly treat the section as a major part of the document in the table of contents (see Figure 12.5 on page 315).

WRITING THE BACK MATTER

The back matter usually consists of some or all of four components.

The elements of the back matter:
Glossary and list of symbols
References
Appendices
Index

The Glossary and List of Symbols

A *glossary* is an alphabetical list of definitions. It is particularly useful if your audience includes readers unfamiliar with the technical vocabulary in your document.

Instead of slowing down your detailed discussion by defining technical terms as they appear, you can use boldface, or some similar method of highlighting words, to inform your readers that the term is defined in the glossary. A footnote at the bottom of the page containing the first boldfaced term will clarify this system for readers. For example, the body of the document might say, "Thus the **positron*** acts as the . . . ," whereas a note at the bottom of the page explains: "*This and all subsequent terms in boldface are defined in the Glossary, page 26."

Although the glossary is usually placed near the end of the document, before the appendices, it can also be placed immediately after the table of contents if the glossary is brief (less than a page) and if it defines terms essential for managers likely to read the body of the document.

A list of symbols is formatted like a glossary, but it defines symbols and abbreviations rather than terms. It too may be placed before the appendices or after the table of contents.

Figure 12.8, based on Ruiu (1994), shows a glossary. Figure 12.9 shows a list of symbols.

References

Many documents contain a list of references (sometimes called a *bibliography* or *works cited*) as part of the back matter.

References, and the accompanying textual citations throughout the document, are called *documentation*. Documentation acknowledges your debt to

---

Glossary

ISDN: integrated-services digital network. A completely digital public telecommunications network that can accommodate both voice and data traffic.

Sonet: synchronous optical network. A family of interfaces that defines the transportation and management of digital traffic on an optical-fiber network in North America. Outside North America, a similar system, called synchronous digital hierarchy (SDH), is used.

Switching fabric: the circuitry in a communications system that routes transmissions from their source to their destination.

Virtual channel: a defined connection through a network in which bandwidth is not allocated permanently, but only when a user's cells require transport.

■ **Figure 12.8 Glossary**

List of Symbols

β	beta
CRT	cathode-ray tube
γ	gamma
Hz	hertz
rcvr	receiver
SNR	signal-to-noise ratio
uhf	ultra high frequency
vhf	very high frequency

■ **Figure 12.9   List of Symbols**

your sources, establishes your credibility as a writer, and helps readers locate and review your sources.

The Appendix, Part A, describes documentation in detail. It explains the kinds of information you need to document and shows the basic forms for three documentation systems:

Publication Manual of the American Psychological Association, 4th ed. (1994). This system, often referred to as APA, is widely used in the social sciences.

Scientific Style and Format: The CBE Manual for Authors, Editors, and Publishers (1994). This system, often referred to as CBE, is widely used in the physical sciences.

MLA Handbook for Writers of Research Papers, 5th ed. (1999). This system, from the Modern Language Association, is widely used in the humanities.

Appendices

An *appendix* is any section that follows the body of the document (and the list of references or bibliography, glossary, or list of symbols). Appendices (or *appendixes*) convey information that is too bulky for the body or that will interest only a few readers. Appendices might include maps, large technical diagrams or charts, computations, computer printouts, test data, and texts of supporting documents.

Appendices, usually labeled with letters rather than numbers (Appendix, Part A, Appendix, Part B, and so on), are listed in the table of contents and are referred to at the appropriate points in the body of the document. Therefore, they are accessible to any reader who wants to consult them.

An appendix that includes or consists of a graphic is titled "Appendix," not "Figure" or "Table," even if it would have been so designated had it appeared in the body of the document.

The Index

An *index* is an alphabetical list of the major items — topics, ideas, functions, and tasks — described and explained in a document. For long documents (100 pages or longer) the index is the most important accessing tool. Readers consult indexes (or *indices*) to locate specific pieces of information. Sometimes they can't find what they are looking for, or it takes them a long time to find it. According to Bonura (1994, p. 4), the most common complaint about technical documents is that the index is poor or lacking altogether.

Characteristics of an Effective Index

An effective index is accurate, comprehensive, precise, and easy to use.

- *Accuracy.* The page numbers listed must be accurate.
- *Comprehensiveness.* The index should include an entry for every term a reader is likely to seek. The average index might list three or four items for each page of text, but that number is only a rough guide.
- *Precision.* An entry followed by 30 page numbers is ineffective. A reader would have to check all of them to locate the relevant one. Useful indexes have subentries and, when appropriate, sub-subentries. For instance, an index to a manual about a word-processing program should have numerous subentries for *tables*, including *creating*, *editing*, and *formatting*; the subentry *formatting* should likewise have sub-subentries such as *justifying*, *sizing*, and *shading*.
- *Ease of use.* Two kinds of cross-references help the reader navigate the index: *see* and *see also* (both of which are usually italicized).
  - *See* directs readers to the proper or main entry: for example, "Drive, *see* disk drive." Use *see* only if the specified entry — *disk drive*, in this case — contains subentries. Otherwise, it is just as easy to list the page numbers for *drive* as well as for *disk drive*.
  - *See also* directs readers to related entries: for example, "Search-and-replace, 17. *See also* Global searching; Editing text." This entry alerts readers to two other entries that contain information related to the search-and-replace function.

Sometimes it is useful to add an explanatory statement at the start of an index. If, for example, you have included names of people in a separate index or chosen to group all the function keys in an entry called *function keys* (rather than alphabetize them separately by the name of the individual function), alert readers to your scheme in a brief headnote.

Planning, Drafting, and Revising an Index

An index is like any other part of a document: it requires planning, drafting, and revising.

- *Planning.* Think of your audience and your purpose. How much do your readers know about the subject? What kinds of information do they want? If you are addressing novices, include entries for basic items. And as with all kinds of technical communication, think in terms of tasks (what readers want to accomplish) in addition to functions (what the item does). If time permits, interview people who fit the reader profile, to learn about the kind of information they expect to find in the index and how they expect that information to be organized.

See Ch. 5 for more about levels of audience.

- *Drafting.* The most popular method of drafting an index, with index cards and a pencil, is the oldest. First, read a section, such as a chapter, and circle the key terms. Then make a card for each term. Some indexers like to introduce subentries at this point; others like to wait they have all the page references to the item. Some indexers like to alphabetize their cards as they go along; others don't want to break their concentration. Some writers work directly on the computer, typing their entries (either in the order in which they occur or alphabetically) during their first pass through the document.

  Word-processing software lets you create an index by locating each reference to an item you designate. The hard part of indexing is determining hierarchy (what is an entry, and what is a subentry) and adding the *see* and *see also* cross-references.

- *Revising.* As you revise the index, consider these six points:
  - Have you entered all items and are your entries easy to understand?
  - Have you used terms consistently throughout?
  - Do you need to create more subentries or sub-subentries? *The Chicago Manual of Style* (1993, p. 722) recommends creating subentries when an entry contains five or six page numbers.
  - Have you consistently used either singular nouns or plural nouns?
  - Do you need to change the way you have indicated page numbers? For instance, in your first draft, you might have *macros, 19–24*, but when you check those pages, you see that they should be listed as 19–22, 23, 24, because the main explanation is on pages 19–22, and macros are merely mentioned on 23 and 24.
  - Do the page numbers make sense? It is possible to transpose numbers inadvertently (14–12 instead of 12–14) or leave out the hyphen between numbers (1214 instead of 12–14).

GUIDELINES

Indexing Effectively

The following guidelines will help you create an effective index:

▶ *Format the index in two columns and use a type size two points smaller than the body text.*

See Ch. 13, pp. 345 and 347, for more on columns and type sizes.

▶ *Alphabetize either letter by letter or word by word.* In the *letter-by-letter format*, ignore spaces between words:

> box drawing
> boxes
> box shading

In the *word-by-word format*, alphabetize all entries that begin with the same word, then go on to the next word:

> box drawing
> box shading
> boxes

▶ *When you use acronyms and abbreviations, consider how most readers will search for the term.* For instance, use "CD-ROM," not "Compact Disk-Read Only Memory."

▶ *Alphabetize symbols and numbers as if they were spelled out.* For instance, "586 microprocessor" is alphabetized under *f* (for "five eighty-six").

▶ *If the document has a glossary, make an index entry for each term listed.*

▶ *Consider using boldface or italics to indicate the pages on which the main discussion of an item appears:*

> dot leaders, **171–172**, 211, 306

Be sure to add a note at the start of the index to explain your use of boldface.

▶ *Find out the appropriate way to express inclusive numbers.* The easiest is to use all the digits: 289–293. To express inclusive numbers for a document with section numbering, use the following form: 3–6 to 3–9 (page 3–6 is the sixth page in section 3).

Some organizations shorten the second number in the number range: 289–93. But abbreviating inclusive numbers can get complicated. For example, most style guides call for "100–104," rather than "100–04" or "100–4." Check your organization's style guide to see how to express inclusive page numbers.

✔ **Revision Checklist**

1. Does the transmittal letter
   ❏ clearly state the title and, if necessary, the subject and purpose of the document?
   ❏ clearly state who authorized or commissioned the document?
   ❏ briefly state the methods you used?
   ❏ summarize your major results, conclusions, and recommendations?
   ❏ acknowledge any assistance you received?
   ❏ courteously offer further assistance?

2. Does the cover include
   - ❑ the title of the document?
   - ❑ your name and position?
   - ❑ the date of submission?
   - ❑ the company name or logo?

3. Does the title page
   - ❑ include a title that clearly states the subject and purpose of the document?
   - ❑ list the names and positions of both you and your principal reader?
   - ❑ include the date of submission of the document and any other identifying information?

4. Does the abstract
   - ❑ list the document title, your name, and any other identifying information?
   - ❑ clearly define the problem or opportunity that led to the project?
   - ❑ briefly describe (if appropriate) the research methods?
   - ❑ summarize the major results, conclusions, and recommendations?

5. Does the table of contents
   - ❑ clearly identify the executive summary?
   - ❑ contain a sufficiently detailed breakdown of the major sections of the body of the document?
   - ❑ reproduce the headings as they appear in your document?
   - ❑ include page numbers?

6. Does the list of illustrations (or list of tables or list of figures) include all the graphics found in the body of the document?

7. Does the executive summary
   - ❑ clearly state the problem or opportunity that led to the project?
   - ❑ explain the major results, conclusions, recommendations, and managerial implications of your document?
   - ❑ avoid technical vocabulary and concepts that a managerial audience is not likely to know?

8. Does the glossary include definitions of all the technical terms your readers might not know?

9. Does the list of symbols include all the symbols and abbreviations your readers might not know?

10. Do the appendices include the supporting materials that are too bulky to present in the document body or are of interest to only a small number of your readers?

11. Does the index
    - ❑ appear in a two-column format in a type size two points smaller than the body text?
    - ❑ use either letter-by-letter or word-by-word alphabetization?
    - ❑ alphabetize acronyms and abbreviations according to how most readers will search for the term?
    - ❑ alphabetize symbols and numbers as if they were spelled out?
    - ❑ include an entry for each item in the glossary?
    - ❑ express inclusive numbers appropriately?

Exercises

1. The following letter of transmittal is from a report written by an industrial engineer to his company president. Write a one-paragraph evaluation that focuses on the clarity, comprehensiveness, and tone of the letter.

Dear Mr. Smith:

The enclosed report, "Robot and Machine Tools," discusses the relationship between robots and machine tools.

Although loading and unloading machine tools was one of the first uses for industrial robots, this task has only recently become commonly feasible. Discussed in this report are concepts that are crucial to remember in using robots.

If at any time you need help understanding this report, please let me know.

Sincerely yours,

2. The following informative abstract is from a report by an electrical engineer to her manager. Write a one-paragraph evaluation. How well does the abstract define the problem, methods, and important results, conclusions, and recommendations?

"Design of a New Computer Testing Device"

The modular design of our new computer system warrants the development of a new type of testing device. The term *modular design* indicates that the overall computer system can be broken down into parts or modules, each of which performs a specific function. It would be both difficult and time consuming to test the complete system as a whole, for it consists of 16 different modules. A more effective testing method would check out each module individually for design or construction errors prior to its installation into the system. This individual testing process can be accomplished by the use of our newly designed testing device.

The testing device can selectively call or "address" any of the logic modules. To test each module individually, the device can transmit data or command words to the module. Also, the device can display the status or condition of the module on a set of LED displays located on the front panel of the device. In addition, the device has been designed so that it can indicate when the module being tested has produced an error.

3. The following table of contents is from a report titled "An Analysis of Corporations v. Sole Proprietorships." Write a one-paragraph evaluation. How effective is the table of contents in highlighting the executive summary, defining the overall structure of the report, and providing a detailed guide to the location of particular items?

CONTENTS

4. The following executive summary is from a report titled "Analysis of Large-Scale Municipal Sludge Composting as an Alternative to Ocean Sludge Dumping." Write a one-paragraph evaluation. How well does the executive summary present concise and useful information to the managerial audience?

Coastal municipalities currently involved with ocean sludge-dumping face a complex and growing sludge management problem. Estimates suggest that treatment plants will have to handle 65 percent more sludge in 2008 than in 1998, or approximately seven thousand additional tons of sludge per day. As the volume of sludge is increasing, traditional disposal methods are encountering severe economic and environmental restrictions. The EPA has banned all ocean sludge-dumping as of next January 1. For these reasons, we are considering sludge composting as a cost-effective sludge management alternative.

Sludge composting is a 21-day biological process in which waste-water sludge is converted into organic fertilizer that is aesthetically acceptable, essentially pathogen free, and easy to handle. Composted sludge can be used to improve soil structure, increase water retention, and provide nutrients for plant growth. At $150 per dry ton, composting is currently almost three times as expensive as ocean dumping, but effective marketing of the resulting fertilizer could dramatically reduce the difference.

Research Projects

Some of the following projects ask you to write a memo. See Chapter 15, page 430, for a discussion of memos.

5. Find a document, such as a textbook or a manual, that contains an index. Study a page of the index and then write a memo to your instructor describing and evaluating the index. For instance, what form of alphabetization is used? Are the entries clear and easy to use? How effectively are *see* and *see also* used? Does the index appear comprehensive and accurate? Attach a photocopy of a page from the index.

6. Choose a journal article of some 8 to 15 pages and compile an index for it. Attach a note describing the audience and indicating what indexing style you have used. For instance, are you alphabetizing letter-by-letter or word-by-word? What style of pagination have you used? Attach a photocopy of the article.

7. Form small groups for this project that compares indexing in an online help system and in a printed manual. Have two people in the group study the indexing principles used in the online help system of one piece of software, such as a word-processing program, a spreadsheet, or a presentations-graphics package. Have two other people in the group study the index in a printed manual for the same product. Meet as a group to compare the two indexes. What are the main similarities and differences? Which is more thorough? Which do you find easier to use? What suggestions would you offer to the indexers? Present your findings in a memo to your instructor.

8. Scholars and librarians are trying to help the Modern Language Association and the American Psychological Association keep up with advances in electronic communication tools. Find two sites on the Internet that offer suggestions on how to document electronic sources according to one of these documentation styles. Download the sections that explain how to cite Web sites, email messages, and FTP documents. Compare and contrast the guidelines on these three types of citations. What are the basic differences between the two documentation styles? Overall, which do you find easier to understand? Which would be more effective in helping the reader find the source? Present your findings in a memo to your instructor.

Case 1: Planning for Better Front and Back Matter

You and the other members of your group work as interns in the office of Leonard Keala, the Senate liaison of the Federal Trade Commission. Keala calls your group in to discuss a problem.

"We've got a situation," he tells you. "Each year we have to submit a bunch of reports to the Senate telling them what we have accomplished in each of our areas of responsibility in the previous year. This year my counterpart in the Senate tossed them back to me. He said they're too hard to read. 'We can't find what we need,' he said. He singled out one of our reports as particularly hard to get into: 'Twenty-First Annual Report to Congress Pursuant to Section 815(a) of the Fair Debt Collection Practices Act.' Would you take a look at that report, figure out what we're doing wrong, and write me a memo recommending how to write these things so we don't get them thrown back at us?"

Study the document (www.ftc.gov/os/statutes/fdcpa/senate99.htm for the html version, www.ftc.gov/os/statutes/fdcpa/senate99.pdf for the pdf

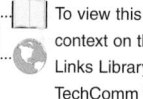

To view this document in context on the Web, follow Links Library for Ch. 12 on TechComm Web (www.bedfordstmartins.com/techcomm).

version). Which elements of front and back matter in the report are effective? Which are ineffective? Which elements are missing and should be presented? Write a memo or report to your supervisor presenting your findings. Where appropriate, excerpt portions of the FTC report to support your claims.

Case 2: Comparing Online Indexes

You work in the technical communication group at United Software. Your group is about to begin writing the online documentation for the first version of your company's newest product: a word-processing program. Your supervisor, Andrea Bonnano, wants to be sure that the online documentation is of high quality. "Let's take a look at how our competitors do their indexes. Would you and the other members of your group study a couple of word processors and write me a memo about their indexes?" As a group, meet and determine the appropriate criteria by which to describe and evaluate the online indexes. Once you have agreed on the criteria, split up and study the indexes. What are the main similarities and differences? Which is more complete? Which is easier to use? What suggestions would you offer to the indexers? Write the memo to Andrea Bonnano presenting your findings and recommendations on how to index your online documentation.

PART FOUR:

DEVELOPING THE VISUAL ELEMENTS

Designing the Document 13

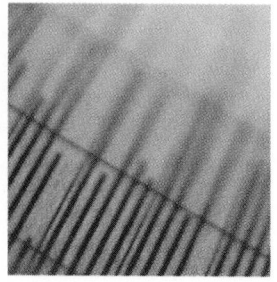

Professors of visual communication Charles Kostelnick and David D. Roberts (1998, p. 4) respond to the idea that designing a document is a mysterious and subjective practice:

Practical design is fairly rational — in the sense that each step of the way you can understand why you're making design decisions, which will enable you to assert control over the process.

See TechComm Web (www
.bedfordstmartins.com
/techcomm) for guidelines,
additional examples, and
links related to topics in this
chapter.

The effectiveness of a document largely depends on how well it is designed. Of course, the information it contains is critical, but readers see the document before they actually read it. In less than a second, the document has made an impression on them, one that might determine how well they read it — or whether they read it at all.

In technical communication, your goal is not to entertain readers with flashy colors and outrageous designs like those in mass-market magazines. But familiarity with design principles can help you make a document that readers will want to read — and that will help them understand and remember what you say.

This chapter covers the basics of document and page design. It describes the goals of document and page design, discusses how to plan the design, and then focuses on designing a document and individual pages. The chapter closes with an analysis of several sample pages.

Most of the concepts discussed in this chapter apply to paper documents and Web pages. To learn about special considerations in designing Web pages, see Chapter 21.

GOALS OF DOCUMENT DESIGN AND PAGE DESIGN

In designing a document, you have five major goals.

Goals of document design:
To make a good impression on readers
To help readers understand the structure and hierarchy of the information

To help readers find the information they need	↓
To help readers understand the information	
To help readers remember the information	

- *To make a good impression on readers.* Your documents should reflect your own professional standards and those of your organization.

- *To help readers understand the structure and hierarchy of the information.* As they navigate through a document, your readers should know where they are and where they are headed. In addition, they should be able to see the hierarchical relationship between one piece of information and another and the importance of what they are reading in relation to the information that preceded it. Design helps to communicate these relationships.

- *To help readers find the information they need.* Usually, people don't read technical documents from cover to cover. Therefore, an effective design helps readers find the information they need — quickly and easily. Design elements (such as tabs, icons, and color), page design, and choice of type are critical in directing a reader's attention to information.

- *To help readers understand the information.* Effective document and page design can clarify information. For instance, if you design a set of instructions so that text describing a step is next to its accompanying graphic, you have made it easier for readers to understand the instructions.

- *To help readers remember the information.* An effective design helps readers create a visual image of the information, enhancing retention. Text boxes, pull quotes, and similar design elements help readers remember important explanations and passages.

GUIDELINES

Understanding Basic Principles of Design

To design an effective technical document, you need a working understanding of five basic design principles. (Each of these principles will be further explained and illustrated throughout the chapter.)

See Roger C. Parker's site on design at www .newentrepreneur.com.

▶ *Use visual contrast to emphasize important information.* Visual cues help readers locate what they need, understand it, and remember it. This principle applies to many aspects of document design. For example, black print is easiest to see against a white background; larger letters stand out among smaller ones; information printed in a color, such as red, grabs attention sooner than information printed in black.

↓

▶ *Arrange information clearly on the page and in the document.* In a well-designed document, the reader should be able to see the relationships between various kinds of information on a page. The principle is simple: things next to each other are related to each other; things distant from each other are not related to each other; the text describing a graphic should be positioned close to the graphic; items in a bulleted list should relate to each other.

▶ *Establish consistent patterns.* Treat the same kind of information in the same way in your design to highlight consistent patterns. For example, all first-level headings should have the same typeface, type size, spacing, and so forth. This common design signals a connection between headings that can make the content easier to understand. Other elements that are used to create consistent visual patterns are colors, icons, rules, and screens.

▶ *Strive for moderation.* Don't fill up every square inch of space with text and graphics; leave enough white space. A cluttered-looking page will be confusing. Use simple graphics to communicate your information and only a few typefaces and colors.

▶ *Achieve a balanced look.* Balance refers to the visual stability of the page; it should not look as if it is tipping to one side or is top-heavy or bottom-heavy. In documents that open to a two-page spread, the concept of balance should apply to both pages.

PLANNING THE DESIGN OF THE DOCUMENT AND THE PAGE

The first step in designing any kind of technical document is planning. Analyze your audience's needs and expectations, and consider your resources.

Analyzing Your Audience's Needs and Expectations

For more about analyzing your audience, see Ch. 5. For more about tables of contents and indexes, see Ch. 12, pp. 313 and 322.

Consider such factors as the audience's knowledge of the subject, their attitudes, their reasons for reading, the way they will be using the document, and the kinds of tasks they will be performing. For instance, if the document is a benefits manual for employees, you know that few people will read it like a novel, from start to finish, but many people will refer to it often. Therefore, you will want to build in as many accessing tools as you can: a table of contents and index, of course, but perhaps tabs or different colors of paper to identify each section. You might include a separate section containing samples of the various forms employees might have to fill out.

Think too about your audience's expectations. Readers expect to see certain kinds of information presented in certain ways. Unless you have a good reason to present information in other ways, you should fulfill their expectations. For instance, tutorial information for complicated software programs is often presented in a small-format book, bound so that it lies flat on the table next to the keyboard.

Determining Your Resources

Once you have considered your readers, think about your resources of time, money, and equipment. Short, informal documents are generally produced in-house; for more ambitious projects, you might need to subcontract some of the job to professionals. If your organization has a technical-publications department, you should consult the professionals there for information on scheduling and budgeting.

- *Time.* What is your schedule? A sophisticated design might require the expertise of professionals at service bureaus and print shops, and their services can require weeks. Creating even a relatively simple design for a newsletter can require many hours.

- *Money.* What is your budget for the project? Can you afford the services of professional designers and print shops? An in-house newsletter should look professional and attractive, but most managers would be unwilling to authorize thousands of dollars for a sophisticated design. They would, however, authorize many thousands to design an annual report.

- *Equipment.* With a pencil and a straight edge you can create a terrific design, but to make that design come to life, you need equipment. Although a word processor is fine for many routine design needs, you need graphics software and desktop-publishing programs for more complicated designs. A good laser printer can produce attractive documents in black and white. For high-resolution color, however, you need a more expensive printer.

Now you can begin to plan the design.

DESIGNING THE DOCUMENT

Before you begin to design individual pages, think through the design of the whole document — how you want the different elements to work together to accomplish your objectives.

There are four major elements to consider in designing the whole document: size, paper, bindings, and accessing tools.

Size

Size refers to two aspects of document design: page size and page count.

- *Page size.* Think about the best page size for your information and about how the document will be used. For a procedures manual that will sit on a shelf most of the time, standard 8.5 × 11-inch paper, punched to fit in a three-ring binder, is an obvious choice. For a software tutorial, you will probably want relatively narrow columns in a document that fits easily on a desk while the user works at the keyboard. Therefore, you might choose a 5.5 × 8-inch size. The physical dimensions of a document can be important if it is to be used in a cramped area or has to fit in a standard-size compartment, such as in a drawer in an airplane cockpit.

  Paper comes precut in a number of standard sizes in addition to 8.5 × 11 inches, such as 4.5 × 6 inches and 6 × 9 inches. Although paper can be cut to whatever size you want, costs increase substantially when you use a nonstandard size. Check with your technical-publications department or a print shop for current prices and availability of paper sizes.

- *Page count.* For the most part, page count is a cost factor and a psychological factor. The cost factor is simple: paper is expensive and heavy, so you want to reduce the number of pages as much as you can, especially if you are printing many copies and mailing them. In the software industry, many companies have replaced 600-page manuals with 100-page versions — in part because of printing, warehousing, and mailing costs. The psychological factor is also easy to understand: everyone wants to spend as little time as possible reading technical communication. Therefore, if you can figure out a way to design the document so that it is 15 pages long rather than 30 — but still easy to read — your readers will appreciate it.

Paper

Paper is made not only in different standard sizes but also in different weights and with different coatings.

The lowest quality paper is newsprint. Because it is extremely porous, it allows inks to bleed through to the other side and picks up smudges and oil. In addition, it can turn yellow in as little as a few weeks. For these reasons, newsprint is generally used only for newspapers, informal newsletters, and similar quick and inexpensive bulk publications.

The most widely used paper is the relatively inexpensive paper stock used in photocopy machines and laser printers. Others include bond (for letters and memos), book paper (a higher grade that permits better print resolution), and text paper (an even higher grade used for more formal documents such as announcements and brochures). A heavier weight is also available for cover pages and reference cards.

Most types of paper can be ordered coated or uncoated. The coating increases strength and durability, and it allows the best print resolution. However, some glossy coated papers produce an annoying glare. To deal with this problem, designers recommend choosing a paper with a slight tint. A bone white, for instance, produces less glare than a bright white.

Paper is also available in different grades. For small jobs (under 5,000 copies of a brief document) you would probably want a high-grade paper. For larger jobs, however, higher grades can increase total costs substantially.

Work closely with printing professionals. They know, for example, about UV-coated paper, which greatly reduces fading. And they know the current costs of recycled paper, which is constantly improving in quality and becoming less expensive.

Bindings

Technical documents of a few pages can be held together with a paper clip or a staple. But longer documents require more sophisticated binding techniques.

Table 13.1 illustrates and describes the four types of bindings commonly used in technical communication.

■ Table 13.1   Common Types of Bindings

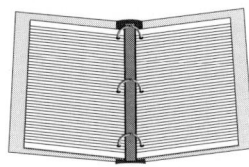

Loose-leaf binders. Loose-leaf binders are convenient when pages must be added and deleted frequently. But a high-quality binder can cost as much as several dollars.

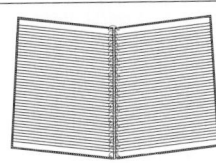

Ring or spiral binders. The wire or plastic coils or combs that hold the pages together let you open the document on a desk or even fold it over, so that it takes up only the space of one page. Neighborhood print shops will bind documents of almost any size in plastic coils or combs for about a dollar each.

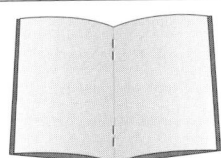

Saddle binding. The document is opened to its middle pages, and a set of large staples is inserted from the outside. Saddle binding is not practical for large documents.

Perfect binding. Pages are glued together along the spine edge, and a cover is attached. Perfect binding, which is used in book publishing, produces the most formal appearance, but it is relatively fragile, and the open document usually does not lie flat.

■ **Table 13.2   Typical Accessing Aids**

Icons. Icons are pictures that symbolize actions or ideas. A garbage can on your computer screen represents the task of erasing a file. An hourglass or a clock tells you to wait while the computer performs a task. Perhaps the most important icon is the stop sign, which alerts you to a warning.

Beware of being too cute in thinking up icons. One computer manual uses a cocktail glass about to fall over to symbolize "tip" (not a good idea). Don't use too many different icons either, or your readers will forget what each one represents.

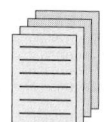

Color. Perhaps the strongest visual attribute is color (Keyes, 1993). Use color to draw attention to important features of the document, such as warnings, hints, major headings, and section tabs; but use it sparingly, or it will overpower everything else in the document.

Use colors logically. Third-level headings should not be in color, for example, if first- and second-level headings are black.

Using a different color of paper for each section of a document is another way to simplify access.

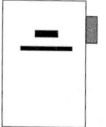

Dividers and tabs. You already know dividers and tabs — the heavy-paper sheets with the paper or plastic extensions — from loose-leaf notebooks. Tabs provide a place for a label, which enables readers to identify and flip to a particular section. Sometimes dividers and tabs are color coded.

For more about using color, see Ch. 14, p. 372.

Accessing Tools

In a well-designed document, readers can move around easily to find the information they seek. Table 13.2 explains six common kinds of accessing aids: icons, color, dividers and tabs, cross-reference tables, headers and footers, and page numbering.

DESIGNING THE PAGE

A page of technical communication is effectively designed if the reader can recognize a pattern — such as where to look for certain kinds of information — but isn't put to sleep by dull monotony.

This section explains how basic design principles apply to page design. It first discusses what learning theory can teach us about designing better pages

See the section on desktop publishing at About.com (www.about.com) for information on design principles and software.

■ **Table 13.2** *(Continued)*

Read . . .	***To learn to . . .***	
Ch. 1	connect to the net	*Cross-reference tables.* These tables refer readers to related material.
Ch. 2	use email	

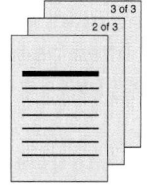

Headers and footers. Headers and footers help readers see at a glance where they are in the text. A header appears at the top of the page, a footer at the bottom. In this book, for example, the headers on the left-hand pages repeat the chapter title; those on the right-hand pages indicate the most recent first-level heading. Sometimes writers build other identifying information into the headers. For example, your instructor might ask you to identify your assignments with a header like the following: "Smith, Progress Report, English 302, page 6."

Page numbering. For one-sided documents, number the pages with arabic numerals in the upper right corner. (The first page of most documents is unnumbered.) For two-sided documents, put the page numbers near the outside margins.

Complex documents often use two number sequences: lowercase roman numerals (*i, ii,* and so on) for front matter, and arabic numerals for the body. The title page is unnumbered; the page following it is *ii.*

Appendices are often paginated with a number and letter combination: Appendix, Part A, begins with page A-1, followed by A-2, and so on; Appendix, Part B, starts with page B-1, and so on.

Sometimes documents that are photocopied and distributed or faxed list the total number of pages in the document (so recipients can be sure they have all of them). The second page is "2 of 3," and the third page is "3 of 3."

Documents that will be updated are sometimes numbered within each section: Section 3 begins with page 3-1, followed by 3-2; Section 4 begins with 4-1. This way, a complete revision of one section does not affect the pagination of subsequent sections.

and then considers page grids, the visual plan for the page; typography; the design of titles and headings; and special design techniques (such as text boxes).

Learning Theory and Page Design

Your job in designing the page is to create visual patterns that help readers find, understand, and remember information. Three principles of learning theory, the result of research into how people learn, can help you design effective pages. These principles are chunking, queuing, and filtering (see Keyes [1993] for a summary of the research on learning theory).

- *Chunking.* People understand information best if it is delivered to them in chunks — small units — rather than all at once. In a business letter,

Chunking, on the right, emphasizes units of related information.

■ **Figure 13.1 Chunking**   a. Without chunking   b. With chunking

which is typed single spaced, chunking involves double spacing between paragraphs, as shown in Figure 13.1.

- *Queuing.* Queuing refers to creating visual distinctions to indicate levels of importance. In a traditional outline, the roman numeral "I" heading is more important than the arabic numeral "1" heading. In designing a page, using more emphatic elements — such as bigger type or boldface type — suggests importance.

   Another visual element of queuing is indentation. Designers start more important information closer to the left margin and indent less important information. (An exception is titles, which are often centered in reports in the United States.) Figure 13.2 shows queuing by size and by alignment.

- *Filtering.* Filtering is the use of visual patterns to distinguish between various types of information. A stop sign in a set of instructions, for

In figure a, the first-level head
at the top is bigger than the
second-level heads.

In figure b, each of the three
headings is a different size, and
the headings and accompanying text are aligned on different
left margins.

■ **Figure 13.2 Queuing**   a. Queuing by size   b. Queuing by size and
                                                              indentation

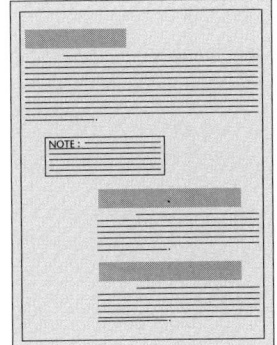 

The text box in this example sets off notes.

a. Filtering using a text box

b. Filtering using a text box and an icon

■ **Figure 13.3   Filtering**

example, often signals safety information. Designers also use typography to produce the same effect. Introductory material might appear in larger type, and notes in italics or another typeface. Figure 13.3 shows filtering.

Page Layout

Every page has two kinds of space: white space and space for text and graphics. The best way to approach page design is to make a grid — a picture of what the page will look like. In making a grid, you will decide how to use white space and determine how many columns to have on the page.

Page Grids

As the phrase suggests, a *page grid* is like a map on which you chart where the text, the graphics, and the white space should go. To devise an effective grid, consider your audience, their purpose in reading, and their reading behavior.

Many writers like to begin with a *thumbnail sketch,* a rough drawing that shows how the different elements — text and graphics — will look on the page. Figure 13.4 shows several thumbnail sketches for a page from the body of a manual.

 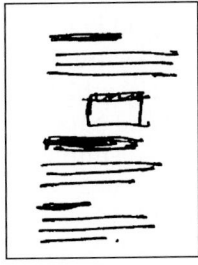  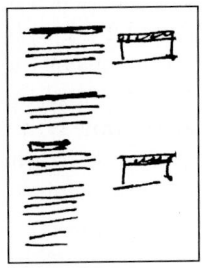

■ **Figure 13.4
Thumbnail Sketches**

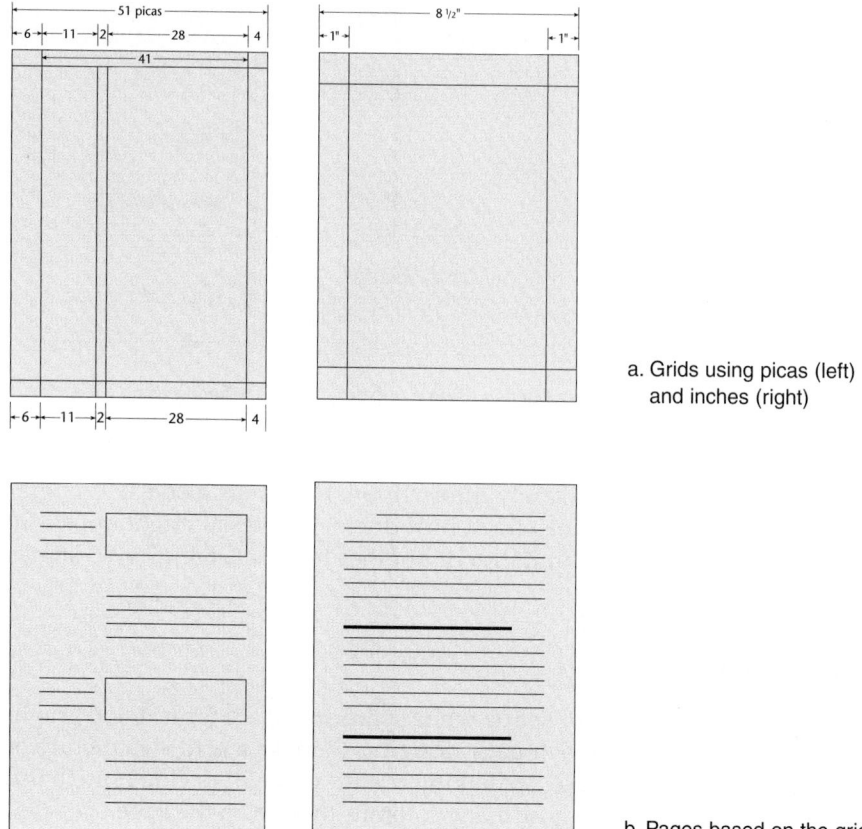

a. Grids using picas (left)
   and inches (right)

■ **Figure 13.5**
Sample Grids Using Picas
and Inches

b. Pages based on the grids

Keep experimenting by sketching the different kinds of pages your document will have: body pages, front matter, and so on. When you are satisfied, make page grids. You can use either a computer or a pencil and paper, or you can combine the two techniques. On a computer, you can use a word-processing program or a draw program or paint program (see Chapter 14, p. 401) to create the grid. Many word-processing programs come with templates you can use as is or modify.

But you don't need sophisticated software to design a grid. You can draw it with a pencil on a sheet of paper of the size you will be using. Some writers like to use graph paper because its printed grid lines are convenient. You can use inches or the metric system for your measurements, or you can use the pica, the unit that printing professionals use, which equals one-sixth of an inch. Some writers experiment with pencil and paper and then transfer their grid to the computer. Figure 13.5 shows two simple grids: one using picas and one using inches.

Experiment with different grids until you think you have a page design that is attractive, able to meet the needs of your readers, and appropriate for the information you are conveying. Figure 13.6 shows some possibilities.

a. Double-column grid

b. Two-page grid, with narrow outside columns for notes

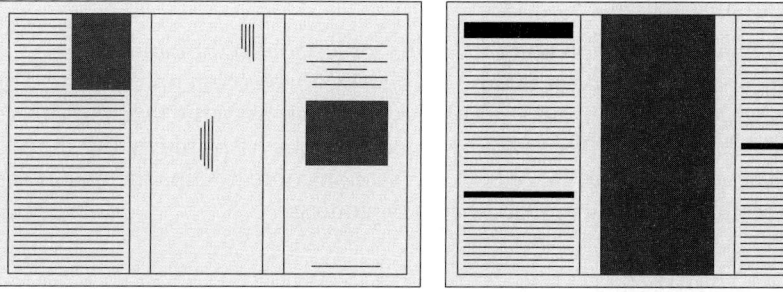

c. Three-panel brochure

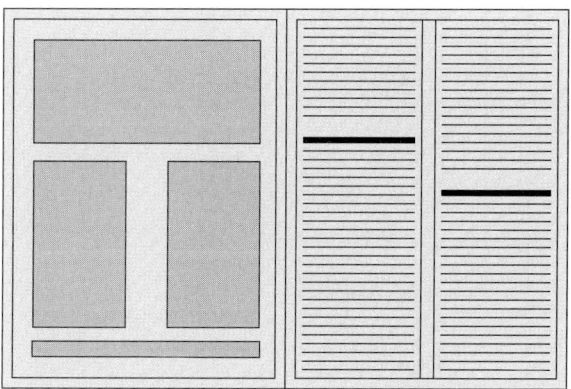

d. Two-page grid, with graphics on the left page and double-column text on the right page

■ **Figure 13.6**
Common Grids

White Space

Sometimes called *negative space,* white space is the area of the paper with no writing or graphics: the space between two lines of text, the space between text and graphics, and, most obvious, the margins. White space directs readers' eyes to a particular element, emphasizing it. In addition, white space helps readers see relationships among elements on the page.

Up to half the area on a typical page is given over to margins. Why so much? Margins serve four main purposes:

- They limit the amount of information on the page, thus making it seem easier to read and use.
- They provide space for binding and allow readers to hold the page without covering up the text.
- They provide a neat frame around the type.
- They provide space for marginal glosses. (Marginal glosses are discussed later in this chapter.)

Figure 13.7 shows common margin widths for left and right-hand pages.

White space can also set off and emphasize an element on the page. For instance, white space around a graphic separates it from the text and draws the reader's eye to it. White space between columns helps the reader to read the text easily. And white space between sections of text helps the reader see that one section is ending and another is beginning.

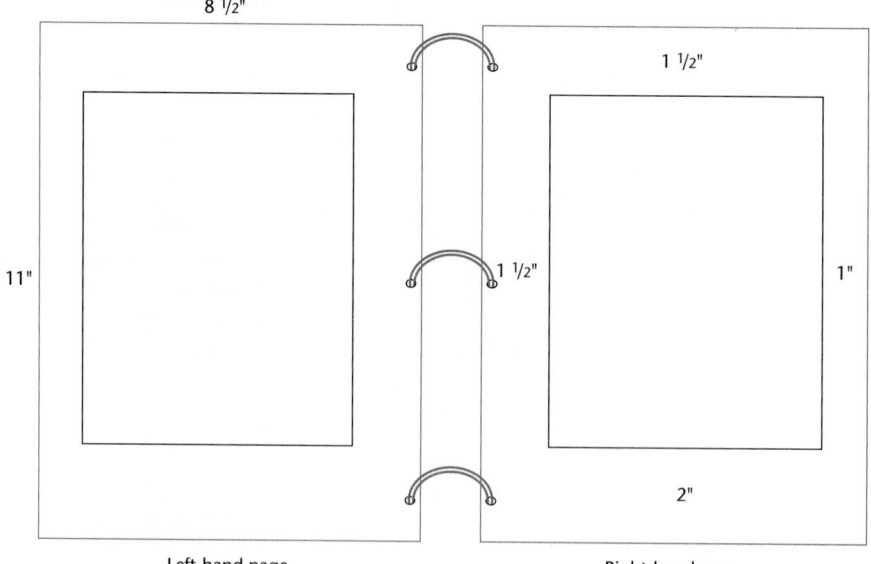

Left-hand page   Right-hand page

■ **Figure 13.7   Typical Margins for a Document That Is Bound like a Book**

Increase the size of the margins when the subject becomes more difficult or when your readers become less knowledgeable about it.

Columns

Unlike the one-column documents you have written as a student, many of the documents you write on the job will have multiple columns. A multi-column design offers four major advantages:

- Text in columns is generally easier to read because the lines are shorter.
- Columns allow you to fit more information on the page, because many graphics can fit in one column or extend across two or more columns. In addition, a multicolumn design can accommodate more words on a page than a single-column design.
- Columns allow you to create a visual pattern: text in one column, accompanying graphic in an adjacent column.
- Columns give you greater flexibility in sizing graphics.

Typography

Typography is the study of type and of the way people read it. When designers and technical communicators study typography, they learn about typefaces, families, case, and sizes. They also consider the white space of typography: line spacing, line length, and justification.

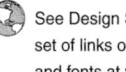 See Design Sphere Online's set of links on typography and fonts at www.dsphere.net/src/stacks/stacks_fonts.html.

Typefaces

A typeface is a set of letters, numbers, punctuation marks, and other symbols, all bearing a characteristic design. Today there are thousands of typefaces, generally bearing the name of their designer or some historical name (such as Times Roman, Avant Garde, Helvetica, Garamond, or Zapf Chancery). More are designed every year. Figure 13.8 shows three contrasting typefaces.

This paragraph is typed in French Script typeface. You are unlikely to see this style of font in a technical document because it is too ornate and too hard to read. It is better suited to wedding invitations and other formal announcements.

This paragraph is Times Roman. It looks like the kind of type used by the *New York Times* and other newspapers in the nineteenth century. It is an effective typeface for text in the body of technical documents.

This paragraph is Univers, which has a modern, high-tech look. It is best suited for headings and titles in technical documents.

■ **Figure 13.8 Typefaces**

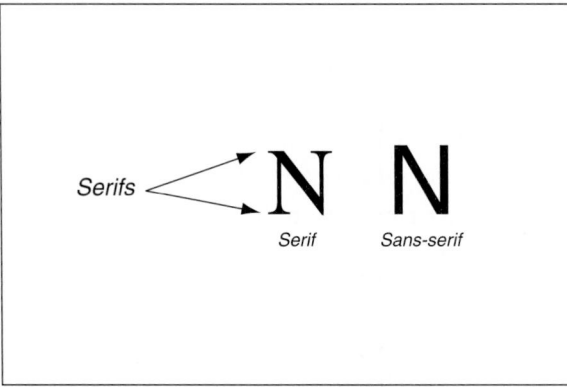

Serif typefaces are often considered easier to read because the serifs — the short extensions on the letters — encourage the movement of the reader's eyes along the line.

Sans-serif typefaces are harder on our eyes because the letters are less distinct from one another than they are in a serif typeface. However, sans-serif typefaces are easier to read on the screen and when printed on dot-matrix printers, because the letters are simpler.

■ **Figure 13.9   Serif and Sans-Serif Typefaces**

Sans-serif typefaces are used mostly for short documents and for headings. They are more common in Europe and are considered easier to read.

As Figure 13.9 illustrates, typefaces are generally classified into two categories: *serif* and *sans-serif.*

Most of the time you will use a standard font such as Times Roman, which is included in your software and which your printer can reproduce. Remember, however, that different typefaces convey different impressions and that some cause less eye fatigue than others.

Type Families

Each typeface belongs to a family of typefaces, which consist of variations on the basic style, such as italic and boldface. Figure 13.10, for example, shows Helvetica.

As with typefaces, be careful not to overload your text with too many different type styles. Used sparingly and consistently, however, the different members of a type family call attention to various kinds of text, such as warn-

Helvetica Light	**Helvetica Bold Italic**
Helvetica Light Italic	**Helvetica Heavy**
Helvetica Regular	***Helvetica Heavy Italic***
Helvetica Regular Italic	Helvetica Regular Condensed
Helvetica Bold	*Helvetica Regular Condensed Italic*

■ **Figure 13.10   Helvetica Family of Type**

ings and notes, or even those characters readers are to type on the computer keyboard (Felker, Pickering, Charrow, Holland, & Redish, 1981). You will use italics for book titles and other elements, and you might use bold type for emphasis and headings, but you can go a long time without needing condensed and expanded versions of typefaces. And you can live a full, rewarding life without using outlined or shadowed versions.

Case

To ensure that your document is easy to read, use uppercase and lowercase letters as you would in any other kind of writing (see Figure 13.11). The average person requires 10 to 25 percent more time to read text using all uppercase letters than to read text using both upper- and lowercase. In addition, uppercase letters take up as much as 35 percent more space than lowercase letters (Haley, 1991).

Individual variations are greater in lowercase words

THAN THEY ARE IN UPPERCASE WORDS.

Lowercase letters are easier to read than uppercase because the individual variations from one letter to another are greater.

■ **Figure 13.11   Individual Variations in Lowercase and Uppercase Type**

Type Sizes

Type size is measured according to a basic unit called a *point*. There are 12 points in a *pica* and 72 points in an inch.

Software and printers with scalable typefaces can produce letters and other characters in sizes, ranging from 0.25 to 999.75 points and in increments of 0.25 points. In most technical documents 10-, 11-, or 12-point type is used for the body of the text:

This paragraph is printed in 10-point type. This size is easy to read, provided it is reproduced on a letter-quality impact printer or laser printer. On most impact printers, however, the resolution isn't high enough for type this small.

This paragraph is printed in 12-point type. If you have a dot-matrix printer, 12-point type is the best size to use because it balances readability and economy.

This paragraph is printed in 14-point size. This size is appropriate for titles or headings.

Type sizes used in other parts of the document include the following:

footnotes	8- or 9-point type
indexes	2 points smaller than body text
slides or transparencies	24- to 36-point type

In general, you should aim for at least a 2 to 4 point difference between the headings and the body. Too many size variations, however, might suggest a sweepstakes advertisement rather than a serious text.

---

a. **Excessive line spacing**

Aronomink Systems has been contracted by Cecil Electric Cooperative, Inc.

(CECI) to design a solid waste management system for the Cecil County

plant, Units 1 and 2, to be built in Cranston, Maryland. The system will consist

of two 600 MW pulverized coal-burning units fitted with high-efficiency elec-

trostatic precipitators and limestone reagent FGD systems. The coal will

contain an estimated 3% sulfur and 10% ash. The station will output approxi-

mately 64 TPH (DWB) of FGD sludge and 24 TPH fly ash at 100% load.

b. **Appropriate line spacing**

Aronomink Systems has been contracted by Cecil Electric Cooperative, Inc. (CECI) to design a solid waste management system for the Cecil County plant, Units 1 and 2, to be built in Cranston, Maryland. The system will consist of two 600 MW pulverized coal-burning units fitted with high-efficiency electrostatic precipitators and limestone reagent FGD systems. The coal will contain an estimated 3% sulfur and 10% ash. The station will output approximately 64 TPH (DWB) of FGD sludge and 24 TPH fly ash at 100% load.

c. **Inadequate line spacing**

Aronomink Systems has been contracted by Cecil Electric Cooperative, Inc. (CECI) to design a solid waste management system for the Cecil County plant, Units 1 and 2 to be built in Cranston, Maryland. The system will consist of two 600 MW pulverized coal-burning units fitted with high-efficiency electrostatic precipitators and limestone reagent FGD systems. The coal will contain an estimated 3% sulfur and 10% ash. The station will output approximately 64 TPH (DWB) of FGD sludge and 24 TPH fly ash at 100% load.

■ **Figure 13.12   Line Spacing**

Line Length

Ironically, the line length most often used on an 8.5 × 11-inch page — about 80 characters — is actually somewhat difficult to read. A shorter line of perhaps 50 to 60 characters — is less tiring to read, especially over the course of a long document (Biggs, 1980). One advantage of a multicolumn format is that you can choose the line length appropriate for the type size and the paper, while keeping in mind the difficulty of the text and the knowledge level of readers.

Line Spacing

Sometimes called *leading* (pronounced "ledding"), *line spacing* refers to the white space between lines or between a line of text and a graphic. If lines are too far apart, the page looks diffuse, the text loses coherence, and the reader tires quickly. If lines are too close together, the page looks crowded and becomes difficult to read. Some research suggests that smaller type, longer lines, and sans-serif typefaces all benefit from a little more line spacing. Figure 13.12 shows three variations on line spacing.

Line spacing is usually determined by the kind of document you are writing. Memos and letters are single spaced; other documents, such as reports and proposals, are double spaced or one-and-a-half spaced.

Line spacing can also be used to separate one section of text from another. Breaks between single-spaced paragraphs usually consist of two spaces. (Double-spaced and one-and-a-half-spaced text has no extra line spacing between paragraphs, but the first line of each paragraph is indented, as in handwritten documents.)

Figure 13.13 shows how line spacing can be used to distinguish one section of text from another and to separate text from graphics.

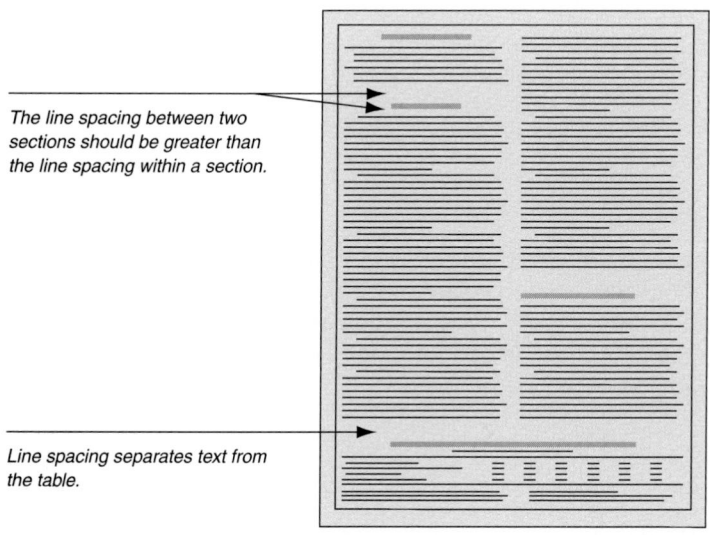

The line spacing between two sections should be greater than the line spacing within a section.

Line spacing separates text from the table.

■ **Figure 13.13**
Line Spacing Used to Distinguish One Section from Another

Line spacing is also used to set off block quotations. If the text is single spaced, set off the quotation above and below with double spacing; if the text is one-and-a-half or double spaced, use triple spacing.

Justification

Justification refers to the alignment of words along the left and right margins. In technical communication, text is often *left-justified*; that is, except for paragraph indentations, the lines begin along a uniform left margin but end on an irregular right border. This irregular right border is often called *ragged right*. Ragged right is most common in typewritten and word-processed text (even though word processors can justify the right margin).

In justified text, also called *full-justified text*, both the left and right margins are justified. Justified text is seen most often in typeset, formal documents, such as books.

The following passage (Berry, Mobley, & Turk, 1994, p. 647) is presented first in left-justified form and then in justified form:

Notice that the space between words is uniform in left-justified text.

Until recently, most software development has adhered to a linear development process, often referred to as the "waterfall" process. Rimbaugh (1992) describes this as several phases performed in sequence, one after another. Using this approach, a product is first designed, then developed, tested, and delivered. Nothing is assembled until the design is complete, and the design is not changed once assembly starts.

In justified text, the spacing between words is irregular, slowing down the reader. Because a big space suggests a break between sentences, not a break between words, the reader can become confused, frustrated, and fatigued.

Until recently, most software development has adhered to a linear development process, often referred to as the "waterfall" process. Rimbaugh (1992) describes this as several phases performed in sequence, one after another. Using this approach, a product is first designed, then developed, tested, and delivered. Nothing is assembled until the design is complete, and the design is not changed once assembly starts.

Notice that the irregular spacing not only slows down reading but also creates "rivers" of white space. Readers are tempted to concentrate on the rivers running south rather than on the information itself.

Should you justify your technical communication? If you are using a standard word processor rather than a sophisticated desktop-publishing system, probably not.

Justification can make the text harder to read in one more way. Some word processors and typesetting systems automatically hyphenate words that do not fit on the line and sometimes hyphenate incorrectly. Hyphenation slows down the reader and can be distracting. Left-justified text does not require as much hyphenation as justified text does.

Designing Titles and Headings

Titles and headings should stand out visually on the page, because they announce the introduction of a new idea.

Designing Titles

Because a title is the most important heading in a document, it should be displayed clearly and prominently. If it is on a cover page or a title page, you might present it in boldface in a large type size, such as 18 or 24 points. If it also appears at the top of the first page, you might make it slightly larger than the rest of the text — perhaps 16 or 18 points for a document printed in 12 point — but smaller than it is on the cover or title page. Titles are often centered on the page between right and left margins.

For more on titling your document, see Ch. 10, p. 251.

Designing Headings

Readers should be able to tell when you are beginning a new idea. The most effective way to distinguish one level of headings from another is to use size variations (Williams & Spyridakis, 1992). A 20 percent size difference between a first-level head and a second-level head will be clear for most readers. Boldface also sets off headings effectively. The least effective way to set off headings is underlining, because in many word processors the underline obscures the *descenders*, the portions of letters such as *p* and *y* that extend below the base line.

For more about using headings, see Ch. 10, p. 252.

Indenting a heading is another way to convey a visual pattern. In general, the more important the heading level, the closer it is to the left margin: first-level headings usually begin at the left margin, second-level headings are indented four or five characters, and third-level headings are indented seven or eight. Indented third-level headings can also be run into the text.

Indenting the text accomplishes two goals:

- It emphasizes the heading and distinguishes it from the text that follows.
- It subordinates the text. You can also decrease the type size for lower-level text. For instance, in a document printed in 12-point type, all third-level and fourth-level text is in 10-point type. Writers sometimes highlight their headings by using *hanging indentation:* extending the heading further out into the left margin than the accompanying text.

In designing headings, use line spacing carefully. A perceivable distance between a heading and the text increases the impact of the heading. Consider these three examples:

Summary

In this example, the writer has skipped a line between the heading and the text that follows it. The heading stands out clearly.

Summary
In this example, the writer has not skipped a line between the heading and the text that follows it. The heading stands out, but not as emphatically.

Summary. In this example, the writer has begun the text on the same line as the heading. This run-in style makes the heading stand out the least.

Other Design Features

Table 13.3 shows five design techniques that are used frequently in technical communication: rules, boxes, screens, marginal glosses, and pull quotes.

■ **Table 13.3   Rules, Boxes, Screens, Marginal Glosses, and Pull Quotes**

rules 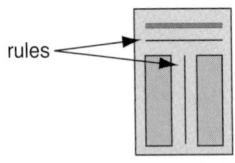	*Rules.* A rule is a design term for a straight line. Using a word processor, you can easily add horizontal or vertical rules to a document. Horizontal rules are often used to separate headers and footers from the body of the page or to divide two sections of text. Vertical rules are used to separate columns on a multi-column page or to identify revised text in a manual.
boxes 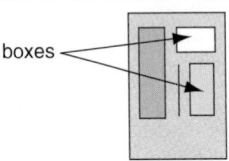	*Boxes.* Adding rules on all four sides of an item creates a box. Boxes can enclose graphics or special sections of text, or form a border for the whole page. Boxed text is often positioned to extend into the margin, giving it further emphasis.
screen 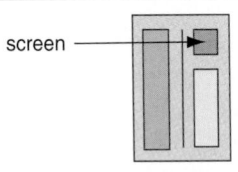	*Screens.* The background shading behind text or graphics for emphasis is known as a screen. The density can range from 1 percent to 100 percent; 5 to 10 percent is usually enough to provide emphasis without making the text illegible. You can use screens with or without boxes.
marginal gloss 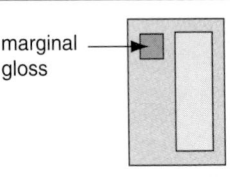	*Marginal glosses.* A marginal gloss is a brief comment on the main discussion, such as a summary statement, in the margin of the document. Marginal glosses are usually set in a different typeface — and sometimes in a different color — from the main discussion and are often separated from it by a vertical rule. Marginal glosses can be helpful in providing a quick overview of the main discussion, but they can also compete with the text for the reader's attention.
pull quote 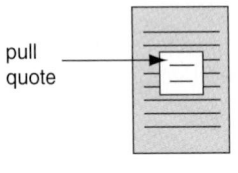	*Pull quotes.* A pull quote is a brief quotation, usually just a sentence, that is pulled from the text and displayed in a larger type size, and generally in a different typeface, often in a box. Newspapers and magazines use pull quotes to attract readers' attention and make them want to read the article. Pull quotes are inappropriate for reports and similar documents, because they look too informal. They are increasingly popular, however, in newsletters.

 To view Figures 13.15 and 13.16 in context on the Web, click on Links Library for Ch. 13 on TechComm Web (www.bedfordstmartins.com/techcomm).

ANALYZING SAMPLE PAGE DESIGNS

Figures 13.14 through 13.16, showing three typical page designs used in technical documents, illustrate the concepts discussed in this chapter.

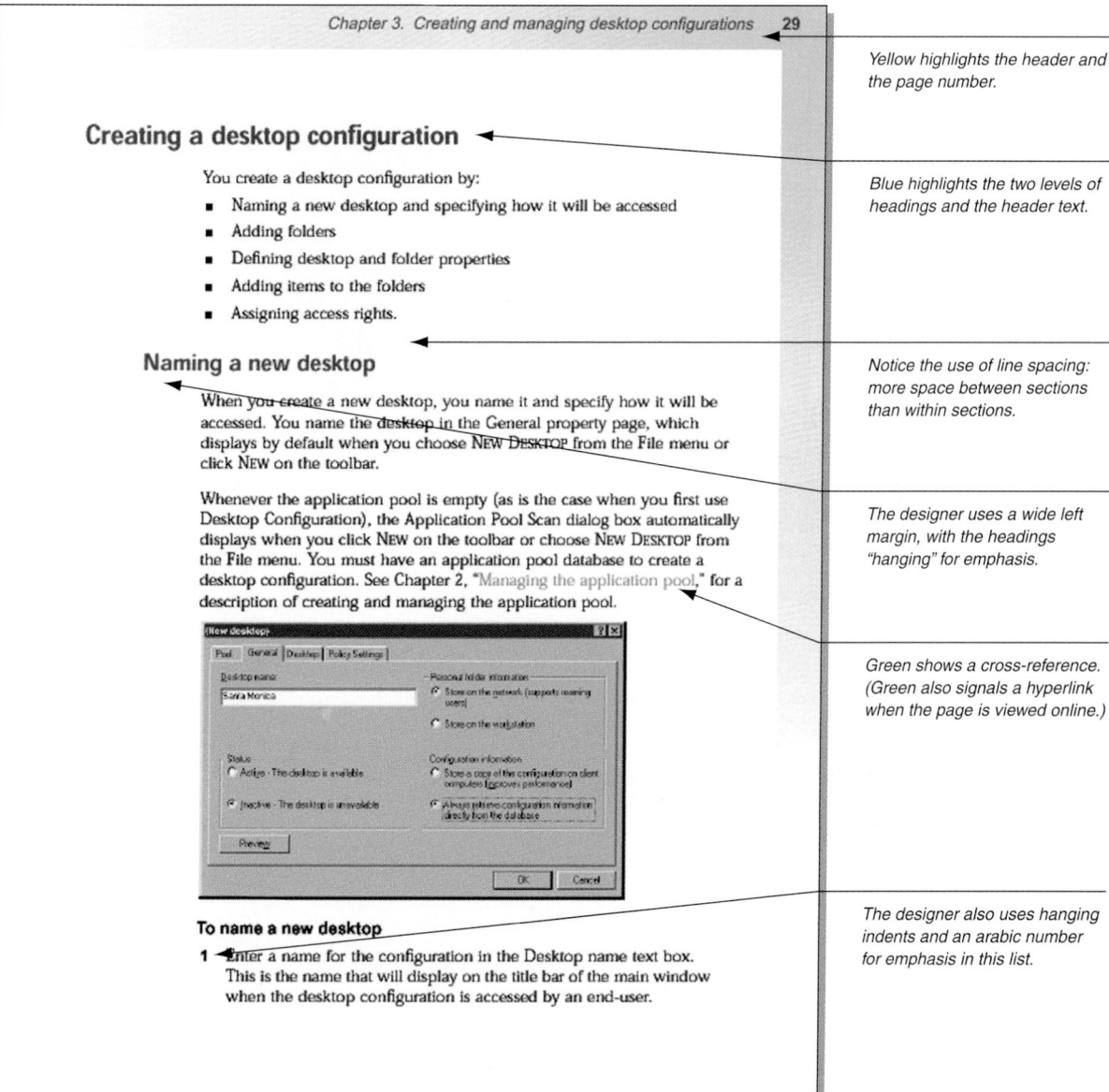

Yellow highlights the header and the page number.

Blue highlights the two levels of headings and the header text.

Notice the use of line spacing: more space between sections than within sections.

The designer uses a wide left margin, with the headings "hanging" for emphasis.

Green shows a cross-reference. (Green also signals a hyperlink when the page is viewed online.)

The designer also uses hanging indents and an arabic number for emphasis in this list.

■ **Figure 13.14**
A Page from a Manual
(Hewlett-Packard, 1997)

This page is designed to emphasize the important content.

The writer uses a gray screen to identify header information. However, the page number — 7 — should be more clearly separated from the title: "Page 7, Oil Spill Program Update."

The body text is set in a serif typeface; the display text — the heading — is set in a sans-serif typeface.

The multicolumn design allows the writer to include the photograph without wasting any space.

Vertical rules separate the text into three columns. The lines are set ragged right.

Containment booms were set up on the mouth of Independence Creek to prevent further contamination of the Missouri River.

Olympic Pipeline Spill and Fire

Three people were killed and 10 injured when a pipeline carrying automotive and jet fuel ruptured, leading to an explosion and fire along Whatcom Creek in Bellingham, Washington, on June 10, 1999. The three fatalities included a fisherman that was apparently overcome by fumes and drowned and two 10-year old boys who died from extensive burns in a Seattle hospital the following morning. Witnesses report that the fuel ignited as the two boys were playing with a cigarette lighter along the creek.

Olympic Pipe Line Company (OPL), the responsible party, estimated that nearly 277,000 gallons of gasoline escaped into Whatcom Creek during the leak. The fuel created a 15-foot thick vapor cloud as it spread downstream. The explosion occurred next to the Bellingham city water treatment facility and disrupted the local water supply. Fires quickly spread about 1.5 miles downstream, destroying one home and damaging a second. Officials report that most of the fuel released from the pipeline was consumed during the intense fires. Local police, fire and OPL employees responded to reports of a gasoline odor just minutes before the explosion occurred.

A *Seattle Times* report of the preliminary investigation describes a series of events that led to the release and explosion. The problem began when computer in a pipeline

needed to contain and remove the spill. An inspection by Scott Hayes, the EPA on-scene coordinator (OSC), confirmed that fuel had escaped the containment measures established by WPC and was entering the Missouri River. Containment booms were set up on the mouth of Independence Creek to prevent further contamination of the Missouri River, although the volume and velocity of the river precluded the use of containment booms on the river itself.

Although vacuum trucks were employed to recover fuel from the containment areas the number of trucks was inadequate for the size of the spill. The six vacuum trucks used at the spill could not remove fuel at the same rate that it was leaking from the pipe. One of the trucks was entirely devoted to removing spilled fuel from a containment trench dug to divert the flow of diesel fuel escaping from the ruptured pipe. Rainy weather conditions made entrainment of fuel beyond the containment booms more difficult. In-situ burning was considered at the spill, but was ruled out due to the potential risks

involved and because more recovery resources became available. Inspection by the OSC and Superfund Technical Assistance and Response Team (START) contractor revealed that containment had been achieved at only one of the three containment areas by the morning of May 12, 1999. Although backup booms were deployed at the other containment sites, nightly storms and inadequate removal equipment resulted in an undetermined amount of fuel escaping from the containment booms. City officials ordered the shut down of industrial operations within the city of Atchison, as hydrocarbons were detected at the water treatment plant.

WPC estimates that 4,565 barrels of diesel were discharged during the incident. Response efforts recovered approximately 2,800 barrels–about 61 percent of the estimated spill amount.

For more information, please contact Scott Hayes, OSC EPA Region 7 at (913) 551-7670.

■ **Figure 13.15**   **A Page from an Online Newsletter (United States Environmental Protection Agency, 1999)**

This page is designed to use space efficiently.

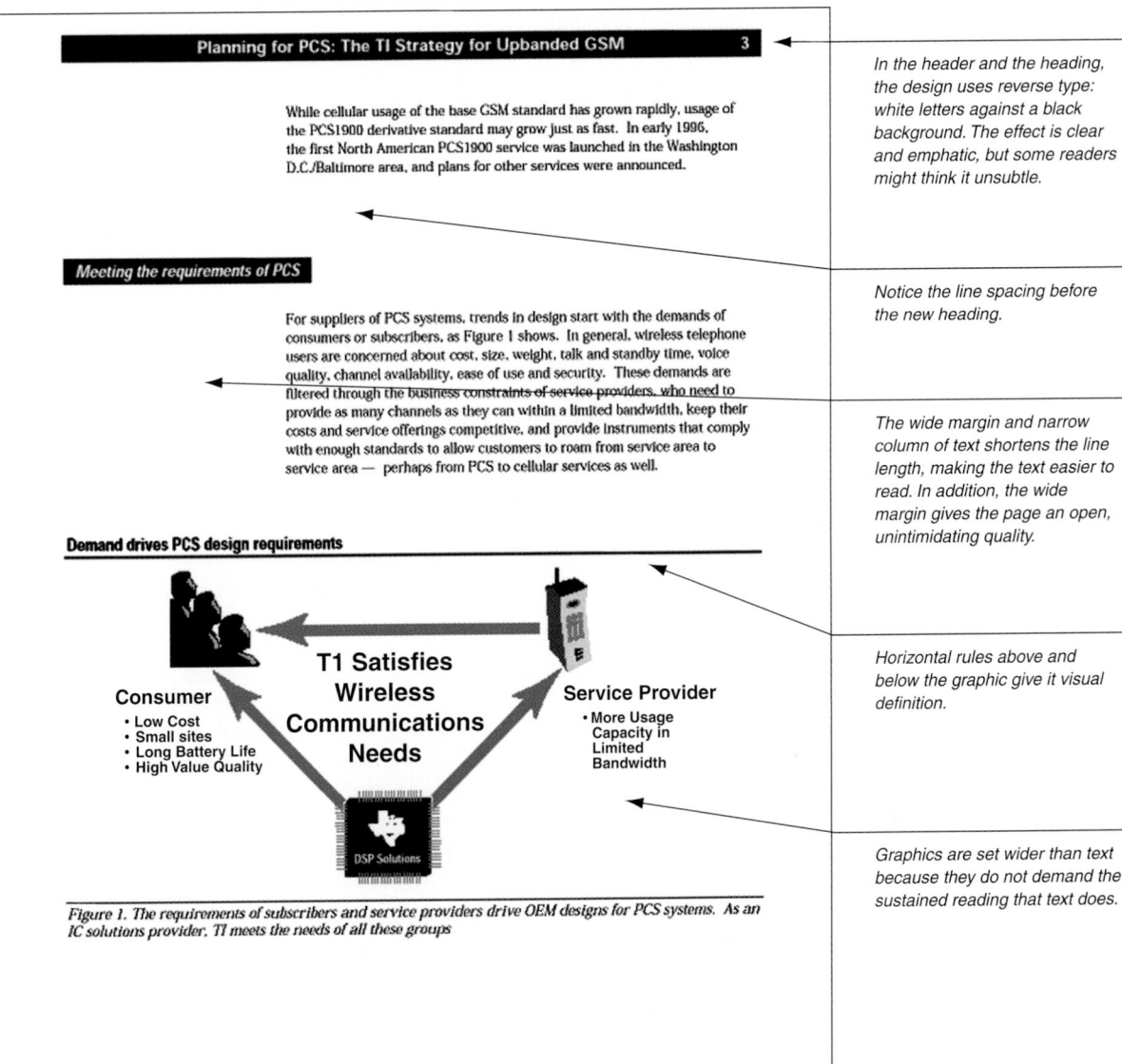

In the header and the heading, the design uses reverse type: white letters against a black background. The effect is clear and emphatic, but some readers might think it unsubtle.

Notice the line spacing before the new heading.

The wide margin and narrow column of text shortens the line length, making the text easier to read. In addition, the wide margin gives the page an open, unintimidating quality.

Horizontal rules above and below the graphic give it visual definition.

Graphics are set wider than text because they do not demand the sustained reading that text does.

Figure 13.16   A Page from an Online Marketing Brochure (Texas Instruments, 1999)

This page is designed to give readers easy access to the information.

Revision Checklist

Did you
- ❏ analyze your audience: their knowledge of the subject, attitudes, reasons for reading, and the kinds of tasks they will be carrying out?
- ❏ think about what your readers will expect to see when they pick up the document?
- ❏ determine your resources in time, money, and equipment?
- ❏ consider the best size for the document?
- ❏ consider the best paper for the document?
- ❏ consider the best binding for the document?
- ❏ think about which accessing tools would be most appropriate, such as icons, color, dividers, tabs, and cross-reference tables?
- ❏ devise a style for headers and footers?
- ❏ devise a style for page numbers?
- ❏ draw thumbnail sketches and page grids that define columns and white space?
- ❏ choose typefaces that are appropriate to your subject?
- ❏ use appropriate styles from the type families?
- ❏ use type sizes that are appropriate for your subject and audience?
- ❏ decide on whether to use left-justified text or full-justified text?
- ❏ choose a line length that is appropriate for your subject and audience?
- ❏ choose line spacing that is appropriate for your line length, subject, and audience?
- ❏ design your title for clarity and emphasis?
- ❏ work out a logical, consistent style for each heading level?
- ❏ use rules, boxes, screens, marginal glosses, and pull quotes where appropriate?
- ❏ use color, if you have access to it, to highlight certain items, such as warnings?

Exercises

1. Your word processor contains a number of templates for such documents as reports, letters, and memos. Study two templates for the same kind of document, for example, memos. What are the main differences between them? For what writing situations — audience, purpose, and subject — is each one most suited?

2. Study the first and second pages of an article in a journal in your field. Describe 10 design features you identify on these two pages. Which design features are most effective for the audience and purpose? Which are least effective?

3. The following excerpt from a report by the Bonneville Power Administration (1993) is underdesigned. Draw three different thumbnail sketches for a new, more interesting design that will better emphasize important information.

2.2.5 Anadromous Fish Columbia River Basin

The Pacific Northwest supports a large number of anadromous fish (species that migrate downriver to the ocean to mature, then return upstream to spawn). The principal anadromous fish runs in the Columbia Basin are chinook, coho, sockeye salmon, and steelhead trout.

These fish are an important resource to the Pacific Northwest, both for their economic value to the sport and commercial fisheries, and for their cultural and religious value to the region's Indian tribes and others.

The development of dam and reservoir projects on the Columbia and Snake Rivers and tributaries has reshaped the natural flows of these rivers. The use of storage reservoirs to capture runoff for later release results in reduced flows during the spring and early summer, when juvenile salmon and steelhead are migrating downstream to the ocean. Water velocities have also been reduced as a result of the increased cross-sectional area of the river due to run-of-river projects. These changes have slowed juvenile fish migration, exposing juvenile salmon and steelhead to predation and disease and impairing their ability to adapt to salt water when they reach the ocean. Additional mortality occurs as fish attempt to pass each dam on their downstream migration to the ocean.

Flow: Flow plays an important part in moving juvenile fish downstream to the ocean. In 1982, the Northwest Power Planning Council established a specific volume of water, known as the water budget, to increase river flows during the April 15 through June 15 period. This coincides with the peak out-migration of spring fish (predominately yearling chinook, steelhead, and sockeye), which depend on adequate river flow, particularly velocity, for a successful migration. The federal hydro system is operated to provide this water each year.

Not only is flow important for moving juvenile fish downstream past the dams, but flow is an important component of ensuring successful spawning and emergence of fall chinook on Vernita Bar (a gravel bar used by spawning fall chinook, located in the Hanford Reach of the Columbia River, downstream from Priest Rapids Dam). In 1988, BPA and the mid-Columbia operators signed a long-term Vernita Bar Agreement, which specifies protection requirements for fall chinook spawning, incubation, and emergence on Vernita Bar.

Spill: Until adequate bypass systems are installed at all the dams, spill remains a necessary means of moving juvenile fish past dams. Planned fish spill now includes the negotiated Spill Agreement, as well as a restricted operation at Bonneville Dam by the Corps. Planned spill also includes spill levels specified by the federal Energy Regulatory Commission for nonfederal projects. Planned spill does not include overgeneration spill (water which is spilled because there is no market for the energy it would produce) and is not changed as a result of the resource additions. Planned fish spills are met under all water conditions.

Research Projects

The following projects ask you to write a memo. See Chapter 15, page 430, for a discussion of writing memos.

4. Form small groups for this collaborative exercise in analyzing design. Photocopy a page from a book or a magazine. Choose a page that does not contain advertisements. Each person works separately for the first part of this project:

   - One person describes the design elements.

   - One person evaluates the design. Which aspects of the design are effective, and which could be improved?

   - One person creates a new design using thumbnail sketches.

   Then, the group members meet and compare notes. Do all members of the group agree with the first member's description of the design? With the second member's analysis of the design? Do all members like the third member's redesign? What have your discussions taught you about design? Write a memo to your instructor presenting your findings. Include the photocopied page with your memo.

5. The design of an organization's Web site is often quite different from the design of its printed materials. One reason is that some design features cannot be achieved in Web pages and some cannot be achieved in print. A second reason is that the audiences for the two media are often quite different; people visiting Web sites tend to be younger and less conservative than people reading print documents. Still, it makes sense for an organization to try to achieve some consistency in the design of its printed materials and its Web site. Choose an organization that maintains a Web site and issues printed information for the public, such as IBM. (Many organizations present printed information, in the form of .pdf files, on their Web sites.) Study a representative screen from the Web site and a page from the printed information. To what extent is the design of the two kinds of communication consistent? How is consistency achieved? Is it effective or does it appear haphazard? Does the difference in the audience and purpose of the two kinds of communication justify the inconsistencies? Present your findings in a memo to your instructor. Along with your memo, include a printed copy of the Web page and a photocopy of the printed page.

This case is available online at TechComm Web (www .bedfordstmartins.com /techcomm). See TechComm Web for additional cases.

C A S E 1
Redesigning a Report

You and the other members of your group work for the National Commission on Libraries and Information Science. Each year your commission publishes a number of reports using the design you see illustrated by the following sample page (National, 1999, p. 1). Your supervisor, Julie Alvarez, asks you to analyze the design and make some recommendations for improving it. "Make up a few sample pages and recommend the one you think works best. Then we'll choose a new design from the samples." Along with the other members of your group, analyze the strengths and weaknesses of the current design. Draw four different thumbnail sketches. Then, using the digital text of this page on TechComm Web (www.bedfordstmartins.com /techcomm), create four designs using your word processor or desktop-publishing software. Print the first page of each of the four designs. Write a memo to Ms. Alvarez recommending one of the four designs and explaining why you think it is the best design.

For a version of this page that you can save as a word-processing file, click on Links Library for Ch. 13 on TechComm Web (www .bedfordstmartins.com/ techcomm).

Moving Toward More Effective Public Internet Access

MOVING TOWARD MORE EFFECTIVE PUBLIC INTERNET ACCESS:
The 1998 National Survey of Public Library Outlet Internet Connectivity

I.   Background

The National Commission on Libraries and Information Science (NCLIS) has a long-standing interest in library statistics. Since 1988 NCLIS has maintained a partnership with the National Center for Education Statistics (NCES) focused on collecting and reporting statistics regarding libraries of various types across the United States. In addition, since 1994 NCLIS has tracked the increasing use of the Internet among public libraries.[1] At first NCLIS supported these studies solely with its own resources. In 1997 and 1998 the survey was funded jointly by NCLIS and the American Library Association (ALA) Office of Information Technology Policy.[2] In each case, the contractors for the research were John Carlo Bertot and Charles R. McClure.[3]

National surveys of U.S. public libraries and the Internet were done in 1994, 1996, 1997, and 1998 during a period in which the Internet has undergone tremendous growth in terms of users, accessible information content, and range of services. The *1998 National Survey of Public Library Outlet Connectivity* was conducted between April and June 1998, against the backdrop of pending universal service support for telecommunications services including Internet access in schools and public libraries. The findings of the 1998 survey provide a very useful baseline of data about public library Internet connectivity prior to the availability of universal service funding.

1.   Telecommunications Act of 1996, Universal Service, and Public Libraries

In the Telecommunications Act of 1996 the U.S. Congress recognized the growth of the Internet and the significance of providing individual citizens with access to Internet information content and services. In that legislation the U.S. Congress acknowledged, "the rapidly developing array of Internet and other interactive computer services available to individuals Americans...[as] an extraordinary advance in the availability of educational and informational resources to our citizens" [47 U.S.C. 230]. The Congress also acknowledged the special role of elementary and secondary schools, public libraries, and health care providers in conveying public access to these resources by providing discounted telecommunications services under a program of universal service originally established as a part of the Communications Act of 1934 [47 U.S.C. 254h].

---

[1] A bibliography of the reports in this series is available in Section IV.
[2] The views, opinions and recommendations expressed in the body of this report (Sections I through IV)) are those of the Commission, not necessarily those of the American Library Association or its members.
[3] Dr. John Carol Bertot, now with the School of Information Science and Policy at the State University of New York, Albany, conducted the 1998 survey and processed the results in collaboration with Dr. Charles R. McClure of the School of Information Studies, Syracuse University. The views, opinions and recommendations in the appendices to this report are theirs and do not necessarily reflect the official position or policy of the research sponsors.

CASE 2
The Brochure with the Inconsistent Design

You work in the corporate communications department at Milton Bradley Company, which produces Scrabble® and many other board games. Your boss, Louise Willis, approaches you one day. "We've got a problem. The redesign of the brochure for the next printing of Scrabble® isn't going well. The designer quit halfway through the job. John was in a real hurry to meet our production deadline, so he hired another firm to finish the design. Would you mind taking a look at the mockup and letting me know whether you think it looks okay or whether it still needs work?"

Study the following copy of the four-page brochure. Does it seem appropriate for the intended audience? Is it clear and attractive? Is there any evidence that two different sets of designers worked on it? If so, would problems be easy to fix or should the company start over? Present your analysis and recommendations in a memo to Louise Willis.

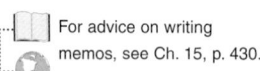
For advice on writing memos, see Ch. 15, p. 430.

Back cover

Front cover

Note: These pages are reduced 50% from the originals.

Special Features

The Official SCRABBLE® Players Dictionary (Second Edition) offers lots of features to help you play faster, better, and more confidently:
• No other dictionary includes all the two-to-eight letter words you'll find here.
• Only words that are permissible in SCRABBLE® can be found here. Proper names, words requiring hyphens or apostrophes, words considered foreign, and abbreviations have been omitted.
• All main entries are given a part-of-speech label, followed by appropriate inflected forms.
• All acceptable variant forms are shown alphabetically.
• All noun plurals are shown, as well as comparatives and superlatives of adjectives and adverbs.

Where to Get It

The Official SCRABBLE® Players Dictionary (Second Edition) costs only $14.95* at your local bookstore.

How to Order It

If the dictionary is out of stock, you can order it directly from the publishers. Send a check or money order in U.S. funds for $15.95,* which includes postage and handling, to:

MERRIAM-WEBSTER INC.
47 Federal Street
P.O. Box 281
Springfield, MA 01102
Attn: Consumer Sales

*Prices subject to change.
©1991, 1992 Milton Bradley Company. All Rights Reserved.
The gameboard shown on the book jacket photograph is
©1948 by Milton Bradley Company

SCRABBLE®
CROSSWORD GAME
by Milton Bradley

JOIN THE CLUB!

◆

The National
SCRABBLE®
Association

TURN THE PAGE TO FIND OUT HOW

Also Inside:

• OFFICIAL SCRABBLE® PLAYERS DICTIONARY OFFER
• FREE TILE REPLACEMENT

Page 2

Page 3

The National
SCRABBLE®
Association

As a member of the National SCRABBLE® Association, you'll receive a membership card; a roster of SCRABBLE® Players Clubs in the U.S. and Canada; 3 special word lists to help improve your play; and a 1-year subscription to the SCRABBLE® News. That's 8 exciting issues with...
• News about clubs and tournaments in your area
• Official National SCRABBLE® Association Tournament Rules
• Special word lists and helpful playing hints
• The latest in strategy
• Challenging quizzes & puzzles
Fill out the form below (print clearly) and send to: NATIONAL SCRABBLE® ASSOCIATION, P.O. Box 700, Greenport, NY 11944

YES! I'd like to join the National SCRABBLE® Association. I enclose a check or money order for $15.00 ($20.00 for Canadian membership; $25.00 outside the 50 states and Canada). Only U.S. funds will be accepted.
☐ *I am interested in starting a club in my area.*

NAME

STREET

CITY

STATE   ZIP

Are any of your tiles missing or defective?

If so, we want to replace them free of charge. Please list them on the form below and send it, with any defective tiles, to: MILTON BRADLEY COMPANY, Attn. Customer Service Department, 443 Shaker Road, East Longmeadow, MA 01028

TILE REPLACEMENT FORM
Please print clearly
NAME   STREET

CITY   STATE   ZIP
GAME:
☐ SCRABBLE   ☐ DELUXE SCRABBLE   ☐ TRAVEL SCRABBLE   ITEM # (from package)

MISSING TILES   DEFECTIVE TILES (please send with form)

IT'S OFFICIAL!

Have you ever had a word challenged that you knew was acceptable—yet it couldn't be found in your dictionary? Now there's a dictionary that was created especially for SCRABBLE® players...

The Official SCRABBLE® Players Dictionary

Second Edition!
Over 5,000 new entries!

Any standard abridged dictionary can be used to settle a SCRABBLE® challenge. But this easy-to-use, 683-page hardcover volume includes well over 100,000 words—and they're all acceptable SCRABBLE® words! That's why it's the official dictionary of first reference for all SCRABBLE® game tournaments sponsored by the National SCRABBLE® Association.

More on next page

14

Creating Graphics

An expert on the visual display of information, Edward R. Tufte (1999) defines an excellent graphic as:

The greatest number of ideas in the shortest time with the least ink in the smallest space.

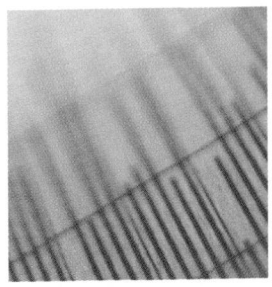

Graphics are the "pictures" in technical communication: drawings, maps, photographs, diagrams, charts, graphs, and tables. Graphics range in appearance from realistic (such as photographs) to highly abstract (such as organization charts). In terms of function, graphics range from the decorative, such as clip art that shows a group of people seated at a conference table, to highly informational, such as a table or a schematic diagram of an electronic device.

Graphics are important in technical communication because they can

 See TechComm Web (www .bedfordstmartins.com /techcomm) for guidelines, additional examples, and links related to topics in this chapter.

- help you communicate some kinds of information that are difficult to communicate with words
- help you clarify and emphasize information
- catch the reader's attention and interest
- help nonnative speakers of English understand information
- help communicate information to multiple audiences with different interests, aptitudes, and reading habits

This chapter begins with an explanation of the characteristics and functions of effective graphics and how to plan, create, and evaluate them. Then it discusses color, the basic kinds of graphics used in technical communication, and how to show motion and to create graphics for multicultural readers. It concludes with a discussion of graphics software.

THE FUNCTIONS OF GRAPHICS

The most obvious reason to use graphics in a document is that they motivate people to study it more closely. Eighty-three percent of what people learn derives from what we see, whereas only 11 percent derives from what we hear

(Gatlin, 1988). Because we are attuned to acquiring information through sight, a document that includes a visual element beyond words on the page is more effective than a document that doesn't. People studying a text with graphics learn about one-third more than people studying a text without graphics (Levie & Lentz, 1982). And graphics help readers retain what they learn; people remember some 43 percent more when a document includes graphics (Morrison & Jimmerson, 1989). Readers like graphics. According to one survey, readers of computer documentation consistently want more graphics and fewer words (Brockmann, 1990, p. 203).

Graphics offer benefits that words alone cannot.

- *Graphics are almost indispensable in demonstrating logical and numerical relationships.* For example, the organization chart used in most businesses is an effective way to represent the lines of authority in the organization. Graphics are also useful in showing numerical relationships of all kinds. If you want to illustrate the trends in the number of nuclear power plants completed each year over the last decade, a line graph would be much easier to understand than a paragraph full of numbers. Graphics can also demonstrate relationships among several variables over time, such as how many 4-cylinder, 6-cylinder, and 8-cylinder cars were manufactured in the United States during each of the last five years.

- *Graphics can communicate spatial information more effectively than words alone.* It's not easy to describe in words what a simple hammer looks like. But in 10 seconds you can draw a diagram of a hammer. Likewise, if you want to show the details of the derailleur mechanism on a bicycle, a diagram of the bicycle with a close-up of the derailleur is more effective than a verbal description.

- *Graphics can communicate steps in a process more effectively than words alone.* A troubleshooter's guide, a common kind of table, can explain how to understand what might be causing a problem in a process and what you might do to fix it. A diagram can explain clearly how acid rain forms.

- *Graphics can save space.* Consider the following paragraph:

  In the Wilmington area, some 90 percent of the population aged 18–24 watches movies or tapes on a VCR. They watch an average of 2.86 tapes a week. Among 35- to 49-year-olds, the percentage is 82, and the average number of movies or tapes is 2.19. Among the 50–64 age group, the percentage is 67, and the number of movies and tapes watched averages 2.5. Finally, among those people 65 years old or older, the percentage is 48, and the average number of movies and tapes watched weekly is 2.71.

Presented as a paragraph, this information is uneconomical and hard to remember. Presented as a table, however, the information is more concise and more memorable.

Age	Percentage Watching Tapes/Movies	Number of Tapes Watched per Week
18–24	90	2.86
35–49	82	2.19
50–64	67	2.50
65+	48	2.71

- *Graphics can reduce the cost of documents intended for international readers.* Translation costs can reach 60 cents per word; used effectively, graphics can reduce the number of words you have to translate (Horton, 1993).

As you plan and draft your document, look for opportunities to use graphics to clarify, emphasize, summarize, and organize information. McGuire and Brighton (1990) point out certain words and phrases that might alert you to an opportunity to create a graphic:

categories	features	numbers	routines
components	fields	phases	sequence
composed of	functions	procedures	shares
configured	if and then	process	structured
consists of	layers	related to	summary of
defines			

If you find yourself writing a sentence such as "The three categories of input modules are . . . ," consider including a diagram to reinforce your idea. If you write "The first step in the procedure is . . . ," think in terms of a flowchart or a logic box. If you write "This structure is related to . . . ," consider a diagram.

CHARACTERISTICS OF AN EFFECTIVE GRAPHIC

To be effective, a graphic must be clear, understandable, and meaningfully related to the larger discussion. Follow these six guidelines:

- *A graphic should have a purpose.* Don't include a graphic unless it will help your reader understand or remember the information. Beware of decorative, content-free clip art: drawings of businesspersons standing with clipboards, shaking hands, and so on. The novelty of meaningless clip art has worn thin.
- *A graphic should be honest.* Graphics can be dishonest, just as words can. You are responsible for making sure the graphic does not lie or mislead

the reader. Following are some common ethical concerns to keep in mind as you create graphics:

– If you did not create the graphic or generate the data, cite your source and, if you wish to publish it, request permission.
– Include all relevant data. For example, if you have a data point that you cannot explain, it is unethical to change the scale to eliminate it.
– Begin the axes in your graphs at zero — or mark them clearly — so that quantities are represented honestly.
– Do not use a table to hide a data point that would be obvious in a graph.
– Show items as they really are. Do not manipulate a photograph of a computer monitor to make the screen look bigger than it is.
– Do not use color or shading to misrepresent an item's importance. A dark-shaded bar in a bar graph, for example, appears larger and nearer than a light-shaded bar of the same size.

Common problem areas are pointed out in the discussions of various kinds of graphics throughout this chapter.

• *A graphic should be simple and uncluttered.* Three-dimensional bar graphs are easy to make with software, but they are harder to understand than two-dimensional ones, as shown in Figure 14.1.

• *A graphic should present a manageable amount of information.* If you present too much information, you risk confusing readers. Consider audience and purpose: what kinds of graphics are readers familiar with, how much do they already know about the subject, and what do you want the document to do? Because people learn best if you present information in small chunks, it is better to create several simple graphics than a single complicated one.

The two-dimensional bar graph is clean and uncluttered, whereas the three-dimensional graph is more difficult to understand because the additional dimension obscures the main data points. The number of uninsured emergency-room visits in February, for example, is very difficult to see in the three-dimensional graph.

■ **Figure 14.1**
Chartjunk and Clear Art

Unnecessary 3-D is one example of chartjunk, a term used by Tufte (1983) to describe the ornamentation that clutters up a graphic, distracting readers from the message.

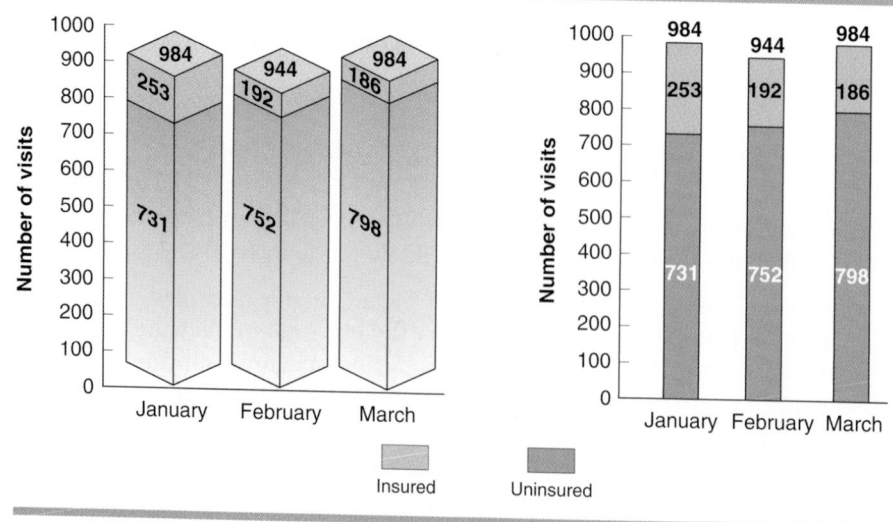

- *A graphic should meet the reader's format expectations.* Through experience, readers learn how to read different kinds of graphics. Follow the conventions — for instance, use diamonds to represent decision points in a computer flowchart — unless you have a good reason not to.

- *A graphic should be clearly labeled.* Every graphic (except a brief, informal one) should have a unique title that is both clear and informative. The columns of a table and the axes and lines of a graph should be labeled fully, complete with the units of measurement. Readers should not have to guess whether you are using meters or yards, or whether you are also including statistics from the previous year.

GUIDELINES

Integrating Graphics and Text

▸ *Place the graphic in an appropriate location.* If readers need the information contained in a graphic to understand the discussion, put the graphic directly after the relevant point in the text — or as soon after it as possible. If the graphic functions merely to support or elaborate on a point that is already clear, include it as an appendix.

▸ *Introduce the graphic in the text.* Whenever possible, refer to a graphic before it appears (ideally, on the same page). Refer to the graphic by number (such as *Figure 7*). Do not refer to "the figure above" or "the figure below"; the graphic might be moved during the production process. If the graphic is in an appendix, tell readers where to find it: "For the complete details of the operating characteristics, see Appendix B."

▸ *Explain the graphic in the text.* State what you want readers to learn from it. Sometimes a simple paraphrase of the title is enough to communicate its significance: "Figure 2 is a comparison of the costs of the three major types of coal gasification plants." At other times, however, you might need to explain why the graphic is important or how to interpret it. If the graphic is intended to make a point, be explicit:

> As Figure 2 shows, a high-sulfur bituminous coal gasification plant is more expensive than either a low-sulfur bituminous or anthracite plant, but more than half of its cost is cleanup equipment. If these expenses could be eliminated, high-sulfur bituminous would be the least expensive of the three types of plants.

Graphics are often accompanied by captions, explanations ranging in length from a sentence to several paragraphs. Captions are useful for readers who are skimming a document; they would overlook the material if it were included in the body text. Captions are usually set smaller than body text and in a different typeface.

See Ch. 13, pp. 344 and 352, for more on white space and rules.

See Ch.12, p. 314, for more on lists of illustrations.

▶ *Make the graphic clearly visible.* Distinguish the graphic from the surrounding text by adding white space or rules (lines) or by enclosing it in a box.

▶ *Make the graphic accessible.* If the document is more than a few pages long and contains more than four or five graphics, consider including a list of illustrations — a table of contents for the graphics — so readers can find them easily.

UNDERSTANDING THE PROCESS OF CREATING GRAPHICS

As with writing textual elements in a document, using graphics in a document involves planning, creating, and revising. The following sections discuss these three steps.

Planning the Graphics

Whether you think first about the text or the graphics, you need to consider the following four aspects of the document as you plan:

• *Audience.* Will your readers understand the kinds of graphics you want to use? Are they familiar with the standard icons in your field? Are they already motivated to read your document, or do you need to enliven the text to hold their attention or use color for emphasis?

• *Purpose.* What point are you trying to make with the graphic? As Figure 14.2 shows, even a few simple facts can yield a number of different points. Your responsibility is to determine what you want to show and how best to show it. Don't rely on your software to do your thinking for you. It can't.

• *The kind of information you want to communicate.* The subject of the document will help you determine what type of graphic to include. If you are writing about the number of languages spoken by citizens in your state, you would probably use tables for the mass of statistical data, maps to show the patterns of language use, and graphs to show statistical trends over time. If you are discussing how the sales of two products have changed over several years, a grouped bar chart or a line graph might be the best type of graphic.

• *Physical conditions.* The physical conditions in which a reader will use the document — amount of lighting, amount of surface space available, exposure to water or grease, and so forth — will influence the kind of graphic as well as its size and shape, the thickness of lines and size of type, and the color.

Rail Line	November		December		January	
	Disabled by electrical problems	Total disabled	Disabled by electrical problems	Total disabled	Disabled by electrical problems	Total disabled
Bryn Mawr	19	27	17	28	20	26
Swarthmore	12	16	9	17	13	16
Manayunk	22	34	26	31	24	33

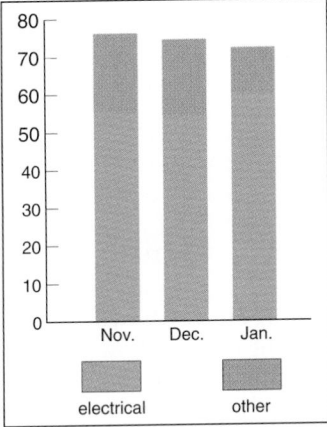

a. Number of rail cars disabled,
    November–January

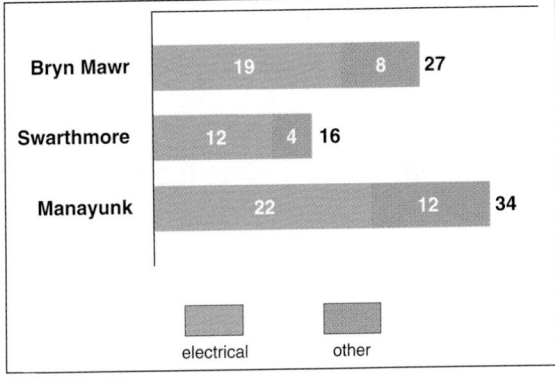

b. Number of rail cars disabled in November

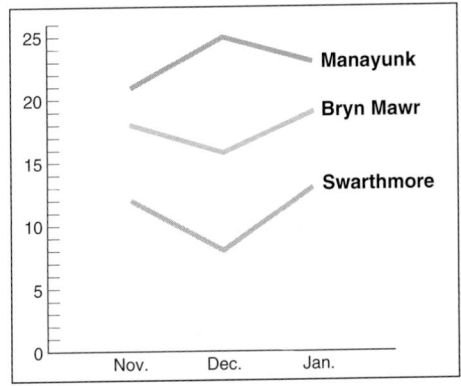

c. Number of rail cars disabled by
    electrical problems November–January

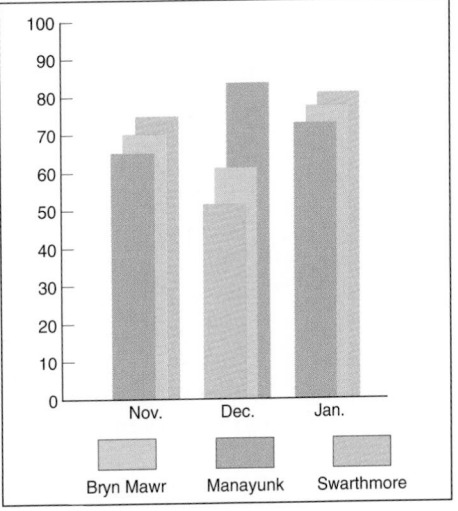

d. Range in percent of rail cars, by line, disabled
    by electrical problems, November–January

■ **Figure 14.2   Different Graphics Emphasizing Different Points**

Each of these four graphs emphasizes a different point. Graph (a) focuses on the total number of cars disabled each month, classified by cause; graph (b) focuses on the three rail lines during one month; and so forth.

Next, as you plan how you are going to create the graphics, consider four important factors:

See Ch. 3, p. 52, for more on planning and budgeting.

- *Time.* Making a complicated graphic can take a lot of time, so you will need to establish a schedule.
- *Money.* A high-quality graphic can be expensive. What is the project budget? How can you use that money effectively?
- *Equipment.* Although some graphics can be made with a straight edge and a pencil, most require other tools, including word-processing software for tables, spreadsheets for tables and graphs, and graphics software for diagrams.
- *Expertise.* How much do you know about creating graphics? Do you have access to the expertise of others?

Creating the Graphics

Time, money, equipment, and expertise are overlapping categories; if you have a large budget, you might be able to hire experts with the latest equipment and take the time to revise the graphics several times. Except for special projects, however, you usually won't have all the resources you would like. You will probably have to choose one of the following four approaches:

For advice about citing graphics in your document, see "Documenting Sources" in the Appendix, Part A (p. 660).

- *Using existing graphics.* For a student paper that *will not be published,* some instructors allow the use of photocopies of existing graphics; other instructors do not. If you are permitted to photocopy existing graphics, you are ethically obligated to cite your source, just as you would for any other information you borrow.

  For a document that *will be published,* whether written by a student or a professional, the use of an existing graphic is permissible if the graphic is the property of the writer's organization or if that organization has obtained permission to use it. If the graphic was created under works-made-for-hire guidelines, the organization has an unlimited right to use it.

  Be particularly careful about graphics you find on the Web. Many people mistakenly think that anything they download from the Web can be used without permission. This is not true. The same copyright laws that apply to printed material also apply to Web-based material, whether words or graphics. However, many Web-based companies would be happy to sell you rights to use their graphics in your documents. Search for "computer graphics," "computer art," and related terms. In addition, there are a number of sources of free clip art and sources of free art, such as buttons, icons, and backgrounds, to use in Web pages.

  Aside from the issue of copyright, think carefully before you use existing graphics:

  - The style of the graphic might not match that of the others you want to use.

For links to clip-art sources, click on Links Library for Ch. 14 on TechComm Web (www.bedfordstmartins .com/techcomm).

- The graphic might lack some features you want or include some you don't.
- If you are photocopying a printed graphic, the lines in the photocopy might lack crispness, and the shading in the original — the shades of gray or the colors — might not reproduce sharply ("void out"), giving the graphic a blotchy appearance.

If you use an existing graphic, assign it your own number and title. The existing number and the title are unlikely to integrate well.

- *Modifying existing graphics.* You can redraw an existing graphic or use a scanner to digitize the graphic and then manipulate it electronically with graphics software. These uses carry the same ethical and legal obligations described in the previous section. Your source citation should indicate that your graphic is "based on" or "adapted from" your source.

- *Creating graphics on a computer.* Although you can draw simple tables and diagrams by hand, in most cases you will want to use a computer to create professional-quality graphics. Computer graphics offer two main advantages over hand-drawn graphics: variety and reusability. Every month, new graphics software with powerful capabilities comes on the market. Three-dimensional effects are now routine, and animation is becoming common. And because they are digital, computer graphics can be stored, reused, and revised. Once you have created a graphic showing this month's budget, you can copy it and easily revise it next month.

  Visit the computer labs at your college or university to test some of the many excellent graphics software packages available. Also, consult the bibliography (page 729) for a list of books about computers and technical communication. But beware: any book about computer graphics is probably out of date as soon as it is published.

  For more information on computer graphics, see page 401.

- *Having someone else create the graphics.* If graphics software is so versatile, why not make the graphics yourself on your own computer? One obvious reason: professional-level software can cost hundreds of dollars. Another less obvious reason: you need artistic ability to create professional-quality graphics. And even if you have the ability, attaining proficiency with a sophisticated graphics package can require dozens or even hundreds of hours of practice.

  In most cases, your choices are constrained by your budget and company policy. Some companies have technical-publications departments with graphics experts, but others subcontract this work. Many print shops and service bureaus have graphics experts on staff or can direct you to them.

Revising the Graphics

As with any other aspect of technical communication, build in enough time and money to revise the graphics. Create a checklist and evaluate each graphic for effectiveness. The Revision Checklist at the end of this chapter

(p. 403) is a good starting point. Show the graphics to people with backgrounds like the intended readers' and ask them for suggestions. Revise the graphics and solicit more reactions.

USING COLOR EFFECTIVELY

See Colorize.com (www
.colorize.com) for articles
about color theory and about
graphics software.

Color is a powerful tool: it draws attention to information you want to emphasize, establishes visual patterns to promote understanding, and adds interest. But it is also easy to misuse. The following discussion is based on Jan V. White's excellent text, *Color for the Electronic Age* (1990).

In using color in graphics and page design, keep these seven principles in mind:

- *Don't overdo it.* Readers can absorb only two or three colors at a time. Use colors for small items, such as portions of graphics and important words. And don't use colors where black will work better.

- *Use color to emphasize particular items.* People interpret color before they interpret shape, size, or placement on the page. Color effectively draws a reader's attention to a particular item or group of items. In Figure 14.3, for example, color adds emphasis to several different kinds of items.

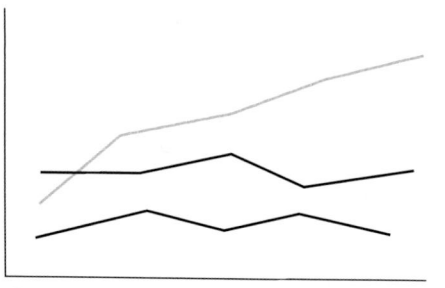

Color draws the reader's attention to the line.

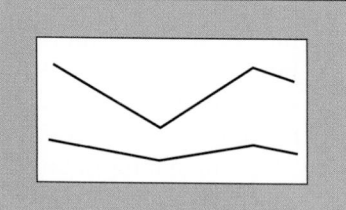

The colored frame focuses
the reader's attention on the
information in the graph.

Color is useful for emphasizing short phrases. Used for longer passages, color loses its impact.

Pixel Driver Renders
600,000 Polygons
per Second

Here the color emphasizes a
row; it could also be used to
emphasize a column or a single
data cell.

Table 2
Israeli Power Plant Sites

Site	Generating	Installed Capacity (MW)
Haifa	4	430
Hadera	4	1400
Tel Aviv	4	530
Ashdod	9	1210
Total	**21**	**3570**

Source: B. Golany, Y. Roll, & D. Rybak. (1994, August). ™Measuring efficiency of power plants in Israel by data envelopment analysis, *IEEE Transactions on Engineering Management 41*, 3, p. 292.

■ **Figure 14.3   Color Used for Emphasis**

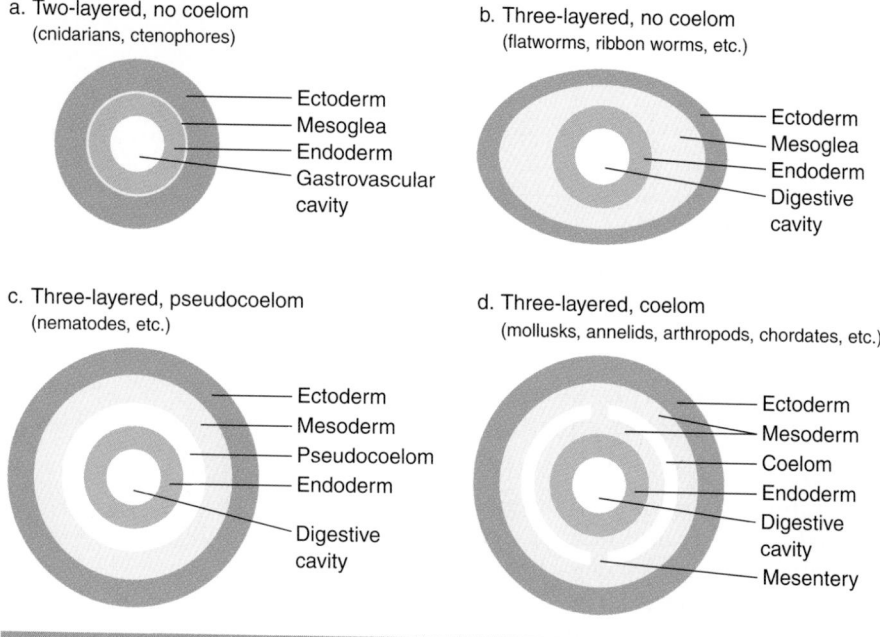

a. Two-layered, no coelom
(cnidarians, ctenophores)

— Ectoderm
— Mesoglea
— Endoderm
— Gastrovascular
 cavity

b. Three-layered, no coelom
(flatworms, ribbon worms, etc.)

— Ectoderm
— Mesoglea
— Endoderm
— Digestive
 cavity

c. Three-layered, pseudocoelom
(nematodes, etc.)

— Ectoderm
— Mesoderm
— Pseudocoelom
— Endoderm
— Digestive
 cavity

d. Three-layered, coelom
(mollusks, annelids, arthropods, chordates, etc.)

— Ectoderm
— Mesoderm
— Coelom
— Endoderm
— Digestive
 cavity
— Mesentery

■ Figure 14.4
**Color Used to Establish
Patterns**

- *Use color to create patterns.* Traffic signs grab attention by using a pattern of black text or images against a yellow background. The same principle applies in graphics and document design. Choose a warm color, such as red, as the background or text color for all safety comments. Choose a cool color, such as green, for tips. Use different-colors of paper for different sections of a manual. In creating patterns, include shape. For instance, use red for safety comments, but place them in octagons resembling a stop sign. This way, you give your readers two visual cues to help them recognize the pattern. Figure 14.4 (Curtis & Barnes, 1989, p. 532) shows a page in a biology textbook that uses color to establish patterns. Color is also an effective way to emphasize text boxes, rules, screens, and headers and footers.
- *Use contrast effectively.* The visibility of a color is a function of the background against which it appears (see Figure 14.5). The strongest contrasts are between black and white and black and yellow.

For more on designing your document, see Ch.13.

Notice that a color washes out if the background color is too similar.

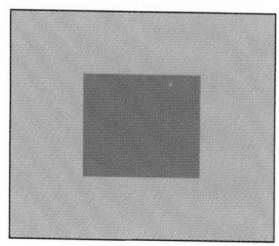

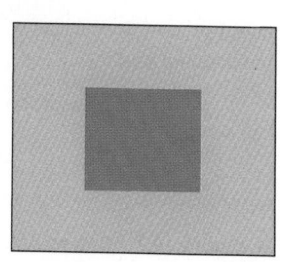

■ Figure 14.5
**The Effect of Background
in Creating Contrast**

Transparencies are used with the room lights on. Therefore, the most popular background is the white provided by the screen, and the most popular color for the text or graphics is black or dark blue. For slides, however, the room lights are either dim or off. Therefore, the most popular background color for slides is dark blue, with yellow or white text and graphics.

New Video Interfaces
- Video Scope
- Video Space Icon
- Video Space Monitor
- Paper Video

New Video Interfaces
- Video Scope
- Video Space Icon
- Video Space Monitor
- Paper Video

■ **Figure 14.6   Contrast in a Transparency (left) and a Slide (right)**

The need for effective contrast applies not only to text but also to the graphics used in presentations. Choose a natural-looking background so that text or graphics produce a vivid contrast, as shown in Figure 14.6.

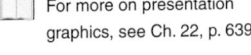
For more on presentation graphics, see Ch. 22, p. 639.

- *Take advantage of any symbolic meaning colors may already have.* In American culture, for example, red signals danger, heat, or electricity; yellow signals caution; and orange signals warning. Using any of these warm colors in ways that depart from these familiar meanings could be confusing. The cooler colors — blues and greens — are more conservative and subtle (Figure 14.7 illustrates these principles.) Keep in mind, however, that different cultures interpret colors differently.

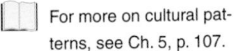
For more on cultural patterns, see Ch. 5, p. 107.

The batteries are red. The warm red contrasts effectively with the cool green of the car body.

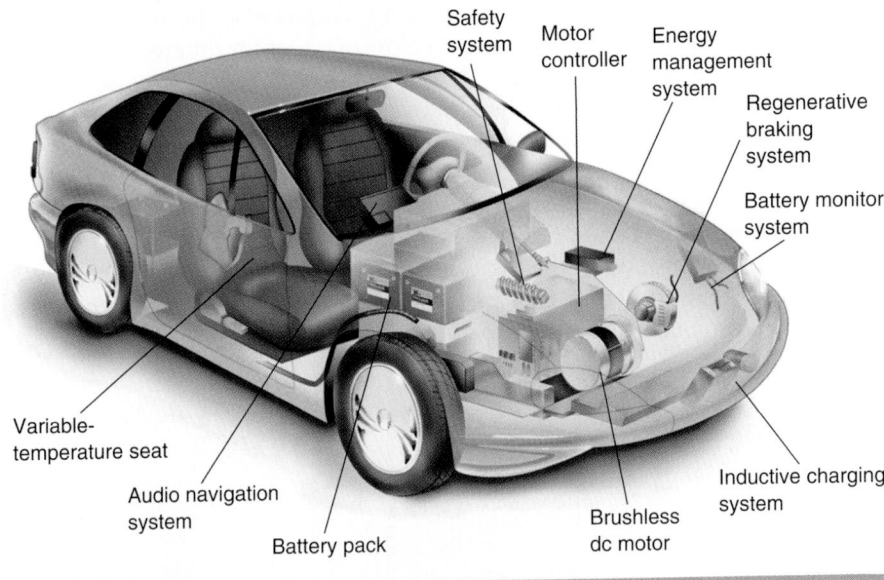

■ **Figure 14.7   Colors Already Have Clear Associations for Readers.**

- *Be aware that color can obscure or swallow up text.*

This text looks bigger because of the white background.	This text looks smaller, even though it is the same size, because of the colored background.

This line of type appears to reach out to the reader.
This line of type appears to recede into the background.

If you are using print against a colored background, you might need to make the type a little bigger, because color makes text look smaller.

Text printed against a white background looks bigger than the same size text printed against a colored background. White letters counteract this effect.

- *Use light colors to make objects look bigger.*

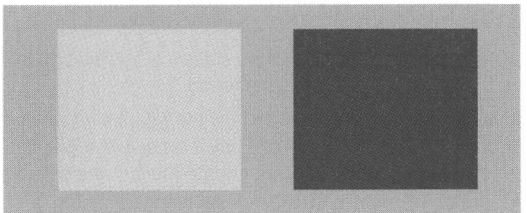

The orange box looks bigger than the blue box even though they are the same size.

CHOOSING THE APPROPRIATE KIND OF GRAPHIC

Graphics used in technical documents fall into two basic categories: tables and figures. Tables are lists of data, usually numbers, arranged in columns. Figures are everything else: graphs, charts, diagrams, photographs, and the like. Generally, tables and figures are numbered separately: the first table in a document is Table 1; the first figure is Figure 1. In documents of more than one chapter (like this book), the graphics are usually numbered within each chapter: Figure 3.2 is the second figure in Chapter 3.

But this broad distinction between tables and figures doesn't help you decide what kind of graphic to use in a particular situation. There is no simple system for choosing, because in many situations several different types would do the job. In general, however, graphics can be categorized according to the kind of information they contain. The discussion that follows is based on the classification system in William Horton's "Pictures Please — Presenting Information Visually," in *Techniques for Technical Communicators* (Barnum & Carliner, 1993).

Technical information can be classified into four categories.

Categories of technical information:

Numerical information

Logical relationships

Process descriptions and instructions

Visual and spatial characteristics

Figure 14.8 presents an overview of the discussion that follows.

Some kinds of graphics convey several kinds of information. A table, for instance, can include both numerical values and procedures.

Purpose	Type of Graphic	What the Graphic Does Best
Illustrating numerical information	Table	Shows large amounts of numerical data, especially when there are several variables for a number of items.
	Bar graph	Shows the relative values of two or more items.
	Pictograph	Enlivens statistical information for the general reader.
	Line graph	Shows how the quantity of an item changes over time. A line graph can accommodate much more data than a bar graph can.
	Pie chart	Shows the relative size of the parts of a whole. Pie charts are instantly familiar to most readers.
Illustrating logical relationships	Diagram	Represents items or properties of items.
	Organization chart	Shows the lines of authority and responsibility in an organization.
Illustrating instructions and process descriptions	Checklist	Lists or shows what equipment or materials to gather, or describes an action.
	Table	Shows numbers of items or indicates the state (on/off) of an item.
	Flowchart	Shows the stages of a procedure or a process.
	Logic box	Shows which of two or more paths to follow.
	Logic tree	Shows which of two or more paths to follow.
Illustrating visual and spatial characteristics	Drawing	Shows simplified representations of objects.
	Map	Shows geographical areas.
	Photograph	Shows precisely the external surface of objects.
	Screen shot	Shows what appears on a computer screen.

▦ **Figure 14.8 Choosing the Appropriate Kind of Graphic (Based on Horton [1993])**

Illustrating Numerical Information

Technical documents often display numerical information. Sometimes they present precise numerical data; at other times, relative values. You may need to present numerical values at one point in time or as they change over time. The basic kinds of graphics for numerical values are tables, bar graphs, pictographs, line graphs, and pie charts.

Tables

Tables convey large amounts of numerical data easily, and they are often the only way to present several variables for a number of items. For example, if you want to show how many people are employed in six industries in 10 states, a table would probably be most effective. Although tables lack the visual appeal of other kinds of graphics, they can handle much more information with complete precision.

As always, keep in mind the needs and aptitudes of your audience. A table can convey much more information than readers need or can understand easily. If you choose to make a large table — one with, say, 20 columns and 30 rows — consider placing it in an appendix and using a summary table for the body of the document.

Figure 14.9 illustrates the standard parts of a table. Tables are identified by a number ("Table 1") and a substantive, informative title, one that encompasses the items being compared, as well as the basis (or bases) of comparison:

Mallard Population in Rangeley, 1990–1994

The Growth of the Robotics Industry in Japan and the United States, 1997

	Valve Readings	Maximum Bypass Cv	Minimum Recirculation Flow (GPM)	Pilot Threads Exposed	Main ΔP at Rated Flow (psid)
Valve #1					
	Initial	43.1	955	+3	4.4
	Final	43.1	955	+3	...
Valve #2					
	Initial	48.1	930	+3	4.5
	Final	48.2	950	+2	...

Number —— **Table 6.   Test Results for Valves #1 and #2** —— *Title*

Stub head —— *Valve Readings*

Row head —— *Valve #1*

Column head

Source statement —— *Source:* "Third Progress Report: Anderson Machine Tools Reconfiguration Project."

■ **Figure 14.9   Parts of a Table**

Most tables are numbered and titled above the data. The number and title are left justified or centered horizontally.

GUIDELINES

Creating Effective Tables

▶ *Indicate the units of measure.* If all data are expressed in the same unit, indicate that unit in the title:

Farm Size in the Midwestern States (in Hectares)

If the data in different columns are expressed in different units, indicate the units in the column headings:

Population (in millions)	Per Capita Income (in thousands of U.S. dollars)

If all the *data cells* in a column use the same unit, indicate that unit in the column head, not in each data cell:

Speed (knots)
15

18

14

You can express data in both real numbers and percentages. A column heading and the first data cell under it might read as follows:

Number of Students (Percentage)
53 (83)

▶ *In the stub — the left-hand column — list the items being compared.* Arrange the items in the stub in some logical order: big to small, important to unimportant, alphabetical, chronological, geographical, and so forth. If the items fall into several categories, you can include the names of the categories in the stub:

Snow Belt States . . . .
    Connecticut . . . .
    New York . . . .
    Vermont . . . .
Sun Belt States . . . .
    Arizona . . . .
    California . . . .
    New Mexico . . . .

For more about screens, see Ch. 13, p. 352.

If the items in the stub are not grouped in logical categories, skip a line after every five rows to help the reader follow the rows across the table. Or use a screen or a colored background for every other set of five rows. Dot leaders, a row of dots that links the stub and the next column, are also useful.

▶ *In the columns, arrange the data clearly and logically.* Line up the numbers consistently by using the decimal tab function:

```
 3,147.4
   365.7
46,803.5
```

In general, don't change units. If you use meters for one quantity, don't use centimeters for another unless the quantities are so dissimilar that readers would have a difficult time understanding them if expressed in the same units:

```
 3.4 hr
12.7 min
 4.3 sec
```

This list would probably be easier for most readers to understand than one in which all quantities were expressed in the same unit.

▶ *Do the math.* If your readers will need to know the totals for the columns or the rows, provide them. If your readers will need to know percentage changes from one column to the next, present them:

Number of Students (Percentage Change from Previous Year)

1998	1999	2000
619	644 (+4.0)	614 (−4.7)

▶ *Use dot leaders if a column contains a "blank" spot: a place where there are no appropriate data:*

```
3,147
. . .
46,803
```

But don't substitute dot leaders for a quantity of zero.

▶ *Don't make the table wider than it needs to be.* The reader should be able to scan across a row easily. As White (1984) points out, there is no reason to make the table as wide as the text column in the document. If a column heading is long — more than five or six words — stack the words:

Computers Sold
Without a DVD Drive

▶ *Minimize the use of rules.* Grimstead (1987) recommends using rules only when necessary: to separate the title and the headings, the headings and the body, and the body and the notes. When you use rules, make them thin rather than thick.

> ▶ *Provide footnotes for any information that needs to be explained.* All the information your readers need to understand the table should accompany the table.
>
> ▶ *Indicate the source of your information if you did not generate it yourself.* A typical source statement is: "Source: Data from *IEEE Spectrum*, vol. 37, no. 8 (2001), p. 34."

In addition to numerical information, tables can also effectively convey textual information, as shown in Figure 14.10 (U.S. Congress, Office of Technology Assessment, 1993, p. 41).

Table 2.3   Types of Electronic Kiosks: Key Characteristics and Selected Applications

Type of Kiosk	Key Characteristics	Selected Applications
Off-line: Stand-alone	For information that does not need updating: no telecommunications costs	GSA's Central Office Building directory
Off-line: Polled	Can update information, and retrieve queries and survey results over a telephone line and modem at night	USPS's "Postal Buddy"; "24-Hour City Hall"
Online	Can process information immediately; can update rules and software in central computer; requires dedicated telephone line and central computer capacity.	Tulare County, CA's "Tulare Touch"; State of California's "InfoCalifornia"
Online: Transactional	Online, but can also collect money via credit or debit cards for bills and services	Long Beach, CA's "Auto Clerk"; State of California's "InfoCalifornia"

KEY: GSA = General Services Administration; USPS = U.S. Postal Service.
Source: Office of Technology Assessment, 1993.

■ **Figure 14.10   Text Table**

Bar Graphs

Like tables, *bar graphs* can communicate numerical values, but they are better at showing the relative values of two or more items. Figure 14.11 shows typical horizontal and vertical bar graphs.

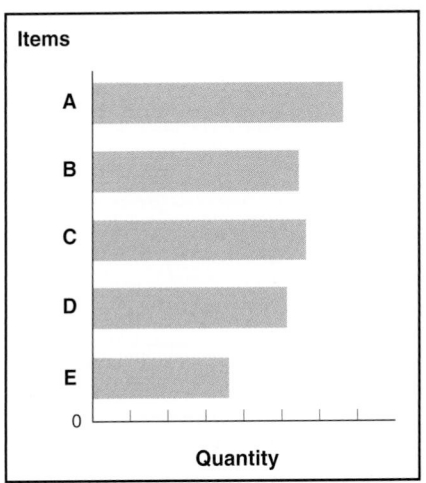

Figure 1. Horizontal graph

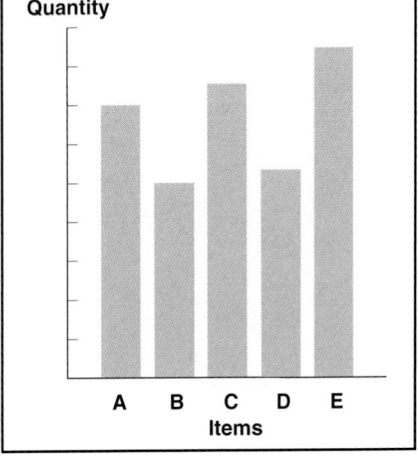

Figure 1. Vertical graph

Horizontal bars are best for showing quantities such as speed and distance. Vertical bars are best for showing quantities such as height, size, and amount. However, these distinctions are not ironclad; as long as the axes are clearly labeled, readers should have no trouble understanding the graph.

■ **Figure 14.11   Structures of Horizontal and Vertical Bar Graphs**

GUIDELINES

Creating Effective Graphs

▶ *Make the proportions fair.* Number the axes at regular intervals. If you are drawing the graph by hand, use a ruler or graph paper.

For vertical bar graphs, choose intervals that make your vertical axis about three-quarters the length of the horizontal axis. If the vertical axis is much longer than that, the differences in the height of the bars will be exaggerated. If the horizontal axis is too long, the differences will be unfairly flattened. Make all bars equally wide, and make the amount of space between them about half the width of a bar. Here are two poorly proportioned graphs:

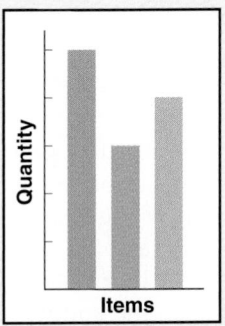

a. Excessively long vertical axis

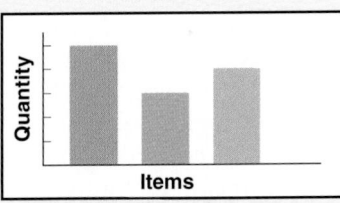

b. Excessively long horizontal axis

▶ *If possible, begin the quantity scale at zero.* Doing so ensures that the bars accurately represent the quantities. Notice how misleading a graph can be if the scale doesn't begin at zero.

Version (a) misrepresents the differences among items a, b, and c.

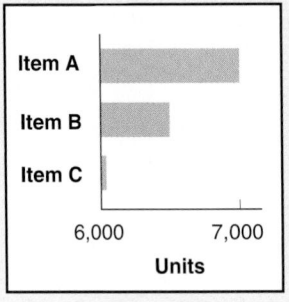 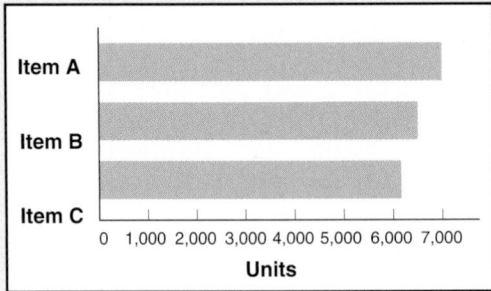

a. Misleading   b. Accurately representative

If it is not practical to start the quantity scale at zero, break the quantity axis clearly at a common point on all bars.

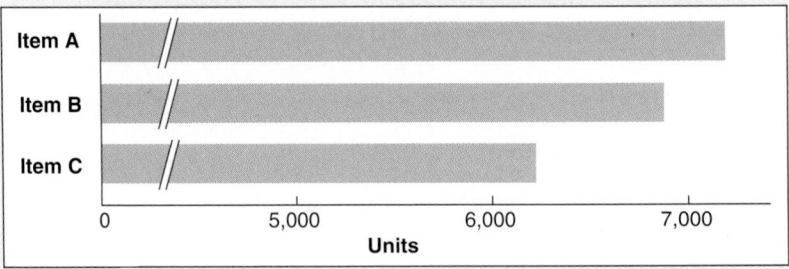

▶ *Use tick marks or grid lines to signal the amounts.* Ticks are the little marks drawn along the axis:

Grid lines are tick marks that extend through the bars. In most cases, grid lines are necessary only if the table has several bars (some of which are too far away from tick marks to allow readers to gauge the quantity easily) and if the actual quantities are not indicated near the bars.

▶ *Arrange the bars in a logical sequence.* For a vertical bar graph, use chronology if possible. For a horizontal bar graph, arrange the bars in order of descending size, beginning at the top of the graph, unless some other logical sequence seems more appropriate.

▶ *Place the title below the figure.* Unlike tables, which are usually read from top to bottom, figures are usually read from the bottom up.

▶ *Indicate the source of your information if you did not generate it yourself.*

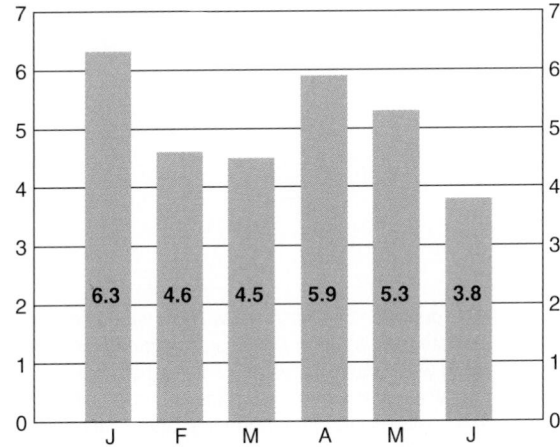

Figure 1. Tri-County Inflation Rate This Year to Date

■ **Figure 14.12**
Effective Bar Graph

Figure 14.12 shows an effective bar graph that uses grid lines.

The four variations on the basic bar graph shown in Table 14.1 can help you accommodate different communication needs.

■ **Table 14.1   Modifying the Basic Bar Graph**

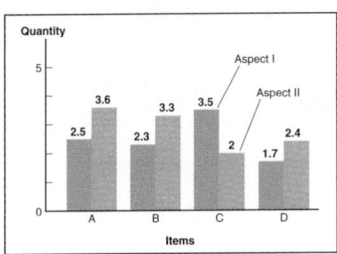

a. Grouped bar graph

The *grouped* bar graph lets you compare two or three quantities for each item. Grouped bar graphs would be useful, for example, for showing the numbers of full-time and part-time students at several universities. One bar could represent full-time students; the other, part-time students. To distinguish between the bars, use hatching (striping), shading, or color, and either label one set of bars or provide a key.

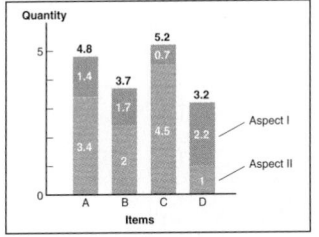

b. Subdivided bar graph

In the *subdivided* bar graph, Aspect I and Aspect II are stacked like wooden blocks placed on top of one another. Although totals are easy to compare in a subdivided bar graph, individual quantities are not.

(Continued)

■ **Table 14.1**
(Continued)

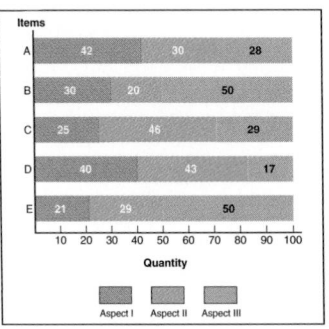

c. 100-percent bar graph

The *100-percent* bar graph, which shows the relative proportions of the elements that make up several items, is useful in portraying, for example, the proportion of full-scholarship, partial-scholarship, and no-scholarship students at a number of colleges.

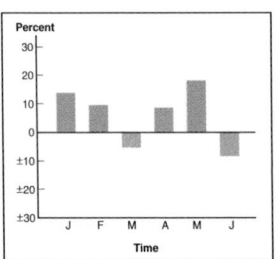

d. Deviation bar graph

The *deviation* bar graph shows how various quantities deviate from a norm. Deviation bar graphs are often used when the information contains both positive and negative values, such as profits and losses. Bars on the positive side of the norm line represent profits, bars on the negative side, losses.

Pictographs

Pictographs — bar graphs in which the bars are replaced by a series of symbols — are used primarily to present statistical information to the general reader. The quantity scale is usually replaced by a statement indicating the numerical value of each symbol. Figure 14.13 shows an example. Represent quantities in a pictograph honestly. Figure 14.14 shows that a picture drawn to scale can appear many times larger than it should.

Clip-art pictures and symbols are available online for use in pictographs. Arrange pictographs horizontally rather than vertically. Pictures of computer monitors balanced on top of each other can look foolish.

Number of Internet Hosts, 1993–1997

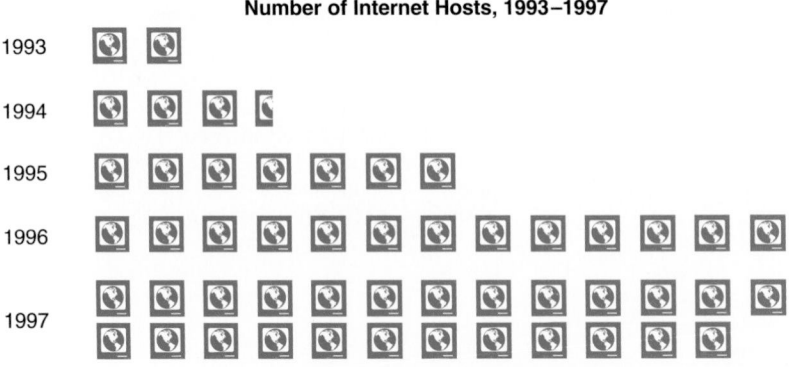

Each symbol represents one million hosts

■ **Figure 14.13**
Pictograph

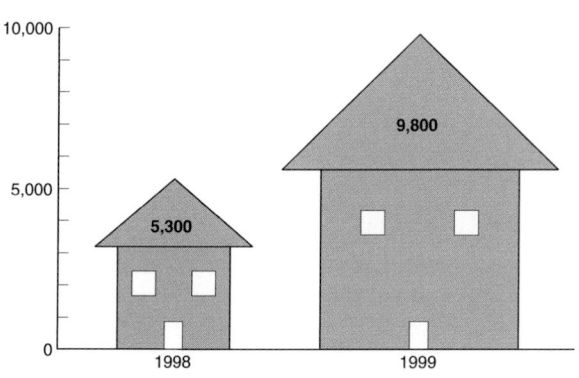

Figure 3. Housing Starts in the Tri-State Area, 1998–1999

Line Graphs

Line graphs are used almost exclusively to show changes in quantity over time, for example, the month-by-month production figures for a product. A line graph focuses the reader's attention on the change in quantity, whereas a bar graph emphasizes the quantities themselves.

An advantage of the line graph for demonstrating change is that it can convey much more data than a bar graph because you can plot three or four lines. If the lines intersect often, however, the graph will be unclear. If this is the case, draw separate graphs. Figure 14.15 shows a line graph.

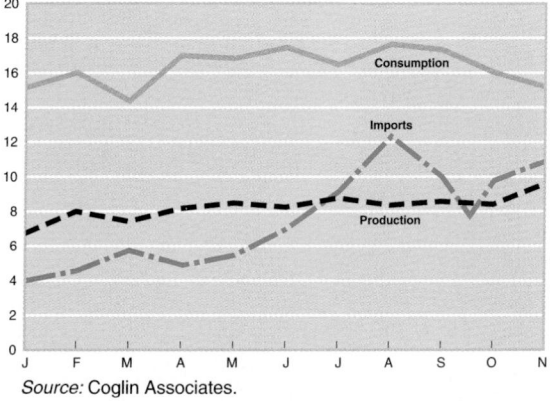

Source: Coglin Associates.

Figure 1. U.S. Petroleum Consumption, Production, and Imports, January–November 1995

■ **Figure 14.15**
Line Graph

The writer has used different colors and patterns to distinguish the lines.

> **GUIDELINES**
>
> ## Creating Effective Line Graphs
>
> ▶ *If possible, begin the quantity scale at zero.* Beginning at zero is the best way to portray the information honestly. If you cannot begin at zero, clearly indicate a break in the axis.
>
> ▶ *Use reasonable proportions for the vertical and horizontal axes.* As with bar graphs, try to make the vertical axis about three-quarters the length of the horizontal axis.
>
> ▶ *Use grid lines — horizontal, vertical, or both — rather than tick marks when readers need to read the quantities precisely.*

Pie Charts

The *pie chart* is a simple but limited design used for showing the relative size of the parts of a whole. Even an untrained reader can recognize and understand a pie chart. Figure 14.16 (Mankiw, 1997, p. 37) shows a typical example.

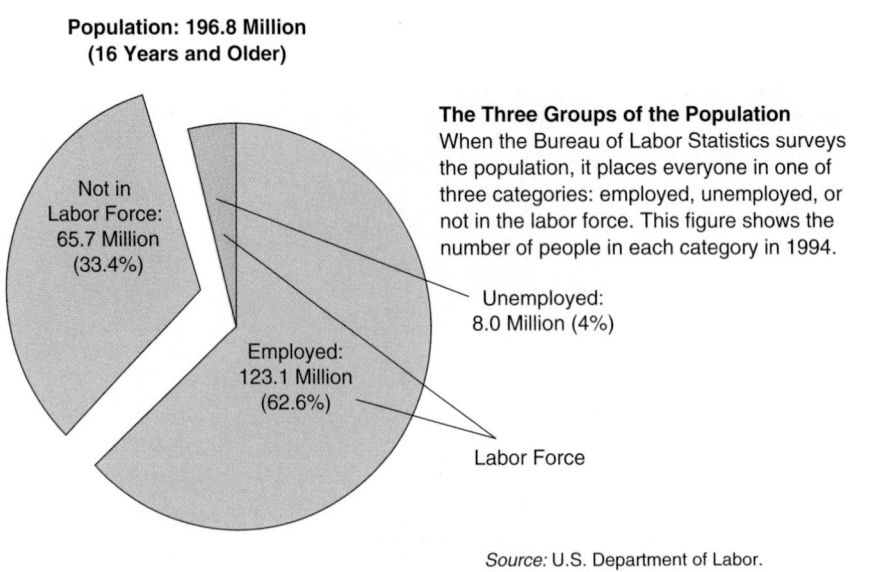

Population: 196.8 Million (16 Years and Older)

Not in Labor Force: 65.7 Million (33.4%)

The Three Groups of the Population
When the Bureau of Labor Statistics surveys the population, it places everyone in one of three categories: employed, unemployed, or not in the labor force. This figure shows the number of people in each category in 1994.

Unemployed: 8.0 Million (4%)

Employed: 123.1 Million (62.6%)

Labor Force

Source: U.S. Department of Labor.

■ **Figure 14.16   Pie Chart**

GUIDELINES

Creating Effective Pie Charts

▶ *Restrict the number of slices to six or seven.* As slices get smaller, judging their relative size becomes more difficult.

▶ *Begin with the largest slice at the top of the pie (noon position) and work clockwise in decreasing-size order, unless you have a good reason to arrange them otherwise.*

▶ *Include a miscellaneous slice for very small quantities that would make the chart unclear.* Explain its contents in a footnote. This slice, sometimes called "other," follows the other sections as you work in a clockwise direction.

▶ *Label the slices (horizontally, not radially) inside the slice, if space permits.* Include the percentage that each slice represents and, if appropriate, the raw numbers.

▶ *To emphasize one slice, use a bright, contrasting color or separate the slice from the pie.* Do this, for example, when you introduce a discussion of the item represented by that slice.

▶ *Check to see that your software follows the appropriate guidelines for pie charts.* Some software adds fancy visual effects that can actually hurt comprehension. For instance, many programs portray the pie in three dimensions, as shown here (U.S. Department of Health and Human Services, 1990, p. 7).

In this three-dimensional pie chart, the "shadows" misrepresent the slices to which they are adjacent by making them appear larger than they should be. In addition, the slices at the bottom of the pie appear larger than they should because they seem closer to the reader. To communicate clearly, make pies two dimensional.

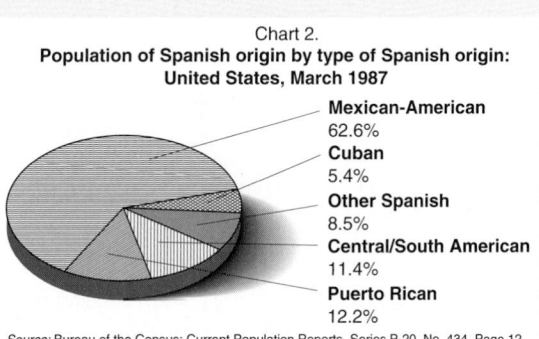

Chart 2.
Population of Spanish origin by type of Spanish origin: United States, March 1987

Mexican-American
62.6%
Cuban
5.4%
Other Spanish
8.5%
Central/South American
11.4%
Puerto Rican
12.2%

Source: Bureau of the Census: Current Population Reports. Series P-20, No. 434, Page 12.

▶ *Don't overdo fill patterns.* Fill patterns are designs, shades, or colors that distinguish one slice from another. In general, use simple, understated patterns or none. In place of fill patterns, include information about the slice, such as raw numbers, percentages, or labels, which reduces the need for legends.

▶ *Check that your percentages add up to 100.* If you are doing the calculations yourself, check your math.

Illustrating Logical Relationships

Graphics can help you communicate logical relationships among items. For instance, in describing a piece of hardware, you might want to show its major components. In writing a proposal, you might want to show alternative courses of action and explain the advantages and disadvantages of each. The two kinds of graphics most useful in showing logical relationships are diagrams and organization charts.

Diagrams

A *diagram* is a visual metaphor that uses symbols to represent items or their properties. In technical communication, common kinds of diagrams include blueprints, wiring diagrams, and schematics. Figure 14.17 (U.S. Environmental Protection Agency, 1991, p. 124) is an example of a schematic diagram.

The artist has used visual symbols, such as the fan and the filter, to represent components working together.

The clip art of the person helps clarify the location and size of the HVAC equipment.

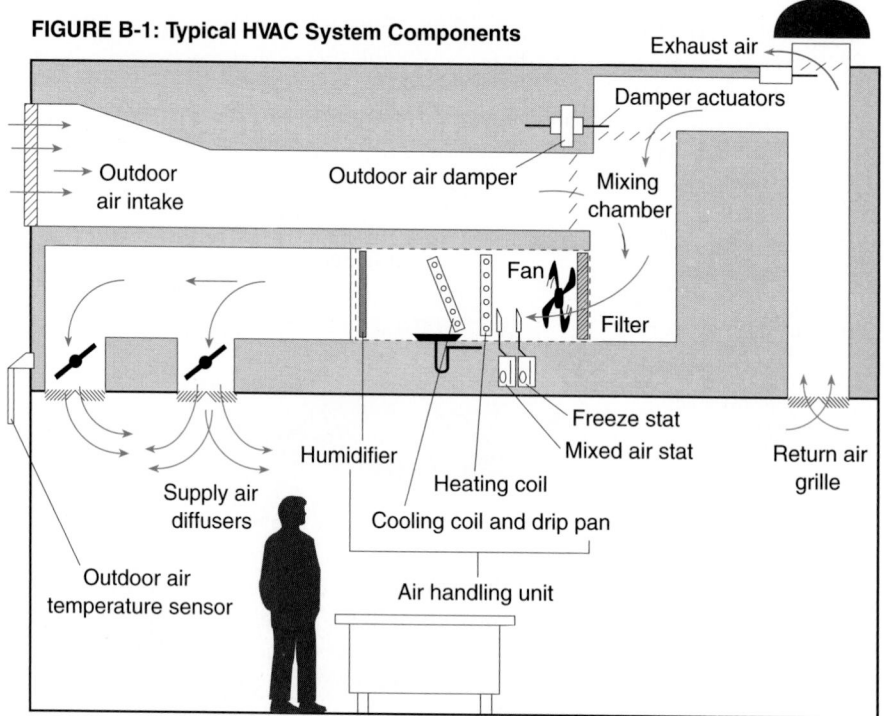

FIGURE B-1: Typical HVAC System Components

Exhaust air

Damper actuators

Outdoor air intake

Outdoor air damper

Mixing chamber

Fan

Filter

Freeze stat

Mixed air stat

Return air grille

Humidifier

Heating coil

Cooling coil and drip pan

Supply air diffusers

Outdoor air temperature sensor

Air handling unit

■ **Figure 14.17   Schematic Diagram**

Although the silhouette of the person looks realistic, the HVAC system itself is drawn to show its function, not its actual appearance.

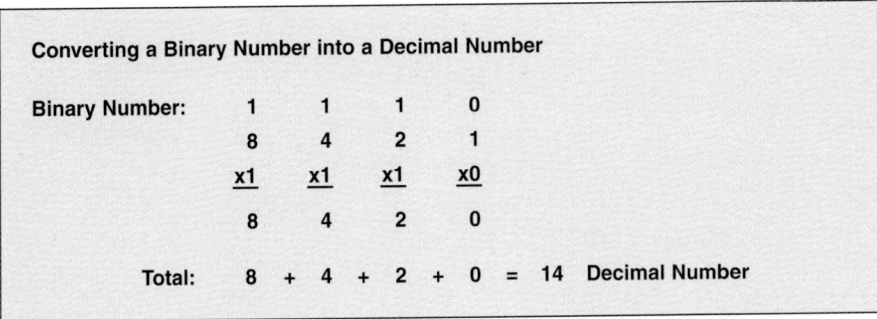

Figure 14.18   Diagram Used to Clarify a Difficult Concept

Here, a series of simple calculations shows how to convert binary numbers into decimal numbers.

Diagrams can also clarify difficult concepts, as shown in Figure 14.18 (Labuz, 1984, p. 184).

A popular form of diagram is the block diagram, in which simple geometric shapes, usually rectangles, suggest logical relationships. In Figure 14.19, the block diagram is much clearer than the prose version of the same information.

Organization Charts

An *organization chart* is a block diagram that portrays the lines of authority and responsibility in an organization. In most cases, the positions are represented by rectangles. Figure 14.20 on page 390 is a typical organization chart.

Illustrating Process Descriptions and Instructions

Graphics often accompany process descriptions and instructions. The following discussion covers some of the graphics used in writing about actions: checklists, tables, flowcharts, logic boxes, and logic trees.

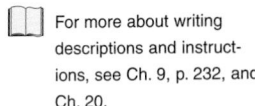
For more about writing descriptions and instructions, see Ch. 9, p. 232, and Ch. 20.

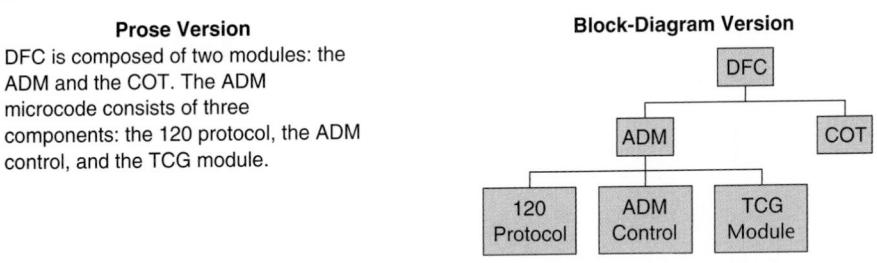

Figure 14.19   Block Diagram and Prose Description

Unlike most other figures, the title of an organization chart generally appears above the chart because the chart is read from the top down.

In this organization chart, different levels are distinguished by color. Levels can also be distinguished by the size of the boxes, the width of the lines that form the boxes, or the typeface. If space permits, the boxes can contain brief descriptions of the positions, duties, or responsibilities.

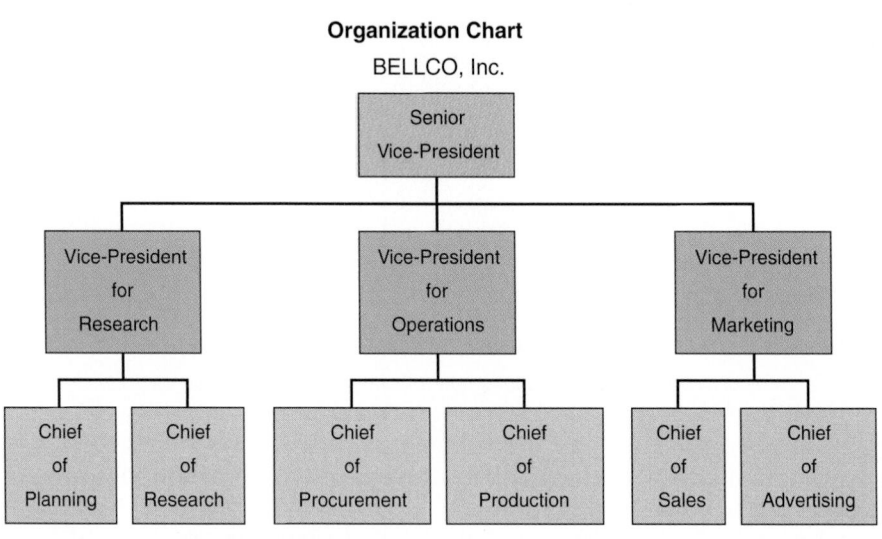

■ **Figure 14.20   Organization Chart**

Checklists

In explaining how to carry out a task, you often need to show the reader what equipment or materials to gather, or describe an action or series of actions to take. A *checklist* is simply a list of items, each preceded by a check box. If readers might be unfamiliar with the items you are listing, include drawings of the items, as shown in Figure 14.21.

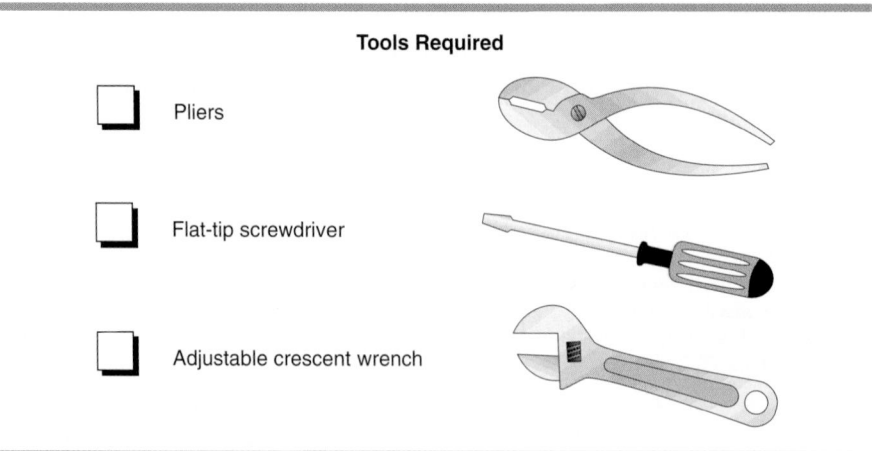

■ **Figure 14.21   Checklist**

Tables

Often you need to indicate that your reader is to carry out certain tasks at certain intervals. A *table* is a useful graphic for this kind of information, as illustrated in Figure 14.22 (McComb, 1991, p. 133).

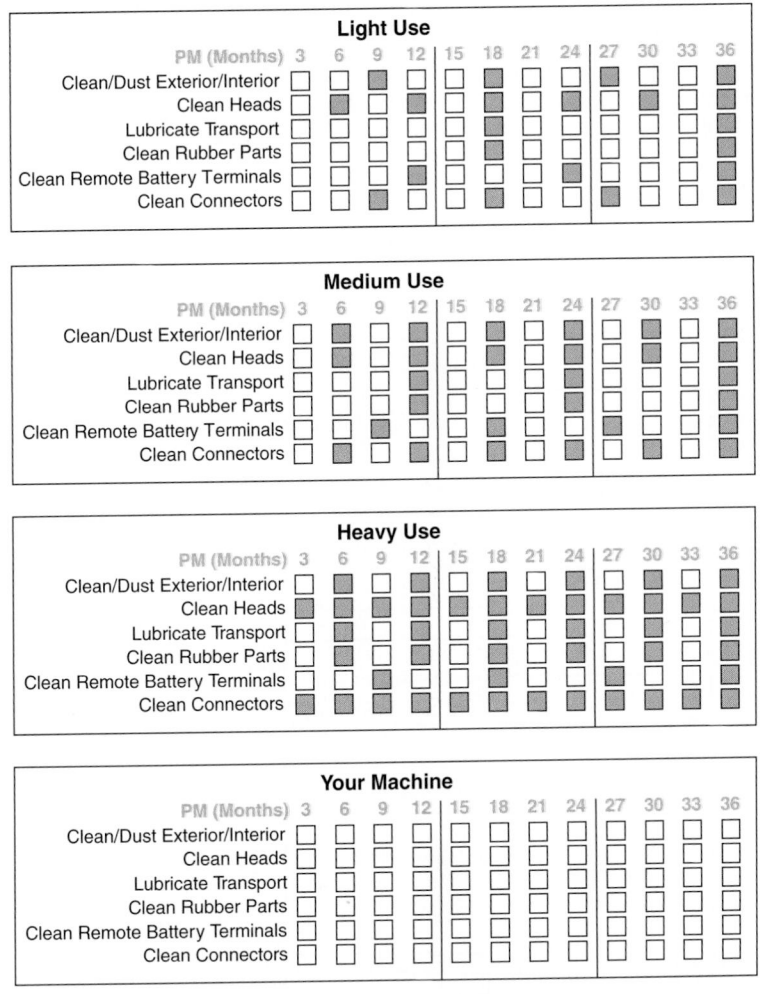

The darkened check boxes indicate when an action is to be performed.

Figure 5-1. Preventative maintenance schedule for VCRs. Determine whether your machine gets light, medium, or heavy use. Then, perform the task indicated by each green box. Record the maintenance in the bottom portion, "Your Machine."

■ **Figure 14.22   A Table Used to Illustrate a Maintenance Schedule**

Flowcharts

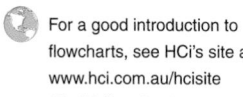

For a good introduction to flowcharts, see HCi's site at www.hci.com.au/hcisite /Toolkit/flowchar.htm.

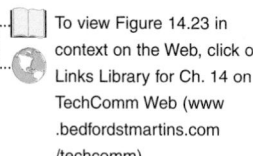

To view Figure 14.23 in context on the Web, click on Links Library for Ch. 14 on TechComm Web (www .bedfordstmartins.com /techcomm).

A *flowchart*, as the name suggests, shows the various stages of a process or a procedure. A flowchart might be used, for example, to show the steps involved in transforming lumber into paper or in synthesizing an antibody. Flowcharts are useful, too, for summarizing instructions. On a basic flowchart, stages are represented by labeled rectangles or circles. If you are addressing general readers, consider pictorial symbols instead of geometric shapes. Flowcharts can portray open systems (those that have a "start" and a "finish") or closed systems (those that end where they began).

Figure 14.23 (U.S. Department of the Treasury, 1999, p. 4) is an open-system flowchart that shows the stages of a procedure. Figure 14.24 (Curtis & Barnes, 1989, p. 251) is a closed-system flowchart that portrays a process.

Figure A. Can You Deduct Business Use of the Home Expenses?*

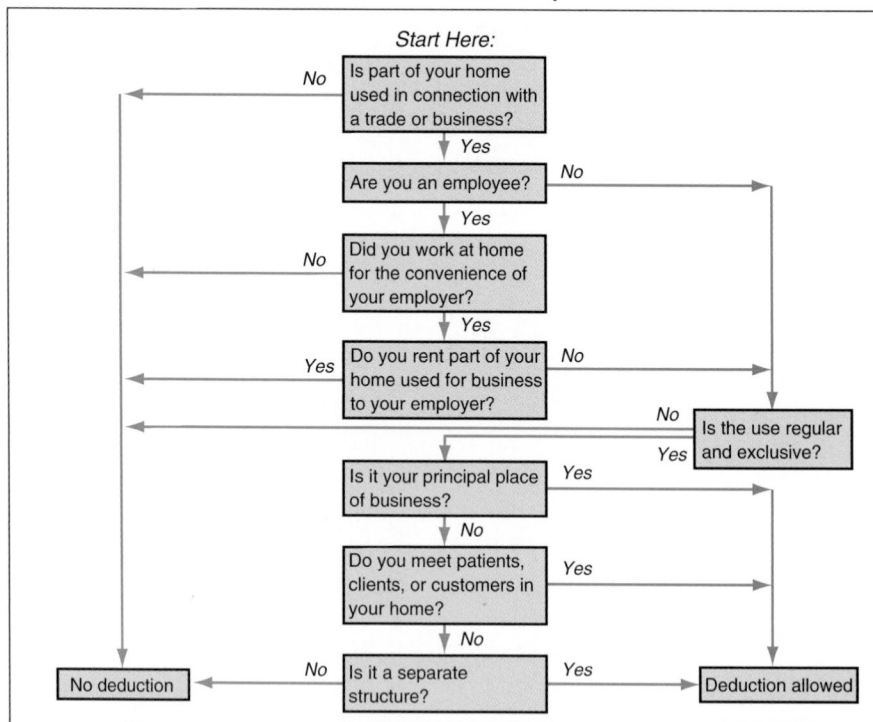

* Do not use this chart if you use your home for the storage of inventory or product samples, or to operate a day-care facility. See *Exceptions to Exclusive Use*, earlier, and *Day-Care Facility*, later.

■ **Figure 14.23   Flowchart Used to Portray a Procedure**

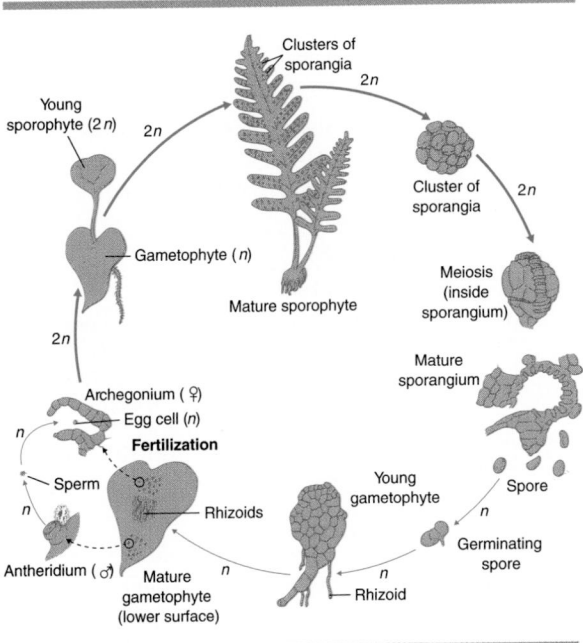

Labels in figure:
Clusters of sporangia 2n
Young sporophyte (2n)   2n
Cluster of sporangia 2n
Gametophyte (n)
Mature sporophyte
Meiosis (inside sporangium)
Mature sporangium
2n
Archegonium (♀)
Egg cell (n)
Fertilization
n
Sperm
n
Rhizoids
Antheridium (♂)
Mature gametophyte (lower surface) n
Young gametophyte n
Spore
n
Germinating spore
n
Rhizoid

■ **Figure 14.24   A Closed-System Flowchart Used to Portray a Process**

Logic Boxes and Logic Trees

Logic boxes guide the reader along one of two or more decision paths. Figure 14.25 on page 394 shows a typical example.

Logic trees are like logic boxes but use a branching metaphor. The logic tree shown in Figure 14.26 on page 394 helps students think through the process of registering for a course.

Illustrating Visual and Spatial Characteristics

To illustrate visual and spatial characteristics, use photographs, screen shots, line drawings, and maps.

Photographs

Photographs are unmatched for reproducing visual detail. If you want to portray realistically the kinds of tire-tread wear caused by various alignment problems, a photograph is the best choice. If you want your readers to recognize a new product, such as a new automobile model, you would probably use a photograph.

Do you want to replace all occurrences of the first word with the second word?			
YES	NO		
Press Y. The computer replaces all occurrences of the first word with the second word.	Press N. The cursor moves to the next occurrence of the word.		
	Do you want to replace that occurrence of the word?		
	YES	NO	
	Press SHIFT/F2.	Press F2.	
	Press F4 to move the cursor to the next occurrence of the word. Then press SHIFT/F2 or F2, as you did above, to replace that occurrence of the word or to leave it as is.		
Do you want to use the search function again?			
YES	NO		
Press F8.	Press F10.		

■ **Figure 14.25   Logic Boxes**

Logic boxes can contain a lot of textual information. However, some readers might have trouble understanding how to read them.

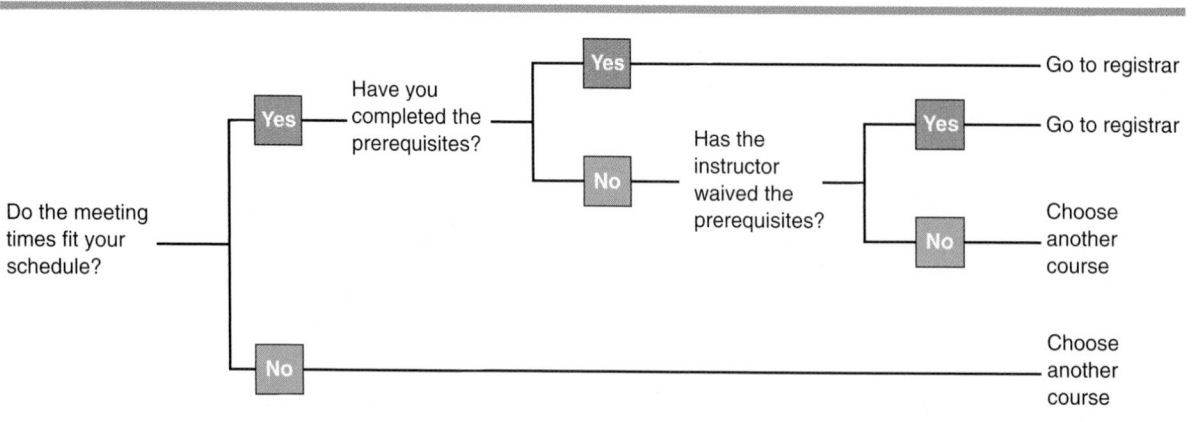

■ **Figure 14.26   Logic Trees**

Logic trees are probably somewhat easier to understand than logic boxes because the tree metaphor is visually more distinct than the box metaphor. However, logic trees cannot handle as much textual information as logic boxes can.

Sometimes, however, a photograph can provide too much information. In an advertising brochure for an automobile, a glossy photograph of the dashboard might be very effective. But in an owner's manual, if you want to show how to find the trip odometer, a diagram will probably work better because it focuses on the item itself.

Sometimes a photograph can provide too little information; the item you want to highlight might be located inside the mechanism or obscured by some other component.

GUIDELINES

Presenting Photographs Effectively

▶ *Eliminate extraneous background clutter that can distract your reader.* Do this by cropping the photograph or by digitizing it and deleting unnecessary detail. Figure 14.27 shows examples of cropped and uncropped photographs.

▶ *Do not electronically manipulate the photograph.* Today it is routine to digitize a photograph, remove blemishes, and crop it (cut it down to a desired size). There is nothing unethical about doing so. However, manipulating a photograph — for example, enlarging the size of the monitor that comes with a computer system — is unethical.

▶ *Help the reader understand the perspective.* Most objects in magazines and journals are photographed at an angle. This perspective provides more information because it shows the depth of the object as well as its height and width.

▶ *If appropriate, include some common object, such as a coin or a ruler, in the photograph to give readers a sense of scale.*

▶ *If appropriate, label components or important features.*

Cropping a photograph lets you direct the reader's attention to the important information. The image on the right is a cropped and enlarged version of the photo on the left.

■ **Figure 14.27   Cropping a Photograph**

Screen Shots

A *screen shot* is a computer graphic of what appears on the computer monitor. Screen shots are often used in software manuals to show the user what the screen looks like at various points during the use of a program. Figure 14.28 is an example of a screen shot.

This screen shot from Software602 shows the spelling/thesaurus feature in action. This screen shot presents a much clearer idea of how the feature works than mere words could.

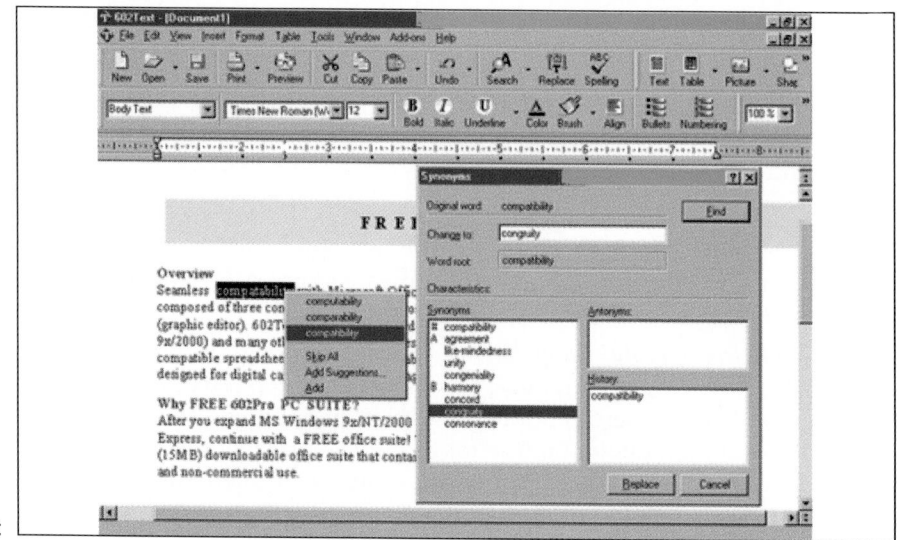

■ **Figure 14.28   Screen Shot**

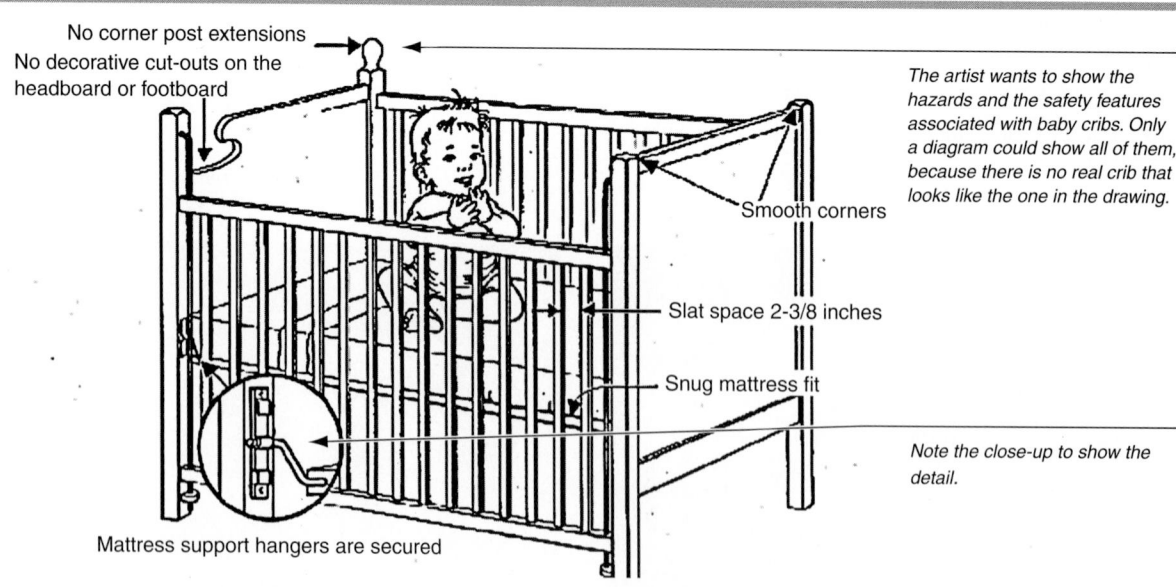

No corner post extensions
No decorative cut-outs on the headboard or footboard

The artist wants to show the hazards and the safety features associated with baby cribs. Only a diagram could show all of them, because there is no real crib that looks like the one in the drawing.

Smooth corners

Slat space 2-3/8 inches

Snug mattress fit

Note the close-up to show the detail.

Mattress support hangers are secured

■ **Figure 14.29   Line Drawing**

Line Drawings

Line drawings are simplified visual representations of objects. Line drawings can have three advantages over photographs:

- Line drawings can focus the reader's attention on desired information better than a photograph can.
- Line drawings can highlight information in a photograph that might be obscured by bad lighting or a bad angle.
- Line drawings are sometimes easier for readers to understand than photographs are.

Figure 14.29 (U.S. Consumer Product Safety Commission, 1999, p. 1) shows the effectiveness of line drawings.

You have probably seen the three variations on the basic line drawing shown in Figure 14.30.

Maps

Maps are readily available as clip art that can be modified with a graphics program. Figure 14.31 on page 398 shows a map derived from clip art.

Showing Motion in Your Graphics

In technical documents you will often want to show motion. For instance, in an instruction manual for military helicopter technicians, you might want to illustrate the process of removing an oil dipstick or tightening a bolt, or you

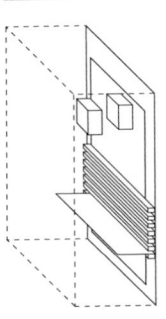 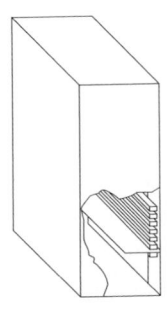 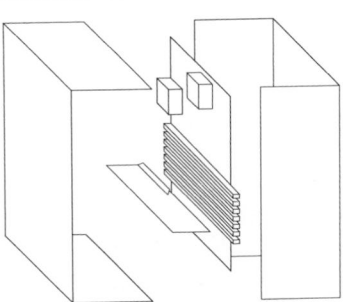

a. Phantom drawings show parts hidden from view by outlining external items that would ordinarily obscure them.

b. Cutaway drawings "remove" a part of the surface to expose what is underneath.

c. Exploded drawings separate components while maintaining their physical relationship.

■ **Figure 14.30   Phantom, Cutaway, and Exploded Views**

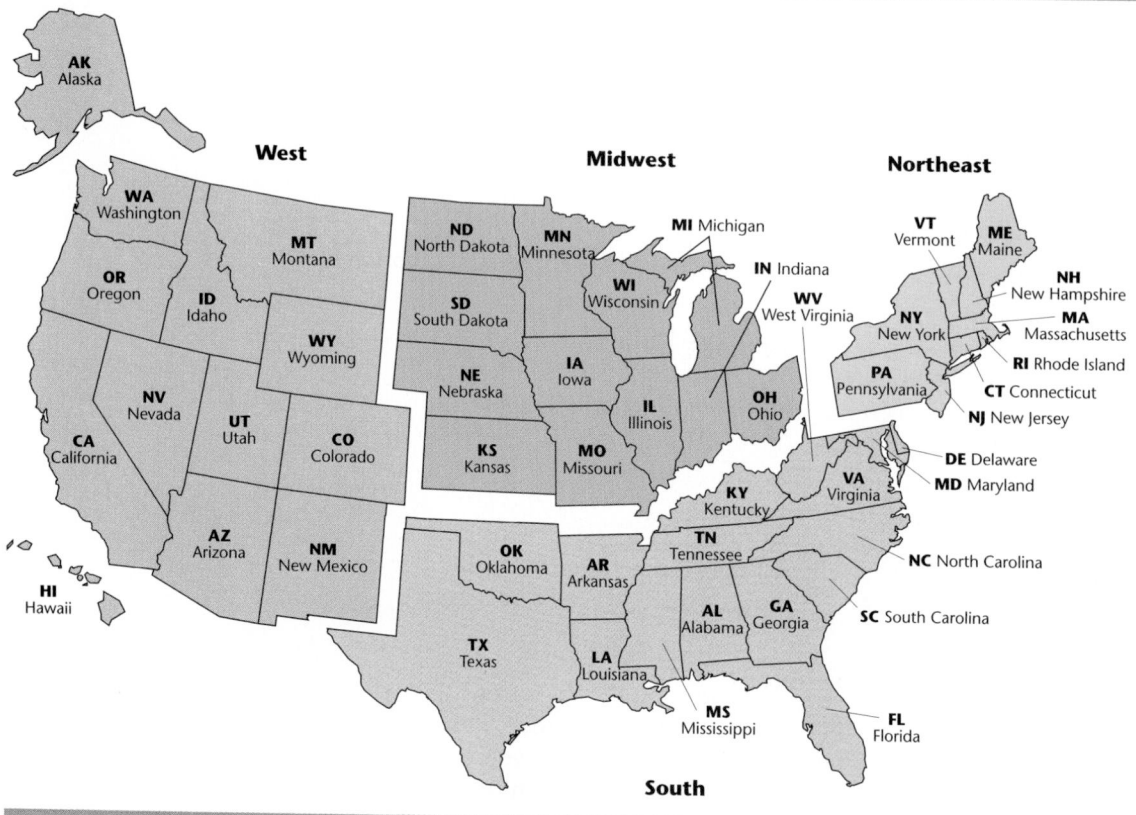

might want to show a warning light flashing. Although document designers frequently use animation or video, printed graphics are still needed to communicate this kind of information.

If the reader is to perform the action, show the action from the reader's point of view. In most cases, you need to show only the person's hands, not the whole body, as shown in Figure 14.32.

Figure 14.33 illustrates four additional techniques you can use to show action.

These symbols are conventional but not universal. If you are addressing multicultural readers, consult a qualified person from that culture to make sure your symbols are clear and inoffensive.

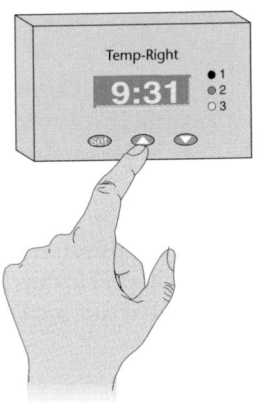

■ **Figure 14.32**
**Showing the Action from
the Reader's Perspective**

*In many cases, you need to
show only the person's
hands, not the whole body.*

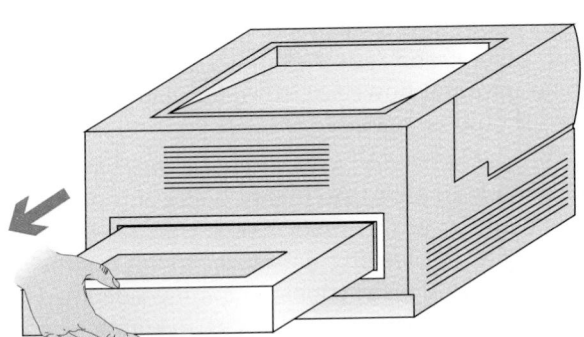

*a. Use arrows or other symbols to suggest the direction in
which something is moving or should be moved.*

c. Shake lines suggest vibration.

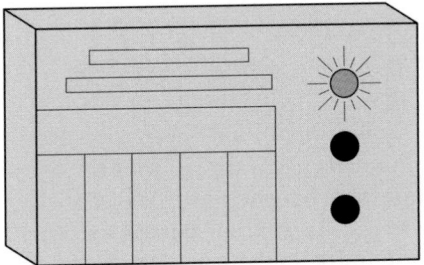

b. Starburst lines suggest a blinking light.

*d. An image of an object both before and after the action
suggests the action.*

■ **Figure 14.33   Showing Motion**

CREATING GRAPHICS FOR MULTICULTURAL READERS

Whether you are writing for people within your organization or outside it, consider the needs of readers who do not speak English as their primary language. If you are writing for people in other countries, remember that many may not speak any English.

See Chs. 5 and 11, pp. 107 and 297, for more on multicultural audiences.

Like words, graphics have cultural meanings. If you are unaware of these meanings, you could communicate something very different from what you intend. The following guidelines are based on William Horton's (1993) article "The Almost Universal Language: Graphics for International Documents."

GUIDELINES

Creating Effective Graphics for Multicultural Readers

▶ *Be aware that reading patterns differ.* In some countries, people read from right to left or from top to bottom. In some cultures, direction signifies value: the right-hand side is superior to the left, or the reverse. You need to think about how to sequence graphics that show action, or where you put "before" and "after" graphics. If you want to show a direction, as in an informal flowchart, consider using arrows to indicate how the chart should be read.

▶ *Be aware of varying cultural attitudes toward giving instruction.* Instructions for products made in Japan are highly polite and deferential: "Please insert the batteries at this time." Some cultures favor spelling out general principles but leaving the reader to supply the details. For people in these cultures, instructions containing a detailed close-up of how to carry out a task might appear insulting. An instructional table with the headings "When You See This . . ." and "Do This . . ." might be inappropriate.

▶ *Deemphasize trivial details.* Common objects, such as plugs on the ends of power cords, come in any number of shapes around the world. Therefore, it is wise to draw them to look generic rather than specific to one country.

▶ *Avoid culture-specific language, symbols, and references.* Don't use a picture of a mouse to symbolize a computer mouse, because the device is not known by that name everywhere. Avoid the casual use of national symbols (such as the maple leaf or national flags), because you might unknowingly make an error in a detail that would insult your readers. Use colors carefully: red means danger to most people from a Western culture, but it is a celebratory color to the Chinese.

▶ *Portray people very carefully.* Every aspect of a person's appearance, from clothing to hairstyle to features, is culture- or race-specific. A photograph of a woman in casual Western attire seated at a workstation would not be effective in an Islamic culture like Saudi Arabia, where only the hands and eyes of a woman may be shown. Horton (1993) recommends using stick figures or silhouettes that do not suggest any one culture, race, or gender.

▶ *Be particularly careful in portraying hand gestures.* Many Western hand gestures, such as the "okay" sign, are considered obscene in other cultures, and long red fingernails would also be inappropriate. Use hands in graphics only when necessary, for example, carrying out a task, and obscure the person's sex and race.

Cultural differences are many and subtle. Sometimes it seems that everything will offend someone. The best advice is to learn as much as possible about your readers and about their culture and outlook, and to make sure your graphics are reviewed by a native of the culture. For more information on this subject, consult these four books:

Edward Hall. (1990). *Understanding cultural differences.* Yarmouth, ME: Intercultural Press.

Nancy L. Hoft. (1995). *International technical communication: How to export information about high technology.* New York: Wiley.

Roger Axtel. (1991). *Gestures: The do's and taboos of body language around the world.* New York: John Wiley.

Scott Jones, Cynthia Kennelly, Claudia Mueller, et al. (1992). *Developing international user information.* Bedford, MA: Digital Press.

UNDERSTANDING GRAPHICS SOFTWARE

Graphics software available for the personal computer fall into two basic categories: spreadsheet business graphics and paint programs and draw programs.

Spreadsheet Business Graphics

Spreadsheet programs — software designed to help businesspeople calculate budgets and evaluate hypothetical business scenarios — can be used to produce business graphics. Once you have typed in the numerical data, such as the profits and losses of the four divisions of your company, you can have the software portray the data in various kinds of graphics and charts.

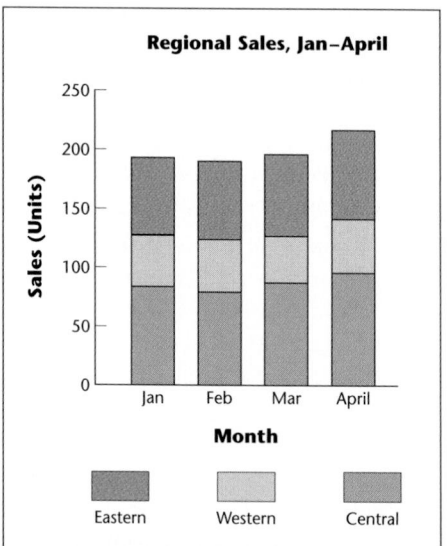

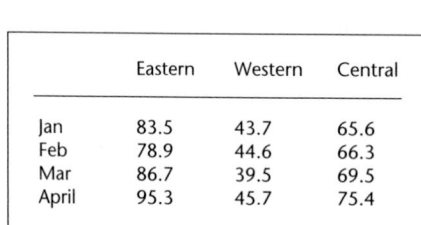

	Eastern	Western	Central
Jan	83.5	43.7	65.6
Feb	78.9	44.6	66.3
Mar	86.7	39.5	69.5
April	95.3	45.7	75.4

a. Spreadsheet data

b. Grouped bar chart created from the spreadsheet data

■ **Figure 14.34   Creating a Graphic from a Spreadsheet**

Spreadsheets generate different kinds of bar graphs, line graphs, and pie charts in both two-dimensional and three-dimensional formats. You then add labels and a title and customize the graphic to suit your needs. Figure 14.34 shows how data entered on a spreadsheet can be displayed as a table and then as a bar chart. Most spreadsheets can display the same numerical data in dozens of styles of graphics.

Be careful when you use spreadsheet business graphics. The software doesn't provide advice on the kind of graphic to make or how to modify the basic presentation — it simply draws whatever you tell it to — and it often makes poor choices: unnecessary 3-D, useless clip art, and lurid color combinations. To use the graphics capabilities effectively, you need to understand the basics of the various kinds of graphics.

Paint Programs and Draw Programs

Paint programs and *draw programs* let you produce and then modify freehand drawings in a number of ways. For instance, you can

- modify the width of lines
- modify the size of shapes
- copy, rotate, and flip images

- fill in shapes with different colors and textures
- add text with a simple text editor

  With both kinds of programs, you can begin in one of three ways:

- by drawing on a blank screen
- by importing an image from a clip-art library
- by scanning an image, which requires hardware — a scanner — and a special software program that translates a graphic on paper into a computerized image

You probably already have a paint program in your computer operating system and a draw program in your word processor. These programs are not professional-level, but they are powerful enough to create many kinds of graphics.

✔   **Revision Checklist**

This checklist focuses on the characteristics of an effective graphic.
1. Does the graphic have a purpose?
2. Is the graphic honest?
3. Is the graphic simple and uncluttered?
4. Does the graphic present a manageable amount of information?
5. Does the graphic meet the reader's format expectations?
6. Is the graphic clearly labeled?
7. For an existing graphic, do you have the legal right to use it? If so, have you cited it appropriately?
8. Does the graphic appear in a logical location in the document?
9. Is the graphic introduced clearly in the text?
10. Is the graphic explained in the text?
11. Is the graphic clearly visible in the text?
12. Is the graphic easily accessible to readers?
13. Is the graphic inoffensive to readers?

Exercises

1. Find out from the admissions department at your college or university the number of students enrolled from different states or from the different counties in your state. Present this information in four different kinds of graphics:

   a. a map
   b. a table
   c. a bar graph
   d. a pie chart

   In three or four paragraphs, explain why each graphic is appropriate for a particular audience and purpose, and how each emphasizes different aspects of the information.

2. Design a flowchart for a process you are familiar with, such as applying for a summer job, studying for a test, preparing a paper, or performing some task at work. Your audience is someone who will be carrying out the process.

3. Create an organizational chart for an organization you belong to or are familiar with: a department at work, a fraternity or sorority, a campus organization, or student government.

4. In two or three sentences, define an audience and a purpose and then make an appropriate drawing of some object you are familiar with, such as a tennis racquet, a stereo speaker, or a weight bench.

5. In two or three sentences, define an audience and a purpose and then draw a diagram illustrating a concept, such as the effects of smoking or the advantages of belonging to the student chapter of a professional organization, that is appropriate to your audience and purpose.

6. Create a graphic for one of the following scenarios:

   a. in a user's manual for a computer, how to insert the floppy disk

   b. in the owner's manual for a bicycle, how to locate the serial number

   c. in a training manual for new wait staff, how to explain the menu to customers

   d. in a set of instructions on how to change the oil in a car, the equipment and materials needed

   e. in a student handbook for your college or university, the process of applying for graduation

7. Create graphics for a set of instructions on how to use an automated teller machine. Your readers are nonnative speakers of English who are unfamiliar with automated teller machines. For each graphic, write a one- or two-sentence summary of what the matching text would be describing.

8. In the *Statistical Abstract of the United States* or an issue of *Forbes* or *Business Week,* find and photocopy a bar graph, a pictograph, a line graph, a pie chart, a diagram, a flowchart, and a photograph. For each graphic, write a brief discussion that responds to the following questions:

   a. Is the graphic necessary?

   b. Is it ethical, or does it misrepresent the information?

   c. Is it professional in appearance?

   d. Does it conform to the guidelines for that kind of graphic?

   e. Is it effectively integrated into the discussion?

9. The following table (U.S. Bureau of the Census, 2000, p. 140) provides statistics on disabilities. Study the table, then perform the following tasks:

   a. Create two different graphics, each of which communicates information about the percentage of people in each group who have a disability.

   b. Create two different kinds of graphics, each of which communicates information about the number of people aged 22 to 44 years who have difficulty with one or more ADLs and those who have a severe disability.

   c. Create two different graphics, each of which communicates information about disability rates for those aged 15 to 21 years and those aged 80 years and over.

No. 222. Disability Status of Persons 15 Years Old and Over: 1997

[In thousands, except as noted (208,059 represents 208,059,000). See headnote, Table 221]

Disability status	Total	15 to 21 years	22 to 44 years	45 to 54 years	55 to 64 years	65 to 79 years	80 years
Total...................	**208,059**	**26,477**	**94,307**	**33,620**	**21,591**	**24,827**	**7,237**
Persons with any disability.........	47,836	2,965	12,157	7,566	7,693	12,141	5,314
Percent of total...............	23.0	11.2	12.9	22.5	35.6	48.9	73.4
With a severe disability..........	30,576	1,396	7,292	4,647	5,208	7,874	4,160
Difficulty with one or more ADLs[1]..	8,661	115	1,364	1,292	1,335	2,616	1,939
Needs personal assistance.......	4,048	90	613	514	527	1,215	1,089
Difficulty with one or more IADLs[1]..	12,912	305	2,291	1,691	1,724	3,925	2,977
Needs personal assistance with one or more ADLs or IADLs[1].........	10,224	262	1,788	1,295	1,345	2,923	2,611

[1]ADLs are activities of daily living and include getting around inside the home, getting in or out of a bed or chair, taking a bath or shower, dressing, eating, and using the toilet. IADLs are instrumental activities of daily living and include going outside the home, keeping track of money and bills, preparing meals, doing light housework, and using the telephone.

10. For each of the following four graphics, write a paragraph evaluating its effectiveness and describing how you would revise it.

a.

	1997	1998	1998
Civil Engineering	236	231	253
Chemical Engineering	126	134	142
Comparative Literature	97	86	74
Electrical Engineering	317	326	401
English	714	623	592
Fine Arts	112	96	72
Foreign Languages	608	584	566
Materials Engineering	213	227	241
Mechanical Engineering	196	203	201
Other	46	42	51
Philosophy	211	142	151
Religion	86	91	72

b.

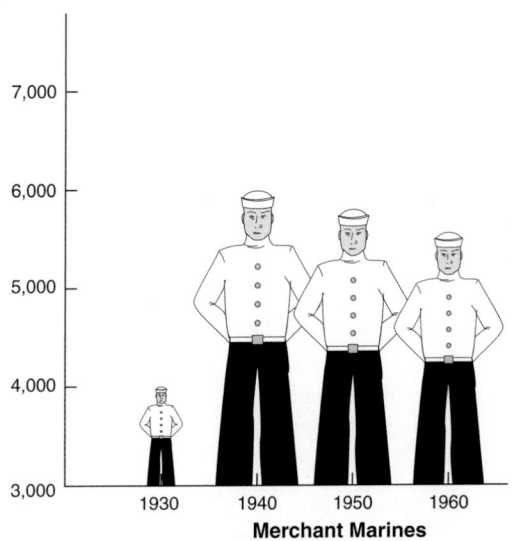

Merchant Marines

c.

Expenses at Hillway Corporation

d.

Costs of the Components of a PC

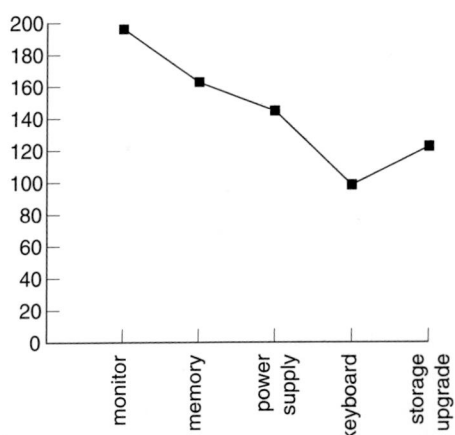

11. The following three graphs illustrate the sales of two products — Series 1 and Series 2 — for each quarter of 1999. Which is the most effective in conveying the information? Which is the least effective? What additional information or revisions would make the best graph better?

a.

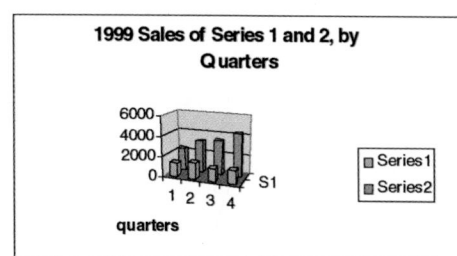

b.

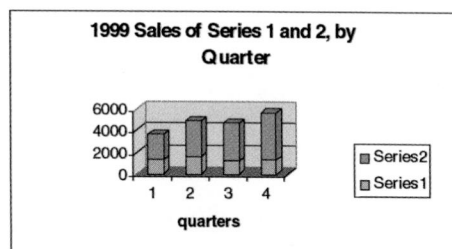

c.

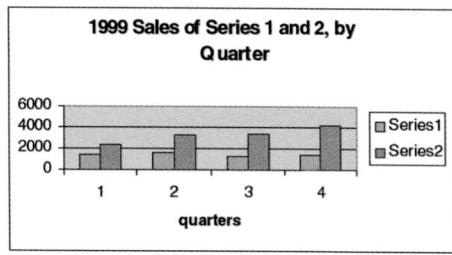

12. In each of the following exercises, express the written information in at least two kinds of graphics and write a one-paragraph statement about which kind works best. (If one kind works well for one audience but not for another, explain.)

a. Following are the profit and loss figures for Pauley, Inc., in early 1999: January, a profit of 6.3 percent; February, a profit of 4.7 percent; March, a loss of 0.3 percent; April, a loss of 2.3 percent; May, a profit of 0.6 percent.

b. The prime interest rate had a major effect on our sales. In January, the rate was 4.5 percent. It went up a full point in February, and another half point in March. In April, it leveled off, and it dropped one full point each in May and June. Our sales figures were as follows for the Crusader 1: January, 5,700; February, 4,900; March, 4,650; April, 4,720; May, 6,200; June, 8,425.

c. This year, our membership can be broken down as follows: 45 percent, electrical engineering; 15 percent, mechanical engineering; 20 percent, materials engineering; 10 percent, chemical engineering; and 10 percent, other.

d. In January of this year we sold 50,000 units of the BG-1, of which 20,000 were purchased by the army. In February, the army purchased 15,000 of our 60,000 units sold. In March, it purchased 12,000 of the 65,000 we sold.

e. The normal rainfall figures for this region are as follows: January, 1.5 in.; February, 1.7 in.; March, 1.9 in.; April, 2.1 in.; May, 1.8 in.; June, 1.2 in.; July, 0.9 in.; August, 0.7 in.; September, 1.3 in.; October, 1.1 in.; November, 1.0 in.; December, 1.2 in. The following rainfall was recorded in this region: January, 2.3 in.; February, 2.6 in.; March, 2.9 in.; April, 2.0 in.; May, 1.6 in.; June, 0.7 in.; July, 0.1 in.; August, 0.4 in.; September 1.3 in.; October, 1.2 in.; November, 1.4 in.; December, 1.8 in.

f. You can access the Rightfile from programs written in six languages. Rightfile classifies these languages as two groups (A and B) and provides separate files for each group. The A group includes C++ and PL/1. The B group includes LISP and RPGII. The object module for the A group is GRTP. The object module for the B group is GRAP.

Research Projects

Some of the following projects ask you to write a memo. See Chapter 15, page 430, for a discussion of memos.

13. Interview three people: an engineer or scientist, a technical communicator, and a graphic artist or illustrator who works in technical communication. Find out about their approaches to creating graphics. Ask about their education in graphics, their use of tools or software, how they integrate graphics into the writing process, and their views of the importance of graphics in the work they do. Write a memo to your instructor discussing what you discover.

14. Form small groups. Go to one of the campus computer centers and study a popular piece of software, such as a spreadsheet or graphics package. Have each group member, working separately, print out three of the basic kinds of graphics, such as graphs, tables, and pie charts, and take notes on the process. Come together as a group and share your experiences in learning to use the software.

What techniques did you try: using a tutorial, reading a manual, trial-and-error, asking someone for help? If appropriate, have the group go back to the lab together and work through some of the questions that arose in the group meeting. Write a memo to your instructor describing how easy or how difficult it was to learn how to produce these graphics and evaluating their quality. How well do the graphics conform to the basic guidelines presented in this chapter? Would you recommend the software to other members of your class? Why or why not?

15. Locate a graphic on the Web that you consider inappropriate for an international audience because it might be offensive or unclear in some cultures. Imagine an audience for the graphic, such as people from the Middle East, and write a brief statement explaining the potential problem. Finally, revise the graphic so that it would be appropriate for its intended audience.

C A S E 1
Evaluating Graphics on a Web Site

You are a student intern working for Marshall Brain at his Web site, How Stuff Works (www.howstuffworks.com). He has asked you to evaluate the effectiveness of the graphics on his site. Select one of the topics, such as How Web Pages Work, then analyze Brain's use of graphics. Are there enough graphics to illustrate the topic? Are the graphics clear and informative? Are they well integrated with the text on the pages? What changes would you recommend? Write a memo to Marshall Brain presenting your conclusions and recommendations. If appropriate, attach copies of excerpts from the Web site.

 To find the Web site for this case, click on Links Library for Ch. 14 on TechComm Web (www.bedfordstmartins.com/techcomm).

CASE 2
Evaluating Graphics Made in a Spreadsheet Program

You and the other members of your group work at Microsoft in the documentation group for the Excel spreadsheet. Your boss, Alan Winston, has written you a memo about revising the program. Here are some excerpts.

"We've gotten some complaints about the charting function in Excel. Some users are unhappy that we don't offer them any help in choosing what kind of chart works best for different kinds of information and readers.

"Would you mind taking a look at the charting function? What kind of instruction do you think we could add that would help our users? Show me an example for one of the chart types, such as bar graphs or line graphs."

Study Excel, then write a memo to Winston, including a sample of the kind of help you think could be incorporated into the software.

PART FIVE
APPLICATIONS

Writing Letters, Memos, and Emails

15

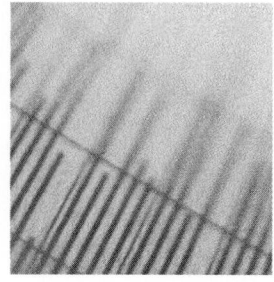

Joel Bowman (1999), a business-communication scholar, on persuasion in business correspondence:

Whenever you are writing to someone who has no compelling reason to do (or think) as you ask in your message — and perhaps even no reason to read or reply to it — you need to write a persuasive message. How persuasive you need to be depends on how obvious it is to your reader that he or she stands to benefit from acting (or thinking) in the manner you suggest.

See TechComm Web (www .bedfordstmartins.com /techcomm) for guidelines, additional examples, and links related to topics in this chapter.

This chapter focuses on the three formats used for correspondence in the working world:

- *Letters.* The oldest of the three formats, letters are still used to communicate important information. Letters are the most formal of the three formats, and always conclude with a signature in ink. Although they are sometimes used to convey formal messages within a writer's own organization, letters are more often addressed to someone in another organization. Today, letters are often sent by fax.

- *Memos.* Because memos are less formal and shorter than letters, they are used most often for communication among those within one organization.

- *Emails.* The least formal of the three, email is replacing many memos because it offers all the advantages of digital communication: the writer can send an email to dozens, or even thousands, of readers instantly and effortlessly; and the recipients can revise and store an email easily.

This chapter begins with a discussion of letters, showing the basic formats, explaining tone and strategy in letter writing, and describing several common kinds of letters. It goes on to discuss the structure and strategy of memos and concludes with email, focusing on the special demands of writing for the screen.

See Online Technical Writing (www.io.com/~hcexres /tcm1603/acchtml/acctoc .html) for more about letter writing.

WRITING LETTERS

Letters are still a basic means of communication between organizations: millions are written each workday. Even as a new employee, you can expect to write letters regularly. As you advance to positions of greater responsibility, you will write letters representing your organization even more frequently. Writing a letter is much like writing any other technical document. First you have to analyze your audience and determine your purpose. Then you have to gather your information, make an outline, write a draft, and revise it.

Projecting the "You Attitude"

Like any other type of technical communication, a letter should be honest, clear, concise, comprehensive, accessible, correct, professional, and accurate. It must convey information in a logical order. It should not contain small talk; the first paragraph should get directly to the point without wasting words. And to help the reader locate information quickly and easily, a topic sentence should appear at the start of each paragraph in the body. Often, letters contain headings, just as reports do.

Moreover, because it is a communication from one person to another, a letter must also convey a courteous, positive tone. The key to an effective letter is the *you attitude*. This term means looking at the situation from your reader's point of view and adjusting the content, structure, and tone to meet the person's needs. If, for example, you are writing to a supplier who has failed to deliver some merchandise on the agreed-upon date, the you attitude dictates that you not discuss problems you are having with other suppliers — those problems don't concern your reader. Rather, you should concentrate on explaining clearly and politely that your reader has violated your agreement and that not having the merchandise is costing you money. Then you should propose ways to expedite the shipment.

Here's another way to look at it: even if the context of the letter is a dispute, always maintain a polite tone. Civilized behavior is good business, as well as a good way to live.

Following are examples of thoughtless sentences, each followed by an improved version that exhibits the you attitude.

BLUNT	You wrote to the wrong department. We don't handle complaints.
BETTER	I have forwarded your letter to the Customer Service Division.
ACCUSING	You must have dropped the engine. The housing is badly cracked.
BETTER	The badly cracked housing suggests that your engine must have fallen onto a hard surface from some height.
SARCASTIC	You'll need two months to deliver these parts? Who do you think you are, the Post Office?
BETTER	Surely you would find a two-month delay for the delivery of parts unacceptable in your business. That's how I feel too.
BELLIGERENT	I'm sure you have a boss, and I doubt if he'd like to hear about how you've mishandled our account.
BETTER	I'm sure you would prefer to settle the account between us rather than have it brought to your supervisor's attention.

CONDE-SCENDING	Haven't you ever dealt with a major corporation before? A 60-day payment period happens to be standard.
BETTER	Perhaps you were not aware of the standard 60-day payment period.
OVERSTATED	Your air-filter bags are awful. They're all torn. We want our money back.
BETTER	You will doubtless be surprised to learn that 19 of the 100 air-filter bags we purchased are torn. We hope you agree that refunding the purchase price of the 19 bags — $190.00 — is the fair thing to do.

When you draft a letter, put yourself in your reader's place. How would you react if you were the recipient? A calm, respectful tone makes the best impression and increases the chances that you will achieve your goal.

Avoiding Letter Clichés

Over the decades, a set of words and phrases has come to be associated with business letters; one common example is *as per your request*. These phrases make the letter sound stilted and insincere. If you would feel awkward or uncomfortable saying these clichés to a friend, avoid them in your letters.

Figure 15.1 is a list of common clichés and their more natural equivalents.

Letter Clichés	Natural Equivalents
attached please find	attached is
cognizant of	aware that
enclosed please find	enclosed is
endeavor (verb)	try
herewith ("We herewith submit . . .")	(None. "Herewith" doesn't say anything. Drop it.)
hereinabove	previously, already
in receipt of ("We are in receipt of . . .")	"We have received . . ."
permit me to say	(None. Permission granted. Just say it.)
pursuant to our agreement	as we agreed
referring to your ("referring to your letter of March 19, the shipment of pianos . . .")	"As you wrote in your letter of March 19, the . . ." (or subordinate the reference at the end of your sentence)
same (as a pronoun: "Payment for same is requested . . .")	(Use the noun instead: "Payment for the merchandise is requested . . .")
wish to advise ("We wish to advise that . . .")	(The phrase doesn't say anything. Just say what you want to say.)
the writer ("The writer believes that . . .")	"I believe . . ."

■ **Figure 15.1**
Letter Clichés and Natural Equivalents

Letter Containing Clichés	*Letter in Natural Language*
Dear Mr. Smith:	Dear Mr. Smith:
Referring to your letter regarding the problem encountered with your new Trailrider Snowmobile. Our Customer Service Department has just tendered its report.	Thank you for writing to us about the problem with your new Trailrider Snowmobile.
It is their conclusion that the malfunction is caused by water being present in the fuel line. It is our unalterable conclusion that you must have purchased some bad gasoline. We trust you are cognizant of the fact that while we guarantee our snowmobiles for a period of not less than one year against defects in workmanship and materials, responsibility cannot be assumed for inadequate care. We wish to advise, for the reason mentioned hereinabove, that we cannot grant your request to repair the snowmobile free of charge.	Our Customer Service Department has found water in the fuel line. Apparently some of the gasoline was bad. While we guarantee our snowmobiles for one year against defects in workmanship and materials, we cannot assume responsibility for problems caused by bad gasoline. We cannot, therefore, grant your request to repair the snowmobile free of charge.
Permit me to say, however, that the writer would be pleased to see that the fuel line is flushed at cost, $30. Your Trailrider would then give you many years of trouble-free service.	However, no serious harm was done to the snowmobile. We would be happy to flush the fuel line at cost, $30. Your Trailrider would then give you many years of trouble-free service. If you will authorize us to do this work, we will have your snowmobile back to you within four working days. Just fill out the enclosed authorization card and drop it in the mail.
Enclosed please find an authorization card. Should we receive it, we shall endeavor to perform the above-mentioned repair and deliver your snowmobile forthwith.	Sincerely yours,
Sincerely yours,	

■ **Figure 15.2   Sample Letters with and without Clichés**

The letter on the right side avoids clichés and shows an understanding of the you attitude. Instead of focusing on the violation of the warranty, it presents the conclusion as good news: the snowmobile is not ruined, and it can be repaired and returned in less than a week for a small charge.

Figure 15.2 shows two versions of the same letter: one written in clichés, the other in plain language.

Understanding the Elements of the Letter

Almost every letter includes a heading, inside address, salutation, body, complimentary close, signature, and reference initials. Some letters also include one or more of the following: an attention line, subject line, enclosure line, and copy line.

In the following paragraphs, the elements of the letter are discussed in the order they would ordinarily appear. In addition, six common types of letters are discussed in detail later in this chapter.

Heading

Most organizations use letterhead stationery with the organization's name, address, phone number, email address, and perhaps a logo printed at the top. This preprinted information and the date the letter is sent (printed two lines below the letterhead) make up the *heading*. (If you are using blank paper rather than letterhead, your address [without your name] and the date form the heading.)

Use letterhead for the first page and do not number it. Use blank paper for the second and all subsequent pages, and carry over the name of the recipient, the page number, and the date in the upper left-hand corner. For example:

Mr. John Cummings
Page 2
July 3, 20XX

Inside Address

The *inside address* consists of the recipient's name, position, organization, and business address:

Ms. Vanessa Stanger
Director of Quality Assurance
Viscount Systems, Inc.
400 Harbor View Drive
San Jose, CA 98213

If the person has a professional title, such as *Professor*, *Dr.*, or — for public officials — *Honorable*, use it. If not, use *Mr.* or *Ms.* (unless you know the recipient prefers *Mrs.* or *Miss*). If the position fits on the same line as the name, add it after a comma; otherwise, drop it to the line below. Spell the name of the organization the way the organization itself does: for example, International Business Machines calls itself IBM. Include the complete mailing address: street number and name, city, state, and zip code.

Attention Line

Sometimes you will be unable to address a letter to a particular person because you don't know (and cannot easily find out) the name of the individual who holds that position in the company. In these cases, use an *attention line*:

Attention: Technical Director

You can also use an attention line if you want to make sure that the organization you are writing to responds, even if the person you address the letter to is unavailable. In this case, put the name of the organization or one of its divisions on the first line of the inside address:

Operations Department
Haverford Electronics
117 County Line Road
Haverford, MA 01765

Attention: Charles Fulbright, Director

Subject Line

On the *subject line*, put either a project number (for example, "Subject: Project 31402") or a brief phrase defining the subject (for example, "Subject: Price Quotation for the R13 Submersible Pump"):

Operations Department
Haverford Electronics
117 County Line Road
Haverford, MA 01765

Attention: Charles Fulbright, Director

Subject: Purchase Order #41763

Salutation

If you decide not to use an attention line or a subject line, put the *salutation*, or greeting, two lines below the inside address. The traditional salutation is *Dear*, followed by the reader's *courtesy title* and last name, followed by a colon (not a comma). If you are fairly well acquainted with the recipient, use *Dear* followed by the person's first name. If you do not know the name, use a general salutation:

Dear Technical Director:
Dear Sir or Madam:

If you are addressing a group of people, use one of the following:

Ladies and Gentlemen:
Gentlemen: (if all the readers are male)
Ladies: (if all the readers are female)

Or tailor the salutation to a particular group:

Dear Members of the Restoration Committee:
Dear Members of Theta Chi Fraternity:

This same strategy is useful for sales letters without individual inside addresses:

Dear Homeowner:
Dear Customer:

Body

The *body* of the letter contains its substance. In most cases, it has at least three paragraphs: an introductory paragraph, a concluding paragraph, and one or more body paragraphs. For information on how to develop a persuasive argument, see Chapter 6.

Complimentary Close

A *complimentary close* follows the body of the letter. Today the convention phrases *Sincerely, Sincerely yours, Yours sincerely, Yours very truly, Very truly yours* are interchangeable. Capitalize only the first word in the phrase and use a comma after it.

Signature

Type your full name on the fourth line below the complimentary close. Sign the letter, in ink, above the typewritten name. Most organizations prefer that you include your position under your typed name. For example:

Very truly yours,

José Santos

José Santos
Personnel Manager

Enclosure Line

If the envelope contains documents other than the letter, include an *enclosure line* that indicates the number of enclosures:

FOR ONE ENCLOSURE   Enclosure

or

Enclosure (1)

FOR MORE THAN ONE ENCLOSURE   Enclosures (2)

Enclosures (3)

In determining the number of enclosures, count only separate items, not pages. A three-page memo and a ten-page report constitute only two enclosures. Some writers like to identify the enclosures:

Enclosure: 2001 Placement Bulletin

Enclosures (2): "This Year at Ammex"

2001 Annual Report

Copy Line

If you want the primary recipient to know that other people are receiving a copy of the letter, include a *copy line*. Use the symbol *c* (for "copy") or *pc* (for "photocopy") followed by the names of the other recipients (listed either alphabetically or according to organizational rank). If you do not want the primary recipient to know about other copies, use *bc* ("blind copy") on the copies *only*, not on the original.

Learning the Format of the Letter

Three common formats are used for letters: modified block, modified block with paragraph indentations, and full block. Figure 15.3 on page 420 illustrates each of these three formats.

Understanding Common Types of Letters

Organizations send out many different kinds of letters. This section focuses on the six types most frequently used in the technical workplace.

Two other types of letters are discussed in this book: the transmittal letter in Ch.12, p. 308, and the job-application letter in Ch. 16, p. 469.

Common types of letters:

Order

Inquiry

Response to inquiry

Sales

Claim

Adjustment

Order Letter

Perhaps the most basic form of business correspondence is the order letter to a manufacturer, wholesaler, or retailer. In writing an order letter, include all the information the recipient will need to identify the merchandise you want — quantity, model number, dimensions, capacity, material, price, and any other pertinent details — and specify the method of delivery. A typical order letter is shown in Figure 15.4 on page 421.

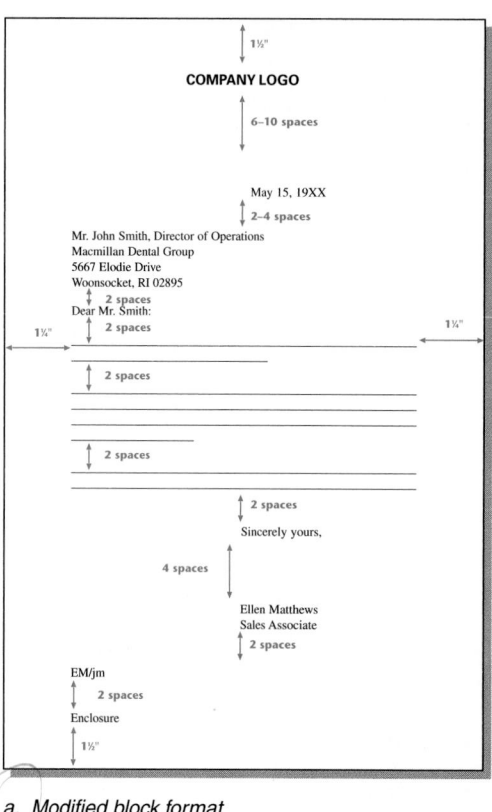

The dimensions and spacing shown here also apply to the other two formats.

a. *Modified block format*

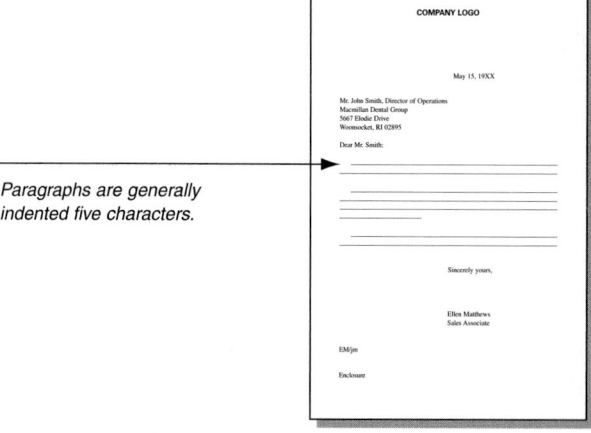

Paragraphs are generally indented five characters.

■ **Figure 15.3**
Common Letter Formats

b. *Modified block with paragraph indentations*

c. *Full block style — everything aligned along the left margin*

WAGNER AIRCRAFT
116 North Miller Road
Akron, OH 44313
www.wagair.com

September 4, 20XX

Mr. Frank DiFazio
Franklin Aerospace Parts
623 Manufacturer's Blvd.
Bethpage, NY 11741

Dear Mr. DiFazio:

Would you please send us the following parts by parcel post? All page numbers refer to your 20XX catalog.

Quantity	Model No.	Catalog Page	Description	Price
2	36113-NP	42	Seal fins	$34.95
1	03112-Bx	12	Turbine-bearing support	19.75
5	90135-QN	102	Turbine disc	47.50
1	63152-Bx	75	Turbine-bearing housing	16.15
				Total Price: $118.35

The writer uses an informal table to describe the parts she orders.

Yours very truly,

Christine O'Hanlon

Christine O'Hanlon
Purchasing Agent

■ **Figure 15.4**
Order Letter

Many organizations use preprinted forms, called purchase orders, to order products or services. A purchase order calls for the same information as an order letter.

Inquiry Letter

Your purpose in writing an inquiry letter is to obtain information. If the recipient is expecting the letter, your task is simple. For example, if a company manufacturing institutional furniture has advertised that it will send its 48-page, full-color brochure to prospective clients, you need write only a one-sentence letter: "Would you please send me the brochure advertised in *Higher Education Today*, May 13, 20XX?" If you want to ask a technical question, or set of questions, about any product or service for sale, your inquiry letter might begin "We are considering purchasing your new X-15 workstations for an office staff of 35 and would like some further information. Would you please answer the following questions?" In this example, the detail about the size of the potential order is not necessary, but it makes the inquiry seem serious and the potential sale substantial.

If the recipient is not expecting the letter, however, your task is more difficult: you must ask a favor. Only careful, persuasive writing will make the recipient want to respond when no direct benefit or potential profit seems likely.

GUIDELINES

Writing an Inquiry Letter

▶ *State why you are writing to the person or organization.* You might use subtle flattery — for example, "I was hoping that, as the leader in solid-state electronics, your company might be able to furnish some information about. . . ." Then explain why you want the information. Obviously, a company will not want to furnish information to a competitor. You have to show that your interests are not commercial — for instance, "I will be using this information in a senior project in agronomy at Illinois State University. I am trying to devise a. . . ." If you need the information by a certain date, mention it: "The project is to be completed by April 15, 20XX."

▶ *List your questions.* Companies are understandably annoyed by thoughtless requests to send "everything you have" on a topic. They much prefer a set of questions showing that you have already done substantial research. "Is your Model 311 compatible with Norwood's Model B12?," for example, is more precise than "Would you please tell me about your Model 311?" If your questions can be answered briefly, leave space for a reply after each question or in the margin.

▶ *Offer something in return, because you are asking someone to do something for you.* In many cases, all you can offer are the results of your research, but if so, say that you would be happy to send a copy of your final report. Express your appreciation. Don't write "Thank you for sending me this information," because it assumes that the reader is both willing and able to meet your request. Instead, write "I would greatly appreciate any help you could give me in answering these questions." Finally, if the answers will be brief, enclose a stamped self-addressed envelope to make it easy for the recipient to reply.

▶ *Always write a thank-you note to the person who has responded to your inquiry letter.*

Figure 15.5 on page 424 shows an example of a letter of inquiry.

Response to an Inquiry

If you are responding to an inquiry letter, keep the following suggestions in mind. If the questions are numbered, number your responses to correspond. If you cannot answer the questions, either because you don't know the answers or because you cannot divulge proprietary information, explain the reasons and offer to assist with other requests. Figure 15.6 on page 425 shows a response to the inquiry letter in Figure 15.5.

Sales Letter

The you attitude is crucial in a sales letter. Your readers don't care why you want to sell your product or service. They want to know why they should buy it. Because you are asking them to spend valuable time studying the letter, you must provide clear, specific information to help them understand what you are selling and how it will benefit them. Be upbeat and positive, but never forget that prospective customers want facts.

GUIDELINES

Writing Effective Sales Letters

1. *Gain the reader's attention.* Unless the opening sentence seems either interesting or important, the reader will toss the letter aside. Use facts, quotations, or questions to identify a problem that will interest your reader. Here are some examples:

   How much have construction costs risen since your plant was built? Do you know how much it would cost to rebuild at today's prices?

14 Hawthorne Ave.
Belleview, TX 75234
November 2, 20XX

Dr. Andrew Shakir
Director of Technical Services
Orion Corporation
721 West Douglas Avenue
Maryville, TN 31409

Dear Dr. Shakir:

I am writing to you because of Orion's reputation as a leader in the manufacture of adjustable X-ray tables. I am a graduate student in biomedical engineering at the University of Texas, and I am working on an analysis of diagnostic equipment for a seminar paper. Would you be able to answer a few questions about your Microspot 311?

1. Can the Microspot 311 be used with lead oxide cassettes, or does it accept only lead-free cassettes?
2. Are standard generators compatible with the Microspot 311?
3. What would you say is the greatest advantage, for the operator, of using the Microspot 311? For the patient?

My project is due on January 15. I would greatly appreciate your assistance in answering these questions. Of course, I would be happy to send you a copy of my report when it is completed.

Yours very truly,

Albert K. Stern

Albert K. Stern

■ **Figure 15.5**
Inquiry Letter

ORION

721 WEST DOUGLAS AVE.   (615) 619-8132
MARYVILLE, TN 31409   www.orioninstruments.com

November 7, 20XX

Mr. Albert K. Stern
14 Hawthorne Ave.
Belleview, TX 75234

Dear Mr. Stern:

I would be pleased to answer your questions about the Microspot 311. We think it is the best
unit of its type on the market today.

1. The 311 can handle lead oxide or lead-free cassettes.
2. At the moment, the 311 is fully compatible only with our Duramatic generator. However,
   special wiring kits are available to make the 311 compatible with our earlier generator
   models — the Olympus and the Saturn. We are currently working on other wiring kits.
3. For the operator, the 311 increases the effectiveness of the radiological procedure while
   at the same time cutting down the amount of film used. For the patient, it reduces the
   number of repeat exposures and therefore reduces the total dose.

I am enclosing a copy of our brochure on the Microspot 311. If you would like additional
copies, please let me know. I would be happy to receive a copy of your analysis when it is
complete. Good luck!

Sincerely yours,

Andrew Shakir, M.D.

Andrew Shakir, M.D.
Director of Technical Services

AS/le

Enclosure

cc: Robert Anderson, Executive Vice-President

■ **Figure 15.6**
Response to an Inquiry

The Datafix copier is better than the Xerox — and it costs less, too. We'll repeat: it's better and it costs less!

If you're like most training directors, we bet you've seen your share of empty promises. We've heard all the stories, too. And that's why we think you'll be interested in what *Fortune* said about us last month.

2. *Describe the product or service you are trying to sell.* What does it do? How does it work? What problems does it solve?

The Datafix copier features automatic loading, so your people don't waste time watching the copies come out. Datafix copies from a two-sided original — automatically! And Datafix can turn out 90 copies a minute — which is 25 percent faster than our fastest competitor. . . .

3. *Convince the reader that your claims are accurate.* Refer to users' experience, or testimonials, or evaluations by reputable experts or testing laboratories.

In a recent evaluation conducted by *Office Management Today*, more than 85 percent of our customers said they would buy another Datafix. The next best competitor: 71 percent. And Datafix earned a "Highly Reliable" rating, the highest recommendation in the reliability category. All in all, Datafix scored higher than any other copier in the desktop class. . . .

4. *Tell the reader how to find out more about your product or service.* Make it easy to proceed to the next step in the sales process. If possible, provide a postcard or a Web address that the reader can use to request more information or arrange for a visit from one of your sales representatives.

Figure 15.7 shows an example of a sales letter.

Claim Letter

A *claim letter* is a polite and reasonable complaint. If you purchase a defective or falsely advertised product, or receive inadequate service, you write a claim letter.

The purpose of a claim letter is to convince the recipient that you are a fair and honest customer who is justifiably dissatisfied. If the letter is convincing, your chances of receiving an equitable settlement are good. Most organizations pay attention to reasonable claims, because they realize that unhappy customers are bad for business. In addition, claim letters help manufacturers identify weak points in their product or service.

DAVIS TREE CARE
1300 Lancaster Avenue
Berwyn, PA 19092
www.davisfortrees.com

May 13, 20XX

Dear Homeowner:

Do you know how much your trees are worth? That's right — your trees. As a recent purchaser of a home, you know how much of an investment you have in that home. But your
property is a big part of your total investment.

Most people don't know that even the hardiest trees need periodic care. Like shrubs, trees
should be fertilized and pruned. And they should be protected against the many kinds of
diseases and pests that are common in this area.

At Davis Tree Care, we have the skills and experience to keep your trees healthy and beautiful. Our diagnostic staff is made up of graduates of major agricultural and forestry universities, and all of our crews attend special workshops to keep current with the latest information
on tree maintenance. Add to this our proven record of 43 years of continuous service in the
Berwyn area, and you have a company you can trust.

May we stop by to give you an analysis of your trees — absolutely without cost or obligation? A few minutes with one of our diagnosticians could prove to be one of the wisest moves
you've ever made. Just give us a call at 555-9187 and we'll be happy to arrange an appointment at your convenience.

Sincerely yours,

Jasmine Brown

Jasmine Brown
President

■ **Figure 15.7**
Sales Letter

GUIDELINES

Writing a Claim Letter

1. *Identify the product or service.* List the model numbers, serial numbers, sizes, and any other pertinent data.

2. *Explain the problem.* State the symptoms clearly and specifically. What function does not work? What exactly is wrong with the service?

3. *Propose an adjustment.* Define what you want the company to do: for example, refund the purchase price, replace or repair the item, improve the service.

4. *Conclude courteously.* You might suggest that you trust the company, in the interest of fairness, to abide by your proposed adjustment.

In a claim letter, the you attitude is as important as the content. You must project a calm, rational tone. A complaint such as "I'm sick and tired of being ripped off by companies like yours" is less effective than "I am very disappointed in the performance of the SureGrip masonry fasteners." There is no reason to show anger in a claim letter, even if the other party has made an unsatisfactory response to an earlier one. Calmly explain what you plan to do, and why. The company will then be more likely see the situation from your perspective. Figure 15.8 is an example of a claim letter that the writer faxed to the reader.

Adjustment Letter

 See Business Communication: Managing Information and Relationship (http://spider.hcob.wmich.edu/bis/faculty/bowman/mir.html#Contents) for excellent advice on adjustment letters.

An *adjustment letter* is a response to a claim letter that tells the customer how you plan to handle the situation. Your purpose, whether you are granting the customer everything proposed in the claim letter, only part of it, or none of it, is to show that your organization is fair and reasonable, and that you value the customer's business.

If you can grant the request, the letter is easy to write. Express your regret about the situation, state the adjustment you are going to make, and end on a positive note by encouraging the customer to continue doing business with you.

If you cannot grant the request, try to salvage as much goodwill as you can. Obviously, your reader is going to be unhappy. If your letter is carefully written, however, it can show that you have acted reasonably. In denying a request, explain your company's side of the matter, thus educating the customer about how the problem occurred and how to prevent it in the future.

ROBBINS CONSTRUCTION, INC.
255 Robbins Place, Centerville, MO 65101 (417) 555-1850
www.robbinsconstruction.com

August 19, 20XX

Mr. David Larsyn
Larsyn Supply Company
311 Elmerine Avenue
Anderson, MO 63501

Dear Mr. Larsyn:

As steady customers of yours for over 15 years, we came to you first when we needed a quiet pile driver for a job near a residential area. On your recommendation, we bought your Vista 500 Quiet Driver, at $14,900. We have since found, much to our embarrassment, that it is not substantially quieter than a regular pile driver.

We received the contract to do the bridge repair here in Centerville after promising to keep the noise to under 90 db during the day. The Vista 500 (see enclosed copy of bill of sale for particulars) is rated at 85 db, maximum. We began our work and, although one of our workers said the driver didn't seem sufficiently quiet to him, assured the people living near the job site that we were well within the agreed sound limit. One of them, an acoustical engineer, marched out the next day and demonstrated that we were putting out 104 db. Obviously, something is wrong with the pile driver.

I think you will agree that we have a problem. We were able to secure other equipment, at considerable inconvenience, to finish the job on schedule. When I telephoned your company that humiliating day, however, a Mr. Meredith informed me that I should have done an acoustical reading on the driver before I accepted delivery.

I would like you to send out a technician — as soon as possible — either to repair the driver so that it performs according to specifications or to take it back for a full refund.

Yours truly,

Jack Robbins

Jack Robbins, President

Enclosure

■ **Figure 15.8**
Claim Letter

> **GUIDELINES**
>
> **Writing a Bad-News Adjustment Letter**
>
> In this more difficult kind of adjustment letter, try to accomplish the following four tasks:
>
> 1. *Meet the customer on some neutral ground.* Consider an expression of regret. You might even thank the customer for bringing the matter to your attention. But be careful about admitting that the customer is right. If you say "We are sorry that the engine you purchased from us is defective," it would bolster the customer's claim if the dispute ended up in court.
>
> 2. *Explain why your company is not at fault.* Most often, you explain to the customer the steps that led to the failure of the product or service. Do not say "You caused this." Instead, use the more tactful passive voice: "Apparently, the air pressure was not monitored."
>
> 3. *State that your company is denying the request for the reasons you have noted.* Note that if you begin with this statement, most readers will not finish reading, and you will not achieve your twin goals of education and goodwill.
>
> 4. *Create goodwill.* Close on a positive note. You might, for instance, offer a special discount on another, similar product. A company's profit margin on any one item is almost always large enough to permit an attractive discount as an inducement to continue doing business.

Figures 15.9 and 15.10 show examples of "good news" and "bad news" adjustment letters. The first is a reply to the claim letter shown in Figure 15.8.

WRITING MEMOS

The memo is a means of communication that is likely to survive even in the age of email. The following discussion explains how to write effective memos.

Writing effective memos:
Include the identifying information.
Clearly state your purpose.
Use headings to help your readers.
If appropriate, summarize your message.
Provide adequate background for the discussion.
Organize the discussion.
Highlight action items.

Larsyn Supply Company
311 Elmerine Avenue
Anderson, MO 63501
www.larsynsupply.com

August 21, 20XX

Mr. Jack Robbins, President
Robbins Construction, Inc.
255 Robbins Place
Centerville, MO 65101

Dear Mr. Robbins:

I was very unhappy to read your letter of August 19 telling me about the failure of the Vista 500. I regretted most the treatment you received from one of my employees when you called us.

Harry Rivers, our best technician, has already been in touch with you to arrange a convenient time to come out to Centerville to talk with you about the driver. We will of course repair it, replace it, or refund the price. Just let us know your wish.

I realize that I cannot undo the damage that was done on the day that a piece of our equipment failed. To make up for some of the extra trouble and expense you incurred, let me offer you a 10 percent discount on your next purchase or service order with us, up to a $1,000 total discount.

You have indeed been a good customer for many years, and I would hate to have this unfortunate incident spoil that relationship. Won't you give us another chance? Just bring in this letter when you visit us next, and we'll give you that 10 percent discount.

Sincerely,

Dave Larsyn, President

■ **Figure 15.9**
"Good News" Adjustment Letter

QUALITY VIDEO PRODUCTS
2077 Highland
Burley, ID 84765
www.qualivid.com

February 3, 20XX

Mr. Dale Devlin
1903 Highland Avenue
Glenn Mills, NE 69032

Dear Mr. Devlin:

Thank you for writing us about the videotape you purchased on January 11, 20XX. I know from personal experience how frustrating it is when a valuable videotape jams.

According to your letter, you used the videotape to record your daughter's wedding. While you were playing it back last week, the tape jammed and broke as you were trying to remove it from your VCR. You are asking us to reimburse you $500 because of the sentimental value of that recording.

As you know, our videotapes carry a lifetime guarantee covering parts and workmanship. We will gladly replace the broken videotape. However, the guarantee states that the manufacturer and the retailer will not assume any incidental liability. Thus we are responsible only for the retail value of the blank tape.

However, your wedding tape can probably be fixed. A reputable dealer can splice tape so skillfully that you will hardly notice the break. It's always a good idea to make backup copies of your valuable tapes.

Attached to this letter is a list of our authorized dealers in your area, any of whom would be glad to do the repairs for you. We have already sent out your new videotape. It should arrive within the next two days.

Please contact us if we can be of any further assistance.

Sincerely yours,

Paul R. Blackwood

Paul R. Blackwood, Manager
Customer Relations

Enclosure

■ **Figure 15.10**
**"Bad News" Adjustment
Letter**

Include the Identifying Information

In almost all memos, five elements appear at the top:

- the organization's logo or an abbreviated letterhead
- the "to" line
- the "from" line
- the "subject" line
- the "date" line

Some organizations also have a "copies" or "c" (copy) heading. Figure 15.11 shows several styles for presenting identifying information.

For more about titles, see Ch. 12, p. 309.

AMRO   MEMO

To: B. Pabst
From: J. Alonso *J. A.*
Subject: MIXER RECOMMENDATION FOR PHILLIPS
Date: 11 June 20XX

Write out the month instead of using the all-numeral format (6/11/XX); multicultural readers might use a different notation for dates and could be confused.

INTEROFFICE

To:	C. Cleveland	c:	B. Aaron
From:	H. Rainbow *H. R.*		K. Lau
Subject:	Shipment Date of Blueprints to Collier		J. Manuputra
			W. Williams
Date:	2 October 20XX		

List the names of persons receiving photocopies of the memo, either alphabetically or in descending order of organizational rank.

NORTHERN PETROLEUM COMPANY
INTERNAL CORRESPONDENCE

Date: January 3, 20XX
To: William Weeks, Director of Operations
From: Helen Cho, Chemical Engineering Dept. *H. C*
Subject: Trip Report — Conference on Improved Procedures for Chemical Analysis Laboratory

Most writers put their initials or signature next to their typed name (or at the end of the memo) to show that they have reviewed the memo and accept responsibility for it.

■ **Figure 15.11   Identifying Information in a Memo**

Some organizations prefer the full name of the writer and the reader; others want only the first initials and the last name. Some prefer job titles; others do not. If your organization does not object, include your job title and your reader's. The memo will then be informative for someone who refers to it after either of you has moved on to a new position, and others in the organization who might not know you.

The subject heading — the title of the memo — deserves special mention. Don't be too concise. Avoid naming only the subject, such as *Tower Load Test;* rather, specify what aspect of the test you wish to address. For instance, *Tower Load Test Results* or *Results of Tower Load Test* would be much more informative than *Tower Load Test,* which does not tell the reader whether the memo is about the date, the location, the methods, the results, or any number of other factors related to the test. The subject line should be accurate, unique, and specific.

Print the second and all subsequent pages of memos on plain paper rather than on letterhead. Include three items in the upper left-hand corner of each page:

- the name of the recipient
- the date of the memo
- the page number

> J. Alders
> April 6, 20XX
> Page 2

Clearly State Your Purpose

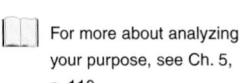

For more about analyzing your purpose, see Ch. 5, p. 110.

The first sentence of the body of a memo should explain its purpose:

> I want to tell you about a problem we're having with the pressure on the main pump, because I think you might be able to help us.

> The purpose of this memo is to request authorization to travel to the Brownsville plant Monday to meet with the other quality inspectors.

> This memo presents the results of the internal audit of the Phoenix branch, an audit that you authorized March 13, 20XX.

> I want to congratulate you on your division's quarterly record.

The best purpose statements are concise and direct. Make sure your purpose statement has a verb that clearly communicates what you want the memo to accomplish, such as *to request, to explain,* or *to authorize.* Some people object to a direct statement of purpose, especially when communicating bad news or asking for something, as in the example about authorization to travel to the Brownsville plant. Instead of beginning with a direct statement of purpose, they would open with a statement of the reasons for the trip, the trip's potential benefits, and so on. They would then conclude the memo with the actual request: "For these reasons, I am requesting authorization to. . . ." Although some readers would rather have the reasons presented first, far more would prefer to know immediately why you have written. There are two basic problems with stating the purpose at the end of the argument:

- Some readers will toss the memo aside if you seem to be rambling on about the Brownsville plant without getting to the point.
- Some readers will suspect that you are trying to manipulate them into doing something they don't want to do.

Placing the purpose statement at the beginning sacrifices subtlety for directness, but it helps your readers understand why you are writing.

Use Headings to Help Your Readers

Headings are useful in all kinds of technical documents. Use them liberally in memos. Headings help your readers in two main ways:

For more about writing coherent headings, see Ch. 10, p. 252.

- *Headings help your readers decide what to read.* If a section is labeled *What Is the Function of a Browser?*, readers who already know can simply skip to the next section.
- *Headings help readers understand the information.* A simple heading such as *Summary* states the function of the section, helping them concentrate on the information without wondering why it is included and how it relates to other information.

If Appropriate, Summarize Your Message

Memos of one page or less do not generally contain summaries. But longer ones, particularly those addressed to multiple readers, can benefit from a summary at the beginning. A summary has three main goals:

For more about writing effective summaries, see Ch. 12, p. 316.

- to help all readers understand the body of the memo
- to enable executive readers to skip the body if they so desire
- to remind readers of the main points

Here are some examples of summaries:

The annual ATC conference was of great value. The lectures on new coolants suggested techniques that might be useful in our Omega line, and I met three potential customers who have since gotten in touch with Marketing.

The analysis of the beam shows that lateral stress caused the failure. We are now trying to determine why the beam did not sustain a lateral stress weaker than that it was rated for.

In March, we completed Phase I (Design) on schedule. At this point, we anticipate no delays that would jeopardize our projected completion date.

The summary should reflect the length and complexity of the memo. It might range from one simple sentence to a long, technical paragraph. If possible, it should reflect the sequence of information in the body.

Provide Adequate Background for the Discussion

Writers of memos sometimes mistakenly assume that their readers already know the background: the events that led to the current situation. If you are in any doubt about whether readers need a background statement, include one. (Those who do not need background will simply skip it.)

Although each background discussion is unique, some basic guidelines are useful. For example, if the memo defines a problem — a flaw detected in a product line — you might discuss how the problem was discovered or present the basic facts about the product line: what the product is, and how long it has been produced and in what quantities. If the memo reports the results of a field trip, you might discuss why the trip was undertaken, what its objectives were, who participated, and so on.

The following background paragraph is from a memo requesting authorization to have a piece of equipment retooled:

Background

The stamping machine, a Curtiss Model 6143, is used in the sheet-metal shop. We bought it in 1996, and it is in excellent condition. However, since we switched the size of the tin rolls last year, the stamping machine no longer performs to specifications.

Organize the Discussion

For more about patterns of organization, see Ch. 8, p. 186; for more about graphics, see Ch. 14.

The discussion is the section in which you present your main arguments. You might divide a detailed discussion into the subsections typical of a more formal report: methods, results, conclusions, and recommendations. Or you might give it headings that pertain specifically to the subject you are discussing. You might also include brief tables or figures but should attach more extensive ones as appendices.

The discussion section of the memo can be developed according to any of the basic organizational patterns, such as chronological, spatial, more-important-to-less-important, and cause and effect.

Highlight Action Items

Some reports require follow-up action, by either the writer or the reader. For example, a memo addressing a group of supervisors might explain a problem, then state what the writer is going to do about it. Or a supervisor might delegate tasks to other employees. In writing a memo, state clearly who is to do what and when. Here are two examples of ways to highlight action items:

Action Items:

I would appreciate it if you would work on the following tasks and have your results ready for the meeting on Monday, June 9.

1. Henderson: recalculate the flow rate.

2. Smith: set up a meeting with the regional EPA representative for sometime during the week of February 13.

3. Falvey: ask Armitra in Houston for his advice.

Action Items:

As we discussed, I will finish these tasks this week:

1. Send the promotional package to the three companies.

2. Ask Customer Relations to work up a sample design to show the three companies.

3. Request interviews with the appropriate personnel at the three companies.

Notice that, in the first example, although the writer is the supervisor of his readers, he uses a polite tone in the lead-in to the list of tasks.

WRITING EMAILS

Email offers four chief advantages over interoffice or U.S. Postal Service delivery of paper documents.

- *Email is fast.* On a local network such as an intranet, delivery usually takes less than a second. On the Internet, email can travel halfway around the world in a few seconds.

- *Email is cheap.* Once the network is in place, it doesn't cost anything to send a message to one person or to a thousand. You don't pay per copy, as you do with photocopies, because you're not using any paper.

- *Email is easy to use.* Once you learn how to use your email software, it is easy to send mail to one person or to an entire group. It is also simple to respond to an email or to forward it to one third party or many.

- *Email is digital.* Email can be read and printed out or deleted, but it can also be stored like any other electronic file. Therefore, it can be used in other documents. For this reason, email is a convenient way for people in different places to collaborate.

Every day, some 130 million US workers send 2.8 billion email messages (ePolicy, 2001). However, there will always be a place for letters and memos, for two reasons:

- In some environments, people do not have access to an intranet or the Internet or do not know how to use it.

- For legal reasons, people sometimes need original hard copies, on letterhead stationery or memo forms, complete with signatures (although the federal government's E-sign program is changing that).

 For statistics on email usage in business, see the ePolicy Institute (www .epolicyinstitute.com). For information on E-sign, see the National Telecommunications and Information Administration site (www.ntia.doc.gov /ntiahome/ntiageneral/esign /esignpage.html).

In writing an email, follow three principles:

- *Use an appropriate level of formality.* In some organizations, managers expect email messages to be as formal as printed documents; in others, they expect them or want them to be quite informal. Find out your organization's email policy. If the emails you read sound just like memos or letters, you know the company considers email a practical way to communicate relatively formal messages. However, if you see writers using *emoticons*, the most popular of which is the smiley face : -) , at the end of a comment that is meant to be taken ironically or as a joke, or abbreviations, such as *BTW* for *by the way*, you know that the organization sees email as a way to foster group cohesiveness. Before you begin to send email messages, read those written by fellow employees.

- *Realize that email, like print, is permanent.* Email, like all traffic on a network, is usually archived, that is, backed up on some kind of tape or disk system. The email that you send a colleague is probably stored somewhere, even if the recipient has deleted it. According to the Electronic Communications Privacy Act of 1986, the organization that established the network may look at all emails on that network without violating employees' privacy rights (Crowe, 1994, p. 31). For this reason, do not write in an email anything that would embarrass you or your organization if the email appeared in the organization's newsletter or in tomorrow's newspaper.

- *Adhere to netiquette guidelines. Netiquette* refers to etiquette on a network.

GUIDELINES

Following Netiquette

See Albion.com's discussion of netiquette at www.albion .com/netiquette/.

▶ *Don't waste bandwidth.* Keep the message brief so that it doesn't clog up the network or the recipient's in-basket. When you reply to another email, don't quote long passages from it. Instead, establish the context of the original email by paraphrasing it briefly or by including a short quotation from it. (When you excerpt a small portion of an email, add a phrase such as <*snip*> at the start and the end of the quotation, to indicate that you have omitted part of the original message.) When you do quote, delete the routing information from the top as well as the signature block from the bottom. And make sure that everyone who is to receive a copy of your email really needs to read it.

▶ *Take some care with your writing.* Email is informal, but messages shouldn't be sloppy. Because text-editing functions on many email systems are much more limited than on a word processor, you need to edit and proofread your emails before sending them.

▶ *Don't flame.* To *flame* is to scorch a reader with scathing criticism, usually in response to something that person has said in a previous message. Flaming is rude. When you are really angry, keep your hands away from the keyboard.

▶ *Use the subject line.* Readers like to be able to decide whether they want to read the message. The subject line helps them decide. Therefore, write specific, accurate, and informative subject lines, just as you would in a memo.

▶ *Make your message easy on the eyes.* Use uppercase and lowercase letters as you do in other forms of correspondence, and skip lines between paragraphs. Don't use italics, underlining, or boldface for emphasis, even if your email software can accommodate them, because you can't be sure your reader's email system can. Instead, use uppercase letters for emphasis. Keep the line length to under 65 characters so that lines do not get broken up if the recipient's monitor has a smaller screen.

▶ *Don't forward a message to another person or to a public forum such as a listserv without the writer's permission.* Doing so is unethical and it might be illegal (the courts haven't decided yet).

▶ *Don't send a message unless you have something to say.* Resist the temptation to write a message that says, in effect, that you agree with another message. If you can add something new, fine, but don't send a message just to be part of the conversation.

Figure 15.12 shows an email that violates some netiquette guidelines. The writer is a technical professional working for a microchip manufacturer. Figure 15.13 on page 440 shows a revised version of this email message.

| To: | Supers and Leads |
| Subject: | |

LATELY, WE HAVE BEEN MISSING LASER REPAIR FILES FOR OUR 16MEG WAFERS. AFTER BRIEF INVESTIGATION, I HAVE FOUND THE MAIN REASON FOR THE MISSING DATA.

OCCASIONALLY, SOME OF YOU HAVE WRONGLY PROBED THE WAFERS UNDER THE CORRELATE STEP AND THE DATA IS THEN COPIED INTO THE NONPROD STEP USING THE QTR PROGRAM. THIS IS REALLY STUPID. WHEN DATE IS COPIED THIS WAY THE REPAIR DATA IS NOT COPIED. IT REMAINS UNDER THE CORRELATE STEP.

TO AVOID THIS PROBLEM, FIRST PROBE THE WAFERS THE RIGHT WAY. IF A WAFER MUST BE PROBED UNDER A DIFFERENT STEP, THE WAFER IN THE CHANGE FILE MUST BE RENAMED TO THE ** FORMAT.

EDITING THE WAFER DATA FILE SHOULD BE USED ONLY AS A LAST RESORT, IF THIS BECOMES A COMMON PROBLEM, WE COULD HAVE MORE PROBLEMS WITH INVALID DATA THAT THERE ARE NOW.

SUPERS AND LEADS: PLEASE PASS THIS INFORMATION ALONG TO THOSE WHO NEED TO KNOW.

ROGER VANDENHEUVAL

The writer does not clearly state his purpose in the subject line and the first paragraph.

The writer has not proofread.

With long lines and no spaces between paragraphs, this email will be difficult to read.

■ **Figure 15.12 Email Message That Violates Netiquette Guidelines**

Using all uppercase letters gives the impression that the writer is yelling at the readers.

The subject line and first para-graph clearly state the writer's purpose.

Double-spacing between para-graphs and using short lines make the email easier to read.

The writer concludes politely.

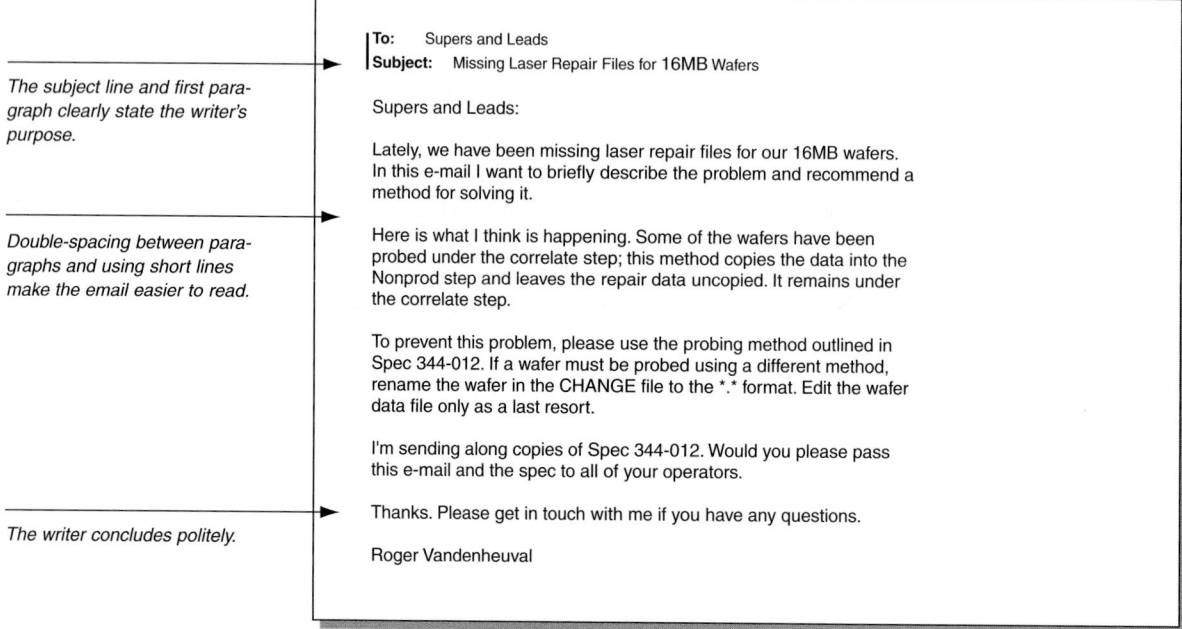

| To: | Supers and Leads |
| Subject: | Missing Laser Repair Files for 16MB Wafers |

Supers and Leads:

Lately, we have been missing laser repair files for our 16MB wafers. In this e-mail I want to briefly describe the problem and recommend a method for solving it.

Here is what I think is happening. Some of the wafers have been probed under the correlate step; this method copies the data into the Nonprod step and leaves the repair data uncopied. It remains under the correlate step.

To prevent this problem, please use the probing method outlined in Spec 344-012. If a wafer must be probed using a different method, rename the wafer in the CHANGE file to the *.* format. Edit the wafer data file only as a last resort.

I'm sending along copies of Spec 344-012. Would you please pass this e-mail and the spec to all of your operators.

Thanks. Please get in touch with me if you have any questions.

Roger Vandenheuval

■ **Figure 15.13   Email That Adheres to Netiquette Guidelines**

The writer diplomatically diagnoses the problem and clearly states the correct method for the procedure. In addition, the writer has edited and proofread the email.

    Revision Checklist

Letter Format

1. Is the first page typed on letterhead stationery?
2. Is the date included?
3. Is the inside address complete and correct? Has the appropriate courtesy title been used?
4. If appropriate, is an attention line included?
5. If appropriate, is a subject line included?
6. Is the salutation appropriate?
7. Is the complimentary close typed with only the first word capitalized? Is the complimentary close followed by a comma?
8. Is the signature clear and legible, and is the writer's name typed beneath the signature?
9. If appropriate, are reference initials included?
10. If appropriate, is an enclosure line included?
11. If appropriate, is a copy line included?
12. Is the letter typed in one of the standard formats?

Types of Letters

1. Does the order letter
   - ❏ include identifying information, such as quantities and model numbers?
   - ❏ specify, if appropriate, the terms of payment?
   - ❏ specify the method of delivery?
2. Does the inquiry letter
   - ❏ explain why you chose the reader to receive the inquiry?
   - ❏ explain why you are requesting the information and to what use you will put it?
   - ❏ specify by what date you need the information?
   - ❏ list the questions clearly and, if appropriate, provide room for the reader's responses?
   - ❏ offer, if appropriate, the product of your research?
3. Does the response to an inquiry letter answer the reader's questions or explain why they cannot be answered?
4. Does the sales letter
   - ❏ gain the reader's attention?
   - ❏ describe the product or service?
   - ❏ convince the reader that the claims are accurate?
   - ❏ encourage the reader to find out more about the product or service?
5. Does the claim letter
   - ❏ identify specifically the unsatisfactory product or service?
   - ❏ explain the problem(s) clearly?
   - ❏ propose an adjustment?
   - ❏ conclude courteously?
6. Does the "good news" adjustment letter
   - ❏ express your regret?
   - ❏ explain the adjustment you will make?
   - ❏ conclude on a positive note?
7. Does the "bad news" adjustment letter
   - ❏ meet the reader on neutral ground, expressing regret but not apologizing?
   - ❏ explain why the company is not at fault?
   - ❏ clearly deny the reader's request?
   - ❏ attempt to create goodwill?

Memos

1. Does the identifying information adhere to your organization's standards?
2. Did you clearly state your purpose at the start of the memo?
3. Did you use headings to help your readers?
4. If appropriate, did you summarize your message?
5. Did you provide appropriate background for the discussion?
6. Did you organize the discussion clearly?
7. Did you highlight items requiring action?

Email

1. Is the tone appropriate?
2. Did you write the message carefully and revise it?
3. Did you avoid flaming?
4. Did you write a specific, accurate subject line?

5. Did you use uppercase and lowercase letters?
6. Did you skip lines between paragraphs?
7. Did you set the line length to under 65 characters?
8. Did you check with the writer before resending his or her message?

Exercises

1. Write an order letter to John Saville, general manager of White's Electrical Supply House (13 Avondale Circle, Los Angeles, CA 90014) to order these items: one SB11 40-ampere battery backup kit, at $73.50; twelve SW402 red wire kits, at $2.50 each; ten SW400 white wire kits, at $2.00 each; and one SB201 mounting hardware kit, at $7.85. Invent any reasonable details about methods of payment and delivery.

2. You are an intern in the Admissions Office at your college or university. Write a letter that can be sent to high-school guidance counselors in your state. The letter is intended to motivate counselors to study the accompanying catalog and interest appropriate students in applying to the school.

3. A beverage container you recently purchased for $8.95 has a serious leak. The grape drink you put in it ruined a $35.00 white tablecloth. Inventing any reasonable details, write a claim letter to the manufacturer of the container.

4. As the recipient of the claim letter described in exercise 3, write an adjustment letter granting the customer's request.

5. You are the manager of a private swimming club. A member has written saying that she lost a contact lens (value $75) in your pool and she wants you to pay for a replacement. The contract that all members sign explicitly states that the management is not responsible for loss of personal possessions. Write an adjustment letter denying the request. Invent any reasonable details.

6. As manager of a stereo equipment retail store, you guarantee that you will not be undersold. If a customer who buys something from you can prove within one month that another retailer sells the same equipment at a lower price, you will refund the difference. A customer has written to you and enclosed an ad from another store showing that it is selling the equipment he purchased for $26.50

less than he paid at your store. The advertised price at the other store was a one-week sale that began five weeks after the date of his purchase. He wants a $26.50 refund. Inventing any reasonable details, write an adjustment letter denying his request. You are willing, however, to offer him a blank cassette tape worth $4.95 for his equipment if he would like to come pick it up.

7. Revise the following letters to increase their effectiveness, adding any reasonable details.

a. to a laboratory-equipment supplier:

> Gentlemen:
>
> Would you please send us the following items:
>
> > one dozen petri dishes
> > one gross pyrex test tubes
> > three bunsen burners
>
> Please bill us.
>
> Sincerely,

b. to the admissions department of a law school:

> Gentlemen:
>
> I am a senior considering going to law school. Would you please answer the following questions about your law school?
>
> 1. How well do your graduates do?
> 2. Is the LSAT required?
> 3. Do you have any electives, or are all the courses required?
> 4. Are computers skills required for admission?
>
> A swift reply would be appreciated. Thank you.
>
> Sincerely yours,

c. from a supplier of home alarm systems:

> Dear Smith Family:
>
> A rose is a rose is a rose, the poet said. But not all home protection alarms are alike. In a time when burglaries are skyrocketing, can you afford the second-best alarm system?

Your home is your most valuable possession. It is worth far more than your car. And if you haven't checked your house insurance policy recently, you'll probably be shocked to see how inadequate your coverage really is.

The best kind of insurance you can buy is the Watchdog Alarm System. What makes the Watchdog unique is that it can detect intruders before they enter your home and scare them away. Scaring them away while they're still outside is certainly better than scaring them once they're inside, where your loved ones are.

At less than two hundred dollars, you can purchase real peace of mind. Isn't your family's safety worth that much?

If you answered yes to that question, just mail in the enclosed postage-paid card, and we'll send you a 12-page, fact-filled brochure that tells you why the Watchdog is the best on the market.

Very truly yours,

8. Form small groups for this exercise on claim and adjustment letters. Have each member of your group study the following two letters. Meet and discuss your reactions to the two letters. How effectively does the writer of the claim letter present his case? How effective is the adjustment letter? Does its writer succeed in showing that the company's procedures for ensuring hygiene are effective? Does its writer succeed in projecting a professional tone? Write a memo to your instructor discussing the two letters. Attach a revision of the adjustment letter to the memo.

Seth Reeves
19 Lowry's Lane
Morgan, TN 30610

April 13, 20XX

Sea-Tasty Tuna
Route 113
Lynchburg, TN 30563

Gentlemen:

I've been buying your tuna fish for years, and up to now it's been OK.

But this morning I opened a can to make myself a sandwich. What do you think was staring me in the face? A fly. That's right, a housefly. That's him you see taped to the bottom of this letter.

What are you going to do about this?

Yours very truly,

SEA-TASTY TUNA
Route 113
Lynchburg, TN 30563
www.seatastytuna.com

April 21, 20XX

Mr. Seth Reeves
19 Lowry's Lane
Morgan, TN 30610

Dear Mr. Reeves:

We were very sorry to learn that you found a fly in your tuna fish.

Here at Sea-Tasty we are very careful about the hygiene of our plant. The tuna are scrubbed thoroughly as soon as we receive them. After they are processed, they are inspected visually at three different points. Before we pack them, we rinse and sterilize the cans to ensure that no foreign material is sealed in them.

Because of these stringent controls, we really don't see how you could have found a fly in the can. Nevertheless, we are enclosing coupons good for two cans of Sea-Tasty tuna.

We hope this letter restores your confidence in us.

Truly yours,

9. Bill, the writer of the following email, is the primary technical supervisor on the production line in a microchip manufacturing facility. He is responding to Larry, a supervisor who reports to him. Larry has sent him the following note: "Bill, I can't seem to find a spec that describes coat tracks. Some of the new hires don't know what they are. What should I tell them?" Bill's reply:

Larry —

Coat tracks are the machines used in the first step of the photo process. The wafers coem to coat to have a layer of a photosensitive resist applyed. This requires several operations to be done by the same machine. First the wafer is coated with a layer of primer or hmds. This ensures that the resist will adhere to the wager. The wafer is then caoted with resist. There are 5 different types of resist in use, each has its won characteristics, and all are used on different levels and part types. I dont really have time to go into all the details now. The photo resist must be applied ina uniform layer as the unfioromity of the resist can effect several other steps to include exsposure on the stepper to the etch rate on a lam. To insure the proper unfiromity and resist volumes the tracks are inspected at 24 hour intervals, all the functions are checked and partical monitors are ran to ensure proper operation and cleanlyness. After the wafer is coated it is soft baked, this rids the wafer of solvents in the primer and resist and also hardens the resist. The wafers are then ready to go to the next step at the p&e or the stepper.

If, in fact, there is no specification that Bill can give to Larry, what would be the best way to communicate the information to him? What impression will his email make on Larry? Why?

10. Evaluate the following email posted to a discussion group. How would you revise it to make it more effective?

To: TECHWR-L List

Subject: Binders v. Updates

On July 13, Barbara Ackley wrote:

>We always prefer using three-ring binders to pub-
>lishing updated manuals. Admittedly, not every
>one is going to keep their binder up to date, but
>the cost differential is just too great for us to
>overlook.

I have to agree with Barbara.

Regards,

John Altow, SysPerfect Corporation

11. Louise and Paul work for the same manufacturing company. Louise, a senior engineer, is chairing a committee to investigate ways to improve the hiring process at the company. Paul, a technical editor, also serves on the committee. The excerpts quoted in Louise's email are from an email written by Paul to all members of the committee in response to Louise's request that members describe their approach to evaluating job-application materials. How would you revise Louise's email to make it more effective?

To: Paul

From: Louise

Sometimes I just have to wonder what you're thinking, Paul.

>Of course, it's not possible to expect perfect
>resumes. But I have to screen them, and last
>year I had to read over 200. I'm not looking for
>perfection, but as soon as I spot an error I make
>a mental note of it and, when I hit a second and
>then a third error I can't really concentrate on the
>writer's credentials.

Listen, Paul, you might be a sharp editor, but the rest of us have a different responsibility: to make the products and move them out as soon as possible. We don't have the luxury of studying documents to see if we can find errors. I suggest you concentrate on what you were hired to do, without imposing your "standards" on the rest of us.

>From my point of view, an error can include a
>misused tradmark.

Misusing a "tradmark," Paul? Is that Error Number 1?

Research Projects

12. 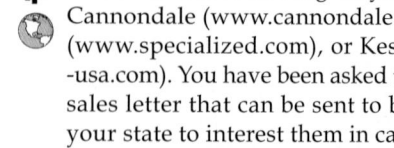 Your group works in the marketing department of one of the following bicycle manufacturers: Cannondale (www.cannondale.com), Specialized (www.specialized.com), or Kestrel (www.kestrel-usa.com). You have been asked to draft a one-page sales letter that can be sent to bicycle retailers in your state to interest them in carrying your products. Use information from your company's Web site in your letter.

13. Because students use email to communicate with other group members when they write collabora-tively, your college or university would like to create a one-page handout on how to use email responsibly. Using a search engine, find three or four netiquette guides on the Internet that focus on using email. Study these guides and then write a one-page student guide to using email to communicate with other students. Somewhere in the guide, be sure to list the sites you studied, so that students can visit them for further information about netiquette.

CASE
Dangerous Wrenches

Your group works for the Customer Service Department at United Tools, a manufacturer of hand tools. Recently, you received the following letter from a hardware store that carries your products.

Handee Hardware, Inc.
Millersville, AL 61304
www.handeehardware.com

December 4, 20XX

United Tools
20 Central Avenue
Dover, TX 76104

Gentlemen:

I have a problem I'd like to discuss with you. I've been carrying your line of hand tools for many years.

Your 9" pipe wrench has always been a big seller. But there seems to be something wrong with its design. I have had two complaints in the last few months about the handle snapping off when pressure is exerted on it. In one case, the user cut his hand seriously enough to require stitches.

Frankly, I'm hesitant to sell any more of the 9" pipe wrenches, but I still have more than two dozen in inventory.

Have you had any other complaints about this product?

Sincerely yours,

You decide to investigate the problem. You learn that in the last two quarters United Tools has sold more than 300 of these wrenches to retailers and received 4 complaints from retailers and retail customers about the handle snapping. No one reported injuries. You go to the company engineers and learn that two months ago they discovered a design flaw that accounts for the problem. At that time, the engineers redesigned the wrench, manufactured samples, tested them thoroughly, and found no problems. Like the old design, the new design exceeds ANSI standards for this type of tool. The old design is no longer sold; all new orders are being filled with the new design.

What additional steps should you take? Are there other people in the company with whom you should meet? Draft a letter, to be sent to the retailers and the retail customers who wrote claim letters to your company, offering an appropriate adjustment. Does the situation you have uncovered within your company merit further action? If so, write a memo or letter to an appropriate officer in the company.

See Ch. 2 for a discussion of ethical and legal considerations.

For additional cases, click on Case of the Month and Archive on TechComm Web (www.bedfordstmartins .com/techcomm).

16

Preparing Job-Application Materials

| Yana Parker (1999), a résumé consultant, advises job applicants: | *Too many people forget that a résumé is a marketing piece — not a career obituary! And it's not a confessional either. Its purpose is to sell the writer's skills.* | 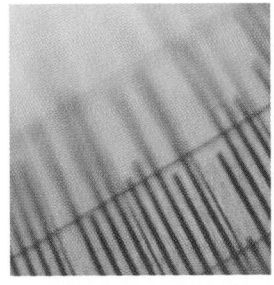 |

For most of you, the first nonacademic test of your technical-communication skills comes when you prepare job-application materials. These materials will inform prospective employers about your academic and employment experience, your personal characteristics, and your reasons for applying for a job with their organization. But they provide much more information, too. Employers have learned that one of the most important skills an employee can bring to a job is the ability to communicate effectively. Potential employers look these materials over carefully for evidence of writing skills. For you, therefore, job-application materials pose a double hurdle: showing employers what you can do and showing them how well you can communicate.

 See TechComm Web (www .bedfordstmartins.com /techcomm) for guidelines, additional examples, and links related to topics in this chapter.

Some students think that once they get a satisfactory job, they will never again have to worry about résumés and application letters. Statistics suggest otherwise. Today, the typical professional changes jobs more than five times, and many organizations require employees to maintain up-to-date résumés. Although this chapter pays special attention to the student's first job hunt, the skills and materials discussed here also apply to established professionals who wish to change jobs or maintain a current résumé. This chapter discusses how to search for jobs and how to write the documents used in a job search: résumés (including electronic ones), application letters, and follow-up letters.

PLANNING THE JOB SEARCH

Planning a job search is a lot of work that will most likely stretch over weeks and months, not days. You have three main tasks in planning the search:

- *Do a self-inventory.* Before you can start thinking of where you want to work, you need to answer some questions about yourself:
  - *What are your strengths and weaknesses?* Are your skills primarily technical? Do you work well with people? Do you work best with supervision or on your own?
  - *What subjects do you like?* Think about what you have liked or disliked about your previous positions and college courses.
  - *What kind of organization would you like to work for?* Profit or nonprofit? Government or private industry? Small or large?
  - *What are your geographical preferences?* If you are free to relocate, where would you like to live? How do you feel about commuting?

- *Learn about the employers.* Don't base your job search exclusively on the information in a job ad. Learn as much as you can about the organization through other means:
  - *Attend job fairs.* Your college and your community probably hold job fairs, where various employers provide information about their organizations. Sometimes a single organization will hold a job fair to find qualified candidates for a wide variety of jobs.
  - *Find out about trends in your field.* Read the *Occupational Handbook*, published by the U.S. Department of Labor, for information about your field and related fields. Talk with professors in your field and with the people at your job-placement office.
  - *Research the companies that interest you.* Visit their Web sites. Scan the index of the *Wall Street Journal* for articles about them. Study their annual reports, which are usually available on the organizations' Web site.

Find the *Occupational Handbook* at www.bls.gov /search/oco_s.asp.

- *Prepare your materials.* You know you will need to write application letters and résumés and that you will go on interviews. Start planning early by obtaining materials from the career-placement office. Talk with friends who have gone through the process successfully; study their application materials. Read some of the books on different aspects of the job search.

  One more very important part of preparing your materials: make a *portfolio,* a collection of your best work. You'll want to give a prospective employer a copy of the portfolio to demonstrate the kind of work you can do. For a technical communicator, the portfolio will include a variety of documents you made in courses and in previous positions. For technical professionals, the portfolio might include proposals and reports as well as computer simulations, Web sites, or presentation graphics. A portfolio is often presented in a loose-leaf notebook, with each item preceded by a descriptive and evaluative statement. Frequently, a portfolio contains a table of contents and an introductory statement.

UNDERSTANDING SEVEN WAYS
TO LOOK FOR A POSITION

Once you have done your planning, you can start to look for a position. There are seven major ways to find a job.

Search for a position through:
A college or university placement office
A professional placement bureau
A published job advertisement
An organization's Web site
A job board on the Internet
An unsolicited letter to an organization
Connections

- *Through a college or university placement office.* Placement offices bring companies and students together. Generally, students submit a résumé — a brief list of credentials — to the placement office. The résumés are then made available to representatives of business, government, and industry, who use the placement office to arrange on-campus interviews with selected students. Those who do best in the campus interviews are then invited by the representatives to visit the organization for a tour and another interview. Sometimes a third interview is scheduled; sometimes an offer is made immediately or shortly thereafter. The advantages of this system are that it is free and it is easy. You merely deliver a résumé to the placement office and wait to be contacted.

- *Through a professional placement bureau.* A professional placement bureau offers essentially the same service as a college placement office, but a placement bureau charges a fee. The fee is often 10 percent of the salary the new employee receives in the first year. Sometimes the fee is paid by the employer; sometimes it is paid by the new employee. Placement bureaus cater primarily to more advanced professionals who are trying to change jobs.

- *Through a published job advertisement.* Published job ads generally offer good opportunities for both students and professionals. Organizations advertise in three kinds of print publications: public-relations catalogs (such as *College Placement Annual*), technical journals, and newspapers. Check the major technical journals in your field and the large metropolitan newspapers, especially the Sunday editions. In responding to an ad, you most likely will need to include with the résumé a job-application letter that highlights the crucial information on the résumé. (See page 469 for more about job-application letters.)

- *Through an organization's Web site.* Many organizations list their job offerings on their Web sites. A job ad on a site usually tells candidates how to apply: by filling out an electronic form, by emailing an application, by sending an electronic résumé, by sending a traditional letter and résumé, or by coming to the organization in person. (See page 466 for more about electronic résumés.)

- *Through a job board on the Internet.* Job boards are sites sponsored by federal agencies, Internet service providers, and private organizations to help introduce employers to prospective employees. Some of these sites offer their services for free; others charge a fee. Some sites merely list positions, to which you respond by regular mail or by email; others let you submit your résumé electronically, so that employers can get in touch with you. Use a search engine to search for "employment," "careers," and "jobs." Or combine one of these terms with the name of your field, as in "careers and forestry." Among the biggest job boards are the following:

    America's Job Bank (sponsored by the U.S. Department of Labor): http://www.ajb.dni.us

    Monster Board: http://www.monster.com

    CareerBuilder: http://careerbuilder.com

    Career Mosaic: http://www.careermosaic.com

    AfterCollege: http://www.aftercollege.com

    Career Magazine: http://www.careermag.com

To find these sites and additional job-related resources on the Web, click on Links Library for Ch. 16 on TechComm Web (www.bedfordstmartins.com/techcomm).

Many of these have links to articles about searching for jobs electronically, including how to research companies, how to write electronic résumés, and how to prepare for interviews.

One caution about using job boards: once you post something to an Internet site, you have probably lost control of it. Here are four questions to ask before you post to a job board:

- Who has access to your résumé? You might wish to remove your home address and phone number from it if anyone can view it.
- How will you know if an employer requests your résumé? Will you be notified by the job board?
- Can your current employer see your résumé? If your employer knows you are searching for a new job, your current position could be jeopardized.
- Can you update your résumé at no cost? Some job boards let you post your résumé for free, then charge you each time you update it.

- *Through an unsolicited letter to an organization.* Instead of waiting for an ad or a notice on a Web site, consider writing unsolicited letters of application to organizations you would like to work for. The disadvantage of this technique is obvious: there might not be an opening. Yet many professionals favor this technique, because there are fewer competitors for those jobs that do exist, and organizations do not advertise all available positions. And sometimes an impressive unsolicited application can prompt an organization to create a position.

Before you write an unsolicited application, make sure you learn as much as you can about the organization: current and anticipated major projects, hiring plans, and so forth. You should be knowledgeable about any organization you are applying to, of course, but when you are submitting an unsolicited application, you need other sources of information to plan your strategy. The business librarian at your college or university will be able to point out additional sources of information, such as the Dun and Bradstreet guides, the *F&S Index of Corporations*, and the indexed newspapers such as the *New York Times*, the *Washington Post*, the *Los Angeles Times*, and the *Wall Street Journal*. You should also study the organization's Web site.

- *Through connections.* A relative or an acquaintance who can exert influence, or at least point out a new position, can help you get a job. Other good contacts include employers from your past jobs and faculty members in your field. Also consider becoming active in the student chapter of your field's professional organization, through which you can meet professionals in your area.

WRITING RÉSUMÉS

This section concentrates on techniques for preparing printed résumés: paper-based résumés that you will mail or carry to a potential employer. Read it even if you know you will be sending electronic résumés, because it provides useful advice about both kinds. The following section discusses electronic résumés.

Many students wonder whether to write their résumé themselves or use a résumé-preparation agency. It is best to write your own résumé, for three reasons:

- *You know yourself better than anyone else does.* No matter how thorough and professional the work of a résumé-preparation agency, you can do a better job communicating important information about yourself.

- *Employment officers know the style of the local agencies.* Readers who realize that you did not write your own résumé might wonder what kinds of deficiencies you are trying to hide.

- *If you write your own résumé, you will be more likely to adapt it to different situations.* You are not likely to return to a résumé-preparation agency and pay an additional fee to make a minor revision.

The résumé communicates in two ways: through its appearance and through its content.

Appearance of the Résumé

Because potential employers normally see your résumé before they see you, it has to make a good first impression. Employers believe — often correctly — that the appearance of the résumé reflects the writer's professionalism. A sloppy résumé implies that you would do sloppy work. A neat résumé implies that you would do professional work. When employers look at a résumé, they see the documents they will be reading if they hire you.

Some colleges and universities advise students to have their résumé professionally printed. A printed résumé is attractive, and that's good, provided, of course, that the information it contains is consistent with its professional appearance. Most employers agree, however, that a neatly word-processed résumé, printed by a laser printer on good-quality paper, is just as effective.

Applicants who photocopy a word-processed résumé are more likely to tailor it to the needs of the organizations to which they apply — a good strategy. People who go to the expense of a professional printing job are far less likely to make several versions; the résumé looks so good that they don't want to tinker with it. This strategy is dangerous because it discourages you from tailoring your résumé to your audience.

Résumés should appear neat and professional. They should have

- *Generous margins.* Leave a one-inch margin on all four sides.
- *Clear type.* Use a good-quality laser printer.
- *Balance.* Arrange the information so that the page has a balanced appearance.
- *Clear organization.* Use adequate white space. The line spacing between items should be greater than the line spacing within an item. That is, there should be more space between your education section and your employment section than between items within either of those sections. You should be able to see the different sections clearly if you stand and look down at the résumé on the floor by your feet.

Use indentation clearly. When you arrange items in a vertical list, indent *turnovers*, the second and subsequent lines of any item, a few spaces. The following list, from the computer-skills section of a résumé, could be confusing:

Computer Experience

Systems: PC, Macintosh, UNIX, Andover AC-256, Prime 360
Software: Lotus 1-2-3, DBase V, PlanPerfect, Micrografx Designer, Adobe PageMaker, Microsoft Word
Languages: Pascal, C++, HTML, XML

When the second line of the long entry is indented, the arrangement is much easier to understand:

Computer Experience

Systems: PC, Macintosh, UNIX, Andover AC-256, Prime 360
Software: Lotus 1-2-3, DBase V, PlanPerfect, Micrografx Designer, Adobe
    PageMaker, Microsoft Word
Languages: Pascal, C++, HTML, XML

For more on page design, see Ch. 13, p. 338.

Content of the Résumé

Although experts advocate different approaches to résumé writing, they all agree on three things:

- *The résumé must be honest.* There are no accurate statistics on how many résumés contain lies, but the figure is probably considerable. You have no doubt read about employees who have been caught including not only exaggerations but outright lies in their résumés. Naturally, they are fired immediately. Many employers today routinely check candidates' credentials. So, for practical as well as ethical reasons, tell the truth.

- *The résumé must be completely free of errors.* Grammar, punctuation, usage, and spelling errors cast doubt on the accuracy of the information in the résumé. Ask for assistance after you have written the draft and proofread the finished product at least twice. Then have someone else proofread it, too.

- *The résumé must provide clear, specific information, without generalizations or self-congratulation.* Your résumé is a sales document, but you are both the salesperson and the product. You cannot say "I am a terrific job candidate," as if you were a toaster or a car. Instead, you have to show the reader by providing the details that will lead the reader to conclude that you are a terrific job candidate. Telling the reader is graceless at best and unconvincing.

A résumé should be long enough to include all pertinent information but not so long that it bores or irritates the reader. Generally, you should keep it to one page. If, however, you have special accomplishments, such as journal articles, patents, or substantial service in student government — a two-page résumé is appropriate. If the information comes to just over a page, either eliminate or condense some of the material to make it fit onto one page, or modify the layout so that it fills a substantial part of a second page. According to a study by Harcourt, Krizan, & Merrier (1991), summarized in Table 16.1 on page 454, most hiring officials prefer that a new college graduate's résumé be brief.

■ **Table 16.1   Preferred Length for a Résumé**

Preferred Length	Hiring Officials (Percent)
No longer than one page	23.6
No longer than two pages	41.8
Depends on the applicant's information	32.7
Other	1.8

Two common styles are the *chronological* résumé and the *analytical* résumé. In a chronological résumé, you use time as the organizing factor for each section, including education and experience. In the analytical résumé, you include a section called *skills*, which is organized according to your talents and achievements.

Recent graduates usually use the chronological résumé because in most cases, they haven't built up the record of skills and accomplishments they need for an analytical résumé. However, if you have professional work experience, you might consider the analytical style.

Elements of the Chronological Résumé

Your résumé should always reflect one particular person: you. Although many people have special skills or backgrounds that can be conveyed in additional sections, most chronological résumés have six basic elements.

A chronological résumé contains:

Identifying information

Job objectives

Education

Employment history

Personal information

References

Identifying Information

Include your full name, address, phone number, and email address. Generally, you should present your name in boldface letters, centered at the top. Use your complete address, with the zip code. Use the two-letter state abbreviations used by the U.S. Postal Service. Also give your complete phone number, with the area code.

If your address during the academic year differs from your home address, list both and identify them clearly. An employer might call during an academic holiday to arrange an interview.

Job Objectives

After the identifying information, add a statement of objectives, in the form of a brief phrase or sentence — for example, "Objective: Entry-level position as a dietitian in a hospital." According to one study, 88.5 percent of managers making personnel decisions consider a statement of objectives important, because it gives the impression that the writer has a clear sense of direction and solid goals (Harcourt, Krizan, & Merrier, 1991). In drafting your statement of objectives, follow these two suggestions:

- *State only the goals or duties explicitly mentioned, or clearly implied, in the job advertisement.* If you unintentionally suggest that your goals are substantially different from the job responsibilities, officials in the organization might infer that you would not be happy working there and not consider you further.
- *Avoid meaningless generalities.* You accomplish little by writing, "Position offering opportunities for advancement in the field of health science, where I can use my communication and analytical skills."

Education

If you are a student or a recent graduate, place the education section next. If you have substantial professional experience, place the employment experience section before the education section.

Include at least the following information in the education section:

- *The degree.* After the degree abbreviation (such as B.S., B.A., A.A., or M.S.), list your academic major (and, if you have one, your minor) — for example, "B.S. in Materials Engineering, minor in General Business."
- *The institution.* Identify the institution by its full name: "Louisiana State University," not "LSU."
- *The location of the institution.* Include the city and state.
- *The date of graduation.* If your degree has not yet been granted, write "Anticipated date of graduation" or "Degree expected in" before the month and year.
- *Information about other schools you attended.* List any other institutions you attended beyond high school, even those at which you did not earn a degree. Employers are generally impressed to learn that a student began at a community or junior college and was able to transfer to a four-year college or university. The listing for other institutions attended should include the same information as the main listing. Arrange entries in reverse

chronological order: that is, list first the school you attended most recently.

Elaborating on Your Education

Stating only the basic information about your educational experience implies that you endured an institution and received a degree. The following guidelines can help you elaborate on your education.

▶ *List your grade-point average.* If your grade-point average is significantly above the median for the graduating class, list it. Or list your average in the courses in your major, if that is more impressive.

▶ *Compile a list of courses.* Include courses that will be of particular interest to an employer, such as advanced courses in your major, or communications courses: technical communication, public speaking, organizational communication, and the like. For example, a list of business courses on an engineer's résumé shows special knowledge and skills. But don't bother listing traditional required courses. Include the substantive titles of listed courses; employers won't know what "Chemistry 450" is. Call it by its official title: "Chemistry 450. Organic Chemistry."

▶ *Describe a special accomplishment.* For a special senior design or research project, for example, include the title and objective of the project, any special or advanced techniques or equipment you used, and — if you know them — the major results: "A Study of Composite Substitutes for Steel — a senior design project to formulate a composite material that can be used to replace steel in car axles." A project discussion makes you seem more like a professional — someone who designs and carries out projects.

▶ *List honors and awards you received.* Scholarships, internships, and academic awards offer evidence of exceptional ability. If you have received a number of such honors, or some that were not exclusively academic, you might list them separately (in a section called *Honors or Awards*) rather than in a subsection of the education section. Ultimately, you must decide where this information will make the best impression.

The education section is the easiest part of the résumé to adapt in applying for different positions. For example, a student majoring in electrical engineering who is applying for a position that calls for strong communications skills can list communications courses in one version of the résumé and advanced electrical engineering courses in another version. As you compose the education section, emphasize those aspects of your background that meet the requirements for the particular job.

Employment

The employment section, like the education section, should convey at least the basic information about each job you have held:

- dates of employment
- the organization's name and location
- your position or title

However, such a skeletal list would be uninformative and unimpressive. As in the education section, you should include carefully selected details. What readers want to know, after they have learned where and when you were employed, is what you actually did. Provide at least a two- to three-line description for each position. For particularly important or relevant jobs, give a more extensive description, focusing on one or more of the following factors:

- *Documents.* What kinds of documents did you write or assist in writing? List, especially, various governmental forms and any long reports, manuals, proposals, or Web sites.
- *Clients.* What kinds of, and how many, clients did you do business with as a representative of your organization?
- *Skills.* What kinds of technical skills did you use on the job?
- *Equipment.* What kinds of technical equipment did you operate or oversee? Mention, in particular, computer skills.
- *Money.* How much money were you responsible for? Even if you considered your bookkeeping position fairly simple, the fact that the organization grossed, say, $2 million a year shows that the position involved real responsibility.
- *Personnel.* How many people did you supervise? Naturally, supervision shows maturity and responsibility.

Whenever possible, emphasize *results*. If you reorganized the shifts of the weekend employees you supervised, state the results:

Reorganized the weekend shift, resulting in a cost savings of more than $3,000 per year.

Wrote and produced (with desktop publishing) a parts catalog that is still used by the company and that increased our phone inquiries by more than 25 percent.

When you describe positions, use the active voice — "supervised three workers" — rather than the passive voice — "three workers were supervised by me." The active voice emphasizes the action of the verb. In thinking about your functions and responsibilities, choose strong action verbs that clearly communicate your activities. Also note that résumés often omit the *I* at the start of sentences. Rather than write, "I prepared bids . . ." many would write, "Prepared bids." Whichever style you use, be consistent.

For more on using strong verbs, see Ch. 11, p. 280.

administered	coordinated	evaluated	maintained	provided
advised	corresponded	examined	managed	purchased
analyzed	created	expanded	monitored	recorded
assembled	delivered	hired	obtained	reported
built	developed	identified	operated	researched
collected	devised	implemented	organized	solved
completed	directed	improved	performed	supervised
conducted	discovered	increased	prepared	trained
constructed	edited	instituted	produced	wrote

■ **Figure 16.1**
Strong Action Verbs Used in Résumés

Figure 16.1 above lists some strong verbs. Here is a sample listing:

June–September 20XX: Millersville General Hospital, Millersville, TX. Student Dietitian. Gathered dietary histories and assisted in preparing menus for a 300-bed hospital. Received "excellent" on all items in evaluation by head dietitian.

In just a few lines, you can show that you sought and accepted responsibility and that you acted professionally. Do not write, "I accepted responsibility"; instead, present facts that lead the reader to that conclusion.

Naturally, not all jobs entail professional skills and responsibilities. Many students find summer work as laborers, sales clerks, short-order cooks, and so forth. If you have not held a professional position, list the jobs you have held, even if they were completely unrelated to your career plans. If the job title is self-explanatory — such as waitperson or service-station attendant — don't elaborate. Every job is valuable. You learn that you are expected to be some-place at a specific time, wear appropriate clothes, and perform specific duties. Also, every job helps pay college expenses. If you can write that you earned, say, 50 percent of your annual expenses through a job, employers will be impressed by your self-reliance. Most of them probably started out with non-professional positions. And any job you have held can yield a valuable reference.

One further suggestion: if you have held a number of nonprofessional as well as several professional positions, group the nonprofessional ones:

Other Employment: Cashier (summer, 1999), salesperson (part time, 1999), clerk (summer, 1998).

This strategy prevents the nonprofessional positions from drawing the reader's attention away from the more important positions.

List jobs in reverse chronological order on the résumé to highlight the most recent employment.

Personal Information

Most résumés do not include such information as the writer's height, weight, date of birth, and marital status; federal legislation prohibits organizations from requiring this information. In addition, most people now feel that such personal information is irrelevant to a person's ability.

However, the personal information section of the résumé is the appropriate place for a few items about your outside interests:

- participation in community-service organizations — such as Big Brothers/Big Sisters — or volunteer work in a hospital
- hobbies related to your career — for example, amateur electronics for an engineer
- sports, especially those that might be socially useful in your professional career, such as tennis, racquetball, and golf
- university-sanctioned activities, such as membership on a team, participation in the college newspaper, or election to a responsible position in an academic organization or a residence hall

Do not include activities that might create a negative impression, such as hunting, gambling, or performing in a rock band. And always omit such activities as meeting people and reading — everybody meets people and reads.

References

Potential employers will want to learn more about you from your professors and previous employers. In providing references, follow three steps:

- *Decide how you want to present the references.* On your résumé, you can list the names of three or four referees — people who have written letters of recommendation or who have agreed to speak on your behalf. Or you may simply say that you will furnish the names of the referees upon request. The length of your résumé sometimes dictates which approach to use. If the résumé is already long, the abbreviated form might be preferable. If it does not fill out the page, the longer form might be the better one. However, each style has advantages and disadvantages that you should consider carefully.

  Furnishing the referees' names appears open and forthright. It shows that you have already secured your referees and have nothing to hide. If one or several of the referees are prominent in their fields, the reader is likely to be impressed. And, perhaps most important, the reader can easily phone the referees or write them a letter. Listing the referees makes it easy for the prospective employer to proceed with the hiring process. The only disadvantage is that it takes up space you might need for other information.

  Writing "References will be furnished upon request" requires only one line and leaves you in a more flexible position. You can still secure referees after you have submitted the résumé. You can also send selected letters of reference to prospective employers according to your analysis of what they want. Supplying different references for different positions is sometimes just as valuable as sending different résumés. However, some readers will interpret the lack of names and addresses as evasive or secretive or perhaps assume that you have not yet asked prospective referees. A greater disadvantage is that if readers are impressed by your résumé and want to learn more about you, they cannot do so quickly and directly.

What do personnel officers prefer? According to Bowman (1999), 70 percent of hiring officials want to see the full references, including the name, title, organization, mailing address, and phone number of each referee. For example:

> Dr. Robert Ariel
> Assistant Professor of Biology
> Central University
> 1910 Westerly Parkway
> Portland, OR 97202
> (503) 555-5746

- *Choose your referees carefully.* Solicit references only from those who know your work best and for whom you have done your best work — for instance, a previous employer with whom you worked closely or a professor from whom you received A's. It is unwise to ask prominent professors who do not know your work well; the advantage of having a famous name on the résumé will be offset by the referee's brief, uninformative letter. Often, a less prominent professor who knows you can write the most informative letter or provide the best recommendation.

- *Give the potential referee an opportunity to decline gracefully.* Sometimes the person has not been as impressed with your work as you think. And if you simply ask "Would you please write a reference letter for me?" the potential referee might accept and then write a lukewarm letter. It is better to follow the first question with "Would you be able to write an enthusiastic letter for me?" or "Do you feel you know me well enough to write a strong recommendation?" If the potential referee shows any signs of hesitation or reluctance, this is the moment to withdraw the request. The scene may be a little embarrassing, but it is better than receiving a half-hearted recommendation.

Other Elements

So far, the discussion has concentrated on the sections that appear on almost everyone's résumé. Other sections are either discretionary or appropriate for only some job seekers.

- *Military experience.* If you are a veteran, include a military-service section. Describe your military service as if it were any other job, citing dates, locations, positions, ranks, and tasks. Often a serviceperson receives regular evaluations from a superior; these evaluations can work in your favor.

- *Language ability.* If you have a working knowledge of a foreign language, your résumé should include a Language Skills section. Language skills are particularly relevant if the potential employer has international interests and you could be useful in translation or foreign service.

- *Willingness to relocate.* If you are willing to relocate, state that fact outright. Many organizations will find you a more attractive candidate if they know you are willing to move around as you learn the business.

Elements of the Analytical Résumé

The analytical résumé differs from the chronological one in that it includes a separate section, usually called *skills* or *skills and abilities*, to convey job skills and experience.

An analytical résumé contains:
Identifying information
Job objectives
Skills
Education
Employment history
Personal information
References

In an analytical résumé, the employment section becomes a brief list of information about the candidate's employment history: name of the company, dates of employment, and name of the position. The important skills section is usually placed prominently near the top of the résumé. Here is an example:

Skills and Abilities

Management
   Served as weekend manager of six employees in the retail clothing business. Also trained three summer interns at a health-maintenance organization.

Writing and Editing
   Wrote status reports, edited performance appraisals, participated in assembling and producing an environmental impact statement by desktop publishing.

Teaching and Tutoring
   Tutored in the University Writing Center. Taught a two-week course in electronics for teenagers. Coach youth basketball.

In a skills section, you choose the headings, the arrangement, and the level of detail. Your goal, of course, is to highlight those skills the employer is seeking.

Figures 16.2, 16.3, and 16.4 on pages 462–464 show three examples of effective résumés.

Many of the job boards listed on p. 450 include samples of résumés.

CARLOS RODRIGUEZ
3109 Vista Street   Philadelphia, PA 19136   (215) 555-3880   crodrig@dragon.du.edu

Objective
Entry-level position in signal processing

Education
Bachelor of Science in Electrical Engineering
Drexel University, Philadelphia, PA
Anticipated June 2001
Grade-Point Average: 3.67 (on a scale of 4.0)
Senior Design Project: "Enhanced Path-Planning Software for Robotics"

Advanced Engineering Courses
Digital Signal Processing   Computer Hardware
Introduction to Operating Systems I, II   Systems Design
Digital Filters   Computer Logic Circuits I, II

Employment
6/98–1/99   *RCA Advanced Technology Laboratory, Moorestown, NJ*
Designed ultra large-scale integrated circuits using VERILOG and VDHL hardware description languages. Assisted senior engineer in CMOS IC layout, modeling, parasitic capacitance extraction, and PSPICE simulation operations.

6/96–1/97   *RCA Advanced Technology Laboratory, Moorestown, NJ*
Verified and documented several integrated circuit designs. Used CAD software and hardware to simulate, check, and evaluate these designs. Gained experience on the VAX and Applicon.

Honors and Organizations
Eta Kappa Nu (Electrical Engineering Honor Society)
Tau Beta Pi (General Engineering Honor Society)
Institute of Electrical and Electronics Engineers, Inc.

References

Ms. Anita Feller	Mr. Fred Borelli	Mr. Sam Shamir
Engineering Consultant	Unit Manager	Comptroller
700 Church Road	RCA Corporation	RCA Corporation
Cherry Hill, NJ 08002	Route 38	Route 38
(609) 555-7836	Moorestown, NJ 08057	Moorestown, NJ 08057
	(609) 555-2435	(609) 555-7849

■ **Figure 16.2
Chronological Résumé of a Traditional Student**

Carlos Rodriguez entered college right after high school and has proceeded on schedule. Because some of his referees are well known, he lists their names at the end of the résumé.

- use italics
- email
- consistency
- verb usage

Alice P. Linder

1781 Weber Road
Warminster, PA 18974
(215) 555-3999
linderap423@aol.com

Objective: A position in molecular research that uses my computer skills

Education: Harmon College, West Yardley, PA
Major: Bioscience and Biotechnology
Expected Graduation Date: June 2002

Related Course Work
General Chemistry I, II, III   Biology I, II, III
Organic Chemistry I, II   Statistical Methods for Research
Physics I, II   Technical Communication
Calculus I, II

Employment Experience: 6/99–present (20 hours per week)
GlaxoSmithKline, Upper Merion, PA
Analyze molecular data on E & S PS300, Macintosh, and IBM PCs.
Write programs in C++, and wrote a user's guide for an instructional
computing package. Train and consult with scientists and deliver in-
house briefings.

8/96–present
Children's Hospital of Philadelphia, Philadelphia, PA
Volunteer in the physical therapy unit. Assist therapists and guide patients
with their therapy. Use play therapy to enhance strengthening progress.

6/88–1/92
Anchor Products, Inc., Ambler, PA
Managed 12-person office in a $1.2 million company. Also performed
general bookkeeping and payroll.

Honors: Awarded three $5,000 tuition scholarships (1995–1998) from the Gould
Foundation.

Additional Information: Member, Harmon Biology Club, Yearbook Staff
Raising three school-age children
Tuition 100% self-financed

References: Available upon request

[handwritten: too much? italics bullets]

[handwritten: OWL—Online Writing Lab]

■ **Figure 16.3
Chronological Résumé of
a Nontraditional Student**

*Alice Linder is a single
mother returning to school.
She is applying for an
internship.*

Alice P. Linder	1781 Weber Road Warminster, PA 18974 (215) 555-3999 linderap423@aol.com
Objective	A position in molecular research that uses my computer skills
Skills and Abilities	*Laboratory Skills* • Analyzed molecular data on E & S PS300, Macintosh, and IBM PCs. Wrote programs in C++. • Have taken 12 credits in biology and chemistry labs. *Communication Skills* • Wrote a user's guide for an instructional computing package. • Trained and consulted with scientists and delivered in-house briefings. *Management Skills* • Managed 12-person office in a $1.2 million company.
Education	Harmon College, West Yardley, PA Major: Bioscience and Biotechnology Expected Graduation Date: June 2002 *Related Course Work* General Chemistry I, II, III Biology I, II, III Organic Chemistry I, II Statistical Methods for Research Physics I, II Technical Communication Calculus I, II
Employment Experience	6/99–present (20 hours per week) *GlaxoSmithKline, Upper Merion, PA* Laboratory Assistant Grade 3 8/96–present *Children's Hospital of Philadelphia, Philadelphia, PA* Volunteer in the physical therapy unit. Assist therapists and guide patients. Use play therapy to enhance strengthening progress. 6/88–1/92 *Anchor Products, Inc., Ambler, PA* Office Manager
Honors	Awarded three $5,000 tuition scholarships (1995–1998) from the Gould Foundation.
Additional Information	Member, Harmon Biology Club, Yearbook Staff Raising three school-age children Tuition 100 percent self-financed
References	Available upon request

■ **Figure 16.4**
Analytical Résumé of a Nontraditional Student

Another version of the résumé in the previous figure.

For a nontraditional student, there are two strategies for presenting information. Alice Linder feels that her nontraditional status is an asset: her maturity and experience will make her a more effective employee than the traditional student. For this reason, she mentions her children and her self-financed tuition on the résumés in Figures 16.3 and 16.4. Others might deemphasize this information because they feel it is irrelevant and draws attention away from important credentials. As a compromise between these two strategies, a student could omit the nontraditional status from the résumé but mention it briefly in the accompanying letter.

Using Tables in Formatting a Résumé

Students and professionals alike spend hours trying to format résumés. The goal is to present the headings in one typeface and size and the body content in another typeface and size. This goal can be hard to achieve when the heading is a long one, such as "Skills and Abilities," because the length cuts into the amount of space left for content. Here is an example of this problem:

Skills and Abilities: *Laboratory Skills*

- Analyze molecular data on E & S PS300, Macintosh, and IBM PCs. Write programs in C++.
- Have taken 12 credits in biology and chemistry labs.

Trying to format "Skills and Abilities" on two lines to leave a wider column for the content is a nuisance.

A good solution is to use a table format. First, create a table with the layout you want. Here is a format that might work for the basic layout:

Skills and Abilities	*Laboratory Skills* • Analyze molecular data on E & S PS300, Macintosh, and IBM PCs. Write programs in C++. • Have taken 12 credits in biology and chemistry labs.

Next, select a table format that hides the grid lines.

Skills and Abilities   *Laboratory Skills*

- Analyze molecular data on E & S PS300, Macintosh, and IBM PCs. Write programs in C++.
- Have taken 12 credits in biology and chemistry labs.

WRITING ELECTRONIC RÉSUMÉS

On the federal government's job board, America's Job Bank, 1.7 million jobs are listed each day. Electronic job searches have become the major way people are hired. The digital revolution is changing the way people look for work.

The key change is that electronic résumés are becoming as popular as traditional paper résumés. The advantage of an electronic résumé is that you can send it in a digital format. There are two main reasons for this growing popularity:

- *Electronic communication is fast and almost free.* You can get your materials to the prospective employer in a matter of seconds, for little or no cost.

- *Electronic résumés can be searched electronically.* Many organizations cannot afford to read every application; Hewlett-Packard, for instance, receives more than 1,000 résumés every day (Robinson, 1997). Large organizations like these often create electronic databases, so that when they need a person with a particular set of skills, they search for keywords electronically.

An electronic résumé can take different forms:

- *A résumé sent as an email message.* You can format your résumé as an email message. The recipient can print it, scan it, or add the information to a database.

- *A résumé attached to an email message.* You can attach the word-processing file to an email message, and the recipient can print it or treat it like an email message. Keep in mind, however, that some email software is unable to read attached files. If the job notice requests "a plain text document sent in the body of the message," do not send an attached file.

- *A printed résumé that will be scanned into an organization's database.* You can send a printed résumé that the recipient will scan and enter into a database. There are several popular database programs for this purpose, such as ResTrac or Resumix. According to Dumas (2001), more than half of medium- and large-sized companies now use scanned résumés extensively. This fact means that even when you submit a printed resume to a company of this size, you should consider how well the document will scan electronically.

- *A Web-based résumé.* You can put your résumé on your own Web site and hope that employers will come to you, or you can post it to a Web board.

Ways of creating and sending résumés will undoubtedly change as the technology changes. For now, you need to know that the traditional printed résumé is only one of several ways to present your credentials, and you need to keep abreast of new techniques for applying for positions. Which form should your résumé take? Whichever form the organization prefers. If you learn of a position from an ad on the organization's own site, it will tell you how to apply.

Content of an Electronic Résumé

Most of the earlier discussion of the content of a printed résumé also applies to an electronic résumé. The résumé must be honest, it must be free of errors, and it must provide clear, specific information, without generalizations.

But if the résumé is to be scanned into a database instead of read by a person, you need to include industry-specific jargon: all the keywords an employment officer might use in a search for qualified candidates. If an employment officer is looking for someone with experience writing Web pages, be sure you include the terms "Web page," "Internet," "HTML," "Java," and any other relevant keywords (Isaacs, 2001). If your current position requires an understanding of programming languages, name the languages you know. Also use keywords that refer to your communication skills, such as "public speaking," "oral communication," and "communication skills." In short, whereas a traditional printed résumé focuses on verbs — tasks you have done — an electronic résumé focuses on nouns. Electronic résumés often contain a section called *keywords,* which is a list of industry-specific jargon.

Format of an Electronic Résumé

Because electronic résumés must be easy to read and scan, they require a very simple design. Consequently, they are not as attractive as paper-based résumés, and they are longer, because they use only a single narrow column of text. Figure 16.5 on page 468 is an example of an electronic résumé.

GUIDELINES

Preparing an Electronic Résumé

▶ *Use ASCII text.* ASCII text includes the letters, numbers, and basic punctuation marks. Avoid boldface, italics, underlining, and special characters such as "smart quotation marks" or math symbols. Also avoid horizontal or vertical lines or graphics. To be sure you are using only ASCII characters, save your file as "text only." Then open it up using your software's text editor, such as Notepad, and check to be sure it contains only ASCII characters.

▶ *Use a simple sans-serif typeface.* Scanners can easily interpret large, open typefaces such as Arial.

▶ *Use a single-column format.* A double-column text will not scan accurately. Align everything on a uniform left margin.

▶ *Use wide margins.* Instead of an 80-character width, set your software for 60 or 65; this way, regardless of the equipment the reader is using, the lines will break as you intend them.

▶ *Use the space bar instead of the tab key.* Tabs will be displayed according to the settings on the reader's equipment, not the settings on yours. Therefore, use the space bar to move text horizontally.

Alice P. Linder
1781 Weber Road
Warminster, PA 18974
(215) 555-3999
linderap423@aol.com

Objective: A position in molecular research that uses my computer skills

Skills and Abilities:
Laboratory Skills. Analyze molecular data on E & S PS300, Macintosh, and IBM
PCs. Write programs in C++. Have taken 12 credits in biology and chemistry labs.

Communication Skills. Wrote a user's guide for an instructional computing package.
Train and consult with scientists and deliver in-house briefings.

Management Skills. Managed 12-person office in a $1.2 million company.

Education:
Harmon College, West Yardley, PA
Major: Bioscience and Biotechnology
Expected Graduation Date: June 2002

Related Course Work:
General Chemistry I, II, III
Organic Chemistry I, II
Physics I, II
Calculus I, II
Biology I, II, III
Statistical Methods for Research
Technical Communication

Employment Experience:
June 1999–present (20 hours per week)
GlaxoSmithKline, Upper Merion, PA
Laboratory Assistant Grade 3

August 1996–present
Children's Hospital of Philadelphia, Philadelphia, PA
Volunteer in the physical therapy unit. Assist therapists and guide patients with their
therapy. Use play therapy to enhance strengthening progress.

June 1988–January 1992
Anchor Products, Inc., Ambler, PA
Office Manager

■ **Figure 16.5   Electronic Résumé**

This is an electronic version of the résumé in Figure 16.4. Notice that the writer uses ASCII text and left justification.

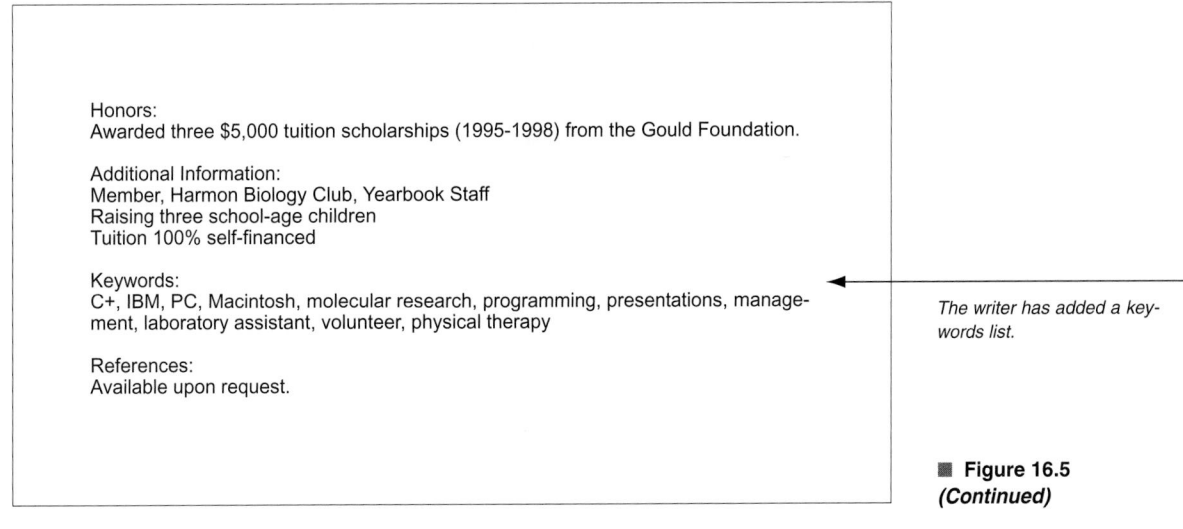

Honors:
Awarded three $5,000 tuition scholarships (1995-1998) from the Gould Foundation.

Additional Information:
Member, Harmon Biology Club, Yearbook Staff
Raising three school-age children
Tuition 100% self-financed

Keywords:
C+, IBM, PC, Macintosh, molecular research, programming, presentations, management, laboratory assistant, volunteer, physical therapy

The writer has added a keywords list.

References:
Available upon request.

■ **Figure 16.5**
(Continued)

If you are mailing a paper résumé that will be scanned, follow these three additional guidelines:

- *Use a good-quality laser printer.* The better the resolution, the better the scanner will work.
- *Use white paper.* Even a slight tint to the paper can increase the chances that the scanner will misinterpret a character.
- *Do not fold the résumé.* The fold line can cause problems for the scanner.

For more information on electronic résumés, use a search engine and look for "careers" or "jobs" or "employment." The search engine will direct you to job boards as well as to Usenet newsgroups.

WRITING JOB-APPLICATION LETTERS

When you send a traditional paper résumé, you include a job-application letter with it. You might also send a letter — as an email — with an electronic résumé.

The job-application letter is crucial because it is the first thing your reader sees. If the letter is ineffective, the reader probably will not bother to read the résumé.

If students had infinite time and patience, they would send a different version of their résumé and a customized letter to each prospective employer, highlighting their suitability for a particular job. But because they don't, they usually make only one or two versions of their résumés. As a result, the typical résumé makes a candidate look only relatively close to the ideal candidate the employer has in mind.

The letter, however, must be customized, because it is addressed to a particular individual at a specific address. Therefore, it makes sense to have the letter appeal as directly and specifically as possible to a particular person.

The Concept of Selectivity

Like the résumé, the job-application letter is a sales document. Its purpose is to convince the reader that you are an outstanding candidate who should be called in for an interview. Of course, you accomplish this purpose through evidence, not empty self-praise.

The job-application letter is not an expanded version of everything in the résumé. The key to a good application letter is selectivity. Choose from the résumé two or three points of greatest interest to the potential employer and develop them into paragraphs. Emphasize results, such as improved productivity or quality or decreased costs. If one of your previous part-time positions called for specific skills that the employer is looking for, that position might be the subject of a substantial paragraph in the letter, even though the résumé devotes only a few lines to it. However, if you try to cover every point on your résumé, the reader will have a hard time forming a clear impression of you, and the letter will not achieve its purpose.

For more about formatting letters, see Ch.15, p. 419.

In most cases, a job-application letter should fill the better part of a page. Like all business letters, it should be single spaced, with double spaces between paragraphs. For more experienced candidates, the letter may be longer, but most students find that they can adequately describe their credentials in one page. Again, selectivity is the key. If you write at length on a minor point, you become boring. Worse still, you appear to have poor judgment. Employers seek candidates who can say a lot in a small space.

Elements of the Job-Application Letter

Among the mechanical elements of the job-application letter, the inside address — the name, title, organization, and address of the recipient — is most important. If you know the correct form of this information from an ad, there is no problem. However, if you are uncertain about any of the information — the recipient's name, for example, might have an unusual spelling — verify it by phoning the organization. Because many people are very sensitive about such matters, you should not risk beginning the letter with a misspelling or an incorrect title.

When you do not know who should receive the letter, do not address it to a department of the company — unless the job ad specifically says to do so — because nobody in that department might feel responsible for dealing with it. Instead, phone the company to find out who manages the department. If you are unsure of the appropriate department or division to write to, address the letter to a high-level executive, such as the president. The letter will be directed to the right person. Also, because the application includes both a letter and a résumé, use an enclosure notation.

The four-paragraph example that will be discussed here is only a basic model, consisting of an introductory paragraph, two body paragraphs, and a concluding paragraph. At a minimum, your job-application letter should include these four paragraphs, but there is no reason it cannot have five or six.

Because this is such an important letter, you should plan it carefully. Select information from your background that best responds to the needs of the potential employer. Draft the letter and then revise it. Let it sit for a while, then revise it again. Spend as much time on it as you can. Make each paragraph a unified, functional part of the whole letter. Supply clear transitions from one paragraph to the next.

The Introductory Paragraph

The introductory paragraph establishes the tone of the letter and captures the reader's attention. It has four specific functions:

For more about developing paragraphs, see Ch. 10, p. 256.

- *It identifies your source of information.* In an unsolicited application, all you can do is ask if a position is available. For most applications, however, your source of information is a published advertisement or an employee already working for the organization. If your source is an ad, identify the publication and its date of issue. If an employee told you about the position, identify that person by name and title.

- *It identifies the position you are interested in.* Often, the organization you are applying to has advertised a number of positions; if you omit the title of the position you are interested in, your reader might not know which one you are seeking.

- *It states that you wish to be considered for the position.* Although the context makes your wish obvious, you should mention it because the letter would be awkward without it.

- *It forecasts the rest of the letter.* Choose a few phrases that forecast the body of the letter, so that the letter flows smoothly. For example, if you use the phrase "retail experience" in the opening paragraph, you are preparing your reader for the discussion of your retail experience later in the letter.

These four points need not appear in any particular order, nor does each need to be covered in a single sentence. The following sample paragraphs demonstrate different ways of providing the necessary information:

Response to a job ad

I am writing in response to your notice in the May 13 *New York Times*. I would like to be considered for the position in system programming. I hope you find that my studies in computer science at Eastern University, along with my programming experience at Airborne Instruments, would qualify me for the position.

Unsolicited

My academic training in hotel management and my experience with Sheraton International have given me a solid background in the hotel industry. Would you please consider me for any management trainee position that might be available?

Unsolicited personal contact

Mr. Howard Alcott of your Research and Development Department suggested that I write to you. He thinks that my organic chemistry degree and my practical experience with Brown Laboratories might be of value to XYZ Corporation. Do you have an entry-level position in organic chemistry for which I might be considered?

The difficult part of the introductory paragraph — and of the whole letter — is to achieve the proper tone: quiet self-confidence. Because your letter will be read by someone who is superior to you professionally, the tone must be modest, but it should not be self-effacing or negative. Never say, for example, "I do not have a very good background in computers, but I'm willing to learn." The reader will take this kind of statement at face value and probably stop reading right there. You should show pride in your education and experience, while at the same time suggesting by your tone that you have much to learn.

The Education Paragraph

For most students, the education paragraph should come before the employment paragraph because the content of the education paragraph will be stronger. If, however, your work experience is more pertinent than your education, discuss your work first.

In devising your education paragraph, take your cue from the job ad (if you are responding to one). What aspect of your education most directly fits the job requirements? If the ad stresses versatility, you might structure your paragraph around the range and diversity of your courses. Also, you might discuss course work in a subject related to your major, such as business or communication skills. Extracurricular activities are often very valuable; if you were an officer in a student organization in your field, you could discuss the activities and programs that you coordinated. Perhaps the most popular strategy for developing the education paragraph is to discuss skills and knowledge gained from advanced course work in your major field.

Whatever information you provide, the key to the education paragraph is to develop one unified idea, rather than to toss a series of unrelated facts. Notice how each of the following education paragraphs develops a unified idea.

EXAMPLE 1   At Eastern University, I have taken a wide range of courses in the sciences, but my most advanced work has been in chemistry. In one laboratory course, I developed a new aseptic brewing technique that lowered the risk of infection by more than 40 percent. This new technique was the subject of an article in the Eastern Science Digest. Representatives from three national breweries have visited our laboratory to discuss the technique with me.

EXAMPLE 2   To broaden my education at Southern University, I took eight business courses in addition to my requirements for the civil engineering degree. Because your ad mentions that the position will require substantial client contact, I believe that my work in marketing, in particular, would be of special value. In an advanced marketing seminar, I used PageMaker® to produce a 20-page sales brochure describing the various kinds of building structures for sale by Oppenheimer Properties to industrial customers in our section of the city. That brochure is now being used at Oppenheimer Properties, where I am serving as an intern.

EXAMPLE 3   The most rewarding part of my education at Western University took place outside the classroom. My entry in a fashion-design competition sponsored by the university won second place. More important, through the competition I met the chief psychologist at Western Regional Hospital, who invited me to design clothing for people with disabilities. I have since completed six different outfits, which are now being tested at the hospital. I hope to be able to pursue this interest once I start work.

Each of these paragraphs begins with a topic sentence — a forecast of the rest of the paragraph — and uses considerable detail and elaboration to develop the main idea. An additional point: if you haven't already specified your major and your college or university in the introductory paragraph, be sure to do so in the education paragraph.

The Employment Paragraph

Like the education paragraph, the employment paragraph should begin with a topic sentence and then elaborate a single idea. That idea might be that you have a broad background or that one job in particular has given you special skills that make you especially well-suited for the available job. Here are several examples of effective experience paragraphs:

EXAMPLE 1   For the past three summers and part-time during the academic year, I have worked for Redego, Inc., a firm that specializes in designing and planning industrial complexes. I began as an assistant in the drafting room. By the second summer, I was accompanying a civil engineer on field inspections. Most recently, I have used CAD to assist an engineer in designing and drafting the main structural supports for a 15-acre, $30 million chemical facility.

EXAMPLE 2   Although I have worked every summer since I was 15, my most recent position, as a technical editor, has been the most rewarding. I was chosen by Digital Systems, Inc., from among 30 candidates because of my dual background in computer science and writing. My job was to coordinate the editing of computer manuals. Our copy editors, who are generally not trained in computer science, need someone to help verify the technical accuracy of their revisions. When I was unable to answer their questions, I was responsible for interviewing our systems analysts to find the correct answer and to make sure the computer novice could follow it. This position gave me a good understanding of the process by which operating manuals are created.

EXAMPLE **3**   I have worked in merchandising for three years as a part-time and summer salesperson in men's fashions and accessories. I have had experience running inventory-control software and helped one company switch from a manual to an online system. Most recently, I assisted in clearing $200,000 in out-of-date men's fashions: I coordinated a campaign to sell half of the merchandise at cost and was able to convince the manufacturer's representative to accept the other half for full credit. For this project, I received a certificate of appreciation from the company president.

The writers of these paragraphs carefully define their duties to convey the nature and extent of their responsibilities.

Although you will discuss your education and experience in separate paragraphs, try to link these two halves of your background. If an academic course led to an interest that you were able to pursue in a job, make that point clear in the transition from one paragraph to the other. Similarly, if a job experience helped shape your academic career, tell the reader about it.

The Concluding Paragraph

The concluding paragraph of the job-application letter, like that of any sales letter, is intended to stimulate action. In this case, you want the reader to invite you for an interview. In the preceding paragraphs you provided the information that should have convinced the reader to give you another look. In the last paragraph, you want to make it easy for him or her to do so. The concluding paragraph contains three main elements:

- *A reference to your résumé.* If you have not yet referred to it, do so now.
- *A polite but confident request for an interview.* Use the phrase *at your convenience.* Don't make the request sound as if you're asking a personal favor.
- *Your phone number and email address.* State the time of day you can be reached. Adding an email address gives the employer one more way to get in touch with you.

Here are two examples of effective concluding paragraphs.

EXAMPLE **1**   The enclosed résumé provides more information about my education and experience. Could we meet at your convenience to discuss further the skills and experience I could bring to Pentamax? You can leave a message for me anytime at (303) 555-5957 or cfilli@claus.cmu.edu.

EXAMPLE **2**   More information about my education and experience is included on the enclosed résumé, but I would appreciate the opportunity to meet with you at your convenience to discuss my application. You can reach me after noon on Tuesdays and Thursdays at (212) 555-4527 or leave a message anytime.

The examples of effective job-application letters in Figures 16.6 and 16.7 correspond to the résumés in Figures 16.2 and 16.3.

3109 Vista Street
Philadelphia, PA 19136

January 19, 20XX

Mr. Stephen Spencer, Director of Personnel
Department 411
Boeing Naval Systems
103 Industrial Drive
Wilmington, DE 20093

Dear Mr. Spencer:

I am writing in response to your advertisement in the January 16 *Philadelphia Inquirer*. Would you please consider me for the position in Signal Processing? I believe that my academic training in electrical engineering at Drexel University, along with my experience with RCA Advanced Technology Laboratory, would qualify me for the position.

My education at Drexel has given me a strong background in computer hardware and system design. I have concentrated on digital and computer applications, developing and designing computer and signal-processing hardware in two graduate-level engineering courses. For my senior-design project, I am working with four other undergraduates in using OO programming techniques to enhance the path-planning software for an infrared night-vision robotics application.

While working at the RCA Advanced Technology Laboratory, I was able to apply my computer experience to the field of the VLSI design. I designed ultra large-scale integrated circuits using VERILOG and VHDL hardware description languages. In addition, I assisted a senior engineer in CMOS IC layout, modeling, parasitic capacitance extraction, and PSPICE simulation operations.

The enclosed résumé provides an overview of my education and experience. Could I meet with you at your convenience to discuss my qualifications for this position? Please write to me at the above address or leave a message anytime at (215) 555-3880. My email address is crodrig@dragon.du.edu.

Yours truly,

Carlos Rodriguez

Carlos Rodriguez

Enclosure (1)

Many of the job boards listed on p. 450 include samples of application letters.

■ **Figure 16.6**
Job-Application Letter

1781 Weber Road
Warminster, PA 18974

January 17, 20XX

Mr. Harry Gail
Fox Run Medical Center
399 N. Abbey Road
Warminster, PA 18974

Dear Mr. Gail:

Last April I contacted your office regarding the possibility of an internship as a laboratory assistant at your center. Your assistant, Mary McGuire, told me then that you might consider such a position this year. With the experience I have gained since last year, I believe I would be a valuable addition to your center in many ways.

The first two paragraphs of the body discuss the applicant's qualifications for the internship position.

At Harmon College, I have earned a 3.7 GPA in 36 credits in chemistry and biology; all but two of these courses had laboratory components. One skill stressed at Harmon is the ability to communicate effectively, both in writing and orally. Our science courses have extensive writing and speaking requirements; my portfolio includes seven research papers and lab reports of more than 20 pages each, and I have delivered four oral presentations, one of 45 minutes, to classes.

At GlaxoSmithKline, where I currently work part time, I analyze molecular data on an E & S PS300, a Macintosh, and an IBM PC. I have tried to remain current with the latest advances; my manager at GlaxoSmithKline has allowed me to attend two different two-day in-house seminars on computerized data analysis using SAS.

Here the applicant explains how her additional experience — as an office manager and a single mother — have enabled her to develop skills that would be of value to anyone in any field.

Having been out of school for more than a decade, I am well aware of how much the technology has changed. However, as the manager of a 12-person office for four years, I believe that I have acquired skills that would benefit Fox Run. In addition, as a single mother of three I know something about time management.

More information about my education and experience is included on the enclosed résumé, but I would appreciate the opportunity to meet with you at your convenience to discuss my application. If you would like any additional information about me or Harmon's internship program, please write to me at the above address, call me at (215) 555-3999, or email me at linderap423@aol.com.

Very truly yours,

Alice P. Linder

Alice P. Linder

Enclosure

■ **Figure 16.7 Job-Application Letter**

The writer discusses her nontraditional background without overemphasizing it. She exploits her situation gracefully, always appealing to the reader's needs, without asking for special consideration.

WRITING FOLLOW-UP LETTERS

After an interview, you should write a letter of appreciation and one of the others listed here. All follow-ups should conform to standard business-letter conventions.

Many of the job boards listed on p. 450 include samples of follow-up letters.

- *The letter of appreciation after an interview.* In a follow-up letter, you thank the representative for taking the time to see you and emphasize your particular qualifications. You can also take this opportunity to restate your interest in the position. The follow-up letter can do more good with less effort than any other step in the job-application procedure, because so few candidates take the time to write it. Here is an example:

Dear Mr. Weaver:

Thank you for taking the time yesterday to show me your facilities and to introduce me to your colleagues.

Your advances in piping design were particularly impressive. As a person with hands-on experience in piping design, I can appreciate the advantages your design will have.

The vitality of your projects and the good fellowship among your employees further confirm my initial belief that Cynergo would be a fine place to work. I would look forward to joining your staff.

Sincerely yours,

Harriet Bommarito

Harriet Bommarito

- *The letter accepting a job offer.* This one is easy: express appreciation, show enthusiasm, and repeat the major terms of your employment. Here is an example:

Dear Mr. Weaver:

Thank you very much for the offer to join your staff. I accept.

I look forward to joining your design team on Monday, July 19. The salary, as you indicate in your letter, is $34,250.

As you have recommended, I will get in touch with Mr. Matthews in Personnel to get a start on the paperwork.

I appreciate the trust you have placed in me, and I assure you that I will do what I can to be a productive team member at Cynergo.

Sincerely yours,

Mark Greenberg

Mark Greenberg

- *The letter of rejection in response to a job offer.* If you decide not to accept a job offer, express your appreciation and, if appropriate, explain why you are declining the offer. Remember, you might want to work for this company some time in the future. Here is an example:

Dear Mr. Weaver:

I appreciate very much the offer to join your staff.

Although I am certain that I would benefit greatly from working at Cynergo, I have decided to take a job with a firm in Pittsburgh, where I have been accepted at Carnegie-Mellon to pursue my master's degree at night.

Again, thank you for your generous offer.

Sincerely yours,

Cynthia O'Malley

Cynthia O'Malley

- *The letter acknowledging a rejection.* Why write back after you have been rejected? To maintain good relations. You just might get a phone call the next week explaining that the person who accepted the job has had to change her plans and that the company is offering you the position. Here is an example of this kind of letter.

Dear Mr. Weaver:

I was disappointed to learn that I will not have a chance to join your staff, because I feel that I could make a substantial contribution. However, I appreciate that job decisions are complex, involving many candidates and many factors.

Thank you very much for the courtesy you have shown me. I have long believed — and I still believe — that Cynergo is a first-class organization.

Sincerely yours,

Paul Goicochea

Paul Goicochea

✔   **Revision Checklist**

Printed Résumé
1. Does the résumé respond to the needs of its readers?
2. Does the résumé have a professional appearance, with generous margins, a symmetrical layout, adequate white space, and effective indentation?
3. Is the résumé honest?
4. Is the résumé free of errors?

5. Does the identifying information section contain your name, address(es), phone number(s), and email address(es)?
6. Does the résumé include a clear statement of your job objectives?
7. Does the education section include your degree, your institution and its location, and your anticipated date of graduation, as well as any other information that will help a reader appreciate your qualifications?
8. Does the employment section include, for each job, the dates of employment, the organization's name and location, and (if you are writing a chronological résumé) your position or title, as well as a description of your duties and accomplishments?
9. Does the personal information section include relevant hobbies or activities, including extracurricular interests? Have you omitted any personal information that might reflect poorly on you?
10. Does the references section include the names, job titles, organizations, mailing addresses, and phone numbers of three or four referees? If you are not listing this information, does the strength of the rest of the résumé offset the omission?
11. Does the résumé include any other appropriate sections, such as military service, language skills, or honors?

Electronic Résumé

In addition to the items mentioned in the checklist for the printed résumé, did you
❏ use ASCII text?
❏ use a simple sans-serif typeface?
❏ use a single-column format?
❏ use wide margins?
❏ use the space bar instead of the tab key?

Job-Application Letter

1. Does the letter respond to the needs of its readers?
2. Is the letter honest?
3. Does the letter look professional?
4. Does the introductory paragraph identify your source of information and the position you are applying for, state that you wish to be considered, and forecast the rest of the letter?
5. Does the education paragraph respond to your reader's needs with a unified idea introduced by a topic sentence?
6. Does the employment paragraph respond to your reader's needs with a unified idea introduced by a topic sentence?
7. Does the concluding paragraph include a reference to your résumé, a request for an interview, your phone number, and your email address?
8. Does the letter include an enclosure notation?

Follow-Up Letter

1. Does the letter of appreciation for a job interview thank the interviewer and briefly restate your qualifications?
2. Does the letter accepting a job offer show enthusiasm and repeat the major terms of your employment?
3. Does the letter of rejection in response to a job offer express your appreciation and, if appropriate, explain why you are declining the offer?
4. Does the letter acknowledging a rejection maintain a positive tone that will help you maintain good relations?

Exercises

1. Using a job board on the Web, list and briefly describe five positions in your field in your state. What skills, experience, and background does each position require? What is the salary range for each position?

2. Locate and provide the URLs of three job boards that provide interactive forms for creating a résumé automatically. What are the strengths and weaknesses of each? Which appears to be the easiest to use? Why?

3. The following résumé was submitted in response to an ad describing the following duties: "Research and develop key technology and system concepts for spectrally efficient digital radio frequency data networks such as digital cellular mobile radio telephones, public safety trunked digital radio systems, and satellite communications." How effective is the résumé? What are some of its problems?

Rajiv Siharath
2319 Fifth Avenue
Waverly, CT 01603
Phone: 611-3356

Personal Data:	22 Years old
	Height 5'11"
	Weight 176 lbs.
Education:	B.S. in Electrical Engineering
	University of Connecticut,
	June, 2003
Experience:	6/01–9/01 Falcon Electronics
	Examined panels for good wiring. Also, I revised several schematics.
	6/00–9/00 MacDonalds Electrical Supply Co. Worked parts counter.
	6/99–9/99 Happy Burger
	Made hamburgers, fries, shakes, fish sandwiches, and fried chicken.
	6/98–9/98 Town of Waverly
	Outdoor maintenance. In charge of cleaning up McHenry Park and Municipal Pool picnic grounds. Did repairs on some electrical equipment.
Backround:	Born and raised in Waverly.
	Third baseman, Fisherman's Rest softball team.
	Hobbies: jogging, salvaging and repairing appliances, reading magazines, politics.
References:	Will be furnished upon request.

4. The following application letter responds to an ad describing the following duties: "Buyer for a high-fashion ladies' dress shop. Experience required." How effective is the letter? How could it be improved?

April 13, 20XX

Marilyn Grissert
Best Department Store
113 Hawthorn
Atlanta, Georgia

Dear Ms. Grissert:

As I was reading the *Sunday Examiner,* I came upon your ad for a buyer. I have always been interested in learning about the South, so would you consider my application?

I will receive my degree in fashion design in one month. I have taken many courses in fashion design, so I feel I have a strong background in the field.

Also, I have had extensive experience in retail work. For two summers I sold women's accessories at a local clothing store. In addition, I was a temporary department head for two weeks.

I have enclosed a résumé and would like to interview you at your convenience. I hope to see you in the near future. My phone number is 555-6103.

Sincerely,

Brenda Sisneros

5. How effective is the following letter of appreciation? How could it be improved?

914 Imperial Boulevard
Durham, NC 27708

November 13, 20XX

Mr. Ronald O'Shea
Division Engineering
Safeway Electronics, Inc.
Holland, MI 49423

Dear Mr. O'Shea:

Thanks very much for showing me around your plant. I hope I was able to convince you that I'm the best person for the job.

Sincerely yours,

Robert Harad

Research Projects

6. In a newspaper or journal or on the Internet, find a job ad for a position in your field for which you might be qualified. Write a résumé and a job-application letter in response to the ad; include the job ad or a photocopy. You will be evaluated not only on the content and appearance of the materials, but also on how well you have targeted them to the job ad.

7. In a newspaper or journal or on the Internet, find a job ad for a position in your field for which you might be qualified. Write a résumé and a job-application letter in response to the ad; include the job ad or a photocopy. Exchange your materials with another group member and meet to discuss your response to his or her materials. Then revise your materials, exchange them again, and meet again to discuss the other person's response to them. Together, draft a memo to your instructor on the success of the exercise. Which aspects of the critique were useful? Which were less successful? When you apply for your first position, what kind of peer review will you use? Why? See Chapter 15, page 430, for a discussion of writing memos.

C A S E
Updating Career-Center Materials

The members of your group are student interns in the Career center at your school. The director of the center, William Karey, would like your group to assist him in updating the materials he distributes to students looking for work. "What I'd like you to do," he tells you, "is to take a look at what is out there on the Web and integrate it with what we already have on résumés and job letters." If possible, secure an electronic copy of available materials from your school's Career Center. Then search the Web for additional information about résumés and job letters. Is the information from the Web consistent with the information from the Career Center? Which sources on the Web seem to be of most use to students at your school? Decide whether you should integrate the new information with the material the Career Center already distributes, put part or all of the new information on the school's Web site, or both. Regardless of your ultimate decision, submit a copy of the existing materials as well as a copy of the new materials.

For additional cases, click on Case of the Month and Archive on TechComm Web (www.bedfordstmartins.com/techcomm).

17

Writing Proposals

Alice Reid (1998), of Delaware Technical and Community College, Wilmington, Delaware, describes the criteria according to which proposals are evaluated:

Any proposal offers a plan to fill a need, and your reader will evaluate your plan according to how well your written presentation answers questions about WHAT you are proposing, HOW you plan to do it, WHEN you plan to do it, and HOW MUCH it is going to cost.

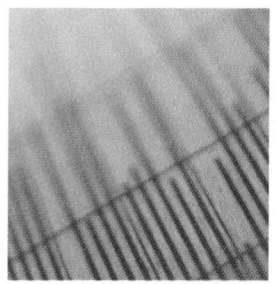

Most projects undertaken by organizations, and most major changes made within organizations, begin with a proposal. A proposal is an offer to carry out research or to provide a product or service.

See TechComm Web (www .bedfordstmartins.com /techcomm) for guidelines, additional examples, and links related to topics in this chapter.

This chapter begins by describing the logistics of proposals, concentrating on external and internal proposals, as well as solicited and unsolicited proposals. Then it explains the "deliverables" of proposals: what you deliver at the end of the project. The chapter then discusses the need for persuasion in proposal writing. Finally, it describes the components of a proposal and presents an example of an internal proposal.

THE LOGISTICS OF PROPOSALS

Proposals can be classified as either external or internal and as either solicited or unsolicited. Figure 17.1 shows the relationship among these four terms.

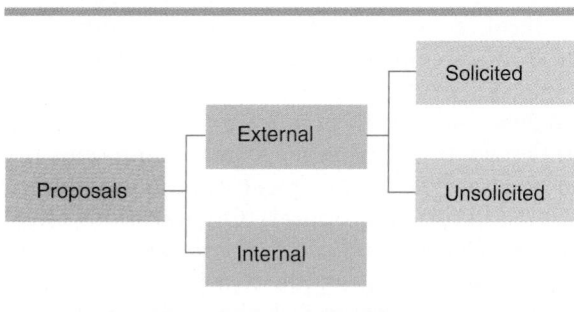

■ Figure 17.1   The Logistics of Proposals

External and Internal Proposals

Proposals are either external (if they are submitted to a reader in another organization) or internal (if they are submitted to a reader in the writer's own organization).

External Proposals

No organization produces all the products or provides all the services it needs. Paper clips and company cars have to be purchased. Offices must be cleaned and maintained. Sometimes projects that require unusual expertise, such as sophisticated market analyses, have to be carried out. Any number of companies would love to provide the paper clips or the cars, and a few dozen consulting organizations would happily conduct the studies. For this reason, it is almost always a buyer's market. To get the best deal, most organizations require that their prospective suppliers compete for the business by submitting proposals, documents created by each supplier to make the case that it deserves the contract.

Internal Proposals

One day, while working on a project in the laboratory, you realize that if you had a new centrifuge you could do your job better and faster. The increased productivity would save your company the cost of the equipment in a few months. You call your supervisor, who tells you to send a memo describing what you want, why you want it, what you're going to do with it, and what it costs; if your request seems reasonable and the money is available, you'll likely get the new centrifuge.

Your memo is an *internal proposal* — a persuasive argument, submitted within an organization, for carrying out an activity that will benefit the organization, generally by increasing productivity or quality or by reducing costs. An internal proposal might recommend that the organization conduct research, purchase a product, or change some aspect of its policies or procedures.

The scope of the proposal determines its format. A simple request might be conveyed orally, either in person or on the phone. A more ambitious request might require an email or a brief memo. The most ambitious requests, however, are generally presented in formal proposals. Organizations often rely on dollar figures to determine which format to use. For instance, employees use a brief form to suggest projects that would cost less than $1,000, whereas they use a formal proposal to suggest projects that would cost more than $1,000.

Solicited and Unsolicited Proposals

External proposals are either solicited or unsolicited. A solicited proposal originates with a request from a customer. An unsolicited proposal originates with the prospective supplier.

Solicited Proposals

When an organization wants to purchase a product or service, it publishes one of two basic kinds of statements:

- An IFB — *information for bid* — is used for standard products. When an agency of the federal government needs office equipment, for instance, it lets suppliers know that it wants to purchase, say, 100 office chairs of a particular type. All other things being equal, the supplier that offers the lowest bid wins the contract.

- An RFP — *request for proposal* — is issued for customized products or services. For example, police cars are likely to differ from the standard consumer model: they might have different engines, cooling systems, suspensions, and upholstery. The police department's RFP might be a long and detailed set of technical specifications. The supplier that can provide the automobile most closely resembling the specifications — at a reasonable price — will probably win the contract. Sometimes the RFP is a more general statement of goals. The customer is, in effect, asking the suppliers to create their own designs or describe how they will achieve the specified goals. The supplier that offers the most persuasive proposal will probably win the contract.

Most organizations issue IFBs and RFPs in newspapers or send them in the mail to past suppliers. Government IFBs and RFPs are published in the journal *Commerce Business Daily*, which is available online. Figure 17.2 on page 486 (CBD-*Net*, 2001) shows a sample entry from that journal.

Unsolicited Proposals

An unsolicited proposal looks like a solicited proposal except that it does not refer to an RFP. Even though the potential customer never formally requested the proposal, in almost all cases the supplier was invited to submit the proposal after people from the two organizations met and discussed the project informally. Because proposals are expensive to write, suppliers are reluctant to submit them without assurances that the potential customer will study them carefully. Thus, the word *unsolicited* is only partially accurate.

External proposals — both solicited and unsolicited — can culminate in contracts of several types: a flat fee for a product or a one-time service; a leasing agreement; or a "cost-plus" contract, under which the supplier is reimbursed for the actual cost plus a profit set at a fixed percentage of the costs.

[Commerce Business Daily: Posted in CBDNet on July 3, 2001]
[Printed Issue Date: July 6, 2001]
From the Commerce Business Daily Online via GPO Access
[cbdnet.access.gpo.gov]

PART: CONTRACT AWARDS
SUBPART: SERVICES
CLASSCOD: R — Professional, Administrative and Management Support
Services
OFFADD: Fleet & Industrial Supply Center, Regional Contracts Dept,
Attn: Bid Officer/Code 210A94, 937 N. Harbor Drive, Suite 212,
San Diego, CA 92132-0212

SUBJECT: R — ON-ORBIT SATELLITE MONITORING SERVICES

POC Contract Negotiator (619) 532-2563
CNT N00244-01-D-0040
AMT $4,465,992.00
LINE 0001-0010
DTD 062101
TO General Dynamics, Government Systems Corporation, 4600 Research Park
Circle, Las Cruces, NM 88001
LINKURL: http://www.neco.navy.mil
LINKDESC: NECO
EMAILADD: glenn_e_brown@sd.fisc.navy.mil
EMAILDESC: Contract Negotiator
CITE: (W-184 SN50Q7T0)

The full description of this RFP appears at this Web address.

■ **Figure 17.2   Extract from CBD*Net***

THE "DELIVERABLES" OF PROPOSALS

When people talk about *deliverables,* they are referring to what the supplier will deliver at the end of the project. Deliverables can be classified into two major categories, as shown in Figure 17.3.

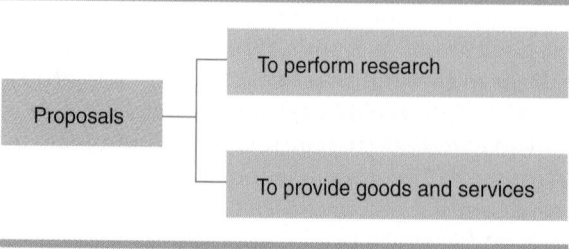

■ **Figure 17.3 "Deliverables" of a Proposal**

Research Proposals

In a research proposal, you are promising to provide a research report of some sort. Here are a few examples:

See Writing Guidelines for Engineering and Science Students (http://fbox.vt .edu:10021/eng/mech /writing/) for sample pro- posals and writing checklists.

> A biologist for a state bureau of land management writes a proposal to the National Science Foundation asking for resources to build a window-lined tunnel in the forest to study tree and plant roots and the growth of fungi. The biologist also wishes to investigate the relationship between plant growth and the activity of insects and worms. The deliverable will be a report sub- mitted to the National Science Foundation and, perhaps, an article published in a professional journal.

> A manager of the technical-publications department at a manufacturing company writes a proposal to her supervisor asking for resources to study whether the company should convert its internal documents from a paper to an electronic format. The deliverable will be a report that contains her recom- mendations.

Also see Online Technical Writing (www.io.com /~hcexres/tcm1603 /acchtml/acctoc.html) for sample proposals.

> A university sociologist writes to his state board of education proposing to study a nearby community of migrant workers to determine how the com- munity uses municipal services and how it regards the providers of those services. The deliverable will be a report, submitted to the state board of education and the municipal government, that presents his findings. If the findings are of interest to people beyond his immediate geographical area, he will also write a journal article.

A research proposal often leads to two other kinds of documents: progress reports and completion reports.

After the proposal has been approved and the researchers have begun work, they often submit one or more *progress reports*. A progress report tells the sponsor of the project how the work is proceeding. Is it following the plan of work outlined in the proposal? Is it going according to schedule? Is it stay- ing within budget?

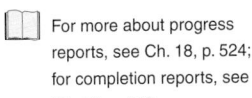

For more about progress reports, see Ch. 18, p. 524; for completion reports, see Ch. 19, p. 540.

At the end of the project, researchers prepare a *completion report*, often called a *final report*, a *project report*, or simply a *report*. A completion report tells the readers the whole story of the research project, beginning with the prob- lem or opportunity that motivated it, the methods that the researchers used in carrying out the project, and the important results, conclusions, and recommendations.

People undertake research projects to satisfy their curiosity and to ad- vance professionally. Organizations often require that their professional em- ployees carry out research and publish in appropriate reports, journals, or books. Government researchers and university professors, for instance, are expected to remain active in their fields. Writing proposals is one way to get the resources — time and money for travel, equipment, and assistants — to carry out the research.

Goods-and-Services Proposals

Whereas a research proposal leads to a report of some kind, a goods-and-services proposal leads to a tangible product (a fleet of automobiles), a service (building maintenance), or some combination of the two (the construction of a building).

A vast network of goods-and-services contracts spans the working world. The U.S. government, the world's biggest customer, spent more than $48 billion in 2000 buying military equipment from organizations that submitted proposals (U.S. Department of Commerce, 2000, p. 360). But goods-and-services contracts are by no means limited to government contractors. One auto manufacturer buys engines from another, and a company that makes spark plugs buys its steel from another company. In fact, most products and services are purchased by contract. The world of work depends on goods-and-services proposals.

PERSUASION AND PROPOSALS

Regardless of whether the supplier is a professor applying for a research grant, an employee proposing a project at the office, or an electronics company seeking a government contract to build a radar device for a new jet aircraft, the proposals will be analyzed carefully and skeptically.

The agency reviewing the professor's proposal wants to see that the professor understands the important research questions pertaining to the subject, presents a feasible plan for carrying out the project, and has a good track record. Because many other professors will be competing for the same grant, the agency wants to make sure it is spending its money wisely.

The supervisor of the employee writing the internal proposal likewise wants to be satisfied that the employee has isolated a real problem and devised a feasible strategy for solving it, and that the employee has a good record of carrying through on similar projects.

The government officials reviewing the radar-device proposals want to be satisfied that the supplier will live up to its promise: to build, on schedule, the best radar device at the best price. With perhaps a dozen suppliers competing for the contract, government officials know only that many companies want the work; they can never be sure — not even after the contract has been awarded — that they have made the best choice.

See Ch. 6 for more on persuasion.

A proposal, then, is an argument. To be successful, it must be persuasive. The writers must convince the readers that the future benefits will outweigh the immediate and projected costs. Basically, proposal writers must clearly demonstrate that they

- understand the readers' needs
- are able to fulfill their own promises
- are committed to fulfilling their own promises

Understanding the Readers' Needs

The most crucial element of the proposal is the definition of the problem or opportunity to which the proposed project is intended to respond. This would seem to be mere common sense: you can't expect to write a successful proposal if you don't show that you understand the readers' needs. Yet people who evaluate proposals — whether government readers, private foundation officials, or managers in small corporations — agree that an inadequate or inaccurate understanding of the problem or opportunity is the most common weakness of the proposals they see.

Readers' Needs in an External Proposal

Sometimes the RFP fails to convey the problem or opportunity clearly. More often, however, the suppliers submitting the proposal are at fault: they might not have read the RFP carefully and simply assumed that they understood the client's needs, or perhaps, knowing they couldn't satisfy a client's needs, they have nonetheless prepared a proposal detailing a project they could complete, hoping that company managers won't notice or that no other supplier will come any closer to responding to the real problem. Suppliers can easily find it opportune to concentrate on what they want to do rather than on what the customer needs.

But most readers will toss a proposal aside as soon as they realize that it doesn't address their needs. When you receive an RFP, study it thoroughly. If you don't understand something in it, contact the organization that issued it. They will be happy to clarify it, because a bad proposal wastes everyone's time.

When you write an unsolicited proposal, analyze your audience carefully. How can you define the problem or opportunity so that your readers will understand it? Keep in mind their needs (even if they are oblivious to them) and, if possible, the readers' backgrounds. Concentrate on how the problem has decreased productivity or quality or on how your ideas would create new opportunities. When you submit an unsolicited proposal, your task in many cases is to convince readers that a need exists. Even when you have reached an understanding with some of your customer's representatives, your proposal will still have to persuade other officials in the company.

Readers' Needs in an Internal Proposal

An internal proposal also must respond to readers' needs. If you propose hiring a new person, you have to make the case that the person is needed and would save or bring in more money than he or she costs. In addition, you have to make sure there is a place for the new person in your current facilities. Writing an internal proposal is both more simple and more complicated than writing an external proposal. It is simpler because you have more access to your readers than you would to external readers. And you can get more information more easily.

However, you might find it more difficult to get an accurate sense of the situation in your organization. Some colleagues might not be willing to tell you directly that your proposal is unlikely to be approved. Another danger is that in identifying a problem, you are often criticizing, directly or indirectly, the person in your organization who instituted the system that needs revising or who failed to take action earlier. Therefore, before you write an internal proposal, it is smart to discuss your ideas thoroughly with as many potential readers as possible. This way, you will more likely find out what the organization really thinks of your idea before you spend time committing it to paper.

Describing What You Plan to Do

Once you have shown that you understand what needs to be done and why, describe what you plan to do. Convince your readers that you can respond to the situation you have just described. Discuss your approach to the subject: indicate the procedures and equipment you would use. If appropriate, justify your choices. For example, if you say you want to do ultrasonic testing on a structure, explain why, unless the reason is obvious.

Present a complete picture of what you would do from the first day of the project to the last. Many inexperienced proposal writers believe they need only convince the reader of their enthusiasm and good faith. Unfortunately, most readers want to see a detailed plan showing that the writer has actually started to do the work.

Of course, no proposal can anticipate and answer every question about what you plan to do. The more planning you have done before you submit the proposal, however, the greater are the chances you will be able to do the work successfully if you get the go-ahead. Providing a complete discussion of your plan suggests to your readers that you are interested in the project itself, not just in winning the contract or in receiving authorization to do the project.

Demonstrating Your Professionalism

Once you have shown that you understand the readers' needs and can offer a well-conceived plan, demonstrate that you are the kind of person — or that yours is the kind of organization — that is committed to delivering what is promised. Many other people or organizations could probably carry out the project. You want to convince readers that you have the pride, ingenuity, and perseverance to solve the problems that inevitably occur in any big undertaking. In short, you want to show that you are a professional.

GUIDELINES

Demonstrating Your Professionalism in a Proposal

In your proposal you can demonstrate your ability to carry out a project by providing four kinds of information:

▶ *Credentials and work history.* Make the case that you know how to make this project work because you have made similar projects work. Who are the people in your organization with the qualifications and experience to carry out the project? What equipment and facilities do you have that will enable you to do the work? What management structure will you use to maintain coordination and keep different activities running smoothly? What similar projects have you completed successfully?

▶ *Work schedule.* Sometimes called a *task schedule,* this schedule, which usually takes the form of a graph or chart, shows when the various phases of the project will be carried out. In one sense, the work schedule is a straightforward piece of information that enables your readers to see how you would apportion your time. But it also reveals more about your attitudes toward your work than about what you will actually be doing on any given day. Events rarely proceed according to plan: some tasks take more time than anticipated, others take less. A careful, detailed work schedule is one way of showing that you have done your homework, that you have attempted to foresee the kinds of problems that might threaten the project.

▶ *Quality-control measures.* Highlight the procedures you have established to evaluate the effectiveness and efficiency of your work on the project. Quality-control procedures might consist of technical evaluations carried out periodically by the project staff. Sometimes the writer will build into the proposal provisions for on-site evaluation by recognized authorities in the field or by representatives of the potential client. Quality control is also measured by progress reports.

▶ *Budget.* Most proposals conclude with a budget, a formal statement of how much the project will cost. A carefully prepared budget is another way of showing that you have done your homework on a project.

WRITING A PROPOSAL

In writing a proposal, you use the same basic techniques of planning, drafting, and revising that you use in any other kind of writing. However, a proposal can be such a big project that two aspects of the writing process — resource planning and collaboration — assume greater importance than they do in smaller documents.

 See Alice Reid's "A Practical Guide for Writing Proposals" (www.members.dca.net /areid/proposal.htm) for a proposal-writing checklist.

As discussed in Chapter 5, planning a project requires a lot of work. You need to see whether your organization can devote resources to writing the proposal and then to carrying out the project if the proposal is successful. Sometimes an organization writes a proposal, wins the contract, and then loses money because it doesn't have the resources to do the project and must subcontract major portions of it.

The resources you need fall into three basic categories:

- *Personnel.* Will the necessary technical personnel, managers, and support people be available?

- *Facilities.* Do you have the facilities to carry out the research and production, or can they be leased? Can you profitably subcontract portions of the job to companies that have the appropriate facilities?

- *Equipment.* Do you have the right equipment? If not, can you buy it or lease it or subcontract the work? Some contracts provide for the purchase of equipment, but others don't.

Don't write the proposal unless you are confident that you can carry out the project if the proposal is successful.

Collaboration is critical in most proposals of more than a few pages because no one person has the time and expertise to do all the work. Writing major proposals calls for the expertise of technical personnel, writers, editors, graphic artists, managers, lawyers, and document-production specialists.

Usually, a project manager coordinates the writing of a large proposal. In most cases, the parts of the proposal are written at different times by different people. These components might not come together into the final package until a few days — or a few hours — before the proposal deadline.

For more on collaboration, see Ch. 4.

THE STRUCTURE OF THE PROPOSAL

For the proposal guidelines of the Society for Human Resource Management, see www.shrm.org/foundation /99guidelines.htm.

Most proposals follow a basic structural pattern. If the authorizing agency provides an IFB, an RFP, or a set of guidelines, follow it to the letter. If guidelines have not been supplied, or if you are writing an unsolicited proposal, use the conventional structure shown here as a starting point. Then modify it according to your subject, your purpose, and the needs of your audience.

Structure of a proposal:
Summary
Introduction
Proposed Program
Qualifications and experience
Budget
Appendices

Summary

For any proposal of more than a few pages, provide a summary. Many organizations impose a length limit — for example, 250 words — and ask the writer to present the summary, single spaced, on the title page. The summary is crucial, because in many cases it will be the only item readers study in their initial review of the proposal.

The summary covers the major elements of the proposal but devotes only a few sentences to each. To write an effective summary, first define the problem in a sentence or two. Next describe the proposed program. Then provide a brief statement of your qualifications and experience. Some organizations wish to see the completion date and the final budget figure in the summary; others prefer that this information be presented separately on the title page along with other identifying information about the supplier and the proposed project.

For more about summaries, see Ch. 12, p. 316.

Figure 17.4 shows an effective summary taken from a proposal submitted by a group of three students (Wells, Tommack, & Tuck, 1997). (An additional example of a summary appears at the end of this chapter, page 505.)

---

page 2

Summary

This proposal recommends that we be authorized to use our time and the Student Special Service's resources to research the possibility of classifying Central State University's American Sign Language classes as a foreign language in the Area I core group. We would then write a recommendation on whether American Sign Language should be considered a foreign language and how we could incorporate it.

The background

Despite the widespread use of sign language in America, Central State University does not give American Sign Language (ASL) any special place among its offered courses. Currently, the ASL courses are only elective credits. Almost half of the states have recognized ASL as a foreign language and grant academic credit for completion of these courses. Six states have universities that independently recognize ASL as a foreign language, while 18 more have passed legislation on the state level recognizing ASL. Central State University is one of 15 states that grant only elective credit to ASL courses.

The research the writer has already done

ASL itself is gaining prominence in mainstream culture and is also being recognized as a foreign language in the educational arena. The CSU English Department, for example, does grant ASL foreign-language status for linguistics majors, but in other departments, ASL is excluded. We propose to research the standards held by Central State University in deeming a language foreign. Our purpose is to determine whether ASL fulfills those requirements. In addition, we propose to research the legislative measures passed in the state of Texas that give ASL statewide recognition as a foreign language. If ASL fulfills CSU's requirements, we will submit a recommendation explaining how it can be implemented. As a safeguard and in the interest of Oklahoma education in general, we propose to investigate the legislative measures needed to attain formal recognition in the same manner as was achieved by the state of Texas.

The proposal

The writers' credentials and the schedule

We have several contacts in the deaf and educational communities in Oklahoma and Texas. With their cooperation we can begin on April 25 and have our research completed and our recommendations prepared by May 12. At that time, we will submit a completion report for your review.

■ **Figure 17.4**
Summary of a Proposal

Note that proposals are often double spaced if they are presented as reports, not memos.

Introduction

The purpose of the introduction is to help the reader understand the context, scope, and organization of the proposal.

GUIDELINES

Introducing a Proposal

Make sure you answer the following seven questions in the introduction to your proposal:

▸ *What is the problem or opportunity?* Be specific. Whenever you can, quantify. Describe the problem or opportunity in monetary terms, because the proposal itself will include a budget of some sort and you want to convince your readers that spending money on what you propose is smart. Don't say that a design problem is slowing down production; say that it is costing $4,500 a day in lost productivity.

▸ *What is the purpose of the proposal?* Even though it might seem obvious to you, the purpose of the proposal is to describe a problem or opportunity and propose a course of action that will culminate in some deliverable. Be specific in explaining what you want to do.

▸ *What is the background of the problem or opportunity?* In answering this question, you probably will not be telling your readers anything they don't already know (except, perhaps, if your proposal is unsolicited). Your goal here is to show them that you understand the problem or opportunity: the circumstances that led to its discovery, the relationships or events that will affect the problem and its solution, and so on.

▸ *What are your sources of information?* Review the relevant literature, either internal reports and memos or external published articles or even books, so that your readers will understand the context of your work.

▸ *What is the scope of the proposal?* If appropriate, indicate what you are proposing to do as well as what you are not proposing to do.

▸ *What is the organization of the proposal?* Indicate the organizational pattern you will use in the proposal.

▸ *What are the key terms that will be used in the proposal?* If you will use any new, specialized, or unusual terms, the introduction is an appropriate place to define them.

Figure 17.5 is the introduction to the proposal about American Sign Language.

page 3

Introduction

Sign language is the fourth most prominent language in the United States. In usage, it ranks below Spanish but above French, German, Russian, and Japanese, which together constitute all the foreign languages taught at CSU. But despite its prominence, Central State University offers few classes in ASL. With a student population of approximately 15,000, only 25 students per semester are given the opportunity to begin study in American Sign Language. These courses offered are not foreign-language credits. We want to research whether this course can or should be included.

Teresa O'Malley, Chair of the Modern Languages Department at CSU, said that her office receives several calls per semester regarding the inclusion of ASL in the core group of foreign languages. She is supportive of ASL's inclusion and cited other state-funded institutions that recognize ASL as a foreign language. The states have various approaches to ASL in the classroom, as shown in Figure 1. The University of Arizona and the University of Delaware, for example, are part of the 12% that have state-funded institutions that independently recognize ASL as a foreign language. An additional 36% of the states have passed legislative bills that formally recognize ASL as a foreign language, such as Texas. CSU is part of the 30% of states that recognize ASL as an elective credit but not a foreign language. Roughly one-half of the state-funded institutions in the United States recognize ASL as a foreign language. CSU does not.

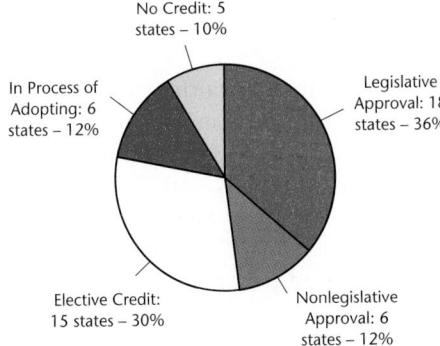

Source: Rhonda Jacobs, *ASL as a Truly Foreign Language,* 1996.

Figure 1: Treatment of ASL in the 50 States

Current research strongly supports ASL's inclusion in the foreign-language category. Both Teresa O'Malley and Suzanne Christopher, the ASL instructor on campus, feel that ASL probably fits into the foreign-language category. Blaine Lee, the Student Special Services Coordinator, and June Yunker, the CSU Interpreter Coordinator, both feel that ASL deserves

The writers describe the problem they wish to study and note that it has both a local and a national dimension.

Some of the primary and secondary research the writers have already done

■ **Figure 17.5**
Introduction to a Proposal

Note that proposals are double spaced if they are presented as reports, not memos.

page 4

foreign-language status. The English Department at Central State does accept Suzanne Christopher's upper-level ASL courses for foreign-language credit in fulfillment of its linguistics-emphasis language requirement. However, this recognition doesn't apply elsewhere on campus. If the English Department feels ASL is foreign enough to be given formal academic credit, then ASL may be a good candidate for formal recognition. Suzanne provided us research by Jacobs (1993) that explores the characteristics of ASL, supporting its qualification as a foreign language, and the various methods of achieving this recognition. In addition, Armstrong (1988) presents strong legal arguments supporting this recognition.

The purpose of this proposal is to request resources to research whether ASL is a foreign language by CSU standards and whether we should petition the university and/or the Oklahoma state legislature to recognize ASL. However, we will not look into significantly expanding the existing ASL curriculum.

In the following sections, we will outline the proposed procedure for our research, our qualifications to do the research, and the budget needed to ensure an effective evaluation of our possibilities.

An advance organizer for the rest of the proposal

■ **Figure 17.5**
(Continued)

Proposed Program

Once you have defined the problem or opportunity, say what you want to do about it. The proposed program demonstrates clearly how much work you have already done. Be specific. You won't persuade anyone by saying that you plan to "gather the data and analyze it." How will you gather the data? What techniques will you use to analyze it? Every word you say — or don't say — will give your readers evidence on which to base their decision. If you know your subject, the proposed program will show it. If you don't, you will inevitably slip into meaningless generalities or include erroneous information that undermines the whole proposal.

If your project concerns a subject written about in the professional literature, show your familiarity with the scholarship by referring to the pertinent studies. However, don't just toss a bunch of references onto the page. For example, don't write, "Carruthers (1996), Harding (1997), and Vega (1996) have all researched the relationship between acid-rain levels and groundwater contamination." Rather, use the recent literature to sketch the necessary background and provide the justification for your proposed program. For instance:

> Carruthers (1996), Harding (1997), and Vega (1996) have demonstrated the relationship between acid-rain levels and groundwater contamination. None of these studies, however, included an analysis of the long-term contamination of the aquifer. The current study will consist of . . .

You might include only one reference to recent research. However if you have researched your topic thoroughly, you might devote several paragraphs or even several pages to recent scholarship.

 For a sample literature review, see Writing Guidelines for Engineering and Science Students (http://fbox.vt .edu:10021/eng/mech /writing/workbooks/proposal .request.html#request).

For more on researching a subject, see Ch. 7.

Whether your project calls for primary research, secondary research, or both, the important point is that the proposal will be unpersuasive if you haven't already done a substantial amount of the research. For instance, say you are writing a proposal to do research on industrial-grade lawn mowers. You are not being persuasive if you write that you are going to visit Sears, JC Penney, and Home Depot to see what kinds of lawn mowers they carry. This statement is unpersuasive for two reasons:

- You need to justify why you are going to visit those three retailers rather than others. Anticipate your readers' questions: Why did you choose these three retailers? Why didn't you choose more specialized dealers?
- You should already have visited the appropriate stores and completed any other preliminary research. If you haven't done the homework, readers have no assurance that you will in fact do it or that it will pay off. If your supervisor authorizes the project and then learns that none of the lawn mowers on the market meets your organization's needs, you will have to go back and submit a different proposal — an embarrassing move.

Unless you can show in your proposed program that you have done the research — and that the research indicates that the project is likely to succeed — the reader has no reason to authorize the project.

Figure 17.6 on pages 498–499 shows the proposed program for the American Sign Language project.

Qualifications and Experience

After you have described how you would carry out the project, show that you can do it. Unless you convince your readers that you can turn an idea into action, your proposal will be unpersuasive.

The more elaborate the proposal, the more substantial the discussion of your qualifications and experience has to be. For a small project, a few paragraphs describing your technical credentials and those of your co-workers will usually suffice. For larger projects, the résumés of the project leader — often called the *principal investigator* — and the other important participants should be included.

External proposals should also discuss the qualifications of the supplier's organization, outlining similar projects the supplier has completed successfully. For example, a company bidding for a contract to build a large suspension bridge should describe other suspension bridges it has built. It should also focus on the equipment and facilities the company already possesses and on the management structure that will ensure the project's successful completion. Although everyone knows that young, inexperienced persons and new firms can do excellent work, when it comes to proposals, experience wins out almost every time.

Proposed Procedure

We will perform the following tasks for your review in determining whether Central State University should recognize ASL as foreign language and how this recognition could be achieved at Central State University (by a university petition) and at all Oklahoma post-secondary institutions (by state legislation):

An advance organizer for this section of the proposal

1. compile a summary of the criteria required by Central State University for a language to be considered foreign
2. research whether ASL fulfills the university-mandated requirements
3. investigate the broad approach of state legislative recognition as used in the state of Texas
4. provide a completion report detailing our findings and recommendations

Describing the proposed program in terms of clear tasks makes the proposal look credible.

Task 1. Compile a summary of the criteria required by Central State University for a language to be considered foreign

We interviewed Teresa O'Malley, Chair of the Modern Languages Department. She did not know the exact criteria for languages to be considered foreign at CSU, but she did know that only languages taught at the intermediate level counted as foreign-language credits. At present, CSU has two ASL courses that could qualify as intermediate courses: CM221: Intermediate American Sign Language, and CM321: Conversational American Sign Language. All beginner-level courses taught are counted as elective credits, including CM121: Beginning Sign Language. She can provide us the exact specifications for a class to count as a foreign-language credit at the end of this week. We request time on Friday morning, April 25, to visit with Teresa O'Malley to discuss the exact qualifications for a language to be considered foreign.

Task 2. Research whether ASL fulfills the university-mandated requirements

We have also met with Suzanne Christopher, the current ASL instructor at Central State. She has done considerable research in American Sign Language and has prepared various proposals to expand the ASL program at the university. Currently, we have several works by Armstrong (1988), Wilcox (1977), and Jacobs (1996) that will assist us in understanding the current status of ASL in university curricula and in making our recommendation. In addition, we have contacted the Oklahoma chapter of the Registry for Interpreters for the Deaf (ORID) and the Oklahoma Association for the Deaf (OAD). Kelly Eastwick, President of ORID, and Janis Seymour, committee member of OAD, are both interested and willing to assist us in our research and deliberations. After we learn the exact criteria for a language to be considered foreign, we request time to meet with Suzanne, Kelly, and Janis to compile our research concerning ASL's qualifications. We request a four-hour reservation of the Centennial Room in the Student Union Building on Monday morning, April 28, to meet with Suzanne Christopher, Kelly Eastwick, and Janis Seymour to prepare our recommendation. After our analysis of the university's requirements and ASL's qualifications, we will submit to you a progress report detailing whether an internal approach to approve ASL is viable.

A further discussion of the research the writers will complete

■ **Figure 17.6**
Proposed Program of a Proposal

Task 3. Investigate the broad approach of state legislative recognition as used in the state of Texas

Regardless of whether ASL as a foreign language is approved at Central State University, we would like to investigate the measures needed to have the state of Oklahoma extend legal recognition of the language. As employees of the state, we see CSU as only one of the several postsecondary institutions in Oklahoma. If ASL is deserving of this status, we wish to extend these benefits to other universities by investigating the feasibility of passing a bill recognizing ASL as a foreign language. If ASL as a foreign language is approved by CSU, it will strengthen our petition to the state. If internal approval is not feasible, petitioning the state will give us one more opportunity. We hope to work on the internal and external research simultaneously.

The alternative method of granting ASL foreign-language recognition is to have a bill passed in the state legislature. According to Figure 1 on page 3, 18 states have legislative approval for ASL to be considered a foreign language in postsecondary schools. Texas is one of these states. We have contacted Sha H. Cowan from the Texas Education Agency, Services for the Deaf Department. He referred us to three educators involved with the legislative measures passed in Texas: Dr. Jean Andrews, Communication Disorders, Lamar University in Beaumont, Texas, and Carol Seeger and Lisa Bissin, deaf ASL instructors at the University of Texas at Austin. We request Friday afternoon, April 25, and Monday afternoon, April 28, to communicate with these individuals and gather research. In addition we request Tuesday, April 29, to follow up with the legislative proposal plan, analyze a feasible Oklahoma legislature approach, contact Oklahoma state representatives, compile our research, and prepare a completion report.

Task 4. Provide a completion report detailing our findings and recommendations

Following the completion of our research, we will submit to you a report summarizing our findings and recommendations. We will explain the criteria required by CSU for a language to be considered foreign, whether ASL fulfills these requirements, and how we could implement a legislative approach to obtain formal recognition for ASL. In this last area, we will report on the strategies used in the state of Texas and provide a list of contact personnel in the Oklahoma state government with whom we will work. This completion report will be on your desk May 12 at 9 A.M.

Figure 2 is our estimated task schedule, beginning on April 25 and ending with our recommendation to you on May 12.

Visit Teresa O'Malley	
Meet with ORID and OAD	
Prepare progress report	
Correspond with Texas	
Prepare completion report	

25 26 27 28 29 30 1 2 3 4 5 6 7 8 9 10 11 12
April   May

Figure 2: Task Schedule

Because the report is the main deliverable, writing the report is one of the tasks.

The writers present their schedule as part of the proposed program. Such schedules are often presented as a separate element in the proposal. See p. 501 for more on task schedules.

■ **Figure 17.6**
(Continued)

page 7

Qualifications and Experience

We believe we have the skills and experience to conduct this research. Each of us brings to the team talents and strengths that, when combined, help us to work effectively and efficiently.

Danny Tommack, our team leader, has the skills to direct the team's efforts and to help us keep our goals in focus. He has developed strong management and communication skills while working at McU Sports. His professional approach in meetings and planning sessions will help us work closely with members of our community and our long-distance contacts in Texas.

Brandon Tuck has been an American Sign Language interpreter for three years. He works for the Department of Student Special Services at Central State University and is a freelance interpreter. He brings to the group a strong knowledge of deafness and sign language, and contacts in the deaf community.

Megan Wells has highly developed research skills she obtained while working for three years in a research laboratory. She has excellent communication and interviewing skills. Her experience at several universities will help us to investigate Central State University and other postsecondary institutions.

■ **Figure 17.7
Qualifications and
Experience Section of a
Proposal**

*The writers describe the
qualifications of each group
member.*

Figure 17.7 shows the qualifications-and-experience section of the ASL proposal.

Budget

Good ideas aren't good unless they're affordable. The budget section of a proposal specifies how much the proposed program will cost.

Budgets vary greatly in scope and format. For simple internal proposals, the writer adds the budget request to the statement of the proposed program: "This study will take me about two days, at a cost of about $400" or "The variable-speed recorder currently costs $225, with a 10 percent discount on orders of five or more." For more complicated internal proposals and for all external proposals, a more explicit and complete budget is usually required.

Most budgets are divided into two parts: direct costs and indirect costs.

- *Direct costs* include such expenses as salaries and fringe benefits of program personnel, travel costs, and necessary equipment, materials, and supplies.

- *Indirect costs* cover the intangible expenses that are sometimes called *overhead*. General secretarial and clerical expenses not devoted exclusively to any one project are part of the overhead, as are other operating expenses such as utilities and maintenance. Indirect costs are usually expressed as a percentage — ranging from less than 20 percent to more than 100 percent — of the direct expenses. In many external proposals, the client imposes a limit on the percentage of indirect costs.

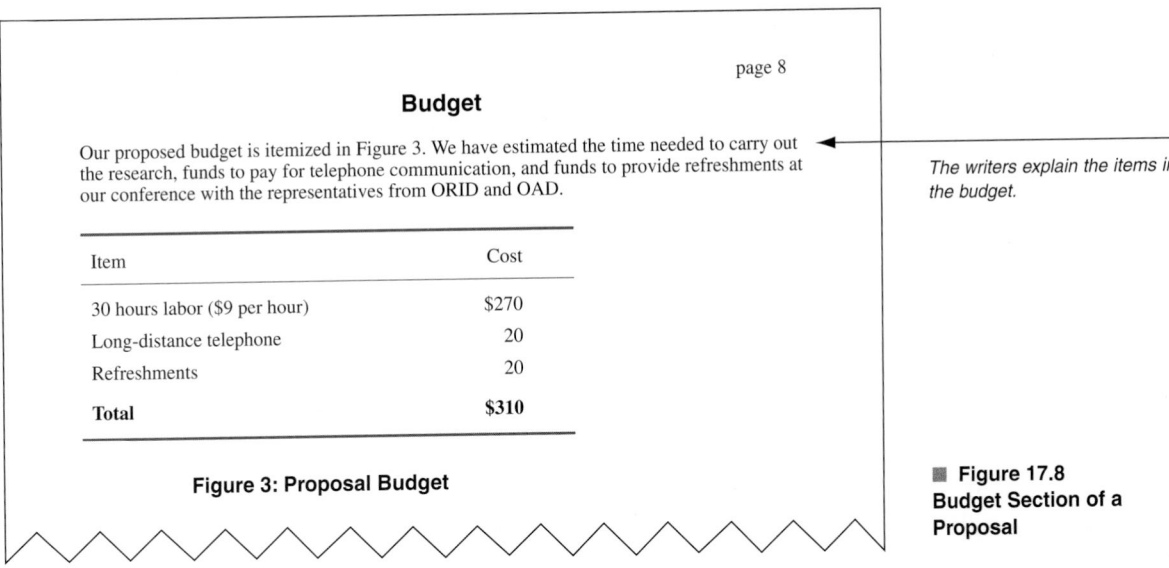

Within the figure:

page 8

Budget

Our proposed budget is itemized in Figure 3. We have estimated the time needed to carry out the research, funds to pay for telephone communication, and funds to provide refreshments at our conference with the representatives from ORID and OAD.

Item	Cost
30 hours labor ($9 per hour)	$270
Long-distance telephone	20
Refreshments	20
Total	**$310**

Figure 3: Proposal Budget

The writers explain the items in the budget.

■ **Figure 17.8 Budget Section of a Proposal**

Figure 17.8 shows the budget section from the ASL proposal.

Appendices

Many types of appendices might accompany a proposal. Most organizations have *boilerplate* descriptions (standard wording that a writer can modify or insert directly into a document) of other projects they have performed. Another popular kind is the supporting letter — a testimonial to the supplier's skill and integrity written by a reputable and well-known person in the field. Two other kinds of appendices deserve special mention: the task schedule and the description of evaluation techniques.

The *task schedule* is almost always drawn in one of three graphical formats:

- table
- bar chart
- network diagram

Tables

The simplest but least informative way to present a schedule is in a table, as shown in Figure 17.9 on page 502. As with all graphics, you should provide a textual reference that introduces and, if necessary, explains it.

Although displaying information in a table is better than writing it out in sentences, readers still cannot "see" the information. They have to read it to figure out how long each activity will last, and they cannot tell whether any of the activities are interdependent. They have no way of determining what would happen to the overall project schedule if one of the activities faced delays.

TASK SCHEDULE

Activity	Start Date	Finish Date
Design the security system	4 Oct. 00	19 Oct. 00
Research available systems	4 Oct. 00	3 Jan. 01
Etc.		

■ **Figure 17.9 Tabular Schedule**

Bar Charts

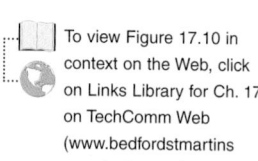

To view Figure 17.10 in context on the Web, click on Links Library for Ch. 17 on TechComm Web (www.bedfordstmartins .com/techcomm).

Bar charts, also called *Gantt charts* after the early twentieth-century American civil engineer who first used them, are a distinct improvement over tables. The basic bar chart shown in Figure 17.10 (SmartDraw, 2001) allows readers to see how long each activity will take and when different activities occur simultaneously. Like tables, however, bar charts do not indicate the interdependency of tasks.

Network Diagrams

There are many types of network diagrams, but all of them show interdependence among various activities, clearly indicating which must be completed before others can begin.

SOFTWARE DEVELOPMENT								
TASKS	November	December	January	February	March	April	May	June
Needs analysis								
Preliminary design								
Project approval			January 5 ●					
GUI designed and created								
Programming								
Alpha testing								
Documentation								
Beta testing								
Final QA analysis								
Web site design								
Release								June 18 ●

■ **Figure 17.10 Bar Chart**

ASI   *Ada Semiconductor Incorporated*
220 Technical Street   Telephone: (208) 555-2200
Boise, Idaho 83706   Fax: (208) 555-2201

www.asisemi.com

MEMORANDUM

DATE: March 4, 1999
TO: Mr. Michael Jensen, Ada Semiconductor, Inc. Manager
           Ms. Shelly McKay, Ada Semiconductor, Inc. Sales and Marketing Manager
           Mr. Brandon Wells, Ada Semiconductor, Inc. Software Support Manager
FROM: Ada Semiconductor Ad Hoc Web Committee:
              Julie Grapatin — Finance Department Rep.
              Tammy T. Moon — Engineering Department Rep.
              Christine Johnson — Software Department Rep.
SUBJECT: Proposal for a Feasibility Study for Hiring a Webmaster

Purpose

We request that Ada Semiconductor, Inc. (ASI), provide the resources to research the feasibility of hiring a Webmaster to create and maintain ASI Web pages.

The purpose of this document

Summary

In 1997, ASI began providing product information, such as data sheets, product selector guides, and technical notes, on the Web. The Web pages were created and maintained by full-time, salaried employees during normal business hours. The area employee with the most HTML expertise was responsible for that area's Web pages; this method required little time and effort.

The background of the problem

However, over the past year the company has grown to two hundred full-time employees and has tripled the number of products being sold. The amount of ASI information posted on the Web has increased by 500 percent. A number of ASI employees are spending a great deal of their time maintaining these pages. Our preliminary research shows that ASI is spending $5092 per month on labor costs for current employees to create and maintain the company Web pages.

The first aspect of the problem: the cost of the current method

And, although some of the pages are very professional and contain accurate information (Attachment 1), most are unprofessional and hard to use. A number of the pages routinely contain inaccurate or hard-to-use information (Attachment 2). Several of our customers have expressed disappointment in the ASI Web pages.

The second aspect of the problem: the quality of the Web pages

ASI's customer base is continuing to grow at a steady pace. With the addition of customers such as IBM, Dell, and Compaq, and new product introductions every quarter, the amount of information that we provide our customers will surely continue to grow.

page 2

A summary of the proposed program

We propose assessing whether all the pages on the Web site are justified, determining whether the ASI site will continue to grow, then researching the following three options:

- contract a Webmaster from an employment service to create and maintain ASI Web pages
- hire a professional, full-time Webmaster as an ASI employee
- hire a college intern to maintain ASI Web pages

This comment gives the proposal credibility.

At the Feb. 2, 1999, ASI staff meeting, Michael Jensen requested this proposal, stating that he felt this might be a good opportunity to research the possibility of hiring a professional Webmaster. The proposal will be presented at the next ASI Management meeting, to be held March 14, 1999.

A summary of the budget and credentials of the writers

Performing the research and presenting the findings in a report to management will require approximately 20 hours over four weeks, for a total cost of $414. Our committee includes three ASI employees, representing the Finance, Engineering, and Software departments. We all have experience creating and maintaining Web pages, as well as other qualifications listed in "Qualifications and Experience," below.

The completion date for the report

If you accept our proposal, we will submit the completion report to you by May 12, 1999. Our completion report will include details regarding our research and a recommendation for the best use of ASI resources to maintain the ASI Web pages.

Introduction

ASI provides valuable data to its external customers via the Internet. We would like to research the best way to continue to provide this information. The following information provides background on this proposed project.

The question-and-answer format is a convenient way to sketch in the background and the problem. Note that the writers provide specific information about the cost of the problem. Note too that they refer to attachments containing more details.

What does it cost ASI to maintain the ASI Web pages?

ASI is currently spending $5092 per month on creating and maintaining our Web pages. As ASI's customer base and product lines expand, the cost of creating and maintaining our Web pages will likely continue to increase. While the cost of maintaining the ASI Web pages is not a problem in itself, the 142 hours of lost production time is a problem (Attachment 3). Based on our preliminary research, we feel that the current system is not the best use of company resources. We think that a trained Webmaster could create and maintain the site, at a lower cost than the current system (Buck Consultants Inc., 1998).

What type of corporate image is ASI projecting with its Web pages?

According to a recent Bay Networks White Paper (1999), "The World Wide Web has emerged as a new paradigm in information access and display, becoming the preferred method for accessing corporate data over the enterprise intranet." ASI conforms to this pattern by providing valuable data to its internal and external customers via the Web. However, we are not presenting a positive image of ASI with these data. Our pages are inconsistent and unprofessional in appearance.

Evidence of the importance of the problem comes from commentary about Web usage and anecdotes about customers who have had problems using the company's Web site.

We are beginning to receive questions and complaints from important customers. For instance, Jeremy White, a buyer from Dell Computers, recently contacted us to report inaccurate information on the SDRAMs that he was interested in ordering. While looking

page 3

at our 128K DRAM data sheet provided on the Web, Claudia Brown, a buyer from Lucent, could not find the voltage limitations (due to the unorganized page format), and became quite frustrated.

Business income could suffer as the result of the problems with some of our Web pages. If our customers cannot find accurate, well-organized information quickly, they may not renew their contracts with us. It would seem to be in the best interest of ASI management to recognize the importance of creating and maintaining professional Web pages. In addition, ASI may gain new customers based on the information we make available on the Web.

What can we do to improve the current system of creating and maintaining the ASI Web pages?
All parties involved agree that the present system is less than fully successful. The ASI employees who maintain the Web pages are most dissatisfied with the current approach because it takes them away from their normal job responsibilities. ASI can reduce the problems with the present system by hiring a full-time Webmaster.

The writers start to describe the proposal itself.

We propose to research options for improving our Web site. Our research would determine whether ASI should continue using current ASI employees to update our site, hire a full-time Webmaster as an ASI employee, contract a Webmaster through a temporary employment office, or hire an intern to provide these services. The following sections of this memorandum include our proposed program, the qualifications and experience of our Ad Hoc Web Committee, and our proposed budget for the project.

Proposed Program
We will perform the following tasks to determine the best method for ASI to maintain its Web pages.

Task 1: Compile a list of ASI Web pages that are currently maintained, including a summary of the information provided on the pages, then determine whether the site is likely to grow.

Task 2: Research the feasibility of contracting a professional Webmaster through a temporary agency to maintain ASI Web pages.

This list functions as an advance organizer for the more detailed discussion that follows.

Task 3: Research the feasibility of hiring a professional Webmaster to work as a full-time ASI employee.

Task 4: Research the feasibility of hiring an intern to create and maintain the ASI Web pages.

Task 5: Write a completion report detailing our findings and recommendations.

Task 1: Compile a list of ASI pages that are currently maintained, including a summary of the information provided on the pages, then determine whether the site is likely to grow.
ASI employees are currently creating and maintaining 20 Web pages. These documents include data sheets, product selector guides, technical notes, and other material. We have already gathered information about the time required to maintain these pages by asking ASI employees to complete an informal survey (Attachment 4). Information from the survey shows that our current employees are spending approximately 142 hours per

This task is necessary to determine, first, whether the company really needs a Webmaster, and second, whether the problem is likely to grow in the future if the Web site expands.

page 4

month (collectively) to create and maintain ASI Web pages. Figure 1 shows the tasks being performed and the number of hours being spent on each task.

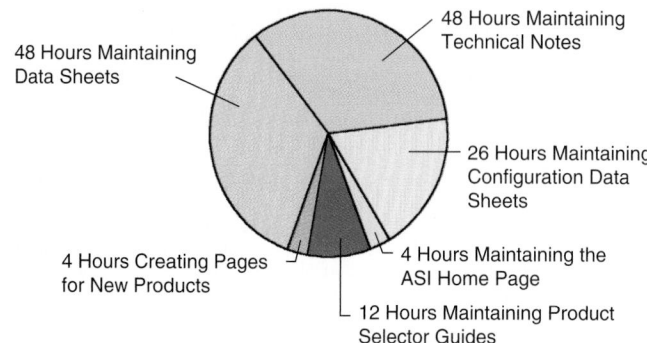

48 Hours Maintaining Technical Notes

48 Hours Maintaining Data Sheets

26 Hours Maintaining Configuration Data Sheets

4 Hours Creating Pages for New Products

4 Hours Maintaining the ASI Home Page

12 Hours Maintaining Product Selector Guides

Figure 1. Breakdown of the 142 hours that ASI employees spend each month creating and maintaining ASI Web pages

These writers have used a spreadsheet to create this pie chart, but the software doesn't offer advice on how to create a correct pie chart. Pie charts should display the largest slice beginning at 12 o'clock, then, moving clockwise, the next-largest slice. See Ch. 14, p. 386, for more about formatting a pie chart.

After compiling the list of ASI Web pages and a summary of each page, we will meet with the key people involved to determine whether all the ASI Web pages are truly needed. This task will also be used to determine whether information is being duplicated on our various Web pages. During these meetings, we will try to determine whether the need to post information on the Web is likely to grow in each area of ASI.

Task 2: Research the feasibility of contracting a professional Webmaster through a temporary agency to maintain ASI Web pages.

If the writers had already met with Human Resources, they would be in a better position to describe the services offered by various contracting agencies. As it stands, the description of Task 2 is thin.

We will meet with the ASI Human Resources Department to determine three professional employment companies to investigate. We will contact each of these companies to determine the standard services their Webmasters provide, associated salary, and contract fees. We will evaluate the companies' Webmaster services according to the skill and experience of the employee, as well as the salary. The evaluation will be documented in the completion report.

Task 3: Research the feasibility of hiring a professional Webmaster to work as a full-time ASI employee.

The description of Task 3 shows that the writers have already done some research on the credentials and salary range of a full-time Webmaster.

During our preliminary research, we found that the salary expectations and the standard services provided by professional Webmasters vary considerably. However, a study of major job boards on the Internet suggests that common requirements include experience working with Web site programming languages (HTML, CGI/PERL, JAVA, JavaScript,

page 5

Active X), development tools (such as NetObjects Fusion and MS FrontPage), operating systems and servers (such as Solaris, Windows NT and 95/98, Apache, and NCSA), peripheral tools (such as Adobe Illustrator and PhotoShop, Version Control Systems), and database environments such as Oracle. In addition, Webmasters need extensive management skills and the ability to work in a team environment and coordinate closely with engineering, marketing, technical communication, and legal staff.

Starting salaries for professionals with these credentials begin at $40,000 and top $100,000. In addition, there would be the standard benefits package of about 34 percent of salary.

Task 4: Research the feasibility of hiring an intern to create and maintain the ASI Web pages.

During our preliminary research, we found that there are qualified students seeking internships in computer-related fields. Although many students work part time on their internships, some work full time for several months. Hiring an intern would be considerably cheaper than retaining a contractor or hiring a full-time employee. We will meet with Human Resources and representatives of local colleges to compile information about interns' skills, experiences, and expectations for salary and benefits.

Task 5: Write a completion report detailing our findings and recommendations.

Following the completion of our research, we will submit a report to you summarizing our findings and recommendations. We will explain what information each of our ASI Web pages provides and discuss future growth of our site. Using the criteria we have listed, we will determine which contractor could provide the best candidate for ASI. We will also compare and contrast the options of contracting a professional Webmaster, hiring a full-time ASI employee as a Webmaster, and using a college intern. We will submit our completion report by May 12, 1999.

page 6

Table 1, below, provides a schedule of the various tasks we will perform to research and compile the completion report.

TABLE 1. SCHEDULE OF TASKS

This Gantt chart is a table. The bars are made by adding a black background to certain cells, then hiding the grid lines of the table. You can also use a spreadsheet to create a Gantt chart.

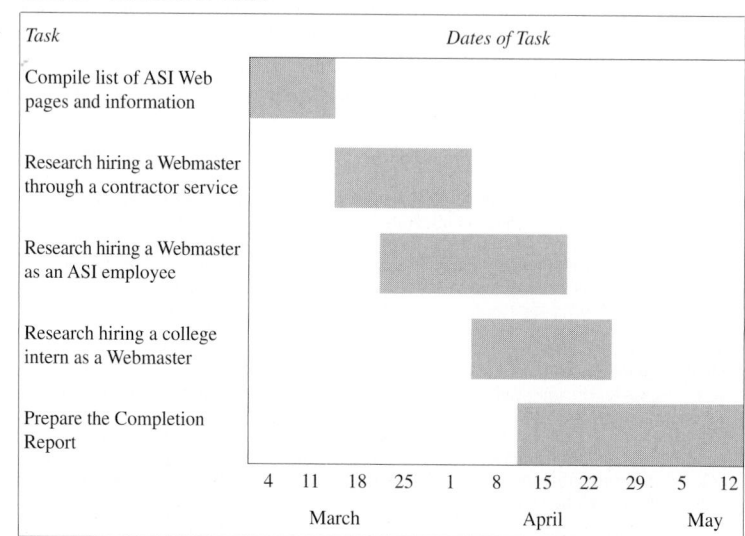

Task	Dates of Task
Compile list of ASI Web pages and information	
Research hiring a Webmaster through a contractor service	
Research hiring a Webmaster as an ASI employee	
Research hiring a college intern as a Webmaster	
Prepare the Completion Report	

4 11 18 25 1 8 15 22 29 5 12

March   April   May

Qualifications and Experience

Our Ad Hoc Web Committee comprises team members with many years of experience in research and a close involvement with maintaining our Web pages. Following is a list of our qualifications:

Julie Grapatin, Finance Department Representative
- Six years' experience in the Finance Department at ASI
- Two years' experience with HTML
- Senior in Finance at Central State University

Tammy T. Moon, Engineering Department Representative
- Engineering Supervisor at ASI for the past 10 years
- Proficient in HTML and Web page design
- Junior in Engineering at Central State University

Christine Johnson, Software Department Representative
- Employed at ASI for the past 7 years
- Excellent technical writing and communications skills
- Expertise in HTML and Web page design
- Bachelor's in Writing and Communication

Budget

Table 2, below, is an itemized budget for the proposed research.

TABLE 2.   ITEMIZED BUDGET

Name	Hours	Rate	Cost
Julie Grapatin	7 hours	19.00	$133
Christine Johnson	9 hours	17.00	153
Tammy T. Moon	8 hours	16.00	128
		TOTAL:	**$414**

References

Baynet Works White Paper (1999). Web Based Network Management. <http://www.baynetworks.com/products/Papers/webbased.html#int> (1999 February 27).

Buck Consultants, Inc. (1998). *National Business Employment Weekly.* <public.wsj.com/careers/resources/documents/19980619-internetpros-tab.htm> (1999 February 28).

page 8

Attachment 1
Sample of a Well-Designed ASI Web Page

A significant strength of this proposal is the writers' ability to create an effective sample Web page. This skill gives them credibility — readers believe that the writers have the technical expertise to evaluate potential Webmasters.

Ada Semiconductor, Inc.

SDRAM Product and Technical Assistance

Information regarding price and availability may be obtained by contacting the North American or International sales representative or distributor nearest you.

Information regarding future product plans and market direction may be obtained by contacting the Marketing Department.

Phone: 208-555-1000

Product Type	Size/Width	Marketing Contact	Technical Contact
SDRAM	16Mb / x16	Mike Jones	Bob Adams Ronda Dodge
	16Mb / x4, x8	Mike Jones	Bob Adams Ronda Dodge Ed Burns
	64Mb / x4, x8, x16	Mike Jones	Bob Adams Ronda Dodge Ed Burns
	64Mb / x32	Mike Jones	Bob Adams Ronda Dodge Ed Burns
	128Mb / x4, x8, x16	Mike Jones	Bob Adams Ronda Dodge Ed Burns
	256Mb / x4, x8, x16	Mike Jones	Bob Adams Ronda Dodge Ed Burns

Email Deb Stevens at Ada Semiconductor, Inc. debstevens@ASI.com if you have questions concerning this product.

The email link to Deb Stevens solicits responses from readers.

©1999, Ada Semiconductor, Inc. All Rights Reserved

Last Updated
2-28-1999

page 9

Attachment 2
Sample of an Unprofessional ASI Web Page

ASI DRAM Contact

Jason Conner

Hi! My name is Jason and I am in charge of Failure Analysis. My home is located in Stanwood, Wa in the beautiful Northwest. People always want to know what I look for when a part fails (So I will tell you). When I first started doing computer work I did reliability testing. I would look at the part and try to figure out why it failed. I would test the part and then deprocess the part. It seemed only natural to become a FA tech. I love computers and am constantly in a stage of learning, things are changing so fast in the industry.

You can email me or fax me or just plain write to me. I like to take pride in doing the best I can.

Email Jason Conner at Ada Semiconductor, Inc. jasonconner@ASI.com if you have questions about failures.

©1999, Ada Semiconductor, Inc. All Rights Reserved

Last Updated
3-28-1999

This sample from the current Web site says all that needs to be said about unprofessional appearance. Rock on, Jason.

Attachment 3
Ada Semiconductor, Inc. Current Labor Cost Estimates for Creating and Maintaining Web Pages

Employee Name	Title	Hourly salary	Hours maintaining Web pages per month	Cost to ASI
Bob Adams	Reliability Engineer	59	24	1,416
Ed Burns	Test Engineer	62	8	496
Jason Conner	Failure Analysis Technician	38	16	608
Ronda Dodge	Process Engineer	61	2	122
Rajiv Gupta	Technical Support	23	16	368
Mike Jones	Sales & Marketing Rep.	58	16	928
Karen Seagate	Software Support Supervisor	42	20	840
Brad Owens	Reliability Engineer	58	4	232
Deanna Peters	Software Support	27	20	540
Deb Stevens	Technical Writer	22	16	352
Monthly Total			**142**	**$5,092**

This information belongs here in an appendix, not in the body of the proposal, because most readers will not be interested in the details. However, because some readers will want to see this information, it should be included in the proposal.

In addition to the questionnaire, the writers should provide at least a summary of the data they gathered, beyond the number of hours spent on maintaining the Web pages.

For example, providing the responses to question 5 would show that the respondents believe the site will likely grow in the future.

Attachment 4
Ada Semiconductor, Inc. Employee Web Page Survey

Ada Semiconductor, Inc.
Interoffice Communication/Survey

To: All Ada Semiconductor, Inc. Employees
From: Julie Grapatin, Finance Department
     Tammy T. Moon, Engineering Department
     Christine Johnson, Software Department
Subject: Employee Survey of ASI Web Site Maintenance
Date: 5 February 1999

We are asking all employees to complete this survey and return it to Tammy Moon by 2/26/99. The results of the survey will help to determine whether ASI will hire a Webmaster to create and maintain ASI Web pages. (Use the back of the survey if you have additional comments.)

1. How many hours do you spend maintaining your pages on the ASI Web site per week?

2. What tasks do you do in working on the ASI Web site?

3. Explain how maintaining the ASI Web site has affected your ability to perform your other duties.

4. If you were a potential client, what would you think about the accessibility, uniformity, and organization of the ASI Web site? Please explain.

5. Do you think that the ASI Web site is likely to stay the same size, or will it need to grow in the coming months and years?

✔   **Revision Checklist**

The following checklist covers the basic elements of a proposal. Guidelines established by the recipient of the proposal should take precedence over these general suggestions.

1. Does the summary provide an overview of
   ❏ the problem or the opportunity?
   ❏ the proposed program?
   ❏ your qualifications and experience?
   ❏ the schedule?
   ❏ the budget?

2. Does the introduction indicate
   ❏ the problem or opportunity?
   ❏ the purpose of the proposal?
   ❏ the background of the problem or opportunity?
   ❏ your knowledge of the professional literature?
   ❏ the scope of the proposal?
   ❏ the organization of the proposal?
   ❏ the key terms that will be used in the proposal?

3. Does the description of the proposed program provide a clear, specific plan of action and justify the tasks you propose performing?

4. Does the description of qualifications and experience clearly outline
   ❏ your relevant skills and past work?
   ❏ the skills and background of the other participants?
   ❏ your department's (or organization's) relevant equipment, facilities, and experience?

5. Is the budget
   ❏ complete?
   ❏ correct?
   ❏ accompanied by an in-text reference?

6. Do the appendices include the relevant supporting materials, such as a task schedule, a description of evaluation techniques, and evidence of other successful projects?

Exercises

The following exercises ask you to write a memo. See Chapter 15, page 430, for a discussion of writing memos.

1. Study the National Science Foundation's Grant Proposal Guide (www.nsf.gov/nsf/nsfpubs/gpg /start.htm). What are the important ways in which the NSF's guide differs from the advice provided in this chapter? What accounts for these differences? Present your findings in a 500-word memo to your instructor.

2. Form groups according to major. Using *Commerce Business Daily* or CBDNET (cbdnet.gpo.gov), find a request for proposals (RFP) for a project related to

your academic field. Study the RFP. What can you learn about the needs of the organization that issued it? How effectively does the RFP describe what the issuing organization expects to see in the proposal? Is the RFP relatively general or specific? What sorts of evaluation techniques does the RFP call for? In your response, include a list of questions that you would ask the issuing organization if you were considering responding to the RFP. Present your results in a memo to your instructor.

3. Using the Web, locate three companies that provide consulting services about proposal writing. In a brief memo to your instructor, list their names and URLs and briefly describe and evaluate their sites and services on the bases of clarity, ease of navigation, and professional appearance. See Chapter 21 for a discussion of Web sites.

Research Project

4. Write a proposal for a research project that will constitute a major assignment in this course. Your instructor will tell you whether the proposal is to be written individually or collaboratively. Start by defining a technical subject that interests you. (This subject could be one that you are involved with at work or in another course.) Using abstract services and other bibliographic tools, compile a bibliography of articles and books on the subject. (See Chapter 7 for a discussion of finding information.) Create a reasonable real-world context. Here are three common scenarios from the business world:

- Our company uses Technology X to perform Task A. Should we instead be using Technology Y to perform Task A? For instance, our company uses traditional surveying tools in its contracting business. Should we be using GPS surveying tools instead?

- Our company has decided to purchase a particular kind of tool to perform Task A. Which make and model of the tool should we purchase, and from which supplier should we buy or lease it? For instance, our company has decided to purchase 10 multimedia computers. Which brand and model should we buy, and from whom should we buy them?

- Our company does not currently perform Function X. Is it feasible to perform Function X? For instance, we do not currently offer day care for our employees. Should we? What are the advantages and disadvantages of doing so? What forms can day care take? How is it paid for?

Here are some additional ideas for topics.

- the need to provide Internet access to students

- the value of using the Internet to form ties with another technical-communication class on another campus

- the need for expanded opportunities for internships in your major

- the need to create an advisory board of people from industry to provide expertise about your major

- the need to raise money to keep the college's computer labs up to date

- the need to evaluate your major to ensure that it is responsive to students' needs

- the advisability of starting a campus branch of a professional organization in your field

- improving parking facilities on campus

- creating or improving organizations for minorities or for women on campus

These topics can be approached from different perspectives. For instance, the first one — on providing Internet access to students — could be approached in several ways:

- Our college currently purchases journals but does not provide Internet access for students. Should we consider reducing the library's journal budget to subsidize Internet access for students?

- Our college has decided to provide Internet access for its students. How should it do so? What vendors provide such services? What are the strengths and weaknesses of each vendor?

- Our college does not offer Internet access to students. Should we make it a goal to do so? What are the advantages of doing so? The disadvantages?

CASE 1
Ethics and Proposals

You work for Devon Electronics, a company that does contract work for the federal government. Your department has submitted a proposal to build a friend-or-foe electronic device to be used in military aircraft. You have just learned that Martha Ruiz, the chief technical person for the project, has accepted a position with another company and will be leaving in two weeks. Your company has not yet decided who would replace her if the project is funded. Your Director of Operations favors delaying a decision until the company learns whether it has won the contract. The government agency is currently reviewing all the proposals and is not scheduled to announce the winner of the contract for three weeks. Should you notify them of Ruiz's impending resignation? If you think you should notify the government agency, draft the letter that you would send. If you think you should not notify the government agency, explain your reasoning in a memo to your instructor.

See Ch. 15, p. 412, for information about letters.

CASE 2
Selecting Project-Management Software

You work in the proposal department of Allied Marine Services, a small manufacturer of electronic equipment used in civilian and military boats and ships. Currently, your company does not use project-management software in creating its Gantt charts; a graphic artist draws them. You think that project-management software might help you complete the charts more quickly and less expensively. You ask your supervisor if he will let you spend a few hours searching the Internet to learn about the software: what is available, what it does, how easy it is to use, and what it costs. Using a search engine, find the sites of three manufacturers of project-management software. Download a few pages from each site. In a memo to your supervisor, describe the different products and state whether you think it is a good idea to download demo copies and try them on your next proposal.

For additional cases, click on Case of the Month and Archive on TechComm Web (www.bedfordstmartins.com /techcomm).

18 Writing Informal Reports

David Beer and David McMurrey (1997, p. 104), technical-communication professors, on the function of the progress report:

Its job is to present to your clients the status of the work you are doing for them. You supply these details to enable your clients to act as manager or executive of the project — to enable them to modify it or even cancel it if the need arises.

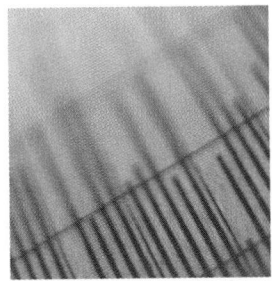

 See TechComm Web (www .bedfordstmartins.com /techcomm) for guidelines, additional examples, and links related to topics in this chapter.

A report is a statement — oral or written — that helps listeners or readers understand, analyze, or take action on some situation, idea, or action. This book classifies reports into two categories: formal and informal. This chapter discusses informal reports; Chapter 19 discusses formal reports.

Here are a few examples of informal reports:

A memo that requests authorization to buy a piece of equipment for $1,200

An email describing a problem in the management of an office and recommending action to solve it

A presentation at a meeting, in which an employee analyzes a set of architectural plans for the company's new offices

A memo describing the projects undertaken in the quality-assurance department of a company over the last quarter

A memo recording what occurred at a recent meeting of a department

A memo responding to a request for information from a supervisor

As this list indicates, informal reports cover many subjects, have many purposes, and take many forms. What exactly is "informal" about an informal report? There is no single answer. In one organization, a particular document might be considered a formal report; in another, an informal one. In general, however, an informal report might be said to be a routine communication of information about an everyday matter.

This is not to say that informal reports are unimportant. An accident report written by police officers at the scene of an automobile accident might be only a page or two long, but it might have enormous implications for the people involved in the accident and for many others.

This chapter discusses the process of writing informal reports. It describes how to write the typical kinds, including trip reports, progress reports, and meeting minutes.

PLANNING, DRAFTING, AND REVISING INFORMAL REPORTS

Writing informal reports involves the same planning, drafting, and revising techniques used in most other kinds of technical communication. This section explains some of the ways in which these techniques apply to informal reports.

For more about analyzing audience and purpose, see Ch. 5.

- *Analyze your audience and purpose.* In some cases, it is easy to identify your audience and purpose. For instance, a set of meeting minutes is addressed to all the members of the committee or department, and its purpose is to serve as the official record of a meeting. A quarterly status report about your department's activities, addressed to the department head, describes the projects you worked on that quarter and might include other comments or recommendations that will help the reader understand your department's situation and needs.

  In other cases, the audience and purpose are not routine. For instance, you might be reporting on an accident at your company's facility. To whom should you address the report: your direct supervisor? your direct supervisor and others? What is your purpose: to describe what happened? to recommend steps that might reduce the chance that such accidents recur? In this situation, you need to analyze the audience and purpose just as carefully as you would for any other kind of document.

For more about conducting research, see Ch. 7.

- *Gather and compile your information.* Sometimes, assembling your information is as simple as printing a database file. At other times it requires sophisticated information-gathering techniques using primary and secondary research.

- *Choose an appropriate format.* The most common format for informal reports is the memo, but in many cases an organization has preprinted forms or templates for you to follow. In some organizations, the title page, a cover, and other elements usually associated with more formal reports are used for all reports. The choice of format is therefore often determined by your organization as well as by your audience and purpose.

For more about drafting and revising, see Ch 3, pp. 53 and 56.

- *Draft the report.* For routine reports, writers sometimes can use sections of previous reports. In a status report, the description of your current project can be copied from the previous report and updated as necessary. Some informal reports are drafted on site: a site study might be "drafted" using a hand-held electronic device or a checklist on a clipboard, as an engineer walks around a site.

- *Revise the report.* Informal reports must be revised like any other kind of technical document. *Informal* does not mean *careless.*

FORMATS FOR INFORMAL REPORTS

Informal reports can be oral — live, on the phone, or in a teleconference — or presented in four basic kinds of documents:

- *Memos.* A memo is a relatively informal medium used for communicating with another person in the same organization. It can range from less than a page to as long as ten pages or more.

- *Forms.* Routine informal reports are often written or typed on preprinted forms or on templates used with a word processor. These forms prompt you to fill in the necessary information. For example, some companies use a template for trip reports: the template has a space in which to write the destination, another for the purpose of the trip, and so forth. Sometimes forms are used as cover sheets for informal reports: the first page is the form, but the subsequent pages are created by the writer.

- *Email.* Email is an increasingly popular format for transmitting informal reports because it is convenient and digital. Writers can use address books and distribution lists to distribute the report easily. And because it is digital, an email message can be pasted into another document. For example, if a supervisor receives four status reports every month, she can easily combine the files into a single report to transmit to her supervisor.

- *Letters.* Letters, the most formal of the documents, are preferred when the writer and the reader work at different organizations.

  The following sections discuss five types of informal reports.

Types of informal reports:

Directives

Trip reports

Field and lab reports

Progress and status reports

Meeting minutes

For more about memos, email, and letters, see Ch. 15.

WRITING DIRECTIVES

In a *directive*, you report on a policy or a procedure you want your readers to follow. If possible, you should explain the reason for the directive; otherwise, it might seem like an arbitrary order rather than a thoughtful, reasonable

Quimby Interoffice

Date: March 19, 20XX
To: All supervisors and sales personnel
From: D. Bartown, Engineering
Subject: Avoiding Customer Exposure to Sensitive Information Outside Conference
             Room B

Recently I have learned that customers meeting in Conference Room B have been allowed to use the secretary's phone directly outside the room. This practice presents a problem: the proposals that the secretary is working on are in full view of the customers. Proprietary information such as pricing can be jeopardized unintentionally.

In the future, would you please escort any customers or non-Quimby personnel needing to use a phone to the one outside the Estimating Department? Thanks very much.

■ **Figure 18.1**
A Directive

The writer uses a polite but informal tone throughout to clearly define the problem and how he would like readers to address it.

request. For brief directives, the explanation should precede the directive itself, to prevent the appearance of bluntness. For longer reports, the directive might precede the detailed explanation, to ensure that readers will not overlook the directive. Of course, the body of the directive should begin with a polite explanatory note:

> I'd like to establish a uniform policy for dealing with customers who fall more than 60 days behind in their accounts. The policy is defined below under the heading "Policy Statement." Following the statement is my rationale.

Figure 18.1 shows an example of a directive.

WRITING TRIP REPORTS

A *trip report* is a record of a business trip written after the employee returns to the office. Most often, it takes the form of a memo, although some organizations use computerized templates. To write an effective trip report, remember that your reader is less interested in an hour-by-hour narrative of what happened than in a carefully structured discussion of what was important. If, for instance, you attended a professional conference, don't list all the presentations; simply attach the agenda or program if you think your reader will be interested. Present any important information you learned, or describe the important questions that didn't get answered. If you traveled to meet a client (or a potential client), focus on what your reader is interested in: how to follow up on the trip and maintain a good business relationship with the client.

Figure 18.2 shows a typical trip report.

Dynacol Corporation

INTEROFFICE COMMUNICATION

To: G. Granby, R&D
From: P. Rabin, Technical Services
Subject: Trip Report — Computer Dynamics, Inc.
Date: September 20, 20XX

This memo presents my impressions of the Computer Dynamics technical seminar of September 18. The purpose of the seminar was to introduce their new PQ-500 line of high-capacity storage drives.

Summary
In general, I was impressed with the technical capabilities and interface of the drives. Of the two models in the 500 series, I think we ought to consider the external drives, not the internal ones. I'd like to talk to you about this issue when you have a chance.

Discussion
Computer Dynamics offers two models in its 500 series: an internal drive and an external drive. Each model has the same capacity (10 G of storage), and they both work the same way. They act just like any other kind of drive, preserving the user's directory structure.

Although the internal drive is convenient — it is already configured for the computer — I think we should consider only the external drive. So many of our employees do teleconferencing that the advantages of portability outweigh the disadvantage of inconvenience. The tech rep from Computer Dynamics walked me through the process of reconfiguring the drive for a second machine; the process will take most of our employees only a few minutes to learn. A second advantage of the external drive is that it can be salvaged easily when we take a computer out of service.

Recommendation
I'd like to talk to you, when you get a chance, about negotiating with Computer Dynamics for a quantity discount. I think we should ask McKinley and Rossiter to participate in the discussion. Give me a call (x3442) and we'll talk.

The headings make the memo easier to read and easier to write, because they are prompts to provide the kind of information readers need.

■ **Figure 18.2**
A Trip Report

Although writer and reader appear to be relatively equal in rank, the writer goes to the trouble of organizing the discussion to make it easy to read and to refer to later.

WRITING FIELD AND LAB REPORTS

A common kind of informal report describes inspections, maintenance, and site studies. These reports, often known as *field reports* or *lab reports*, include the same type of information that high-school lab reports do — the problem, methods, results, and conclusions — but they deemphasize methods and can include recommendations.

GUIDELINES

Responding to Readers' Questions in a Field or Lab Report

Be sure to answer the following questions:

▶ What is the purpose of the report?

▶ What are the main points covered in the report?

▶ What were the problems leading to the decision to perform the procedure?

▶ What methods were used?

▶ What were the results?

▶ What do the results mean?

▶ What should be done next?

The report in Figure 18.3 illustrates possible variations on this standard report structure.

WRITING PROGRESS AND STATUS REPORTS

A *progress report* communicates to a supervisor or sponsor the current status of an ongoing project. A *status report* is an update on the entire range of operations of a department or division of an organization. For example, the director of marketing for a manufacturing company might submit a monthly status report.

For more on proposals, see Ch. 17; for completion reports, see Ch. 19, p. 540.

As its name suggests, a progress report is an intermediate communication between the proposal (the argument that a project be undertaken) and the completion report (the comprehensive record of a completed project).

Progress reports let you check in with your audience. Supervisors are vitally interested in the status of their projects, because they have to integrate them with other present and future commitments. Sponsors (or customers) have the same interest, plus an additional one: they want the projects to be done right — and on time — because they are paying for them.

LOBATE CONSTRUCTION
3311 Industrial Parkway
Speonk, NY 13508

Quality Construction Since 1957

April 11, 20XX

Ms. Christine Amalli, Head
Civil Engineering
New York Power
Smithtown, NY 13507

Dear Ms. Amalli:

We are pleased to report the results of our visual inspection of the Chemopump after Run #9, a 30-day trial on Kentucky #10 coal.

The inspection was designed to determine if the new Chemopump is compatible with Kentucky #10, the lowest-grade coal that you anticipate using. In preparation for the 30-day test run, the following three modifications were made by your technicians:

• New front-bearing housing buffer plates of tungsten carbide were installed.
• The pump-casting volute liner was coated with tungsten carbide.
• New bearings were installed.

Our summary is as follows. A number of small problems with the pump were observed, but nothing serious and nothing surprising. Normal break-in accounts for the wear. The pump accepted the Kentucky #10 well.

The following four minor problems were observed:

• The outer lip of the front-end bell was chipped along two-thirds of its circumference.
• Opposite the pump discharge, the volute liner received a slight wear groove along one-third of its circumference.
• The impeller was not free-rotating.
• The holes in the front-end bell were filled with insulating mud.

The following three components showed no wear:

• 5-1/2" impeller
• suction neck liner
• discharge neck liner

The word "visual" describes the methods.

The writer states the purpose of the inspection.

This writer has chosen to incorporate the terms "summary" and "conclusion" in the body of the letter rather than use headings as a method of organization.

■ **Figure 18.3**
A Trip Report

Because the writer and the reader work for different companies, the letter is the appropriate format for this brief informal report.

page 2

Our conclusion is that the problems can be attributed to normal break-in for a new Chemopump. The Kentucky #10 coal does not appear to have caused any extraordinary problems. In general, the new Chemopump seems to be operating well.

We would recommend, however, that the pump be modified as follows:

1. Replace the front-end bell with a tungsten carbide-coated front-end bell.
2. Replace the bearings on the impeller.
3. Install insulation plugs in the holes in the front-end bell.

Further, we recommend that the pump be reinspected after another 30-day run on Kentucky #10.

If you have any questions, or would like to authorize these modifications, please call me at 555-1241. As always, we appreciate the trust you have placed in us.

Sincerely,

Marvin Littridge

Marvin Littridge
Director of Testing and Evaluation

The writer concludes politely.

■ **Figure 18.3**
(Continued)

Regardless of whether the project is proceeding smoothly or has encountered difficulties, you need to explain clearly and fully what happened and how it will affect the overall project. Your tone should be objective, neither defensive nor casual. Unless ineptitude or negligence has caused the problem, you're not to blame. Regardless of the news you are delivering — good, bad, or mixed — your job is the same: to provide a clear and complete account of your activities and to forecast the next stage of the project.

Progress reports are crucial because they enable managers to make informed decisions when things go wrong. You might be tempted to cover up problems and hope that you can solve them before the next progress report. This course of action is unwise and unethical. Chances are that problems will multiply, and you will have a harder time explaining why you didn't alert your readers earlier.

GUIDELINES

Reporting Your Progress Honestly

Withholding bad news is unethical because it misleads your readers. Because they are the sponsors or supervisors of the project, they have a right to know how it is going. If you find yourself faced with any of these three common problems, consider responding in the following ways:

▶ *The deliverable — the document or product you will submit at the end of the project — won't be what you thought it would be.* The progress report is your opportunity to explain that even though you thought you would be able to recommend to your contracting company which earth mover under $200,000 to purchase, you will not be able to do so. Then explain why. Or explain that the deliverable will not meet one of the specifications in the proposal.

▶ *You won't meet your schedule.* Explain why you are going to be late and state when the project will be complete.

▶ *You won't meet the budget.* Explain why you need more money, and state how much more you will need.

For more about deliverables, see Ch. 17, p. 486.

Organizing Progress and Status Reports

The time pattern and the task pattern, two organizational patterns frequently used in progress or status reports, are illustrated in Figure 18.4 on page 528. A status report is usually organized according to task; by its nature, a status report covers a specified period of time. If the purpose of a status report is to describe a department's work in, say, July, it makes sense to focus on each task or project carried out by that department during July.

In the time pattern, you describe the work completed during the reporting period; then you sketch in the work that remains. Some writers include a section on present work, which enables them to focus on a long or complex task still in progress.

■ **Figure 18.4 Organizational Patterns in Progress and Status Reports**

The Time Pattern	*The Task Pattern*
Discussion	Discussion
A. Past Work	A. Task 1
B. Future Work	1. Past work
	2. Future work
	B. Task 2
	1. Past work
	2. Future work

The task pattern allows you to describe, in order, what has been accomplished on each task. Often, a task-oriented structure incorporates the chronological structure, shown here.

Concluding Progress and Status Reports

In the conclusion of a progress or status report, evaluate how the project is proceeding. In the broadest sense, there are two possible messages: things are going well, or things are not going as well as anticipated.

GUIDELINES

Projecting an Appropriate Tone in a Progress or Status Report

▶ *If the news is good, convey your optimism but avoid overstatement.*

OVERSTATED   We are sure the device will do all that we ask of it, and more.

REALISTIC   We expect that the device will perform well and that, in addition, it might offer some unanticipated advantages.

Beware of promising early completion. Such optimistic forecasts rarely prove accurate, and it is always embarrassing to have to report a failure to meet the promised deadline.

▶ *Don't panic if the preliminary results are not as promising as you had planned or if the project is behind schedule.* Readers know that the most levelheaded and conservative proposal writers cannot anticipate all problems. As long as the original proposal was well planned and contained no wildly inaccurate computations, don't feel responsible. Just do your best to explain unanticipated problems and the current status of the project. If you suspect that the results will not match earlier predictions — or that the project will require more time, personnel, money, or equipment — say so, clearly. If your news is bad, at least give the reader as much time as possible to deal with it effectively.

Find other samples of progress reports at Online Technical Writing (www.io .com/~hcexres/tcm1603 /acchtml/progrep.html) and at Writing Guidelines for Engineering and Science Students (http://fbox.vt .edu:10021/eng/mech /writing/).

If appropriate, append supporting materials, such as computations, printouts, schematics, diagrams, charts, tables, or a revised task schedule. Be sure to cross-reference these appendices in the body of the report, so that a reader can consult them at the appropriate stage in the discussion.

Sample Progress Report

The progress report shown below (Grapatin, Moon, & Johnson, 1999) was written for the project proposed on page 505 in Chapter 17. (The completion report for this study is on page 563 in Chapter 19.)

To see another sample progress report, for a study on American Sign Language, click on Additional Sample Documents for Ch. 18 on TechComm Web (www.bedfordstmartins.com /techcomm).

ASI   *Ada Semiconductor Incorporated*
220 Technical Street   **Telephone:**   **(208) 555-2200**
Boise, Idaho 83706   **Fax:**   **(208) 555-2201**

www.asisemi.com

M E M O R A N D U M

DATE: March 19, 1999
TO: Mr. Michael Jensen, Ada Semiconductor, Inc. Manager
           Ms. Shelly McKay, Ada Semiconductor, Inc. Sales and Marketing Manager
           Mr. Brandon Wells, Ada Semiconductor, Inc. Software Support Manager
FROM: Ada Semiconductor Ad Hoc Web Committee:
               Julie Grapatin — Finance Department Rep.
               Tammy T. Moon — Engineering Department Rep.
               Christine Johnson — Software Department Rep.
SUBJECT: Progress Report on the Feasibility Study for Hiring a Webmaster

The writers identify the document as a progress report.

PURPOSE
This is a progress report on the feasibility study for hiring a Webmaster to create and maintain ASI Web pages.

SUMMARY
The ASI Ad Hoc Web Committee has been studying our current Web site and researching the following three options:

- contract for a Webmaster through an independent temporary employment agency
- hire a full-time Webmaster to work at ASI
- hire a college intern to perform Webmaster duties

At present, the research effort is ahead of schedule and on budget, and we expect to submit the completion report by the May 12 deadline.

The summary briefly explains the purpose of the project and answers the question, "How is the project going, and will it be completed on schedule and under budget?"

INTRODUCTION
On March 4, we received approval to investigate the best use of ASI resources to maintain ASI's Web site. Our proposal was based on a request initiated by Michael Jensen, ASI Manager. The results of our research will be presented in the form of a completion report at the next ASI management meeting, scheduled for May 12.

Our proposal included the following activities:

- Task 1: Compile a list of all ASI Web pages currently being maintained, summarize the information provided on each page, and determine whether the site is likely to grow in the future.

Most of the information in the introduction is taken directly from the proposal.

page 2

- Task 2: Research the feasibility of contracting a professional Webmaster through independent temporary employment agencies.

- Task 3: Research the feasibility of hiring a professional Webmaster to work as a full-time ASI employee.

- Task 4: Research the feasibility of hiring an intern to create and maintain ASI Web pages.

- Task 5: Write and submit a completion report detailing the committee's findings and recommendations.

The introduction concludes with a brief summary of the project's current status.

Task 1 is complete and included in this report. Tasks 2–4 are in various stages of completion, but all are within 85 percent of being completed. Task 5 will be completed after the results of Tasks 2–4 are analyzed for final recommendations.

RESULTS OF RESEARCH

In this section, the writers are presenting information that will go into the completion report.

Task 1: Compile a list of all ASI Web pages currently being maintained, summarize the information provided on each page, and determine whether the site is likely to grow in the future.

- One home page. This is the first page that external customers see when going to www.asisemi.com. It links to all other ASI pages.

- One product selector guide page. This page, which lists all the current ASI products, includes part density, memory configuration, features, part numbers, package types, access time, and availability.

- One literature-request form page. This is the form that Web customers use to order additional product literature.

- One newsletter page. This page announces new ASI information such as special offers and product qualifications.

- Four product guide pages (DRAM, SRAM, Flash, and Modules). These pages contain boot block selector guides, compatibility guides, and ordering information.

- Six technical notes pages. These pages contain pertinent information based on the application and design of ASI products; topics are separated by product type.

- Six product and technical assistance pages. These pages contain names of the marketing contact and technical contact(s) and are separated by product type.

On the basis of the surveys we distributed to the ASI employees who maintain the Web pages, as well as follow-up interviews with them, there is a clear consensus that the site will need to grow at the rate it has expanded in the last year, if not at a greater rate. ASI continues to introduce new products, and the hit counters show an increasing number of visitors to our site. Last year, our site was visited approximately 1,500 times per month. This year, the

page 3

figure is more than 4,000. It seems clear that the Web is likely to remain our primary means of furnishing product information to our customers for the foreseeable future.

Task 2: Research the feasibility of contracting for a professional Webmaster through independent temporary employment agencies.
Originally we had planned to investigate three professional contractor services, but after meeting with Paul Mitchell from ASI's Human Resources department, we revised our plan. Paul informed us that although ASI works with several local temporary employment agencies when filling production positions, ASI prefers to contract software professionals through Roberts, Inc.

A Webmaster hired through Roberts, Inc. will be qualified to perform production-level HTML scripting, to implement basic features for online Web content, and to provide tools to enhance efficiency in creating HTML Help, Web, and other online deliverables. In addition, Roberts will screen the Webmaster (referred to as Web Author at Roberts) to ensure that he or she has at least one year's demonstrated programming experience as well as experience in graphics, layout, and design. All candidates will provide a current portfolio. The hourly rate for a Webmaster ranges from $25 to $45, based on expertise and experience. In addition, through Roberts we have the option of hiring a Webmaster for a specific project without future commitment. This option gives us the flexibility to rehire the same Webmaster if his or her job performance has been satisfactory, or contract for a different Webmaster.

Task 3: Research the feasibility of hiring a professional Webmaster to work as a full-time ASI employee.
We are currently working with ASI's Human Resources department to develop a Webmaster position description (PD). The PD is 85 percent complete. Once it is completed, the hiring qualifications and appropriate salary and benefits levels can be set. At present, it appears the position will be graded out at the technical skill level (TSL) of 9 (the scale is 1 to 15). The beginning annual salary of a TSL 9 employee is $50,025 ($23.85 per hour). The aggregate total annual cost of a TSL 9 employee for ASI (this figure includes such tangibles as benefits, retirement plan, and salary) is currently determined to be $67,002.

Task 4: Research the costs associated with hiring an intern to create and maintain ASI Web pages.
ASI could arrange for an intern through the Computer Science or English department at Central State University (CSU). The internships in both departments consist of 50 hours for each credit earned. Most internships are 150 hours/3 credits. ASI would need to do the following:

• Sign a contract stating ASI will supervise the intern.

• Provide a supervisor for the intern.

• Provide the intern with instruction and direction for creating and maintaining Web sites.

• Evaluate the intern's work at the end of 150 hours (end of semester).

The writers explain clearly why they had to modify this task.

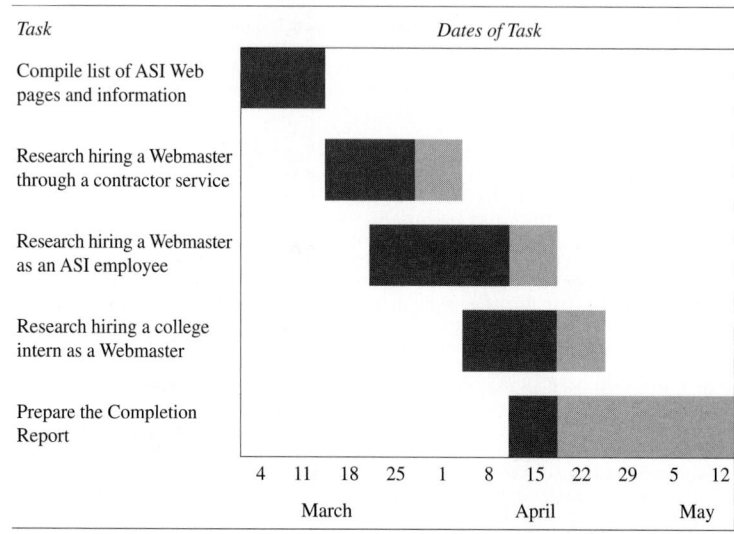

page 4

UPDATED TASK SCHEDULE

Note: Black bars represent completed tasks; gray bars represent tasks not yet complete.

This Gantt chart shows the progress toward completing each of the project's tasks.

Task   *Dates of Task*

Compile list of ASI Web pages and information

Research hiring a Webmaster through a contractor service

Research hiring a Webmaster as an ASI employee

Research hiring a college intern as a Webmaster

Prepare the Completion Report

4 11 18 25 1 8 15 22 29 5 12

March   April   May

The conclusion offers a summary and interpretation of the current status of the project.

CONCLUSION

The Ad Hoc committee has been successful in completing one task and is ahead of schedule for completing the remaining four tasks on or before May 12. The committee researched ASI's existing Web pages for function and content. There are a total of 20 pages; the content of each is worthwhile and requires vigilant maintenance. In addition, we will almost certainly want to add new ASI Web pages in the near future, which supports our proposal to hire a Webmaster. Additionally, the committee has researched the costs and intangibles associated with contracting a Webmaster, hiring a full-time, salaried ASI employee, or arranging for a university intern (paid or unpaid). Recommendations will be available in the completion report on or before May 12.

The writers end with a polite offer to provide additional information.

Please get in touch with Julie Grapatin at extension 3098 if you have any questions or wish to discuss the project further.

WRITING MEETING MINUTES

The minutes, an organization's official record of a meeting, are distributed to all those who belong to the committee or any other unit represented at the meeting. In writing a set of minutes, your goal is to be clear, comprehensive, objective, and diplomatic. Your job is not to interpret what happened. You have three main goals:

For more about conducting meetings, see Ch. 4, p. 71.

- to record the logistical details of the meeting
- to provide an accurate record of the meeting
- to ensure that the minutes reflect positively on the participants and the organization

Recording the Logistical Details of the Meeting

According to *Robert's Rules of Order* (Robert & Patnode, 1994), the authoritative reference work on parliamentary procedure, you should record the following information:

- the name of the group or committee that met
- the location, date, and time of the meeting
- the type of meeting (regular or special)
- the presence of the chair and secretary, or their substitutes
- the time at which the meeting was adjourned

Another respected source, the *Gregg Reference Manual* (Sabin, 1999), suggests that you include the names of those who attended and those who did not and a separate list of guests.

Also record what action was taken about the minutes of the previous meeting. For example, the minutes were read (or distributed) and approved (or amended and approved). You should record any changes to the previous minutes.

Providing an Accurate Record of the Meeting

Because meetings rarely follow the agenda perfectly, you might find it challenging to provide an accurate record of the meeting. Record the major topics discussed at the meeting as well as any actions taken. For example, write down the names of reports read or approved, motions made (and whether they were approved, defeated, or tabled), and resolutions adopted. Record the outcomes of votes. Record the names of the people who made motions, read reports, and so forth; for example, "Barry Young presented a report on the June activities of the Safety Department." If the conversation is going too fast for you to keep an accurate record, interrupt the discussion to request a clarification.

Robbins Junior High School
Weekly Planning Committee Minutes
Minutes of the February 14, 20XX Meeting

The meeting was called to order by Ms. De Grazia at 3:40 P.M. in the conference room. In attendance were Mr. Sipe, Ms. Leahy, Mr. Zaerr, Mr. Simon, and Principal Barson. Ms. Evett was absent.

The minutes of the February 7, 20XX meeting were read. The following correction was made: In paragraph 2, "800 hours" was replaced with "80 hours." The minutes were then unanimously approved.

There was one topic: authorization for the antidrug presentations by motivational speaker Alan Winston. Principal Barson reported on his discussion with Peggy Giles of the School District. She offered positive comments about Winston's presentations at other schools in the district last year. Principal Barson has also been in contact with the three other principals who invited Mr. Winston last year; they all spoke highly of his presentations.

Principal Barson moved that the committee authorize Winston's visit, to be scheduled in late May. The motion was seconded by Louis Simon.

Mr. Sipe asked whether the individual school or the district was to pay for the expenses (approximately $2,800) for the visit. Principal Barson replied that the school and the district would split the costs evenly, as had been done last year. The school has more than $4,000 surplus in operating expenses.

Mr. Zaerr expressed serious concerns about the effect of the visit on the teaching schedule. The visit would disrupt one whole day for all three grades in the school. Principal Barson acknowledged this but replied that in his weekly meeting with department chairs earlier in the week, they gave their approval to the idea. Since student participation would be voluntary, teachers were to offer review sessions to those students who elected not to attend Winston's presentation.

There being no more discussion, Ms. De Grazia called for a vote on the motion. The motion carried 4–0, with one abstention.

Principal Barson asked the committee if they would assist him in planning and publicizing Winston's visit. The committee agreed. Ms. De Grazia asked if there was any new business. There was none.

Ms. De Grazia adjourned the meeting at 4:20 P.M.

Zenda Hill
Recording Secretary

■ **Figure 18.5**
Set of Minutes

Reflecting Positively on the Participants and the Organization

Your task in recording the minutes includes separating the substance of the meeting from the emotional exchanges of participants. Do not write: "The motion to add a new position in the QA Department was defeated 7 to 6 after a heated argument in which Bob Minor complained that 'Alice states that she supports my department, but when it comes to action, she's all talk.' " Instead, write: "After considerable discussion, the motion to add a new position in the QA department was defeated by a vote of 7 to 6." Sometimes the smartest thing a recording secretary can do is choose not to record what was said.

Figure 18.5 is an example of an effective set of minutes.

✔   **Revision Checklist**

1. Did you choose an appropriate format for the informal report?

2. Does the directive
   ❑ clearly and politely explain your message?
   ❑ explain your reasoning, if appropriate?

3. Does the trip report
   ❑ clearly explain the purpose of the trip?
   ❑ highlight the important information?
   ❑ append appropriate supporting materials?

4. Does the field or lab report
   ❑ clearly explain the important information?
   ❑ use, if appropriate, a problem-methods-results-conclusion-recommendations organization?

5. Does the progress or status report
   ❑ clearly announce that it is a progress or status report?
   ❑ use an appropriate organization?
   ❑ clearly and honestly report on the subject and forecast the problems and possibilities of the future work?
   ❑ append supporting materials that substantiate the discussion?

6. Do the minutes
   ❑ provide the necessary housekeeping details about the meeting?
   ❑ explain the events of the meeting accurately?
   ❑ reflect positively on the participants and the organization?

Exercises

Some of the following exercises ask you to write a memo. See Chapter 15, page 430, for a discussion of writing memos.

1. As the manager of Lewis Auto Parts Store, you have noticed that some of your salespeople smoke in the showroom. You have received several complaints from customers. Write a directive in the form of a memo defining a new policy: salespeople may smoke in the employees' lounge but not in the showroom.

2. There are 20 secretaries in the six departments at your office. Although they are free to take their lunch hours whenever they wish, sometimes several departments have no secretarial coverage between 1:00 and 1:30 P.M. Write a directive in the form of a memo to the secretaries explaining why this lack of coverage is undesirable and asking for their cooperation in staggering their lunch hours.

3. If you have attended a lecture or presentation in your area of academic concentration, write a trip report to your instructor assessing its quality.

4. The three reports below could be improved in tone, substance, and structure. Revise them to increase their effectiveness, adding any reasonable details.

a.

KLINE MEDICAL PRODUCTS

Date:	1 September 20XX
To:	Mike Framson
From:	Fran Sturdiven
Subject:	Device Master Records

The safety and efficiency of a medical device depends on the adequacy of its design and the entire manufacturing process. To ensure that safety and effectiveness are manufactured into a device, all design and manufacturing requirements must be properly defined and documented. This documentation package is called by the FDA a "Device Master Record."

The FDA's specific definition of a "Device Master Record" has already been distributed.

Paragraph 3.2 of the definition requires that a company define the "compilation of records" that makes up a "Device Master Record." But we have no such index or reference for our records.

Paragraph 6.4 says that any changes in the DMR must be authorized in writing by the signature of a designated individual. We have no such procedure.

These problems are to be solved by 15 September 20XX.

b.

Diversified Chemicals, Inc.
Memo

Date:	August 27, 20XX
To:	R. Martins
From:	J. Speletz
Subject:	Charles Research Conference on Corrosion

The subject of the conference was high-temperature dry corrosion. Some of the topics discussed were

1–thin film formation and growth on metal surfaces. The lectures focused on the study of oxidation and corrosion by spectroscopy.

2–the use of microscopy to study the microstructure of thick film formation on metals and alloys. The speakers were from the University of Colorado and MIT.

3–one of the most interesting topics was hot corrosion and erosion. The speakers were from Penn State and Westinghouse.

4–future research directions for high-temperature dry corrosion were discussed from five viewpoints.

    1–university research
    2–government research
    3–industry research
    4–European industry research
    5–European government research

5–corrosion of ceramics, especially the oxidation of Si_3N_4. One paper dealt with the formation of Si ALON, which could be an inexpensive substitute for Si_3N_4. This topic should be pursued.

c.

FREEMAN, INC. INTEROFFICE

To:	C. F. Ortiz
From:	R. C. Nedden
Subject:	Testing of Continuous Solder Strip Alternative for Large-Scale Integrated Terminals
c:	J. A. Jones
	M. H. Miller

We ordered samples of continuous solder strips in three thicknesses for our testing: 1.5 mil, 2.5 mil, and 4.0 mil. Then we manufactured each thickness into terminals to test for pull strength.

The 1.5 mil material had an average pull strength of 1.62 pounds, which is above our goal of 1.5 pounds. But 30 percent of these terminals did not meet the goal. The 2.5 mil material had an average pull strength of 2.4 pounds, with a minimum force of 1.65 pounds. The 4.0 mil material had an average pull strength of 2.6 pounds per terminal, with a minimum of 1.9. Even though there was 60 percent more solder available than with the 2.5 mil material, the average pull strength increased by only 8 percent.

We concluded from this that the limit to the pull strength of the terminal is dependent on the geometry of the terminal, not on the amount of solder.

Research Projects

5. Write a progress report about the research project you are working on in response to exercise 4 on page 516 in Chapter 17. If the proposal was a collaborative effort, collaborate with the same group members on the progress report.

6. You and another student are co-chairs of the Associated Student Organization (ASO) at your college. Student organizations that receive university funding, including the ASO, are required to submit formal minutes of each of their meetings. You have been approached by the dean of students, who oversees student organizations. The dean is concerned that ASO minutes are unprofessional and could reflect poorly on the organization. The dean has asked you to revise the most recent set of minutes (the one you see printed here) and attach a memo describing the changes you have made and explaining your decisions. Submit to your instructor the revised minutes, along with the memo.

Minutes of the Associated Student Organization

May 13, 20XX

The meeting began at 3:10, even though it was scheduled to begin at 3:00. We had to wait for enough people to come to constitute a quorum.

The first issue we discussed was the proposed raise in fees for the General Parking Permit. This year it is $15, which it has been for the last six years. The university's plan to raise it to $30 is unreasonable, especially given the 8% increase in tuition and the 12% increase in dormitory fees. A resolution was passed unanimously to protest this increase.

The discussion turned next to the lineup for next semester's musical events. Bob commented that there were too many country acts, and not enough alternative acts. He said, "A little country goes a long way." Marty got insulted at this and replied, "What do you want, some idiot in leather who spits up fake blood?" He added that country acts are popular with the whole community, whereas a lot of the alternative acts draw very poorly. With the current policy of up-front deposits required by Elite Productions, we have to be sure we don't book any weak acts, which could deplete our whole budget for the year. Bob said he didn't mean to be insulting about country acts, and the two agreed to keep talking about the issue.

The meeting adjourned at 4:15.

7. You are one of three members of the Administrative Council of your college's student association. Recently, the three of you have concluded that your weekly meetings have become chaotic, largely because you do not use rules for parliamentary procedure and because controversial issues have arisen that have attracted numerous students (the meetings are open to all students). You have decided that it is time to consider adopting some parliamentary procedures to make the meetings more orderly and more effective. Look on the Web for models of parliamentary procedure. Is there one that you can adopt? Could you combine elements of several to create an effective model? Find or put together a brief set of procedures, being sure to cite your sources. In a memo to your instructor, discuss the advantages and disadvantages of the model you propose and submit it along with the procedures.

For additional cases, click on Case of the Month and Archive on TechComm Web (www.bedfordstmartins.com /techcomm).

CASE
Amending a Proposal

You are the chief technician for United Paper Company, an Oregon manufacturer of paper and paper products. The company has decided to replace two of its older rolling machines with newer models, each of which will cost more than $300,000. Last month, you formally proposed to your supervisor that you be permitted to visit two of the leading manufacturers of this equipment to study their products; your proposal was approved. You have carried out more research and are preparing your first progress report on this project. You have described the two competing machines and explained how each would fit in your company's manufacturing processes, and you have completed arrangements for visiting the two companies. With a budget of $6,000, you will visit American Equipment in Hawthorne, CA, and Consolidated Industrial Equipment, in Newark, NJ. Your supervisor had balked a little at the cost of the two trips, but he agreed that they were necessary, given the importance of the new machines to the company.

Studying your papers in preparing your progress report, your heart sinks. You discover that you made a serious omission in your proposal: there is a third manufacturer, Southern Printing Equipment, in Atlanta, Georgia. You look back through your materials and discover that Southern's equipment is fully competitive with that of the other two manufacturers. You remember that when you wrote the proposal you were having some problems at home. Your spouse had been laid off, and with your oldest child beginning college next year, you had spent some sleepless nights.

You realize now that adding Southern to your itinerary would delay the completion date, and you would have to request an increased budget. What would be the costs and benefits to you and to the company of explaining the oversight to your supervisor? Of not explaining it? What should you do? If you think it is best to explain the oversight, write a memo to your supervisor. If you think it is best not to mention it, write a memo to your instructor, explaining your thinking.

Writing Formal Reports 19

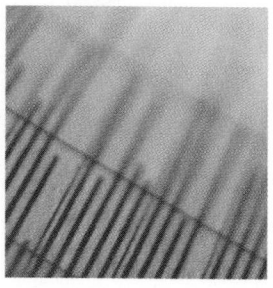

The technical writing instructors J. C. Mathes and Dwight W. Stevenson (1991, p. 87) on planning a report:

Your first concern is . . . providing information that readers need in order to respond or act so that the report accomplishes its purpose. The information that most readers need, however, is minimal — far less than the information you have available and indeed considerably less than you put into your report. Thus, you plan your basic structure according to reader needs rather than in terms of your investigation.

See TechComm Web (www .bedfordstmartins.com /techcomm) for guidelines, additional examples, and links related to topics in this chapter.

For more about proposals, see Ch. 17; for progress reports, see Ch. 18, p. 524.

A formal report can be the final link in a chain of documents that begins with a proposal and continues with one or more progress reports. This last, formal report is often called a *final report, project report,* or *completion report*. The sample document beginning on page 563 is the final report in the series about hiring a Webmaster presented in Chapters 17 and 18.

A formal report can also be a freestanding document, one that was not preceded by a proposal or by progress reports. For instance, your supervisor might ask you to find out employee attitudes toward comp pay: compensating employees who work overtime with time off rather than with overtime pay. Such a project would require that you research the subject and write a single, complete report.

This chapter begins with a discussion of the types of formal reports — informational, analytical, and recommendation — and a problem-solving model for writing them. It discusses an important type of recommendation report, the feasibility report, and the typical components of formal reports. Finally, it presents a sample report written by a group of students.

UNDERSTANDING THE TYPES OF FORMAL REPORTS

One of the challenges in talking about reports is that there is no standard terminology. There are *lab reports* and *site studies, progress reports, fact-finding reports, environmental-impact statements, analytical reports, completion reports, annual reports, informational reports, interim reports,* and more. Some of these terms refer to the topic of the report (*environmental-impact statements, lab re-*

ports), some to the phase of the investigation (*progress reports, annual reports*), and some to the function or purpose of the report (*informational reports, analytical reports*). Some of these terms have special meanings in particular fields and in particular organizations. When you are asked to write a report, talk to more-experienced people at your organization to determine what the word *report* means in that context.

This section describes three kinds of reports, all classified according to their main purpose.

 See Business Communication: Managing Information and Relationships (http://spider.hcob.wmich.edu/bis/faculty/bowman/mir.html#Contents) for a clear and thorough discussion of business reports.

Informational Report: | Presents results

Analytical Report: | Presents results + draws conclusions

Recommendation Report: | Presents results + draws conclusions + makes recommendations

Informational Reports

An informational report presents facts, often referred to as *results*, so that readers can understand a situation. For instance, the circulation department at a library might present an informational report to the library planning board, providing statistics about the number of patrons who visit the library, the kinds of items they borrow, the number and kinds of items the library purchases, the number and kinds of items stolen, and so forth.

If this report is routine and covers a particular period, such as the last quarter, then you would probably write an informal status report. However, if the report is nonroutine — requested by the planning board because it believes the library is not being run well or because the government body that funds the library requested it — you would probably write a formal informational report.

For more on status reports, see Ch. 18, p. 524.

Here are examples of the kinds of questions an informational report addresses:

What is the status of Project X? What is the current status of the work on the project to reinforce the levees on the river? How many open-heart surgeries did the hospital perform last year, and what were the outcomes?

How do we do Function X? What procedures do we currently use to assess the performance of our clerical personnel? What are our quality-control procedures for the laptops we manufacture?

What are the most popular ways of doing Function X? What are our major competitors' policies on returns? What is the ratio of surgical to laser operations for treating gallstones in the United States? What are the three top-selling brands and models of ink-jet printers?

What do our people think of Situation X? How satisfied are our employees with our medical benefits? What do our production personnel think we could do to improve the quality of our products?

An informational report usually involves the primary- and secondary-research techniques discussed in Chapter 7: using the library and the Internet, conducting interviews, and distributing questionnaires.

Analytical Reports

Like informational reports, analytical reports provide information, but then they go one step further: they analyze it and present *conclusions*. A conclusion is an interpretation of the results, an answer to the question, "What do these results mean?" For instance, in writing the report to the planning board of the library, the staff in the circulation department might present all the statistics but highlight the fact that thefts of compact discs have increased substantially. With further analysis, the writers can conclude that theft is high for one type of music disc but not for other types. An analytical report, therefore, tries to describe *why* something happens or *how* it happens or what it *means*.

Like an informational report, an analytical report can be routine (for example, a status report) or special.

Here are some examples of the kinds of questions an analytical report addresses:

What is the best way to do Function X? What is the most effective method for tracking manuscripts being edited by different people? Which therapy for people with spinal-cord injuries has the best outcome?

What causes Situation X? Why is there a high turnover in the Information Systems Department? Why is our competitor's new product doing so much better than ours?

What are the results of Situation X? If we create a Web site for our company, how much increase in business can we expect? If our accounting functions were performed by an outside firm, would we save money?

Could we do Function X? Do we currently have the infrastructure to compete with Company A in bidding on these government contracts? Could we afford to offer tuition reimbursement for all college classes our employees wish to take?

For more on causal reasoning, see Ch. 6, p. 123.

Analytical reports usually include an informational element. For instance, if you work for a software manufacturer that provides customer support, you might commission an analytical report to learn why your customer-support costs are high. In the informational section — the results — you would describe your customer-support costs, focusing on which products, and which aspects of those products, are causing your customers the most problems. For example, most of the problems occur when your customers try to perform a specific type of calculation using your spreadsheet program. In the analytical section — the conclusion — you use these data to explain why this calculation is difficult to perform.

Analytical reports involve analyzing the data to determine cause-and-effect relationships. Until you understand the problem, you can't fix it. An analytical report helps you — and your readers — understand the problem.

Recommendation Reports

A recommendation report goes a step further than presenting information and analyzing it: it advocates a certain course of action. (Sometimes, however, the course of action it advocates is to do nothing.) What is the difference between drawing conclusions and presenting recommendations? If your conclusion is that Product A meets our needs better than Product B or Product C, the recommendation would seem obvious: purchase Product A.

Maybe, but maybe not. A report might recommend that, even though a high number of compact discs is being stolen, the library should take no action: we can't afford any of the available antitheft devices, none of the devices does a good enough job, all the devices create unacceptable problems, or next year there will be a better device on the market. A report can even recommend that, even though Product A is better for us than B or C, we should buy Product B, because B does 90 percent of what A does but costs only 50 percent as much as A.

Here are some examples of the kinds of questions a recommendation report addresses:

What should we do about Problem X? What should we do about the fact that the number of calls to our Technical Support lines is very high? How should we celebrate our company's one hundredth anniversary next year?

Should we do Function X? Although we currently have the infrastructure to compete with Company A in bidding on these government contracts, could we increase our business more effectively in some other way? Although we cannot afford to offer tuition reimbursement for all college classes our employees wish to take, can we reimburse them for classes highly relevant to their work?

Should we use Technology A or Technology B to do Function X? Should we buy copiers or should we lease them? Should we buy several high-output copiers or a larger number of low-output copiers?

We currently use Method A to do Function X. Should we be using Method B? Currently, we have a Technical Publications Department; should we instead subcontract that department's work? Currently, we sort our bar-coded mail by hand; should we buy an automatic sorter instead?

This brief look at informational, analytical, and recommendation reports might suggest that there is a clear demarcation among them and that the process of producing them is neat and orderly. The working world, however, is messy. Often, a routine informational report reveals — or confirms — a problem or an opportunity within the organization. Then a report is commissioned to analyze the situation or present recommendations. As problems and opportunities arise and events occur, people make the best decisions they can with the information they have at the moment. Sometimes they conclude that a recommendation report outlines a wise course of action, only to realize later that the plan won't work; they have to start over, gathering more information and analyzing it all again. Even if the recommendations are implemented and the situation appears to improve, in time a new problem might arise, and the whole process begins again. Writing reports, like the writing process itself, is more cyclical than linear.

You should also be aware of how organizational culture affects report writing:

- *In some organizations, only certain people are authorized to draw conclusions and make recommendations.* For instance, bench chemists at a pharmaceuticals manufacturer might be entitled to present results but not to draw conclusions and make recommendations. Those steps are left to more senior people.

- *Some organizations like to segment the report-writing process so that different collaborative groups work on results, conclusions, and recommendations.* One group writes the informational report, then hands the project over to a different group, which writes the analytical report. A third group writes the recommendation report. Each group, in effect, checks the work of the group or groups that preceded it. The rationale for this method is that it can reduce the chances that inaccurate or incomplete results are used to derive conclusions, and that invalid conclusions are used to formulate unwise recommendations.

A PROBLEM-SOLVING MODEL FOR PREPARING FORMAL REPORTS

See NASA's guide to report writing (ltid.grc.nasa.gov /Publishing/editing/vidcover .htm).

Although the process of researching and writing every report is unique, many writers find that a problem-solving approach helps them to put together an effective report. A problem-solving approach has six basic steps. Figure 19.1 shows a basic problem-solving model and its relationship to various types of reports.

Analyze Your Audience

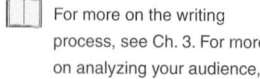
For more on the writing process, see Ch. 3. For more on analyzing your audience, see Ch. 5, p. 92.

The process used in writing most documents — planning, drafting, and revising — is equally effective for reports. Analyzing your audience during the planning stage is critical, because it helps you determine how much information and what kinds of information you need to provide and where to put it. If readers are already knowledgeable, they do not need much additional de-

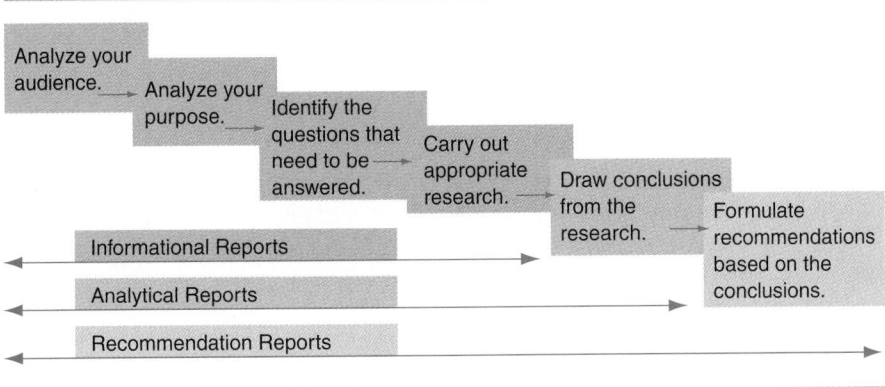

For more on appendices and glossaries, see Ch. 12, pp. 320 and 321. For more on patterns of organization, see Ch. 8.

■ Figure 19.1   **Problem-Solving Models for Formal Reports**

tail to understand the project. That information could go in appendices rather than in the body of the document. This way, readers can refer to it, but they don't have to read it.

Before you start to write, think about these five audience-related questions:

- *How well do your readers know your field or the subject of your report?* The better the readers know the subject, the less explanation you need to provide in the body of the report. If some of them are managers who do not know the subject well, however, consider adding detailed explanations as appendices. Less knowledgeable readers can also benefit from glossaries.

- *Why are the readers reading the report?* If readers merely need to understand the subject, you can provide less information than if they plan to use your report as the basis for some further action. For instance, if some of your readers will duplicate your methods, you need to provide complete details about those methods. Again, it might be appropriate to put the complete description of the methods in an appendix.

- *Are your readers negative, neutral, or positive about the project?* The attitude of the readers might affect the organizational pattern, the amount of detail, and the vocabulary you choose.

- *What is your standing within the organization?* Writers with strong technical credentials and experience do not need to justify their assertions, conclusions, or recommendations as fully as writers who lack an extensive track record.

- *How routine is the project?* Some fields have accepted methods for carrying out certain projects. For example, if in assessing hurricane damage to a physical structure, you always perform a visual inspection first, there is no need to explain how or why you have done so. But if the project is unusual or unique, readers will want to know why you used the approach you did.

Analyze Your Purpose

For more on determining your purpose, see Ch. 5, p. 110.

When you plan a formal report, it is especially important to analyze your purpose carefully, if only because a formal report is likely to be a long and important document. You want to make sure you get the purpose right.

In the broadest sense, you have one of three basic purposes:

- to present information
- to present information and analyze it
- to present and analyze information and make recommendations

However, you also have a purpose or set of purposes that relates more directly to your subject. Think about why you are doing the research, why your readers have requested it, and what they are going to do with it once you present it to them. Is the report routine, or is there a specific problem or opportunity that your readers need to examine in detail? If, for instance, you work for a library and know that your research will help your readers understand the broad problem of theft of library materials, your purpose might be to help them understand how antitheft systems work and what they can be expected to do, or it might be to help the library trustees determine which of the systems (if any) would solve the library's problems. Of course, you can combine several purposes in one report.

Before you proceed too far in your planning, make sure your principal reader agrees with your understanding of your purpose. Writing a memo explaining to your supervisor the basic audience, purpose, and scope of the report can prevent you from going off in the wrong direction.

Identify the Questions That Need to Be Answered

After you have analyzed your audience and purpose, start thinking in detail about the subject of the investigation. Obviously, you have already thought about it a great deal. But now it is time to think specifically about the critical questions you will have to answer. Although sometimes it is your responsibility to determine these questions, often they are established before you begin.

Study your purpose statement carefully. If you are writing an informational report on library antitheft systems, here are some questions you might consider:

- What are antitheft systems?
- What technologies do they use?
- How effective are they in reducing thefts?
- What kinds of library materials do they protect?

- What have been the experiences of libraries like ours that have used them?
- Are there reliable benchmark tests of the different systems?
- How easy are they to install and maintain?
- How often do they break down?
- Do they pose any health risks to library workers or patrons?
- Is there some sort of financial support — grants, for instance — to help finance the purchase of the antitheft systems?

State these questions as precisely as you can.

VAGUE	How do library antitheft systems work?
SPECIFIC	What different technologies are used in the library antitheft systems currently available?

The problem with the vague question is that it could refer to the different technologies or to the steps that library personnel have to follow to operate these devices.

VAGUE	What are some of the disadvantages of library antitheft systems?
SPECIFIC	Do any of the technologies pose any health risks to our workers or patrons?

The specific question doesn't address all the concerns suggested by "some of the disadvantages," but it is a start. You could easily pose a half-dozen more questions about other disadvantages.

Carry Out Appropriate Research

The questions you need to answer will determine the kinds of research you should carry out. Often, you will consult company records, interview people, distribute questionnaires, perform experiments, make observations, and consult books, journals, and Internet sources. You should try to do as much research as time and other resources permit, and you should analyze the information to be sure it is valid and current.

For more on conducting research, see Ch. 7.

Keep in mind, as well, that the needs and preferences of your audience will help you determine the kinds of research to perform. For example, if you know your readers are very concerned about the attitudes of your organization's workers toward policy changes, you will naturally want to interview workers and distribute questionnaires.

Draw Conclusions from the Research

There is no foolproof way to draw valid conclusions from data. You have to think about the data carefully, looking for a body of evidence that suggests a

For more about causal reasoning and avoiding logical fallacies, see Ch. 6, pp. 123 and 129.

causal relationship between two factors. Some kinds of evidence are more valuable than others; benchmark tests from disinterested groups are obviously more persuasive than claims from manufacturers about their own products. Look for a number of different kinds of evidence that all point in the same direction; one or two pieces of evidence are generally insufficient.

As you draw your conclusions, pay particular attention to the possible problems with the logic of your argument. Just as washing your car does not cause rain, new cost-cutting measures implemented in your company are not necessarily responsible for decreased costs. Those decreased costs might have nothing to do with the new measures.

Formulate Recommendations Based on the Conclusions

In most cases, if you have carefully formulated the questions you need to answer, carried out appropriate research, and drawn valid conclusions, the recommendation will flow directly and inevitably from those conclusions. For instance, if you have identified why customers are having problems with one of your software products and concluded that the best course of action is to send all customers new disks containing a software patch, your recommendation would probably be to go ahead with this plan as soon as possible.

But this is not necessarily the recommendation you will present. For example, you might know that your company does not have the funds to go ahead with this plan right away. Or maybe your company is planning a new upgrade next quarter, and you feel that your plan would be more trouble than it is worth to your customers.

Sometimes managers decide not to implement recommendations, even though the recommendations seem perfectly valid and reasonable. This fact can surprise and dismay the report writers, who have spent many hours on the project. In some cases, managers simply change their minds about priorities, making the recommendations irrelevant. In other cases, the situation changes, making the recommendations irrelevant. For example, you might be unaware that your company is being acquired by another company, and management wants to get authorization to carry out the recommendations from its new supervisors next month.

GUIDELINES

Confronting Political Realities in Writing Reports

For more on ethics, see Ch. 2.

Managers sometimes wish to pursue a course of action that is inconsistent with your findings. For example, you might conclude that the best course of action is to implement Plan A, but management favors Plan B and has no intention of implementing Plan A. Should you just tell them what they want to hear by recommending Plan B, adjusting your results and conclusions to lead to this recommendation? In most instances, the answer is simple: no. Recommending Plan B would be unethical, because it would entail lying or misleading. On the other hand, an appropriate course of action might be to recommend each plan equally, provided that the following conditions are present:

- There is nothing unsafe or unethical about Plan B.
- Plan B is not clearly inferior to Plan A.
- You do not have to lie or mislead to recommend Plan B.

This course of action would give you the opportunity to make the case for Plan A, while still making it easy for management to choose Plan B if they remain unconvinced of the superiority of Plan A.

UNDERSTANDING FEASIBILITY REPORTS

One kind of recommendation report is written so often that it deserves special discussion. A *feasibility report* documents a study that evaluates at least two alternative courses of action. For example, should we expand our product line to include a new item, or should we make changes in an existing product?

A feasibility report is an argument that answers three kinds of questions:

- *Questions of possibility.* We would like to build a new rail line to link our warehouse and our retail outlet, but if we cannot raise the money, the project is not possible at this time. Even if we have the money, do we have authorization from government authorities? If we do, are the soil conditions adequate for the rail link?

- *Questions of economic wisdom.* Even if we can raise the money to build the rail link, is it wise to do so? If we use all our credit on this project, what other projects will we have to postpone or cancel? Is there a less expensive — or a less financially risky — way to achieve the same goals?

- *Questions of perception.* If your company's workers have recently accepted a temporary wage freeze, they might view the rail link as unnecessary. The truckers' union might see it as a threat to truckers' job security. Some members of the general public might also be interested parties because any large-scale construction might affect the environment. Even though your plan might be acceptable according to its environmental-impact statement — the study required by the government — some citizens might disagree with the statement or oppose the project on aesthetic grounds. Whether or not you agree with their objections, going ahead with the project could create bad publicity.

For more about considering opposing viewpoints, see Ch. 6, p. 126.

The following discussion explains the six steps that are particular to preparing feasibility reports.

To prepare a feasibility report:
Identify the problem or opportunity.
Establish criteria for responding to the problem or opportunity.
Determine the options.
Study each option according to the criteria.
Draw conclusions about each option.
Formulate recommendations based on the conclusions.

Of course, you must begin by analyzing your audience and purpose, as you would for any technical document.

Identify the Problem or Opportunity

What is not working, or not working as well as it might? What situation presents an opportunity for our organization to decrease its costs or improve the quality of its product or service? Without a clear statement of your problem or opportunity — that is, if you don't answer these kinds of questions — you cannot plan your research.

For example, you know that a number of employees in your organization smoke, and that these employees are absent and ill more often than nonsmoking employees. Your supervisor has asked you to investigate whether your company should offer employees a free smoking-cessation program. In talking with your supervisor, you learn that the only way the company can offer the program is if the company's insurance carrier will pay for the program or reduce premiums for those people who successfully complete it. This fact tells you that the first thing you need to do is talk with the insurance agent; if the agent says the carrier will pay for the program or reduce the premiums, then you can proceed with your investigation. If the agent says no, the problem might shift; now the problem is to determine whether another insurance company offers your organization better coverage or whether there is some other way to encourage employees to stop smoking.

Establish Criteria for Responding to the Problem or Opportunity

For more about establishing criteria, see Ch. 8, p. 193.

Criteria are standards against which you measure your options. Criteria can take two forms: *minimum specifications* and *evaluative criteria*. For example, if you need to find an apartment to live in when you return to school in the fall, your minimum specifications might be that it has to cost less than $400 per month and be located within one mile of campus. Your evaluative criteria might be that the kitchen is separated from the living room by a wall and that laundry facilities are available. In other words, minimum specifications define what you need, while evaluative criteria define what you want. These examples of evaluative criteria can be further refined: by laundry facilities, do you mean a washer and dryer in the apartment or a laundromat located on the premises? Which would be preferable?

When you solve problems in the working world, you use the same process of establishing what you *need* and what you *want*. For example, in selecting photocopiers for your department, a minimum specification might be that each copy cost less than two cents to produce and that the photocopier be able to handle oversized documents. Evaluative criteria might include that the photocopier do double-sided copying and stapling. You could get by without these features, but they are desirable; therefore, you might rate a photocopier offering these features more highly than one without the features, all other things being equal. That is the function of evaluative criteria: to enable you to make distinctions among a variety of similar objects, objectives, actions, or effects.

Until you can establish your criteria, you have no way of knowing what your options are. Sometimes you inherit your criteria; your supervisor tells you how much money you can spend, for instance, and that figure becomes part of your minimum specifications. Sometimes you derive your criteria from your research; you study your photocopying needs and determine the standards for evaluating different copiers.

Determine the Options

After you establish your criteria you can determine your options. *Options* are potential solutions. For example, your options in choosing an apartment are the apartments that have vacancies and that meet your minimum specifications and, if possible, many of your evaluative criteria. Likewise, your options for the photocopier project are the different photocopiers that meet your minimum specifications and evaluative criteria.

Determining your options is a critically important phase in the problem-solving process because if you fail to identify all the appropriate options, the rest of the analysis will be flawed. For instance, if you fail to consider the photocopiers made by Xerox, the results, conclusions, and recommendations you present later, in your report, might well be invalid, because Xerox might manufacture the best copier for your needs.

For this reason, you have to be especially careful that you carry out your research thoughtfully and thoroughly. You must use the appropriate methods of primary and secondary research (see Chapter 7) to discover what companies manufacture the kind of copier you are looking for.

Study Each Option
According to the Criteria

Once you have established your criteria and determined your options, you study each option according to the criteria, a process that might include both primary and secondary research. In the case of the photocopier project, secondary research would include studying articles about photocopiers in technical journals and specification sheets from the different manufacturers. Primary research might include observing product demonstrations as well as interviewing representatives from different manufacturers and managers who have purchased different brands.

Studying the options usually involves collecting objective and subjective information. An objective piece of information would be the number of copies per minute that a photocopier produces. A subjective piece of information would be an assessment of ease of use or print quality.

To make the analysis of the options as objective as possible, investigators sometimes create a *matrix*, a method for systematically entering an evaluation for each option according to each criterion. A matrix is really just a table (or a spreadsheet), as shown in Figure 19.2.

Does using a matrix ensure an objective analysis? Not at all. Subjectivity can creep in at three stages:

Option	Criterion 1: Pages per Minute	Criterion 2: Ability to Duplex	Criterion 3: Extra Paper Bins	Criterion 4: Color Printing	Total Points
Option 1: Ricoh Model 311	9	8	7	9	33
Option 2: Xerox Model 4500	8	9	7	2	26
Option 3: Savin Model 12X	10	8	8	9	35
Option 4: Sharp Model S350	7	8	8	9	32

■ **Figure 19.2   A Matrix**

To use a matrix, you assign a value (0–10 is a common range) for each criterion for each option. Then you add up the values for each option and compare the totals. In this case, option 3 scores the highest, with 35 points; option 2 scores the lowest, with 26 points.

- *Determining which criteria to examine.* The Xerox Model 4500 — Option 2 — did very poorly on criterion 4; if criterion 4 were removed from the analysis, or if many other criteria were added, the Xerox might score much higher.

- *Deciding the range of values for each criterion.* If one of your criteria is whether the copier can staple, how do you decide what value to assign to a machine that does stapling? To a machine that does not? If you give 10 points to a machine that does stapling and 0 points to a machine that doesn't, you have probably eliminated the nonstapling machine from contention. However, if the ability to staple is not very important, it might be more reasonable to assign a value of 8 points to nonstapling machines. To help readers understand your thinking, you should explain your ranking system. For instance, to accompany Figure 19.2, you should explain that you give a copier 9 points if it can print duplex and 8 points if it cannot, because printing duplex is an unimportant criterion. However, you give a copier 9 points if it can print in color and 2 points if it cannot, because printing in color is a very important criterion.

- *Assigning values to criteria.* If, for example, one of your criteria is ease of operation, you might give one machine a 9, whereas someone else might give it a 3. Other criteria are equally difficult to assess objectively. For example, what value do you give to the cost criterion if one machine costs $12,500 per year to operate and another costs $14,300?

Clearly, evaluating options according to your criteria is subjective. Still, for you as the writer, the principal advantage of a matrix is that it helps you do a methodical analysis. For your readers, a matrix makes your analysis easier to follow by clarifying the presentation of your methods and results.

Draw Conclusions about Each Option

Whether you use a matrix or some other, less formal means of recording your evaluations, your next step is to draw conclusions about the options you have studied — to interpret your results and write evaluative statements about the options.

For the study of photocopiers, your conclusion might be that a particular model made by Ricoh is the best copier: it meets all your minimum specifications and the greatest number of your evaluative criteria, or it scores highest on your matrix. Depending on your readers' preferences, present your conclusions in any one of three ways.

- *Rank all the options:* the Ricoh copier is the best option, the Savin copier is second best, and so forth.

- *Classify all the options in two categories:* acceptable and unacceptable.

- *Present a compound conclusion:* the Ricoh offers the most technical capabilities; the Savin is the best value.

Formulate Recommendations Based on the Conclusions

 See Writing Guidelines for Engineering and Science Students (http://fbox.vt.edu:10021/eng/mech/writing) and Online Technical Writing (www.io.com /~hcexres /tcm1603/acchtml/acctoc .html) for samples and discussions of reports.

The earlier discussion of recommendations (p. 548) applies equally to feasibility reports, but with one additional note: don't be afraid to recommend that your organization take no action. Research projects can yield mixed or bad news: none of the options would be an unqualified success or none would work at all. Don't feel that a negative recommendation reflects negatively on you. If the problem being studied were easy to solve, it probably would have been solved before you came along. Give the best advice you can, even if that advice is to do nothing. You would not want to recommend a course of action that will not live up to the organization's expectations.

GUIDELINES

Preparing Feasibility Reports

To prepare a feasibility report, follow these eight steps:

1. *Analyze your audience.* What do they need to know?

2. *Analyze your purpose.* What do you want to accomplish?

3. *Identify the problem or opportunity.* What is not working, or what improvements could you make?

4. *Establish criteria.* Decide on minimum specifications and evaluative criteria for responding to the problem or opportunity.

5. *Determine your options.* What are the available courses of action?

6. *Study each option according to your criteria.* Work out a systematic way to evaluate each option to reduce the subjectivity of the analysis.

7. *Draw conclusions about each option. Conclusions* are statements of what your data mean.

8. *Formulate recommendations based on your conclusions. Recommendations* are statements of what we ought to do next.

ORGANIZING THE BODY OF THE FORMAL REPORT

Every report should reflect the characteristics of its audience, purpose, and subject. One basic structure can serve as a good starting point for thinking about how to structure a report. A typical formal report contains many of the elements shown in Figure 19.3.

More than most other technical documents, reports are read by multiple audiences: managers who are not technically competent in the field and need only an overview of the project; technical personnel who are competent in the field and need detailed information; technical personnel in related fields, attorneys, or representatives of regulatory agencies.

front matter	body	back matter
• title page	• introduction	• glossary
• abstract	• methods	• list of symbols
• table of contents	• results	• references
• list of illustrations	• conclusions	• appendices
• executive summary	• recommendations	

Front matter and back matter are discussed in Chapter 12.

The body of the report is discussed in this chapter.

■ **Figure 19.3   Elements of a Typical Report**

To accommodate all these audiences, a formal report generally consists of modular components that remain independent yet work together to form the whole. For instance, in a typical formal report, an executive summary precedes the body of the document. Some readers will skip the executive summary; others will read nothing but the executive summary.

Some writers like to draft these components in the order in which they will be presented. They like to compose the introduction first because they want to be sure they have a clear sense of direction before drafting the discussion and findings. Other writers prefer to put off writing the introduction until they have completed the other elements of the body. They reason that, in writing the discussion and the findings, they will inevitably have to make substantive changes; if they wrote the introduction first they would have to revise it. In either case, analyze your audience and purpose, brainstorm, and carefully outline before you begin to draft.

Introduction

The introduction enables readers to understand the technical discussion that follows. Start by analyzing who its readers are and then considering these standard questions:

- *What is the subject of the report?* If the report follows a proposal and progress report, you can probably copy this information from one of these documents and modify it as necessary.

- *What is the purpose of the report?* The purpose of the report is not the purpose of the project. The purpose of the report is to present information to enable readers to understand a subject, to affect readers' attitudes toward a subject, or to enable readers to carry out a task.

- *What is the background of the report?* You might be able to copy this information from a previous document. However, in describing the background, you are addressing readers who might not have read your previous documents and who therefore might be ignorant of the project; do not assume that they have been following along. Even those readers who have read the previous documents might need a refresher.

For more about purpose statements, see Ch. 5, p. 110.

- *What are your sources of information?* Briefly describe your primary and secondary research, to prepare your readers for a more detailed discussion of your sources in subsequent sections of the report.

- *What is the scope of the report?* Indicate the topics you are including as well as those you are not.

- *What are the most significant findings?* Summarize the most significant findings of the project.

For more on introductions, see Ch. 8, p. 206.

- *What are your recommendations?* In a short report containing a few simple recommendations, you might include those recommendations in the introduction. In a lengthy report containing numerous complex recommendations, you might briefly summarize those recommendations in the introduction, and then refer readers to the more detailed discussion in "Recommendations."

- *What is the organization of the report?* Indicate the organizational pattern you will use , so readers can understand where you are going and why.

- *What key terms are you using in the report?* If you are introducing new terms, the introduction is an appropriate place to provide definitions. If you need to define many terms, consider placing them in a glossary and refer readers to it in the introduction.

See OCLC (www.oclc.org) for examples of research reports.

Figure 19.4 shows the introduction to the body of a formal report — in this case, a feasibility report. The subject of the report (Corder, 1993) is an investigation to determine whether the Quality Assurance (QA) lab at a microchip manufacturer should purchase a scanning electron microscope (SEM).

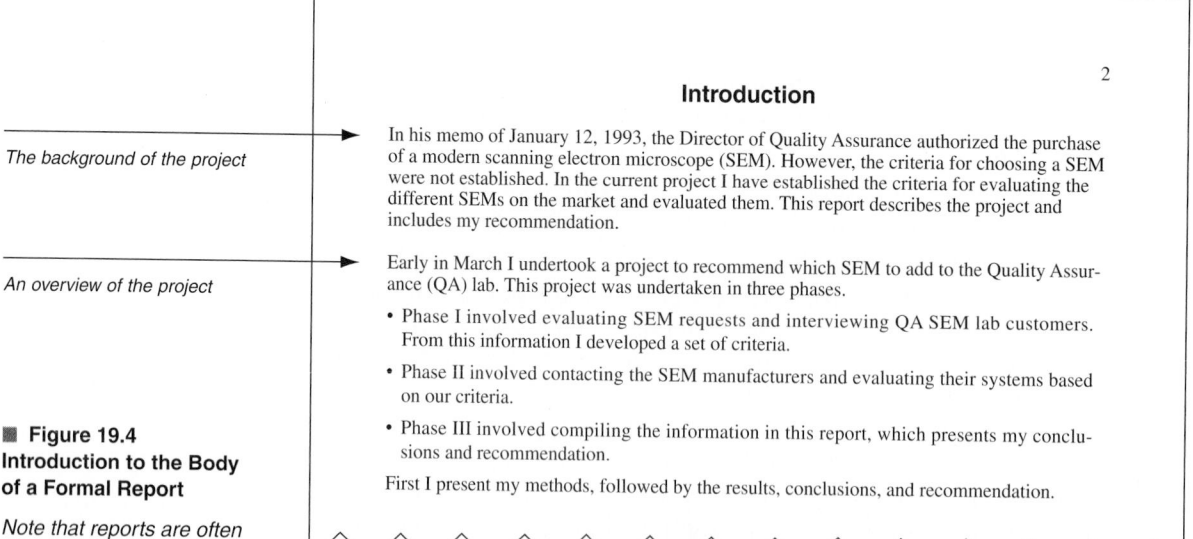

**■ Figure 19.4
Introduction to the Body
of a Formal Report**

*Note that reports are often
double-spaced.*

The figure shows:

2

Introduction

The background of the project →

In his memo of January 12, 1993, the Director of Quality Assurance authorized the purchase of a modern scanning electron microscope (SEM). However, the criteria for choosing a SEM were not established. In the current project I have established the criteria for evaluating the different SEMs on the market and evaluated them. This report describes the project and includes my recommendation.

An overview of the project →

Early in March I undertook a project to recommend which SEM to add to the Quality Assurance (QA) lab. This project was undertaken in three phases.

- Phase I involved evaluating SEM requests and interviewing QA SEM lab customers. From this information I developed a set of criteria.

- Phase II involved contacting the SEM manufacturers and evaluating their systems based on our criteria.

- Phase III involved compiling the information in this report, which presents my conclusions and recommendation.

First I present my methods, followed by the results, conclusions, and recommendation.

3

Methods

To carry out this project, I performed four tasks:

1. I studied all the internal QA records of the requests that the SEM lab has received over the last year. For each request I noted the department of the requester and the kind of information requested. For example, some requesters needed to determine the location of a fault. Others needed to determine the nature of the fault or needed an image of the fault.
2. I interviewed the 12 requesters to find out how well the lab is meeting their needs and what further information would help them do their jobs. I used a standard script (see Appendix B) that combined closed-ended and open-ended questions.
3. I wrote to the six North American SEM manufacturers (see Appendix C) to get specifications, warranties, and prices on the available systems that would fulfill the requirements I established on the basis of the requests and the interviews.
4. I analyzed the information from the SEM manufacturers and the independent reviews of the equipment in the professional journals (see References).

■ **Figure 19.5**
Methods Section of a Formal Report

Notice how specifically the methods are described.

The report is titled "QA Lab SEM Evaluation: Project Completion Report." Marginal notes have been added.

Methods

The methods section answers the question "What did you do?" In drafting the methods section, provide the appropriate kind and amount of information for your audience.

In analyzing your audience, consider your readers' knowledge of the field, their perception of you, and the uniqueness of the project, as well their reasons for reading the report and their attitudes toward the project. Provide enough information to enable readers to understand what you did and why you did it that way. If others will be using the report to duplicate your methods, be sure to include sufficient detail.

Figure 19.5 shows the methods section from the report on the SEM.

Results

Results are the data you have discovered or compiled. Present the results objectively, without comment; save the interpretation of the results — your conclusions — for later. If you combine results and conclusions, your readers might be unable to follow your reasoning and might not be able to tell whether your conclusions are justified by the evidence.

The methods section answers the question "What did you do?" The results section answers the question "What did you see?"

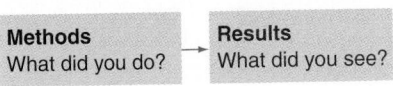

Methods
What did you do?

Results
What did you see?

The needs of your audience will help you decide how to structure the results. How much they know about the subject, what they plan to do with the report, what they expect your recommendation to be — these and many other factors will affect how you present the results.

For instance, suppose that your company is considering installing an intranet, a companywide computer network. In the introduction, you have discussed the company's current system for internal communication, explaining its disadvantages and limitations. In the methods section, you have described how you established the minimum specifications and evaluative criteria you applied to the available intranet systems, as well as the procedures you followed in your research. In the results section, you provide the details of each intranet system you are considering, as well as the results of your evaluation of those systems.

Figure 19.6 is an excerpt from the results section of the scanning electron microscope report.

Conclusions

For more on drawing conclusions and evaluating evidence, see Ch. 6, pp. 124 and 125.

Conclusions are the implications — the meaning — of the results. To draw conclusions, you need to think carefully about your results, weighing whether they point clearly to a single meaning.

Just as the results section answers the question "What did you see?" the conclusions section answers the question "What does it mean?"

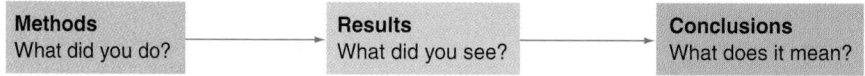

The conclusions section of the scanning electron microscope report is shown in Figure 19.7 on page 561.

Recommendations

Recommendations are suggestions for taking particular actions. Just as the conclusions section answers the question "What does it mean?" the recommendations section answers the question "What should we do now?" As discussed earlier in this chapter, recommendations do not always flow directly from conclusions. Always consider recommending that the organization take no action, or no action at this time.

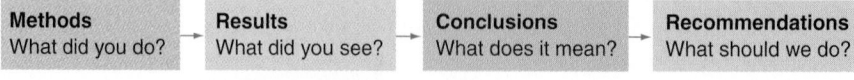

5

Results

Job Requests and Interviews
I began by collecting all SEM job requests submitted to the QA SEM lab over a period of one year. In analyzing these job requests, I found that the largest proportion (some 75 percent) of the job requests came from either the failure-analysis engineers in QA and product engineering, or from the R&D design engineers. Most of the requests were similar: find a specific bit or location on an individual die, and image the area to find defects or failure mechanisms.

I simultaneously began interviewing the 12 requesters. I asked them if our lab provided them adequate images to define failure mechanisms, and if there were other functions that would make their identification of defects and locations simpler, quicker, or better in some other way.

The answers to these questions were that the images and data provided by the QA SEM lab are of sufficient resolution and contain adequate information. However, many requesters expressed a desire to have the images on the network and available for manipulation with an image processor, such as Semicaps. According to the requesters, this would accomplish three things for the engineers:

- It would allow them to include images in their reports through Harvard Graphics or other similar software.
- It would allow the engineers, at their local workstation, to make adjustments to the contrast, brightness, gamma, and other aesthetic characteristics of the image.
- It would allow the images to be archived electronically, rather than saving file cabinets full of Polaroid photos, as is currently done.

Determining Criteria
Using the information I obtained through the analysis of requests and the interviews of requesters, I developed a set of criteria for evaluating the various SEMs available on the market. The features I specified as criteria for the QA lab are the following:

- Turbomolecular vacuum pump (TMP)
- Lanthanum hexaboride electron gun (LaB6 Cathode)
- Digital imaging
- Solid-state backscattered electron detector (BSE)
- 100 mm or larger specimen stage
- Eucentric stage controls
- Minimal integral image processing

Turbomolecular Pump
Of the two major types of vacuum pumps on SEMs — diffusion and turbomolecular — turbomolecular will better meet our needs.

■ **Figure 19.6**
Results Section of a Formal Report

6

Diffusion pumps use a heated oil vapor to remove air molecules. These pumps are quite efficient and relatively inexpensive to operate, but simple errors can cause the oil to back-stream into the specimen chamber and even into the column. Backstreaming oil causes many problems once inside the SEM column.

Turbomolecular pumps use a series of very high speed, rotating turbines to remove air molecules. These pumps are very dependable, very fast, and very efficient. Most important, however, they cannot backstream oil into the column; therefore, they are a much safer type of vacuum pump for a SEM.

Lanthanum Hexaboride Electron Gun
The lanthanum hexaboride electron gun (LaB6) is the best of the three common types of electron guns used in SEMs.

- The most common is a tungsten gun. Tungsten is the material used for filaments in light bulbs. Although the value of tungsten guns is well documented, they have problems. A tungsten filament (the consumable portion of the gun) is very inexpensive, at about $25 each, but typical lifetimes are only 20–25 hours of use.

- The LaB6 gun uses an LaB6 crystal instead of a tungsten filament. There are several advantages to this. First, LaB6 puts out more electrons at lower accelerating voltages than does tungsten. Second, LaB6 filaments last several hundreds of hours. The trade-off is that LaB6 filaments cost close to $600 each. The extended use and the enhanced signal, however, more than make up for the extra cost of replacement.

- The field-emission (FE) gun uses a slightly different technique. FE guns put out even more electrons at lower accelerating voltages than LaB6, but they are much more delicate to operate. FE guns require a higher vacuum, therefore, more vacuum pumps. FE guns are temperamental and difficult to use. They are also typically very expensive to operate, up to 50 percent more expensive than TMPs or LaB6s. The difference in low accelerating voltage does not justify the use of an FE gun in the QA lab.

Digital Imaging
Digital imaging capability, a requirement for a new SEM, is a standard feature on most new SEMs.

Other Criteria
All modern SEMs meet our other four criteria:

- Solid-state backscattered electron detector (BSE)

- 100 mm or larger specimen stage

- Eucentric stage controls

- Minimal integral image processing

■ **Figure 19.6** *(Continued)*

7

Contacting SEM Manufacturers
I contacted the SEM manufacturers and asked them to provide brochures and information on the SEMs they have available.

We had assumed that because 10 years have passed since we purchased our JEOL 840, many advances had been made in SEM technology. That assumption proved false. The technology of producing images with a scanning electron beam has not changed significantly. No dramatically different columns or detectors have been developed. The major changes include only two additions:

- Nearly all SEMs now produce digital video images, as opposed to the analog imaging of our 840.

- Nearly all SEMs now include much higher degrees of computerization and integral image processing.

The computerization of SEMs is very much like the Semicaps software already mentioned. The difference is that the image processors that are integral to many SEMs are not compatible with systems we already have in use in QA. Nearly any image processor can manipulate a digital image, and most SEMs built today produce digital images. Therefore, "new" SEM technology is fairly old.

■ **Figure 19.6**
(Continued)

9

Conclusions

Based on its technical capabilities, price, and service record, the JEOL 5100 SEM is the best option for us.

In addition to Amray and JEOL, I interviewed representatives from Hitachi, Olympus, ISI, Cambridge, and Electroscan. Each of these companies offer SEMs that meet the basic criteria I established. I have evaluated the differences that each of these companies highlights. I found none of these features to be significant for our application. In addition, our experience with most of these companies is quite limited.

On the other hand, we have considerable experience with both Amray and JEOL. We have a good idea of their reliability; we have established service contracts with both companies. Therefore, we have established relationships with their field-service engineers. In addition, we have some stock of spares from each company, many of which are applicable to their new models.

With all these factors considered, Amray and JEOL were the only manufacturers whom I invited to visit our site for more thorough discussions and to provide system quotes.

Amray quoted a model 1810 SEM. The total price from Amray is $100,000. JEOL quoted a model 5100 SEM. The total price of the JEOL system is $115,500. While the JEOL appears 15 percent higher in price, we have learned from experience that the Amray will require a retrofitted Eddie Fjeld stage before it is operational. The Eddie Fjeld stage suitable for the Amray model 1810 is available for $28,000, installed. Therefore, the actual price for the Amray 1810 is $128,000.

Beyond the price difference, however, the SEM technicians, the equipment engineers, and I all have greater confidence in the JEOL service organization on the basis of our excellent relationship with them over the years.

In the last paragraph the writer describes the unanimous opinion of his colleagues.

■ **Figure 19.7   Conclusions Section from a Formal Report**

Notice that the writer explains his reasoning carefully.

10

Recommendation

I recommend that we purchase a JEOL 5100 SEM configured with the features listed in the report and on the quotation. The total system price will be $115,000. Installation will be another $1,000, for a total project cost of $116,000.

■ **Figure 19.8**
Recommendations Section from a Formal Report

GUIDELINES

Writing Recommendations

As you draft your recommendations, consider the following four factors:

▶ *Content.* Be clear and specific. If the project has been unsuccessful, don't simply recommend that your readers "try some other alternatives." What alternatives do you recommend and why?

▶ *Tone.* When you recommend a new course of action, you run the risk of offending whoever formulated the earlier course. Do not write that acting on your recommendations will "correct the mistakes" that have been made. Instead, write that acting on your recommendations "offers great promise for success." A restrained, understated tone is not only more polite but also more persuasive: it indicates that you are interested only in the good of your company, not personal rivalries.

▶ *Form.* If the report leads to only one recommendation, use traditional paragraphs. If the conclusion of the report leads to more than one recommendation, consider a numbered list.

▶ *Location.* Consider including a summary of the recommendations — or, if they are brief, the full list of recommendations — after the executive summary or in the introduction as well as at the end of the body of the report.

 To see the completion report for the American Sign Language study, click on Additional Sample Documents for Ch. 19 on TechComm Web (www.bedfordstmartins.com /techcomm).

Figure 19.8 shows the recommendations section of the scanning electron microscope report.

SAMPLE FORMAL REPORT

The following example (Grapatin, Moon, and Johnson, 1999) is the completion report on the project proposed in Chapter 17 on page 505. The progress report for the same project appears in Chapter 18 on page 529. Presentation graphics from an oral presentation on the project appear in Chapter 22 on page 643.

Note: Reports are often double spaced.

ASI   *Ada Semiconductor Incorporated*
220 Technical Street   Telephone: (208) 555-2200
Boise, Idaho 83706   Fax: (208) 555-2201

www.asisemi.com

MEMORANDUM

DATE: May 12, 1999
TO: Michael Jensen, President and CEO
FROM: Ad Hoc Web Committee Team Members:
        Julie Grapatin — Finance Department Rep.
        Tammy T. Moon — Engineering Department Rep.
        Christine Johnson — Software Department Rep.
SUBJECT: Completion Report for a Feasibility Study for Hiring a Webmaster

Transmittal "letters" can be presented as memos.

Attached is our completion report for the feasibility study about hiring a Webmaster. We researched the three hiring options presented in our proposal dated March 4, 1999: (1) contract/outsource, (2) permanent, full-time ASI employee, and (3) college intern.

The purpose of the study

First, we researched ASI's current site. We compiled an inventory of the Web pages ASI is currently using, and we determined whether the ASI site is likely to grow. Finally, we researched the three hiring options available to ASI.

The methods

We concluded that the best course of action is to contract through Roberts, Inc. to hire a Webmaster to set up our site, then hire our own full-time employee to maintain the site. We recommend that these actions be taken immediately.

The major conclusion and recommendation

Thank you for the opportunity to perform this research. We look forward to working on future projects for ASI. If you have any questions or comments, please feel free to contact us.

A polite offer to provide more information

The title indicates the subject and purpose of the report.

**FEASIBILITY STUDY FOR HIRING A WEBMASTER:
A COMPLETION REPORT**

Prepared for: **Michael Jensen**
                       President and CEO
                       Ada Semiconductor, Inc.

Prepared by: **Ad Hoc Webmaster Committee**
                       Julie Grapatin — Finance Department Representative
                       Tammy T. Moon — Engineering Department Representative
                       Christine Johnson — Software Department Representative

May 12, 1999

ABSTRACT

"Feasibility Study for Hiring a Webmaster: A Completion Report"

Prepared by: **Ad Hoc Webmaster Committee**
Julie Grapatin — Finance Department Representative
Tammy T. Moon — Engineering Department Representative
Christine Johnson — Software Department Representative

Ada Semiconductor, Inc. management authorized a study to determine whether ASI should hire a Webmaster to improve and maintain its Web site. The current Web site is disorganized, cumbersome to navigate, and inaccurate. Additionally, creating and maintaining ASI's Web site takes an ever-increasing amount of time, preventing employees from devoting enough attention to the tasks they were hired to perform. We researched three different methods of hiring a Webmaster: (1) a contract/outsource hire, (2) a permanent, full-time ASI employee, or (3) a college intern. We recommend that ASI hire a Webmaster from Roberts, Inc. to restructure and update the existing ASI Web site. While a Webmaster from Roberts is working on the Web site, ASI should search for a Webmaster to hire as a full-time ASI employee. This employee, who should be fully competent in Web design and programming languages, e-commerce, and site functionality, would supervise the company's Web site and overall Internet presence, create internal policies regarding the site, maintain the site, and publicize it. ASI should also attempt to participate in Central State University's intern program (when skilled interns are available) to assist the ASI Webmaster. By combining the above options, ASI can provide a well-developed, user-friendly, accurate, and standardized Web site.

Keywords: Ada Semiconductor, Inc., ASI, Webmaster, Roberts

The abstract briefly summarizes the purpose of the study and its methods, then focuses on the technical aspects of the subject: the duties the permanent employee would be expected to perform.

A keywords list ensures that if the report is searched electronically, it will register "hits" for each of the terms listed.

i

CONTENTS

Note that the typeface and design of the headings in the contents page match the typeface and design of the headings in the report itself.

For more about using styles to create a table of contents automatically, see Ch. 12, p. 313.

ii

Executive Summary

Three years ago (1996), Ada Semiconductors, Incorporated's (ASI) employees began creating, designing, and maintaining an ASI Web site. The ASI Web site provides product information such as data sheets, product selector guides, and technical notes. The ASI online market has grown rapidly and is responsible for a large share of our sales base. Over the past three years, ASI has grown from 78 employees to 200 employees and has tripled the number of products being sold; in this period, the amount of ASI information posted on the Web has increased by 500%.

Within each department, the employee with the most hypertext markup language (HTML) expertise is responsible for that department's Web site. ASI is currently spending $5,092 (142 employee hours) per month creating and maintaining the ASI Web site.

There are two major problems with the current approach. First, the employees performing these functions are not professional Webmasters, which is clearly reflected by the unprofessional appearance and usability of the ASI Web site. Second, some employees who once spent only a small fraction of their work day performing this function now find they are sacrificing large amounts of valuable production time.

We researched three different methods of hiring a Webmaster: (1) a contract/outsource hire, (2) a permanent, full-time ASI employee, and (3) a college intern. Our committee recommends hiring a Webmaster through Roberts, Inc. to restructure and update the existing ASI Web pages so that they are consistent, accurate, and professional. This would ensure that the information posted on the ASI Web site is updated quickly. The cost to ASI would be approximately $7,506, and the project would take 4 weeks (180 hours of development, production, and editing time) to complete. This step would be the most expedient way to improve the Web site and bring it up to ASI's high standards.

We also recommend that ASI hire a full-time Webmaster as an ASI employee to maintain the site. This employee would start at a salary of $50,000 (cost with benefits: $67,000). A full-time employee would offer significant benefits over a contract employee, including greater accessibility for ASI employees, greater visibility for the position, and a greater understanding of ASI product line and corporate goals.

Additionally, ASI might benefit from participating in the Central State University (CSU) intern program. When a qualified intern is available, he or she could be used to assist the ASI Webmaster at minimal cost to ASI. By combining the above options, ASI can provide a well-developed, user-friendly, professional, and accurate Web site.

The executive summary describes the project with a focus on the managerial aspects, particularly the recommendations. Note the writers' emphasis on the problem at ASI and their use of dollar figures.

1

In some organizations, all first-level headings begin a new page.

The background and purpose of the report

The specific problems with the current situation

Introduction

On March 4, 1999, we received approval to research the feasibility of hiring a Webmaster to create and maintain the ASI Web site.

This report presents the findings of our study. We studied our current site to determine whether all our pages are necessary, and we determined whether the ASI site is likely to continue to grow. Next, we researched three options for hiring a Webmaster: (1) contracting/outsourcing, (2) hiring a permanent, full-time ASI employee, and (3) hiring a college intern.

In 1997, ASI began providing product information on the Web, including data sheets, product selector guides, and technical notes. The Web site was created and is maintained by full-time, salaried employees during normal business hours. Over the past three years, the company has grown from 78 full-time employees to 200 full-time employees, and the amount of ASI information posted on the Web has increased by 500 percent.

The current situation has two major drawbacks:

- *The current site is unprofessional.* According to a recent Bay Networks White Paper, "The World Wide Web has emerged as a new paradigm in information access and display, becoming the preferred method for accessing corporate data over the enterprise intranet" (Bay Networks, 1999). In business today, the Web browser has become the universal interface to corporate data. ASI does provide valuable data to its internal and external customers via the Internet, but the presentation is unprofessional. ASI's Web site is disorganized, inconsistent, cumbersome to navigate, and inaccurate, a problem that could have serious business implications.

  We are increasingly receiving disturbing reports about our site. For example, a buyer from Dell Computers recently contacted us to report inaccurate information on the SDRAMs that he was interested in ordering. While looking at our 128K DRAM data sheet provided on the Web, a buyer from Lucent stated she could not find the voltage limitations. In fact, the information was there, but it was difficult to find. See Appendix B on page 13 for examples of unprofessional pages from our site.

- *The time spent maintaining the site is taking ASI employees away from their regular duties.* ASI is currently spending $5,092 per month on creating and maintaining the Web site. As ASI's customer base and product lines expand, the cost of creating and maintaining our Web pages will likely continue to increase. Although the cost of maintaining the ASI Web pages is not yet a problem in itself, the 142 hours of lost production time is a problem. Based on our research, we feel that the current system is not the best use of company resources. We think that a trained Webmaster could create and maintain the ASI Web pages at a lower cost than the current system (Buck Consultants Inc., 1998). A professional Webmaster, even at a relatively high salary, would be a good value because he or she would work much more efficiently than our current employees in maintaining the site.

Figure 1 shows the amount of time currently devoted to maintaining the site.

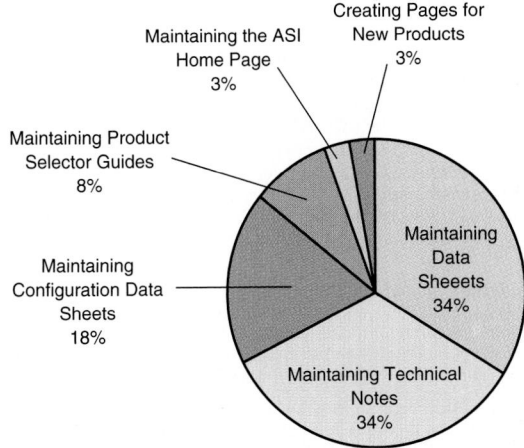

Figure 1.   Breakdown of the 142 hours that ASI employees spend each month creating and maintaining the ASI Web site.

The following sections describe our research methods, our principal findings, and our recommendations. Our principal recommendation is to retain Roberts to create a Web site for ASI, then hire a full-time ASI employee to maintain it.

The organization of the report, and a statement of the main recommendation

3

In many corporate reports, much of the research consists of interviewing and checking company records.

Research Methods

We performed the following research to determine the best methods for creating and maintaining ASI's Web site:

1. To assess the current ASI site, we surveyed and interviewed the eight ASI employees who are currently creating and maintaining ASI Web pages. We asked each of the eight to comment on whether they thought the ASI site would stay the same size in the future or would need to grow. We summarized each of the Web pages and checked for duplication of information.

2. To determine ASI's position on retaining contract software professionals, we met with Paul Mitchell, ASI's Human Resource Supervisor. Mitchell told us that ASI's policy is to contract software professionals through Roberts, Inc. We researched Roberts's Web page and solicited from them a proposal to update our current site. That proposal appears in Appendix C on page 14.

3. To determine the job requirements and compensation of a full-time Webmaster, we met with Paul Mitchell, ASI's Human Resource Supervisor, and gathered additional data. Mitchell provided the current market expectations for determining the position description, educational requirements, experience, and salary for a Webmaster. We also researched several competitors' Webmaster positions posted on the Web.

4. To learn more about the internship program at Central State University, we met with the CSU internship coordinator, Karen Pearson, who described the program and helped us understand the process of hiring a college intern. The information she provided us is presented in Appendix D on page 15.

5. We analyzed the information from our research, drew conclusions, and formulated the recommendations presented in this report.

4

Results

In this section, we present our findings regarding the current ASI Web site, as well as each of the three options for hiring a Webmaster.

An advance organizer for this section

The Current ASI Web Site

We met with the eight key ASI people involved with each of the ASI Web pages: Bob Adams, Ed Burns, Jason Conner, Ronda Dodge, Mike Jones, Brad Owens, Deanna Peters, and Deb Stevens. Based on interviews with each of these people, we determined that all of the current ASI Web pages are justified. We also found that none of the information on the existing pages is duplicated on other pages. Additionally, each of the eight ASI employees felt strongly that we will need to expand the site, not only to advertise and support our newest product, but also to provide more detailed information for each of our products. Last year, our site was visited approximately 1,500 times per month. This year, the figure is more than 4,000. It seems clear that the Web is likely to remain our primary means of furnishing product information to our customers for the foreseeable future. Customers are demanding complete information on Web sites; to remain competitive, ASI will need to devote additional resources to its site.

Following is a list of the pages in the current ASI Web site and a description of what is contained in each:

The list of the pages and their functions increases the writers' credibility.

- Home Page — this is the first page external customers see when going to www .asisemi.com. This page contains links to all other ASI pages.

- Product Selector Guide Page — this page, which lists all the current ASI products, includes part density, memory configuration, features, part numbers, package, types, access time, and availability.

- Literature Request Form Page — this is the form Web customers use to order additional product literature.

- Newsletter Page — this page announces new ASI information such as special offers and product qualifications.

- Four Product Guide Pages (DRAM, SRAM, Flash, and Modules) — these pages contain selector guides, compatibility guides, and ordering information.

- Six Technical Notes Pages — these pages contain pertinent information based on the application and design of ASI products; topics are separated by product type.

- Six Product and Technical Assistance Pages — these pages contain names of the marketing contact and technical contact(s) and are separated by product type.

5

The writers worked with the company's human resources supervisor, who has had a positive experience contracting a particular vendor.

The writers went to the trouble to request a proposal from Roberts.

The writers have put the details of the different options in an appendix.

At this company, the human resource supervisor oversees the writing of job descriptions.

Contracting/Outsourcing a Webmaster

Paul Mitchell, ASI Human Resource Supervisor, recommended that we contact Roberts, Inc. to research hiring a temporary Webmaster. According to Mitchell, Roberts has successfully completed several contract jobs for ASI in the past two years. With five years' experience creating Web sites, Roberts can provide a professional to help restructure and maintain our Web site using the latest Web site development tools. Roberts is currently designing and implementing Web sites for the Microsoft intranet system (Roberts, 1999).

A Webmaster hired through Roberts would be qualified to perform production-level HTML scripting, implement basic features for online Web content, and provide tools to enhance efficiency in creating HTML Help, as well as Web and other online deliverables.

In addition, Roberts would screen the Webmaster to ensure that he or she has demonstrated HTML programming experience as well as experience in graphics, layout, and design. All candidates would provide a current portfolio upon request. The hourly rate of a Webmaster ranges from $25 to $45, based on expertise. In addition, we found that through Roberts we could hire a Webmaster for the duration of a specific project. This would give us the flexibility to rehire the same contractor if that individual is meeting our needs, or to decline renewal if the individual is not meeting our needs.

Roberts provided a proposal (Appendix C on page 14), which includes the estimated cost of restructuring the ASI Web site. Additionally, Roberts agreed to continue to provide a Webmaster for an hourly rate of $35 for simple site maintenance.

Further information on the advantages and disadvantages of this hiring option are detailed in Appendix A on page 11.

Hiring a Permanent, Full-Time ASI Employee

Our committee contacted Paul Mitchell, ASI's Human Resource Supervisor. We provided Mr. Mitchell the information we researched for our proposal concerning ASI's need to create a Webmaster position. Using the criteria set forth in our proposal, Mr. Mitchell researched current market expectations for determining the ASI Webmaster position description and salary. In addition, Mr. Mitchell researched the educational requirements and suggested experience for a Webmaster. Mr. Mitchell drafted a position description, which we then revised. Following is the revised position description:

General Position Description

Successful candidates should be self-motivated software engineers with an interest in Internet technologies, open protocol standards, database, and Web application development tools. The position provides overall software development and support for ASI infrastructure and consulting services to external clients.

Preferred Educational Requirements

• Two years of graduate-level education or master's, or equivalent graduate degree (such as LL.B or J.D.).

6

- Education in a major field of study such as computer science, information science, information systems management, mathematics, statistics, operations research, or engineering. Course work that required the development or adaptation of computer programs and systems and provided knowledge equivalent to a major in the computer field will be considered. (Transcripts must be provided.)

Preferred Experience

- Two years of experience in Web design with demonstrated ability in graphic design, layout, and Web-site creation.

- Three to five years experience in a responsible position in a team environment.

- Knowledge of Web design, e-commerce, and site-functionality trends.

- Understanding of Web programming.

Additional Skills

- Ability to work with little direct supervision.

- Ability to take detailed and well-organized notes to summarize meeting results, agreements, action items, and problems.

- Ability to communicate effectively, orally and in writing, with different constituencies in ASI and outside the company.

Mean Salary

The starting salary would be $50,000 per year.

Mr. Mitchell reminded our committee that hiring an ASI employee of this caliber might take several months, and it would be costly and time consuming. He also gave the committee a breakdown of the total cost of an ASI employee in terms of the benefits and salary outlay for a period of one year.

Further information on the advantages and disadvantages of this hiring option are detailed in Appendix A on page 11.

Hiring a College Intern Webmaster

ASI has the option of arranging for a college intern through the English Department and/or the Computer Information Systems (CIS) Department at Central State University (CSU). ASI would be interested in hiring an intern who is an English major with a technical communication emphasis and/or a CIS major. Internships consist of 50 hours for each credit earned. Most internships are 150 hours/3 credits. ASI would need to:

- sign a contract stating ASI will supervise the intern
- provide a supervisor for the intern

7

- provide the intern with input and direction
- evaluate the intern's work at the end of the 150 hours (end of semester)

In discussing internships with the coordinators in the English and Computer Science departments, we concluded that an intern would be an inappropriate choice for the Webmaster position because an intern could stay with ASI for only a limited period of time, and because the chance of securing an intern with the level of expertise and experience ASI required is remote. However, we concluded that college interns could provide valuable assistance to a contract or regular ASI Webmaster.

Further information on the advantages and disadvantages of this hiring option are detailed in Appendix A on page 11.

Conclusions and Recommendations

As suggested by customer complaints and internal records, the present system for maintaining the ASI Web site is less than successful. ASI employees do not have the time or the skills to maintain the Web site in a manner that will increase Internet sales and promote a professional image. Our conclusion is that, over the long term, ASI can solve the problems with the current system only by hiring a full-time Webmaster.

Based on the results of our research, however, we recommend that ASI hire a Webmaster through Roberts, Inc. on an interim basis to restructure and update the existing ASI Web pages so that they are consistent, accurate, and professional. This would ensure that the information posted on the ASI Web site is updated quickly and brought up to ASI's high standards. The process would take 4 weeks (180 hours of development, production, and editing time) to complete and cost approximately $7,506.

However, having Roberts provide a Webmaster long term would not be the best option. ASI's resources would be best spent hiring a full-time Webmaster as an ASI employee. Through the interviewing process, ASI would be assured of hiring an expert in HTML and Web design and could look for someone with industry knowledge. ASI could also determine the salary. We therefore recommend that ASI immediately launch a search for a full-time Webmaster. This employee, who should be fully competent in Web design and programming languages, e-commerce, and site functionality, would supervise the company's Web site and overall Internet presence, create internal policies regarding the site, maintain the site, and publicize it.

Finally, we recommend that an appropriate ASI employee be asked to maintain contact with the Central State University internship coordinators and, when appropriate, oversee the hiring of a CSU student intern to assist in maintaining the ASI site.

We feel these three steps would be the best solution for eliminating ASI's current problems with creating and maintaining its Web site.

The writers present and justify their major conclusions and recommendations. Notice that they explain their reasoning clearly.

9

References

Baynet Works White Paper (1999). Web Based Network Management. <http://www
.baynetworks.com/products/Papers/Webbased.html#int> (1999 February 27).

Buck Consultants, Inc. (1998). *National Business Employment Weekly*. <public.wsj.com
/careers/resources/documents/19980619-internetpros-tab.htm> (1999 February 28).

Roberts, Inc. (1999). Roberts — Intranet Information Development. <www.robertsinc
.com/whatwedo/intra.htm> (1999 April 15).

10

Appendices

Appendix A: Advantages and Disadvantages of Hiring Options

Contractor/Outsourcing	*ASI Permanent Employee*	*College Intern*
Advantages	**Advantages**	**Advantages**
• A Webmaster provided by Roberts would be able to begin work immediately; ASI's Web site would be restructured and updated within 4 weeks.	• Profit sharing would provide increased incentive.	• ASI would not provide employee benefits (health insurance, vacation, etc.).
• Roberts provides high-quality professionals who are experts in HTML design, etc.	• The employee would work on site: (1) this would allow for greater coordination efforts between departments; (2) the supervisor could observe employee's work performance; and (3) managers and peers would have constant access for facilitating position objectives.	• Payment of intern would be low: approximately $10/hour.
• Webmasters are hired per project; if the Webmaster does not perform satisfactorily, ASI could terminate him or her.	• Permanent status would give equal standing among peers.	
• ASI would not provide an employee-benefits package (health insurance, vacation, etc.).	• Permanent status would allow for promotion within the company.	
	• Permanent status would give year-to-year benefit of experience gained and lessons learned ("corporate knowledge" retained over a long period of time).	
	• A permanent employee would work under company's policies governing employees: – Supervisor would enforce performance of duties as described in the position description. – Documented poor performance would be grounds for termination. – The employee would not receive overtime pay. – A work schedule would be enforced.	

This table makes it easy for readers to see the advantages of each of the three options.

11

Appendix A: Advantages and Disadvantages of Hiring Options (*Continued*)

Contractor/Outsourcing	*ASI Permanent Employee*	*College Intern*
Advantages	**Advantages** – Annual and sick leave policies would be enforced. – Professional standards and safety rules would be enforced.	**Advantages**
Disadvantages • Long term, this arrangement would be expensive (approximately $75,000 – 100,000 annually). • Continuation of service by a specific Webmaster who is performing well would not be guaranteed. • A Webmaster hired through Roberts may not have semiconductor-specific knowledge.	**Disadvantages** • Salary and benefits would total approximately $67,000. • Profit sharing would be diluted. • Tenured employees (7 years continuous employment) would be protected from most termination actions. • The process of hiring a permanent employee would be costly and time consuming.	**Disadvantages** • Supervision by a manager would be required. • Continuation of service would not be guaranteed, since only 150 hours is required for the internship. • An intern may not have semiconductor-specific knowledge.

Appendix B: Examples of ASI Web Pages
[Here the writers provide three annotated samples of ASI Web pages, showing the inconsistencies that undermine the professional image the company wishes to project.]

Appendix C: Roberts Proposal
[Here the writers provide the full proposal from Roberts, describing the Webmaster services the company would provide to ASI.]

Appendix D: CSU College Intern Information
[Here the writers provide a brochure from the CSU College Intern Office describing the policies related to hiring an intern.]

✔   **Revision Checklist**

1. In planning your informational, analytical, or recommendation report, did you
   ❑ analyze your audience?
   ❑ analyze your purpose?
   ❑ identify the questions that need to be answered?
   ❑ carry out appropriate research?
   ❑ draw valid conclusions about the results (if appropriate)?
   ❑ formulate recommendations based on the conclusions (if appropriate)?

2. In planning to write your feasibility report, did you
   ❑ analyze your audience?
   ❑ analyze your purpose?
   ❑ identify the problem or opportunity?
   ❑ establish criteria for responding to the problem or opportunity?
   ❑ determine your options?
   ❑ study each option according to your criteria?
   ❑ draw valid conclusions about each option?
   ❑ formulate recommendations based on your conclusions?

3. Does the introduction
   ❑ explain the subject of the report?
   ❑ explain the purpose of the report?
   ❑ explain the background of the report?
   ❑ describe your sources of information?
   ❑ indicate the scope of the report?
   ❑ briefly summarize the most significant findings of the project?
   ❑ briefly summarize your recommendations?
   ❑ explain the organization of the report?
   ❑ define key terms used in the report?

4. Does the methods section describe your methods in sufficient detail? Have you justified your methods where necessary, explaining, for instance, why you chose one method over another?

5. Are the results presented
   ❑ clearly?
   ❑ objectively?
   ❑ without interpretation?

6. Are the conclusions
   ❑ presented clearly?
   ❑ drawn logically from the results?

7. Are the recommendations
   ❑ clear?
   ❑ objective?
   ❑ polite?
   ❑ in an appropriate form (list or paragraph)?
   ❑ in an appropriate location?

Exercises

1. An important element in carrying out a feasibility study is determining the criteria by which to judge each option. For each of the following topics, list five minimum specifications and five evaluative criteria you might apply in assessing the options.

   a. buying a computer printer
   b. selecting a major
   c. choosing a company to work for
   d. buying a car
   e. choosing a place to live while you attend college

2. In the Links Library for Chapter 7 on TechComm Web (www.bedfordstmartins.com/techcomm), find a site that links to government agencies and departments. Find a government report on a subject that interests you. Determine whether it is an informative, an analytical, or a recommendation report. In what ways does the structure of the report differ from the structure described in this chapter? In other words, does it lack some of the elements described in this chapter, or does it have additional elements? Are the elements arranged in the same order in which they are described in this chapter? In what ways do the differences reflect the audience, purpose, and subject of the report?

Research Projects

Some of the following projects ask you to write a memo. See Chapter 15, page 430, for a discussion of writing memos.

3. Interview a local professional who writes completion reports. What process does he or she use in creating the report, and what does he or she see as the most important aspects of writing a completion report? To what extent do the information and suggestions provided by the professional match those provided in this chapter? Present your findings in a memo to your instructor. See Chapter 7, page 171, for a discussion of interviewing.

4. Write the completion report for the research project you proposed in response to exercise 4 on page 516 in Chapter 17. Your instructor will tell you whether the report is to be written individually or collaboratively, but work with a partner in reviewing and revising your report. You and your partner will work together closely at the end of the project as you revise your reports, but keep in mind that a partner can be very helpful during the planning phase, too, as you choose a topic, refine it, and plan your research.

5. Secure a completion report for a project subsidized by a city or federal agency, a private organization, or a university committee or task force. (Be sure to check your university's Web site; universities routinely publish strategic planning documents and other sorts of self-study reports. Also check www .nas.edu, which is the site for the National Academy of Sciences, the National Academy of Engineering, the Institute of Medicine, and the National Research Council, all of which publish reports on the Web.) In a memo to your instructor, analyze the report. Overall, how effective is the report? How could the writers have improved the report? If possible, submit a copy of the report along with your memo.

To find government agencies and sites for this case, click on Links Library for Ch. 7 on TechComm Web (www. bedfordstmartins .com/techcomm).

C A S E

Writing a Recommendation Report

Form small groups for this project. Select a government department or agency. Imagine that you work for this organization and that you have been asked by a supervisor to review all the information on the site and to write a recommendation report about the site. Your supervisor wants to know "what we have on the site, how well it communicates to our viewers, and what changes (if any) we should make to it." Write the recommendation report.

Writing Instructions and Manuals   20

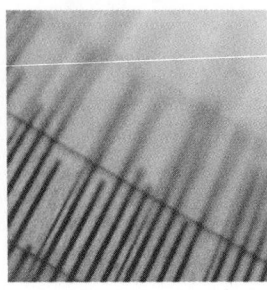

Elna Tynes, a technical communicator, on the role of instructions in selling the products they accompany:

When companies make products and sell them, they generally think that the product is what they're selling. Nope. The product is a symbol for all or part of the solution to their problem, and the information associated with the product is what actually sells the product and tells the user how to solve the problem. (Qtd. in Mead, 1998, p. 375)

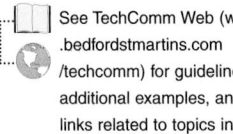

See TechComm Web (www .bedfordstmartins.com /techcomm) for guidelines, additional examples, and links related to topics in this chapter.

The customer-support staff at Dell Computers can tell you that no matter how hard their technical communicators try, their manuals will never please everyone. A customer once called asking how to install batteries in her new laptop. When told that the instructions were on page 1 of the manual, she replied, "I just paid $2,000 for this [deleted] thing, and I'm not going to read a book" ("Befuddled," 1994, p. B1). Because instructions and manuals have acquired a bad reputation over the years, many people don't even bother trying to read them. Perhaps that is why many technical communicators use the phrase RTFM, which (roughly translated) stands for Read the Fine Manual. But many people apparently don't. This situation is regrettable, because instructions and manuals are fundamentally important in carrying out procedures and using products safely and effectively.

Effective instructions and manuals are challenging to write. You must make sure that your audience will want to read the document and be able to understand it easily. You also must make sure that in performing the tasks, your readers won't damage any equipment or, more important, injure themselves or other people.

Chapter 9 (p. 232) discussed process descriptions, which explain how a process occurs. This chapter discusses instructions, which are process descriptions written to help the reader perform a specific task. For instance, a set of instructions shows how to install a water heater in your home. By contrast, a process description might describe how the water heater burns natural gas to heat the water inside.

This chapter also discusses manuals, which are larger documents consisting primarily of instructions. Often manuals are printed and bound, like books. Manuals can be classified according to function. One common type is a user's manual. Like a set of instructions, its function is to instruct by explaining, for example, how to use a software program, maintain inventory, or operate a piece of machinery. Other types include installation manuals, maintenance manuals, and repair manuals.

One other aspect of instructional writing is also discussed in this chapter: writing instructions and manuals for multicultural readers.

See Online Technical Writing (www.io.com/~hcexres /tcm1603/acchtml/acctoc .html) and Writing Guidelines for Engineering and Science Students (http://fbox.vt .edu:10021/eng/mech /writing/) for examples of instructions.

UNDERSTANDING THE ROLE OF INSTRUCTIONS AND MANUALS

Instructions and manuals are central to technical communication. If you are a technical professional, such as an engineer, you will probably be asked to write or contribute to instructions and manuals often. If you are a technical communicator, you will write them more often than any other kind of major technical document.

Whereas just a decade or two ago little attention was paid to the quality of instructions and manuals, today the goal is to make products, procedures, and systems safe and "user-friendly." Safety is an important issue largely because of liability suits. A company can lose millions if a court finds that its instructions and manuals failed to explain how to use a product properly or to alert users about dangers related to using a product. User-friendliness is critical, because if the product is hard to use, it could fail in the marketplace. Ads for all kinds of products stress clear, simple, easy-to-use manuals.

Instructions and manuals are no longer an afterthought, but an integral part of the planning process. At the most progressive companies, the people who create the documentation — the documents that come with the product — are part of the research and development team.

For more about ethical and legal considerations, see Ch. 2.

ANALYZING YOUR AUDIENCE

When instructions and manuals are ineffective, chances are that the writer has inaccurately assessed the audience. Performing a function, such as assembling a backyard swing or maintaining a conveyor belt, is easy for the expert but not necessarily for the person reading the documentation. Someone who doesn't know what a self-locking washer is, or what the calibrations on a timing control mean, might not be able to complete the process if such things are left unexplained.

Before you start to write a set of instructions or a manual, think carefully about the background and skill level of your audience. If you are writing to people who are experienced in the field, use technical vocabulary and concepts. But if you are addressing general readers, define technical terms and provide

For more about analyzing your audience, see Ch. 5, p. 92.

For more about defining terms, see Ch. 9, p. 220.

detailed directions. Don't be content to write "Make sure the tires are rotated properly." Instead, define proper rotation and explain how to achieve it.

Consider, too, the language skills of your readers. If you are addressing multicultural readers, include instructions in their native languages or take other measures described later in this chapter.

The best way to make sure you have assessed your audience effectively is to find people whose backgrounds resemble those of your intended readers and test the effectiveness of the documentation. This process, called *usability testing*, is discussed in Chapter 3, page 60.

PLANNING FOR SAFETY

Your most important responsibility in writing documentation is to make sure you do everything you can to ensure the safety of your readers. Even though some kinds of tasks do not involve safety risks, many do. Therefore, planning for safety is critically important.

Plan for safety by:

Writing clear safety information

Designing safety information

Placing safety information in the appropriate location

Writing Clear Safety Information

Be clear and concise. Avoid complicated sentences.

COMPLICATED   It is required that safety glasses be worn when inside this laboratory.

SIMPLE   You must wear safety glasses in this laboratory.

SIMPLE   Wear safety glasses in this laboratory.

Sometimes a phrase works better than a sentence: "Safety Glasses Required."

Because a typical set of instructions or manual can contain dozens of comments — both safety and nonsafety — experts have devised terms to indicate the seriousness of the advice. Unfortunately, the use of this terminology is not consistent.

For instance, the American National Standards Institute (ANSI) and the U.S. military's MILSPEC publish definitions that differ significantly, and many private companies have their own definitions that don't conform to either ANSI or MILSPEC. The following explanation of four common terms, presented here from most to least serious, points out the significant differences between ANSI and MILSPEC.

- *Danger*. MILSPEC does not use this term, but for ANSI and many companies *danger* alerts the reader to an immediate and serious hazard that will likely be fatal.

  DANGER. EXTREMELY HIGH VOLTAGE. STAND BACK.

  Often, safety warnings are written in all-uppercase letters.

- *Warning*. For MILSPEC, *warning* is the most serious level, indicating an action that could result in serious injury or death. For ANSI, it also suggests the potential for serious injury or death. Among different companies, however, the meaning of *warning* ranges from serious injury or death to serious damage to equipment.

  WARNING: TO PREVENT SERIOUS INJURY TO YOUR ARMS AND HANDS, MAKE SURE THE ARM RESTRAINTS ARE IN PLACE BEFORE YOU OPERATE THIS MACHINE.

- *Caution*. For MILSPEC, *caution* warns of the potential for both equipment damage and long-term health hazards. For ANSI, it indicates the potential for minor or moderate injury. Among companies, *caution* signals the potential for anything from moderate injury to serious equipment damage or destruction.

  Caution: Do not use nonrechargeable batteries in this charging unit; they could damage the charging unit.

- *Note*. A *note* is a tip or suggestion to help the readers carry out the procedure successfully.

  Note: Two kinds of washers are provided: regular washers and locking washers. Be sure to use the locking washers here.

If your organization does not already have guidelines for safety labeling, you might consider using the following definitions:

- *Danger*. Likelihood of serious injury, including death.
- *Warning*. Potential for minor, moderate, or serious injury.
- *Caution*. Potential for damage to equipment.
- *Note*. A suggestion on how to carry out a task.

Designing Safety Information

Whether printed in a document or on machinery or equipment, safety information should be prominent and easy to read. Many organizations use visual symbols to represent levels of danger, but these symbols are not standard-

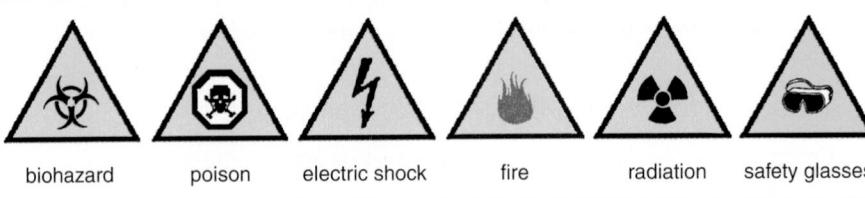

■ **Figure 20.1   Common Symbols in Safety Information**

ized. If your organization doesn't have a set of symbols that you can use in your document, devise a different design for each kind of comment. For instance, you could present warnings in 18-point type, boldfaced, within a box. The more critical the safety comment, the larger and more emphatic it should be.

Safety information is often printed in color: text, for example, against a yellow, orange, or red background. Symbols are printed in color, too: flames, for example, in red.

Figure 20.1 shows common symbols for safety information. Figure 20.2 (HCL, 2001) shows a safety label that would be affixed to a container used for a dangerous chemical.

 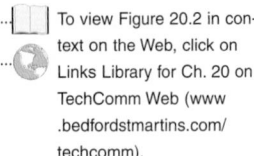

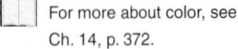For more about color, see Ch. 14, p. 372.

To view Figure 20.2 in context on the Web, click on Links Library for Ch. 20 on TechComm Web (www .bedfordstmartins.com/ techcomm).

ACETONE
(Dimethyl Ketone, CAS 67-64-1)

DANGER !   EXTREMELY FLAMMABLE

Acute: **CAUSES IRRITATION OF EYES, SKIN AND MUCOUS MEMBRANES.**
Chronic: **EXPOSURE TO LIQUID MAY CAUSE DERMATITIS.**
Keep away from heat, sparks and flame. Avoid contact with eyes, skin, and clothing. Keep container closed. Use with adequate ventilation. Wash thoroughly after handling.

FIRST AID:
IMMEDIATELY CALL POISON CONTROL CENTER OR HOSPITAL EMERGENCY ROOM.

IF CONTACTED: Immediately flush eyes with plenty of water for at least 15 minutes. Wash skin with soap and plenty of water. GET MEDICAL ATTENTION for eyes. Wash clothing before reuse.
IF INHALED: Remove to fresh air. If not breathing, give artificial resuscitation.
IF SWALLOWED: Give water to dilute. CONSULT POISON CONTROL CENTER OR HOSPITAL EMERGENCY ROOM. Never give anything by mouth to an unconscious or convulsive person.

                                                                              530-1

■ **Figure 20.2   Safety Label**

Placing Safety Information in the Appropriate Location

What is an appropriate location? This question has no easy answer because you cannot control how your audience reads your document. Be conservative: put safety information wherever you think the reader is likely to see it, and don't be afraid to repeat yourself. Of course you wouldn't want to repeat the same piece of advice in front of each of 20 steps, because readers would stop paying attention to it. But a reasonable amount of repetition — such as including the same safety comment at the top of each page — might be effective. If your company's procedure format calls for a safety section near the beginning of the document, place the information there and repeat it just before the appropriate step in the step-by-step section.

The Occupational Safety and Health Administration Guidelines (Chapter XVII, Sections 1910.145 and 1926.155) describe proper standards for placing safety messages on products and manuals. These standards address the following questions: Is the message prominently displayed so that users see it? Is the message large enough and clearly legible under operating conditions? Are the graphics and the words of the message clear and informative? Figure 20.3, from an operator's manual for a John Deere lawnmower, shows one company's approach to placing safety warnings on machinery.

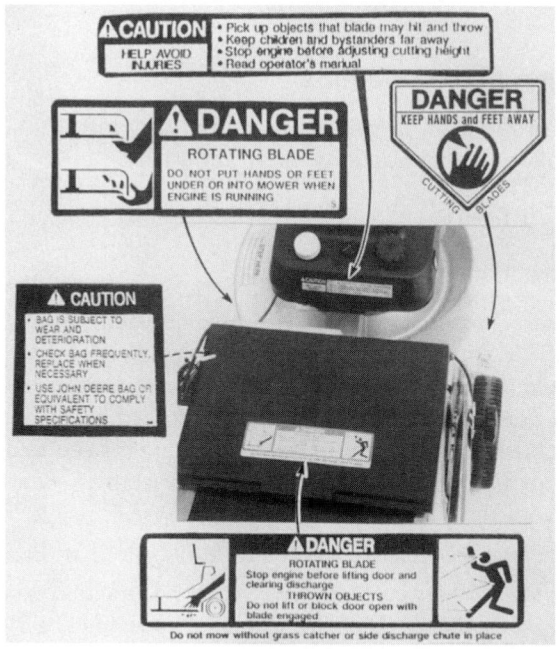

■ **Figure 20.3**
Safety Information on Machinery

Notice that the safety warnings combine words and graphics.

DRAFTING EFFECTIVE INSTRUCTIONS

Instructions can be brief — a small sheet of paper — or extensive, up to 20 pages or more. Brief instructions might be produced by one or two people: a writer, or a writer and a graphic artist. Sometimes a subject-matter expert — an expert in the technical subject being written about — is added to the team. For more extensive instructions, other people, such as marketing and legal personnel, might be added. The team could consist of as many as 10 or even 20 professionals working with a budget of many thousands of dollars.

For more about process descriptions, see Ch. 9, p. 232.

As with process descriptions, you will probably find it easiest to write instructions sequentially, that is, in the order in which the tasks appear. Many writers perform the task they are explaining as they write, a process that helps them detect errors and omissions.

Regardless of the size of the project, most instructions are structured like process descriptions. The main structural difference is that the conclusion of a set of instructions is not a summary but an explanation of how to make sure the reader has followed the instructions correctly. Most sets of instructions contain three elements.

> **Elements of a set of instructions:**
>
> General introduction
>
> Step-by-step instructions
>
> Conclusion

Drafting General Introductions

The general introduction provides the preliminary information that readers will need to follow the instructions safely and easily.

> **GUIDELINES**
>
> **Drafting Introductions for Instructions**
>
> A general introduction answers these three questions:
>
> ▶ *Why should the reader carry out this task?* Sometimes the reason is obvious: you don't need to explain why a backyard barbecue grill should be assembled. At other times, however, you do need to explain, as in the case of many preventive-maintenance chores such as changing radiator antifreeze every two years.
>
> If appropriate, answer two more questions:
>
> – *When should the reader carry out this task?* Some tasks, such as rotating tires or planting crops, need to be performed at particular times or at particular intervals.

— *Who should carry out the task?* Sometimes you need to describe or identify the person or persons who are to carry out a task. Some kinds of aircraft maintenance, for example, may be performed only by those certified to do it.

▶ *What safety measures or other concerns should the reader understand?* In addition to the safety measures that apply to the whole task, mention any tips that will make the job easier:

NOTE: For ease of assembly, leave all nuts loose. Give only 3 or 4 complete turns on bolt threads.

▶ *What items will the reader need?* List necessary tools, materials, and equipment in the introduction so that readers will not have to interrupt their work to hunt for something. If you think readers might not be able to identify these items easily, include drawings next to the names (see Chapter 14 for more about graphics).

Following is a list of tools and materials from a set of instructions on replacing broken window glass:

You will need the following tools and materials:

Tools	Materials
glass cutter	putty
putty knife	glass of proper size
window scraper	paint
chisel	hand cleaner
electric soldering iron	work gloves
razor blade	linseed oil
pliers	glazier's points
paintbrush	

Drafting Step-by-Step Instructions

The heart of a set of instructions is the step-by-step information on how to carry out the task or procedure.

 For examples of instructions, see Learn2.com (www .learn2.com) and the Knowledge Hound (www .knowledgehound.com).

G U I D E L I N E S

Drafting Steps in Instructions

▶ *Number the instructions.* For long, complex instructions, use two-level numbering, such as a decimal system.

1
   1.1
   1.2
2
   2.1
   2.2
etc.

▶ *Present the right amount of information in each step.* Each step should define a single task the reader can carry out easily, without having to refer to the instructions.

TOO MUCH
INFORMATION 1. Mix one part of the cement with one part water, using the trowel. When the mixture is a thick consistency without any lumps bigger than a marble, place a strip of the mixture about 1" high and 1" wide along the face of the brick.

TOO LITTLE
INFORMATION 1. Pick up the trowel.

RIGHT
AMOUNT OF
INFORMATION 1. Mix one part of the cement with one part water, using the trowel, until the mixture is a thick consistency without any lumps bigger than a marble.

          2. Place a strip of the mixture about 1" high and 1" wide along the face of the brick.

▶ *Use the imperative mood.* For example, "Attach the red wire. . . ." The imperative is more direct and economical than the indicative mood ("You should attach the red wire . . ." or "The operator should attach the red wire . . ."). Avoid the passive voice ("The red wire is attached . . ."), because it can be ambiguous: is the red wire already attached? Finally, make sure your sentences are grammatically parallel.

▶ *Include graphics.* When appropriate, add a photograph or a drawing to show the reader what to do. Some activities — such as adding two drops of a reagent to a mixture — do not need an illustration, but they might be clarified by charts or tables. Figure 20.6 on page 594 shows the extent to which a set of instructions can integrate words and graphics. See Chapter 14 for more about graphics.

▶ *Do not omit the articles (a, an, the) to save space.* Omitting the articles can make the instructions unclear and hard to read. In the sentence "Locate midpoint and draw line," for example, the reader cannot tell if "draw line" is a noun (as in "locate the draw line") or a verb and its object (as in "draw the line").

Drafting Conclusions

Instructions often conclude with *maintenance tips*. Another popular conclusion is a *troubleshooter's guide*, usually in the form of a table, that identifies common problems and explains how to solve them.

Here is a portion of the troubleshooter's guide in the operating instructions for a lawnmower.

Problem	Cause	Correction
The mower does not start.	1. The mower is out of gas.	1. Fill the gas tank.
	2. The gas is stale.	2. Drain the tank and refill it with fresh gas.
	3. The spark plug wire is disconnected from the spark plug.	3. Connect the wire to the plug.
The mower loses power.	1. The grass is too high.	1. Set the mower to a "higher cut" position. See page 10.
	2. The air cleaner is dirty.	2. Replace the air cleaner. See page 11.
	3. There is a buildup of grass, leaves, or trash in the underside of the mower housing.	3. Disconnect the spark plug wire, attach it to the retainer post, and clean the underside of the mower housing. See page 8.

A LOOK AT SAMPLE INSTRUCTIONS

Figure 20.4 on page 592 (Learn2.com, 1999a) is a portion of the first screen of a Web-based set of instructions on how to jump-start a car. Figure 20.5 on page 593 (Learn2.com, 1999b) shows an excerpt from the step-by-step section of the instructions. Figure 20.6 on page 594 (Hewlett-Packard, 1998, p. 29) shows a set of instructions contained in a user manual.

 To view Figures 20.4 and 20.5 in context on the Web, follow the links in Chapter 20 on TechComm Web (www .bedfordstmartins.com /techcomm).

The pun suggested by this cartoon is a little strained. Graphics can be informal but still present information effectively. This graphic falls short of that mark.

Learn2 Jump-Start a Car

Send to a friend

the **steps**

<u>Intro</u>:
Before you begin
<u>Step 1</u>:
Make sure the battery is not damaged
<u>Step 2</u>:
Attach the cables
<u>Step 3</u>:
Start the car
<u>Step 4</u>:
Disconnect the cables

The steps in the instructions are presented as separate Web pages.

That jumpin' jive..

Car batteries can lose their charge for more than a few reasons. Leaving the lights on overnight has got to top the list, though, and it's such an unexpected thing that most of us are caught unprepared. It's times like these when many of us rely on the kindness of strangers to jump-start our cars, but sometimes YOU need to be the kindly stranger- or at least the one who knows which cable goes where.

About 5 minutes

Here is the list of materials.

before you begin

When you're stranded in the cold rain and snow, asking passing motorists if they'll provide a jump start for your car, you'll find that there are two kinds of motorists: those who have jumper cables and those who do not.

A car with the same voltage battery as your own, fully charged

A set of jumper cables (see below)

A wire brush (optional)

Gloves (optional)

■ **Figure 20.4   Excerpt from the Introduction to a Set of Instructions**

These instructions use informal language and graphics but still provide a lot of useful information.

Step 1: Make Sure The Battery is Not Damaged

Step 1 is clearly labeled.

Batteries have an electrolyte solution inside them. If that solution is frozen, don't try to jump-start the car. If your battery has removable vent caps, you can look inside to see if the liquid is frozen (replace the caps after looking). It's not easy to tell otherwise.

- If there are cracks in the battery casing, don't try to jump-start it. Curse a few times, call a cab and go buy a new one.
- If there is whitish (or greenish or yellowish) residue around the battery terminals, clean it off with a wire brush. Wear gloves, if possible, because that stuff can be nasty if it touches your skin. When this stuff forms, it's a sign that you may need a new battery.

Intro:
Before you begin
Step 1:
Make sure the battery is not damaged
Step 2:
Attach the cables
Step 3:
Start the car
Step 4:
Disconnect the cables

AutoTrader.com

Find the used car you want in your area:

Make:
Acura ▼

ZIP:
[   ]
submit

This is a good example of an informal graphic that conveys useful information.

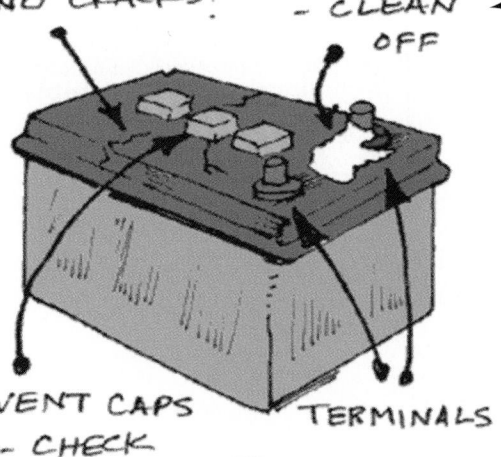

■ Figure 20.5   Excerpt from Step-by-Step Instructions

Even though the style of the text is informal, it is clear and informative.

The writers use color to emphasize the heading, the notes, and those parts of the graphics that show action.

The five graphics are keyed to the five explanatory paragraphs on the right side. The design would be clearer if the paragraphs were next to the graphic each describes.

►Loading the Optional 500-Sheet Tray

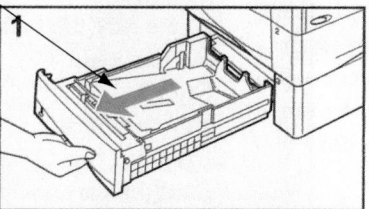

The optional 500-sheet tray is Tray 3 for the HP LaserJet 4000/4000 N printers, or Tray 4 for the HP LaserJet 4000 T/ 4000 TN printers. This tray is adjusted differently than Tray 2 (page 27). Tray 2 and the optional 500-sheet tray can be interchanged in the printer. For supported sizes of paper, see page A-4.

Note
To avoid paper jams, do not load trays while the printer is printing.

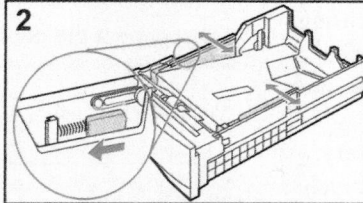

1 Pull the tray completely out of the printer.

2 If the tray is not already set to the desired paper size, squeeze the release on the left guide and adjust the left and right guides to match the width of the paper.

3 Squeeze the release on the back of the rear paper guide and slide it to the desired paper size.

4 Load the tray. For correct orientation, see page 36.

5 Make sure the paper is flat in the tray at all four corners and below the tabs on the guides. The guides should touch the paper without bending it.

Continued on the next page.

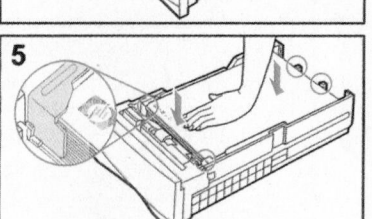

EN   Chapter 2: Printing Tasks   29

■ **Figure 20.6   A Set of Instructions from a Manual**

DRAFTING EFFECTIVE MANUALS

Most of what has been said in this chapter about instructions also applies to manuals. For example, when writing a manual, you have to analyze your audience, explain procedures clearly, and include graphics.

However, because manuals are usually more ambitious projects than instructions, they require more planning. A bigger investment is at stake, not only in the costs of writing and producing the manual but also in the potential effects of the published manual. A good manual reduces the possibility of injuries and liability, but it should also attract customers and reduce costs (because the organization needs fewer customer-support people). A poorly written manual is expensive to produce because it must be revised more often and because it alienates customers.

Drafting a manual, like drafting a book, is so complex that the following discussion can provide only an introduction. If you are involved in producing a manual, consult the additional resources in the bibliography (page 730).

Writing a manual is almost always a collaborative project. A full-size manual can require too many skills for one person to write: technical skills in the subject area, writing skills, graphics skills, production skills, even a knowledge of contracts and law to prevent lawsuits. In addition, other people's perspectives can prevent little problems from becoming big problems.

For more about collaboration, see Ch. 4.

There are three stages to writing a manual.

- *Planning.* The most important stage in manual writing is planning. Again, start by analyzing your audience and purpose.

   The documentation for a sophisticated procedure or system almost always addresses a multiple audience. You might decide to produce a set of manuals — one for the user, one for the manager, one for the installer, one for the maintenance technicians, and so on. Or you might decide on a main manual combined with related documents, such as brochures, flyers, and workbooks.

   Although your primary purpose might vary according to the kind of manual, you also have a secondary purpose: to motivate your readers. The mere fact that the manual might run 200 pages or more means that you have to persuade people to read it. In many cases, your readers are uncomfortable with the new product, procedure, or system that the manual documents. Your job is to make the task of learning these things less overwhelming. A well-designed manual, with plenty of white space and graphics, will help considerably.

- *Drafting.* Drafting a manual is much the same as drafting any other kind of technical document. Sometimes one person collects information from all the people on the documentation team and creates a draft based on that information. Sometimes various subject-matter experts write their own drafts, which a writer or a writing team then revises. In the drafting stage, you also need to create and integrate graphics and to design the whole manual as well as each page.

For more about revising, see Ch. 3, p. 56.

- *Revising.* Because manuals are longer and more complex than most other kinds of technical documents, revision is more complicated. It involves a series of different kinds of checks — on the accuracy of the technical information, structure, organization, emphasis, style, and correctness.

Drafting Front Matter

The front matter, which helps readers understand the content and the organization of the manual and the best ways to use it, consists of everything before the body. Most manuals have a cover or title page and a table of contents. In addition, most manuals contain a preface and a section about how to use the manual. (Sometimes these last two items are simply combined in an introduction.)

For more about covers, see Ch. 12, p. 309.

To decide whether to use a cover or just a title page, consider the manual's size and intended use. Manuals that will receive some wear and tear, such as maintenance manuals for oil-rigging equipment, need a hard cover, usually of a water-resistant material. Manuals used around an office usually don't need hard covers, unless they are large enough to require extra strength.

For more about creating headings, see Ch. 10, p. 252.

The title page, usually designed by a graphic artist, contains the title of the manual, plus the company's name, address, and logo. An extensive table of contents is also important because people don't read a manual straight through, but refer to it for specific information. To be effective, table of contents headings should focus on the tasks readers want to accomplish.

The other introductory information, which might precede or follow the table of contents, takes a number of forms. The first page of text in the manual, for instance, can be called "Introduction" or "Preface" or have no heading at all. Another popular strategy is to use several "about" phrases, as in "About Product X," or "About the Product X Documentation Set."

This introductory information should answer five basic questions for your reader:

- Who should use this manual?
- What product, procedure, or system does the manual describe?
- What is the manual's purpose?
- What are the manual's major components?
- How should the manual be used?

These questions need not be answered in separate paragraphs or even in separate sentences.

Some manuals need to answer an additional question: what does the typography signify? If the typography signifies different kinds of information (Courier typeface might represent the text the reader is supposed to type) and you want your readers to understand your conventions, define them in the front matter.

Figure 20.7 (Group 1 Software, 1994, pp. vii–ix, xi–xii) shows selected pages from the front matter of a manual. For more information on front matter, see Chapter 12, page 308.

Preface

About This Manual

The Purpose of This Guide

This User's Guide provides you with the information you need to effectively use the AccuMail System. With this guide, you will learn about fundamental name and address matching concepts, AccuMail's many features and functions, how to define and run jobs with AccuMail, and more.

This guide is divided into three sections:

Part I: Before You Begin	Provides basic mailing and functional concepts of the AccuMail System, and important instructions for installing and configuring your AccuMail software. This section also provides definitions of the terms used in this guide, and an overview of basic mouse and keyboard functions.
Part II: Using AccuMail	Provides instructions for performing the various tasks needed to work with jobs. This section also covers step-by-step instructions for working with filters, databases, mailing lists, and other AccuMail features.
Part III: Appendices	Provides a glossary, index, and appendices regarding special issues such as error messages, optimizing for speed, troubleshooting, and calling AccuMail from your own application.

Who Should Use This Guide?

This guide was developed for three types of users:

- Novice AccuMail users
- Experienced AccuMail users
- Programmers wanting to call AccuMail from their own applications.

> **NOTE:** This book assumes some familiarity with the Windows and DOS environments.

User's Guide   vii

■ **Figure 20.7**
Pages from the Front
Matter of a Manual

A conventions section helps readers understand the typographical styles that will be used in the manual.

Conventions Used in This Guide

This guide uses the following documentation conventions:

NOTE:	Indicates important information or warnings.
①	Indicates a numbered step in a process.
Italics	Indicates words or phrases that require emphasis or that appear in the glossary. The glossary words only appear in italics the first time they appear in the manual. Italics are also used for all dialog box and window titles referenced in this manual.
Bold	Indicates any of the following: menu name or menu option; icon; name of a list box, text box, check box, or radio button; group label and fields.
"Quoted Text"	Indicates one of the following (1) text that appears on a screen, window, or dialog box, such as a message or list box option or (2) text that you are to type (you do not type the quotation marks).
<Key>	Indicates a keyboard key such as <Return> or <Shift>. When you are to hold two or more keys down at the same time, they are shown in a single set of brackets, such as <Ctrl-C>.
[OK]	Indicates the name of a button on a dialog box.
ALL CAPITALS	Indicates file names, directory names, and job names. This typeface is also used to reference drives and DOS prompts.
[text]	Indicates user-specific names, directories, drives, and other information.

AccuMail

■ **Figure 20.7**
(Continued)

Summary of Chapters

The following table summarizes the chapters in this guide. Novice users will find Chapter 4, "Example Job" and all of Part I most useful, while experienced users will find the more advanced chapters of Parts II and III valuable.

Chapter	Topics Discussed
Part I: Before You Begin	
Chapter 1. AccuMail Basics	This chapter provides an overview of AccuMail's features and functions, including sections on address standardization, CASS, and other issues to consider before using AccuMail.
Chapter 2. Setting Up AccuMail	This chapter describes how to install AccuMail. It includes software and hardware requirements and separate installation and log-on instructions for the DOS, and Windows LAN versions.
Chapter 3. Getting Started	This chapter begins with an overview of the AccuMail interface, identifying all components of windows and dialog boxes used in the AccuMail system. It then describes navigation basics such as using the arrow and tab keys, entering and highlighting text, and using the mouse.
Part II: Using AccuMail	
Chapter 4. Example Job	This chapter functions as a tutorial for the novice user. It describes certain key concepts such as a job and the basic features of the **File** menu. It demonstrates the necessary steps to create and run a job, and to open, close, and copy a job. It includes sections on selecting lists, defining and matching fields, selecting databases, setting job defaults, and printing the CASS Form 3553.
Chapter 5. Looking for an Address	This chapter explains how to perform single-address lookups, browse the USPS National Database, and print envelopes and labels.
Chapter 6. Using Filters	This chapter explains how to select certain records based on criteria by using filters that you specify.

■ **Figure 20.7**
(Continued)

Related References

The front matter often directs readers to other documents.

The following pamphlets and handbooks can be obtained from the United States Postal Service:

- *Domestic Mail Manual (DMM)*: USPS Document ID DMM
- *Addressing Conventions*: USPS Document ID #MI DM-940-89-3
- *Postal Addressing Standards:* USPS Document ID #28
- *A Guide to Business Mail Preparation:* USPS Document ID #25
- *Addressing for Success:* USPS Notice #221
- *Automation Plan For Business Mailers:* USPS Document ID #67.

For further information, please contact:

National Address Information Center
United States Postal Service
6060 Primary Pkwy Suite 101
Memphis, TN 38188-0001

NOTE: The USPS may charge a fee for some or all of these publications.

■ **Figure 20.7**
(Continued)

About This Manual

If You Need More Help

Group 1 Software provides free telephone hotline support on all Group 1 microcomputer products for 90 days from the date of purchase. When calling Technical Support:

- Have your phone near the computer.

- Have the AccuMail program up on the screen.

- Know the name of the person and the company to whom AccuMail is registered.

- Know the AccuMail version number that you are using.

- Know any specific system configuration information (it is best if you have a printout of your CONFIG.SYS and AUTOEXEC.BAT files readily available).

- Be prepared to describe where you are in the program and what you are trying to accomplish.

- Have all product documentation within reach.

The **Technical Support Hotline** number for customers in the United States and Canada is **(301) 731-2316** between 9:00 a.m. and 7:00 p.m. Eastern Standard Time. The **facsimile number** for Technical Support is **(301) 306-4373**. Please address your facsimile cover sheet to Microproducts Customer Support.

> **NOTE:** To be eligible for this service, you MUST complete and return the Product Registration Card.

Your Comments Are Welcome

We appreciate and welcome your comments concerning this guide! If you have suggestions, please let us know. For your convenience, a Documentation Comment Form is provided with your new system. If the Documentation Comment Form is missing, however, you may address your comments to:

Group 1 Software, Inc.
Documentation Department
4200 Parliament Place, Suite 600
Lanham, MD 20706-1844

---

xii   AccuMail

A manual helps readers understand how to get technical support.

The company solicits comments and suggestions.

■ **Figure 20.7**
(Continued)

Drafting the Body

The body of a manual might look like the body of a traditional report, or it might look radically different. Its structure, style, and graphics will depend on its purpose and audience. For instance, the body of a manual might include summaries and diagnostic tests to help readers determine whether they have understood the discussion. A long manual might have more than one "body"; that is, each chapter might be a self-contained unit with its own introduction, body, and conclusion.

GUIDELINES

Drafting the Body of a Manual

▶ *Structure the body according to how the reader will use it.* If the reader is supposed to carry out a process, arrange it chronologically, beginning with the first step in the process and continuing on to the end. If the reader is supposed to understand a concept, move from more important elements to less important elements. Consider one of the organizational patterns discussed in Chapter 8 (p. 186), but be open to combining or altering it to meet the needs of your audience.

▶ *Write clearly.* Simple, short sentences work best. Use the imperative to give instructions.

▶ *Be informal, if appropriate.* For some kinds of manuals, especially those intended for readers unfamiliar with the subject, an informal style that uses contractions and everyday vocabulary is effective. One caution: safety warnings and information about serious subjects such as disease or war usually require a formal style.

See Ch. 14 for information on graphics.

▶ *Use graphics.* Graphics break up the text and encourage comprehension. Whenever readers are to perform an action with their hands, include a drawing or photograph showing the action. Where appropriate, use tables and figures.

Figure 20.8 (Microsoft, 1998, p.74) shows a page from the body of a user manual for a software program.

Drafting Back Matter

For more about glossaries and indexes, see Ch. 12, pp. 320 and 322.

The manual's audience and purpose will determine the items in the back matter, but three are common: a glossary, an index, and appendices. A *glossary* is an alphabetized list of definitions of important terms in the document. An *index* is common for most manuals of 20 to 30 pages or more.

The word *appendices* refers to a range of elements. For instance, procedures manuals often have flowcharts or other graphics that illustrate the

---

74   **Getting Started**

Customizing the Start Menu and the Taskbar

You can rearrange items on the Start menu by dragging items to another location on the Start menu.

You can customize the Start menu to help you work more efficiently. You can add folders or files that you open frequently, so that you can open them quickly from the Start menu at any time. Or you can create your own groups of files and *programs*. You can also add items to or remove them from the Start menu. For example, you can reduce the size of the Start menu by removing a program that you no longer use.

As in this book, the manual uses the margin for notes.

If you remove an item from the Start menu, you're not uninstalling the program or removing it from your computer. For more information, see "Adding and Removing Programs" in Chapter 3, "Using Your Desktop."

The tone of the writing is informal; note the contractions.

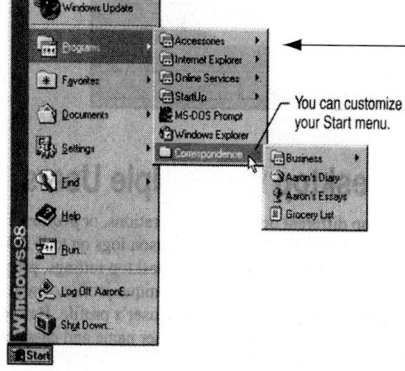

You can customize your Start menu.

A screen shot helps the reader see what is supposed to appear on the screen.

▶ **To customize the Start menu**

1. Click the **Start** button, point to **Settings**, and then click **Taskbar & Start Menu**.

   The Taskbar Properties dialog box appears.

2. Click the **Start Menu Programs** tab.

3. Click **Add** or **Remove**, and then follow the instructions that appear on the screen.

The instructions are conveyed in numbered steps and are expressed in the imperative.

■ **Figure 20.8   A Page from the Body of a Manual**

Note the generous use of white space in this design.

processes described in the body of the manual. Sometimes these graphics can be removed from the manual and taped to the office wall for frequent reference. User's guides often have diagnostic tests and reference materials: error messages and sample data lists for the computer system, troubleshooting guides, and similar items.

Drafting Revisions of Manuals

In high-tech industries, a new generation of a product might come along as often as every 18 months. As products evolve, technical communicators need to revise the manuals that go with them. When a new version of the product is released, you can take one of two approaches to revising.

- *Publish a "new" manual.* Although a new manual will most likely contain elements from the old manual, it might never mention the old manual. The advantage of a new manual is that it can reflect the look of the new product. The disadvantage is that people who are switching from the old product to the new one might have to spend time finding the information that applies specifically to the new product.

- *Publish a "revised" manual.* Revised manuals contain a lot of information from the old version, but the new information is flagged so that users can find it easily. The new information is usually marked with an icon in the margin or with a *change bar:* a vertical rule in the margin. Sometimes a section in the front matter describes the changes to this version of the product. The advantage of a revised manual is that it is easier for previous users to use. The disadvantage is that it can look like a patch job.

WRITING INSTRUCTIONS AND MANUALS FOR MULTICULTURAL READERS

When a company wishes to provide a set of instructions or a manual to multicultural readers, it can either translate the document into the reader's native language or try to make the English easy to understand.

Translation is sometimes the best or only alternative, but companies often use a simplified form of English, particularly when readers need to acquire a basic competence for long-term use of English. For instance, thousands of non–English-speaking aircraft technicians train for months in the United States to learn how to maintain U.S.-made civilian and military aircraft purchased by their countries. The manufacturers use a simple form of English because it would be prohibitively expensive to translate the many volumes of documentation into all the languages used by their customers.

The different forms of simple English include Simplified English, Controlled English, and Fundamental English. Some of these forms have been devised by scholars, others by companies such as Caterpillar and Kodak. They share three basic characteristics:

- *A limited vocabulary.* Some forms use vocabularies of as few as 500 words. Every word has only one meaning: *right* is the opposite of *left*, not of *wrong;* and *correct* is the opposite of *incorrect* (Sanderlin, 1988).

- *A limited grammar and sentence structure.* Sentences are short and simple. The imperative mood dominates. Here is an example of a sentence that has been revised from standard English into a simple English.

 For more about Simplified English, see Ch. 11, p. 297.

ORIGINAL	After visually inspecting the gap to ensure the gap is no more than 0.025 inches, replace the housing.
REVISED	Use the gauge to make sure the gap is no bigger than 0.025 inches. Then replace the housing.

- *A reliance on graphics.* As discussed in Chapter 14 (p. 365), graphics are used extensively in communicating with speakers whose first language is not English. If you can use a drawing, a diagram, or a photograph instead of words, do so. But be careful — graphics can project a cultural bias that interferes with communication or offends your readers.

The amount of instruction required to learn to use different forms of Simplified English ranges from 30 hours to 3 months. For more information, see Sanderlin (1988).

✔ **Revision Checklist**

Instructions

1. Does the introduction to the set of instructions
   - ❏ state the purpose of the task?
   - ❏ describe safety measures or other concerns that the readers should understand?
   - ❏ list necessary tools and materials?

2. Are the step-by-step instructions
   - ❏ numbered?
   - ❏ expressed in the imperative mood?
   - ❏ simple and direct?

3. Are appropriate graphics included?

4. Does the conclusion
   - ❏ include any necessary follow-up advice?
   - ❏ include, if appropriate, a troubleshooter's guide?

Manuals

1. Does the manual include, if appropriate, a cover?

2. Does the title page provide all the necessary information to help readers determine whether they are reading the appropriate manual?

3. Is the table of contents clear and explicit? Are the items phrased clearly to indicate the task readers are to carry out?

4. Does the other front matter clearly indicate
   - ❏ the product, procedure, or system the manual describes?
   - ❏ the purpose of the manual?
   - ❏ the major components of the manual?
   - ❏ the best way to use the manual?

5. Is the body of the manual organized clearly?
6. Are appropriate graphics included?
7. Is a glossary included, if appropriate?
8. Is an index included, if appropriate?
9. Are all other appropriate appendix items included?
10. Is the writing style clear and simple throughout?

Exercises

The following exercises ask you to write a memo. See Chapter 15, page 430, for a discussion of writing memos.

1. Study a set of instructions from Learn2.com (www.learn2.com) or the Information Hound (www.informationhound.com). Write a memo to your instructor evaluating the quality of the instructions. Attach a printout of representative pages from the instructions.

2. You work in the customer-relations department of a company that makes plumbing supplies. The head of product development has just handed you the draft of installation instructions for a sliding tub door (see page 607). She has asked you to comment on their effectiveness. Write a memo to her, evaluating the instructions and suggesting improvements.

3. Write a memo to your instructor evaluating the front matter of a manual. Does it explain the purpose of the manual and how to use it? Is its tone welcoming and professional? How might it be improved? Attach to your memo copies of the pages you are evaluating.

Research Projects

4. Write a brief manual for a process familiar to you. Consider writing a procedures manual for a school activity or a part-time job, such as your work as the business manager of the school newspaper or as a tutor in the Writing Center.

5. Write instructions for one of the following activities or for a process used in your field. Include appropriate graphics. In a brief note preceding the instructions, indicate your audience and purpose. Exchange these materials with a partner. Observe your partner and take notes as he or she attempts to carry out the instructions. Then revise your instructions and share them with your partner; discuss whether the revised instructions are easier to understand and apply, and in what ways they are easier to understand and apply. Submit your instructions to your instructor.

a. how to change a bicycle tire
b. how to delete the contents of the cache in your browser
c. how to light a fire in a fireplace
d. how to make a cassette-tape copy of a compact disc
e. how to find a listserv discussion group and subscribe to it
f. how to locate, download, and install a file from CNET shareware.com (shareware.cnet.com), FileDudes! (filedudes.com), or a similar download site

INSTALLATION INSTRUCTIONS

CAUTION: SEE BOX NO. 1 BEFORE CUTTING ALUMINUM HEADER OR SILL

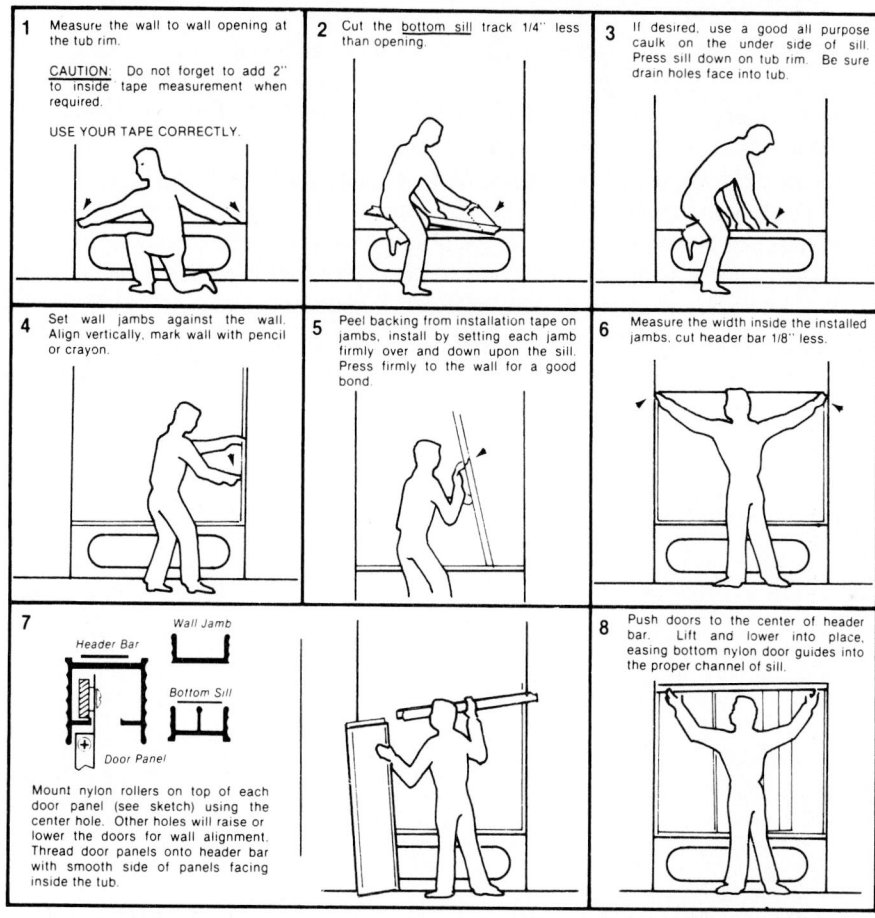

1 Measure the wall to wall opening at the tub rim.

CAUTION: Do not forget to add 2" to inside tape measurement when required.

USE YOUR TAPE CORRECTLY.

2 Cut the bottom sill track 1/4" less than opening.

3 If desired, use a good all purpose caulk on the under side of sill. Press sill down on tub rim. Be sure drain holes face into tub.

4 Set wall jambs against the wall. Align vertically, mark wall with pencil or crayon.

5 Peel backing from installation tape on jambs, install by setting each jamb firmly over and down upon the sill. Press firmly to the wall for a good bond.

6 Measure the width inside the installed jambs, cut header bar 1/8" less.

7 Wall Jamb

Header Bar

Bottom Sill

Door Panel

Mount nylon rollers on top of each door panel (see sketch) using the center hole. Other holes will raise or lower the doors for wall alignment. Thread door panels onto header bar with smooth side of panels facing inside the tub.

8 Push doors to the center of header bar. Lift and lower into place, easing bottom nylon door guides into the proper channel of sill.

TRIDOR MODEL ONLY:

To reverse direction of panels, raise panels out of bottom track and slide catches past each other thereby reversing direction so that shower head does not throw water between the panels.

HARDWARE KIT CONTENTS
TUDOR MODEL
4 nylon bearings
4 ball bearing screws # 8-32 × 3/8"
TRIDOR MODEL
6 nylon bearings
6 ball bearing screws # 8-32 × 3/8"

This set of instructions accompanies exercise 2 on page 606.

To find the Web sites for this case, click on Links Library for Ch. 20 on TechComm Web (www.bedfordstmartins .com/techcomm).

CASE
Writing a Set of Instructions

The instructor in your technical-communication class would like your help in familiarizing students with the process of downloading software from the Internet. She has asked you to write a set of instructions that students in the class can use to download an updated version of an Internet browser from Microsoft (www.microsoft.com) or Netscape (home.netscape.com). Visit one of these sites, download the software, install it, and configure it either for a home computer or a computer in one of your labs. Then write a clear set of instructions for someone who is basically familiar with using a computer but has not downloaded and installed software. Include instructions on how to modify the size of the cache.

Creating Web Sites 21

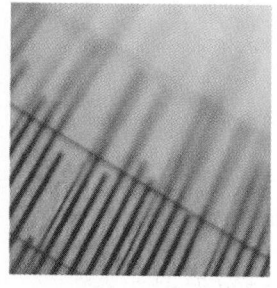

Amy Gahran (1999), the editor of *Contentious*, a Web magazine for writers and editors who create content for online media, on analyzing the purpose of your site:

Too many Web sites don't succeed because they don't have clear goals. Often, the people behind a site simply dive in before defining what they really want the site to achieve, or what the site can realistically achieve. While it is possible to "learn as you go" with online publishing, that's not necessarily the best way to grow your audience or enhance your reputation.

See TechComm Web (www .bedfordstmartins.com /techcomm) for guidelines, additional examples, and links related to topics in this chapter.

Since its inception in 1991, the World Wide Web has grown much faster than anyone predicted. The Web is now a gigantic publishing medium, made up of millions of Web pages and sites housed on computers all around the world. Nobody has precise figures about Web usage because it changes too quickly and because there is no Internet headquarters. According to the *Computer Industry Almanac* (2001), by the year 2005, more than a billion people will have used the Internet.

Who makes Web pages? Everyone, it seems. Governments, research institutions, publishers, interest groups, all kinds of large and small businesses, and students in many kinds of courses. A basic, primarily text-based site on the Web requires considerably less than one megabyte of storage space on an Internet server; many Internet service providers make this space available as part of their service.

Because the Web is such an important publishing medium for organizations, all professionals need to understand the basics of creating a Web site. This chapter begins with a discussion of the process of making Web sites and then discusses how to design a Web site and its individual pages. Next, it considers important issues in designing sites for people with disabilities and for multicultural readers. Finally, it discusses some of the complex ethical issues involved and analyzes sample Web pages.

UNDERSTANDING THE PROCESS OF CREATING WEB SITES

The process of building and maintaining a Web site is far from linear. You will find yourself going forward, then doubling back and rewriting. In fact, a Web site is never really finished, for you will probably be adding, deleting, and revising information as long as the site exists. The needs of your audience might change, and the information you present almost certainly will.

One major difference between a Web site and a paper document is that when you create a paper document, you are in complete control of its look: you design it, print it, and make the copies. When you create a Web site, however, you cannot control its appearance completely. What the reader sees is determined by his or her hardware and software. Such factors as the size of the monitor, the browser used to download files from the server, and the settings on the reader's computer and browser will affect the appearance of your site.

Another important difference between a Web site and a paper document is that the Web makes hyperlinks possible. Because a Web site may contain numerous links to other pages within that site or to other sites on the Internet, you need to pay particular attention to navigation issues: helping readers know where they are and know how to get where they want to go.

The following sections discuss the process of creating Web sites.

To create a Web site:
Analyze your audience and purpose.
Design the site and its pages.
Create and code the content.
Revise and test the site.
Launch the site.
Register the site with search engines.
Maintain the site.

Analyzing Your Audience and Purpose

Your first goal, as always, is to understand who will be viewing the site and why you are creating it. Who are your readers? Why would they visit the site? What kinds of information do they seek? How much do they already know about your subject? Are they looking mainly for links to other sites? Do they need to download information to their own computers? What are your specific goals in launching the site: to project a positive image for your organization? to publicize your products or services? to sell?

For more about audience and purpose, see Ch. 5.

You need to think about all these issues as you start to plan. In addition, you need to think about four additional questions that can affect the whole design of your site:

- *What kind of equipment do your readers have?* If they have fast Internet connections, you can use more and bigger graphics without causing annoying delays as the information downloads. If they have slow connections, you should use only a few graphics and keep them small. In addition, you will want to create a number of small pages rather than a few large ones, because small pages load more quickly.

- *Do your readers want to print out the information on your site?* If so, you need to create a version that prints as a single, unified document, not as many small pages. For more about providing printable versions of your site, see page 625.
- *Do your readers have any disabilities?* If many of your readers are elderly, you need to design the site to accommodate vision impairment, hearing impairment, and perhaps motion impairment. For more about designing sites for people with disabilities, see page 626.
- *Do your readers speak English?* If not, you should consider creating the site in other languages. For more about designing sites for multicultural readers, see page 627.

Depending on your answers to these questions, you may need to include extra time in your schedule and extra labor in your budget.

Designing the Site and Its Pages

Figure 21.1 shows the basic structure of a simple Web site. Almost all sites consist of a home page — the main page of the site — and other pages that are linked to it. A page refers to a file; one file might fit on one screen or extend to a number of screens.

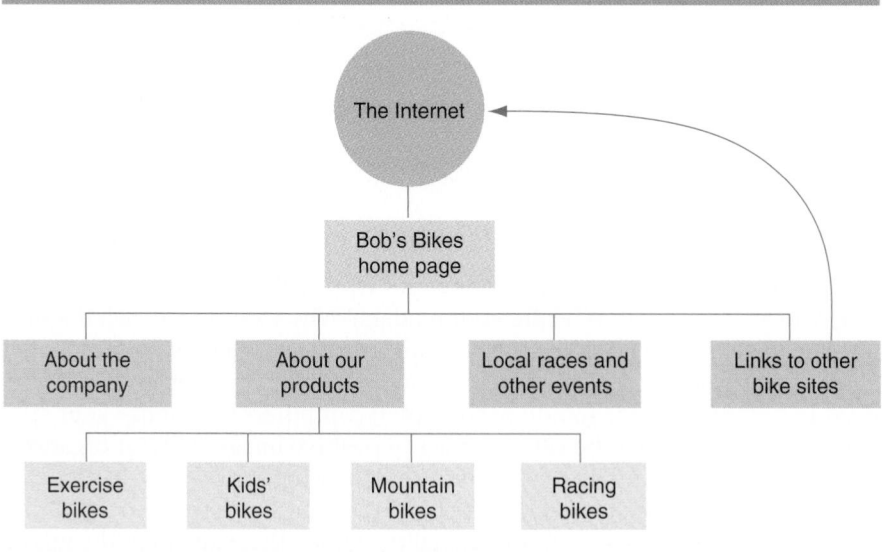

■ **Figure 21.1**
Structure of a Basic Web Site

Each box in the diagram represents a page on the Web site for Bob's Bikes. The pages in the bottom row are linked only to the "About our products" page.

The main characteristic of a well-designed site is that readers can easily find the information they need. Your job is to figure out the kinds of information they will need and how they will look for it. Consider your audience and purpose. If you are creating a site for a small insurance agency, you might conclude that your readers will want to visit your site for five reasons. Your site should be designed so that readers can figure out how to fulfill each of these purposes from the home page:

- to understand the types of insurance you offer
- to find out your rates
- to follow links to other sources of information about insurance
- to email questions to you
- to make appointments to meet with you

For simple sites such as this one, a shallow design, like the one shown in Figure 21.2, might work best. Larger sites often call for a deeper design, as shown in Figure 21.3.

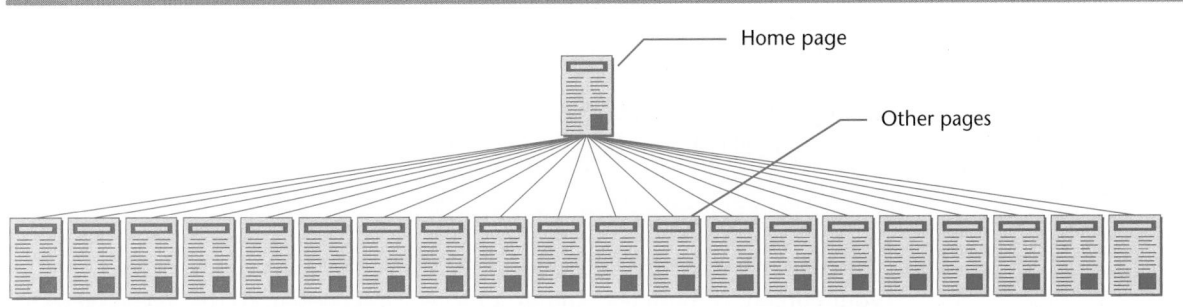

■ **Figure 21.2   A Shallow Site Design**

If the home page clearly displays links to the second-level pages, a reader can easily navigate the site. Electronic phone books and bibliographies are often presented using a shallow design.

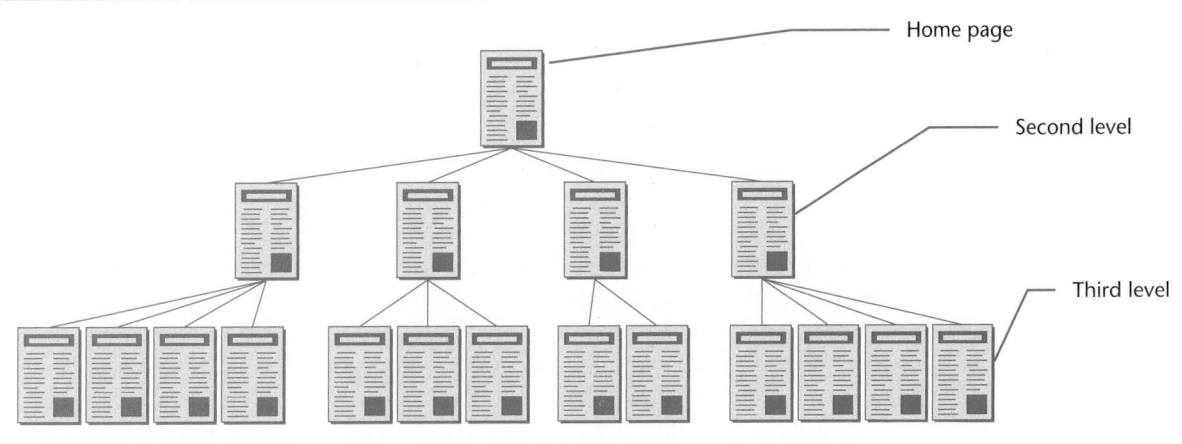

■ **Figure 21.3   A Deeper Site Design**

A deeper site lets you classify and subclassify information so that readers are not overwhelmed with links on a home page. Unfortunately, information on lower-level pages may get overlooked. Many site designers try to structure their sites so that a reader does not have to click more than twice to get from any one place on the site to any other.

In designing the site as a whole, try to give all pages a consistent appearance. The content and function of different pages might vary greatly, but the typography, types of graphics, and colors should be consistent from page to page. The site's navigation elements should also appear in the same place on each page, creating a pattern that will help readers anticipate where to find the information they seek.

For more specific advice about designing the site and its pages, see page 618, or consult one of the many excellent online tutorials on designing Web sites. Three well-known ones are the following:

Click on the link to Mike Markel's Web Design Tutorial on TechComm Web (www.bedfordstmartins.com /techcomm) for a tutorial on designing Web pages.

- *Getting Started with HTML,* by Dave Raggett (http://www.w3.org /MarkUp/Guide/)
- *HTML Goodies,* by Joe Burns (http://www.htmlgoodies.com/primers /basics.html)
- *Webmonkey* (http://www.hotwired.lycos.com/webmonkey/design/)

Search for "HTML tutorial" or "Web tutorial" to find many dozens of tutorials.

Creating and Coding the Content

Once you have created a design for your site, you will need to create or gather your content — both text and graphics — and code it, so that it can be transmitted on the Web.

Gathering your content can be a bigger job than you think. If you simply code print documents and put them on the Web, your site will be ineffective. Readers on the Web tend to jump from place to place rather than read consecutively. To make paper documents effective on the Web, you need to revise or even rewrite them.

In addition, you also need to collect or make the graphics for the site. A graphic the size of a standard sheet of paper would probably be a poor choice for a Web site, because readers with slower modems or phone lines would have to wait several minutes for it to download.

The next step is to code your information into a digital format that can be transmitted on the Web. For text, the current standard is HTML (*hypertext markup language*), which is now evolving into XHTML (*extensible hypertext markup language*). Other new standards, such as WML (*wireless markup language*), will make Web pages more compatible with evolving technologies, such as personal digital assistants (PDAs) and cell phones. Figure 21.4 shows a simple Web page; Figure 21.5 shows the HTML code that produced it.

How do you code material for use on the Web? To add HTML tags to text, use one of three techniques:

- *Save your word-processed files as HTML.* Unfortunately, most word processors add erroneous code. If you use word-processed code, open the file in a text editor and remove the faulty code.
- *Use a Web-editor program.* Sometimes called a *Web-authoring program,* a Web editor automates most of the coding. For example, instead of typ-

Some Basic HTML Codes

HTML can be a complicated markup language, but most of the *tags* you will need for a simple page are easy to understand.

The default typeface on the Web is Times Roman. You can use tags to change the color, the size, or the typeface. The title of this page is formatted as H1, the largest of six heading sizes.

You can use most of the design features that you use all the time on your word processor. For instance, you can easily make a bulleted list, called an *unordered list:*

- first bulleted item
- second bulleted item

Or a numbered list, called an *ordered list:*

1. first numbered item
2. second numbered item

You can insert a graphic:

You can add a hyperlink to another file, such as the Web site of this textbook.

You can make a table:

Column head	Column head
data	data

Tables are useful in HTML because they let you create columns of text:

Here you place the text and graphics for the left-hand column of the screen. If you eliminate the grid lines, your reader sees only the text and graphics.	Here you place the text and graphics for the right-hand column of the screen.

To help chunk information, use a *horizontal rule,* which you see here.

To view XHTML versions of Figures 21.4 and 21.5, click on Additional Sample Documents for Ch. 21 on TechComm Web (www.bedfordstmartins.com/techcomm).

■ **Figure 21.4   A Simple Web Page**

ing the tags for italics, you simply select the text you want to italicize, then click a button. Sophisticated Web editors, such as PageMill® or Front-Page®, contain numerous other features, including design templates, for creating complex sites and pages. Like word processors, however, Web editors often introduce unnecessary or erroneous codes.

- *Enter the tags by hand in a text editor such as Notepad.* For simple sites, using the text editor that comes with your computer operating system is the easiest way to enter tags. Creating the tags shown in Figure 21.5, for example, required only a few minutes in Notepad. Learning basic HTML tags will allow you to make changes to a file yourself.

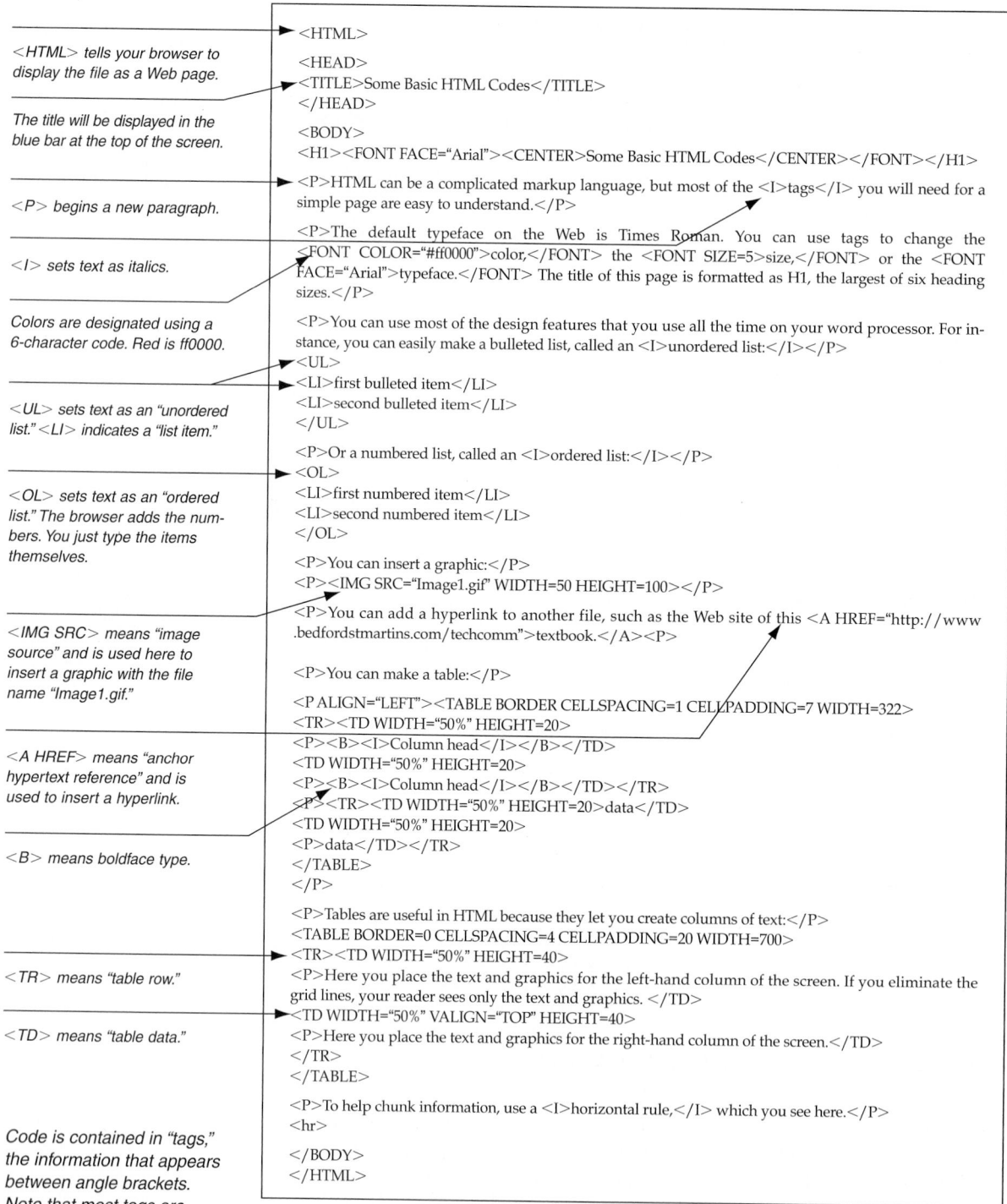

<HTML> tells your browser to display the file as a Web page.

The title will be displayed in the blue bar at the top of the screen.

<P> begins a new paragraph.

<I> sets text as italics.

Colors are designated using a 6-character code. Red is ff0000.

<UL> sets text as an "unordered list." <LI> indicates a "list item."

<OL> sets text as an "ordered list." The browser adds the numbers. You just type the items themselves.

<IMG SRC> means "image source" and is used here to insert a graphic with the file name "Image1.gif."

<A HREF> means "anchor hypertext reference" and is used to insert a hyperlink.

<B> means boldface type.

<TR> means "table row."

<TD> means "table data."

Code is contained in "tags," the information that appears between angle brackets. Note that most tags are used in pairs.

```
<HTML>

<HEAD>
<TITLE>Some Basic HTML Codes</TITLE>
</HEAD>

<BODY>
<H1><FONT FACE="Arial"><CENTER>Some Basic HTML Codes</CENTER></FONT></H1>

<P>HTML can be a complicated markup language, but most of the <I>tags</I> you will need for a simple page are easy to understand.</P>

<P>The default typeface on the Web is Times Roman. You can use tags to change the <FONT COLOR="#ff0000">color,</FONT> the <FONT SIZE=5>size,</FONT> or the <FONT FACE="Arial">typeface.</FONT> The title of this page is formatted as H1, the largest of six heading sizes.</P>

<P>You can use most of the design features that you use all the time on your word processor. For instance, you can easily make a bulleted list, called an <I>unordered list:</I></P>
<UL>
<LI>first bulleted item</LI>
<LI>second bulleted item</LI>
</UL>

<P>Or a numbered list, called an <I>ordered list:</I></P>
<OL>
<LI>first numbered item</LI>
<LI>second numbered item</LI>
</OL>

<P>You can insert a graphic:</P>
<P><IMG SRC="Image1.gif" WIDTH=50 HEIGHT=100></P>

<P>You can add a hyperlink to another file, such as the Web site of this <A HREF="http://www.bedfordstmartins.com/techcomm">textbook.</A><P>

<P>You can make a table:</P>

<P ALIGN="LEFT"><TABLE BORDER CELLSPACING=1 CELLPADDING=7 WIDTH=322>
<TR><TD WIDTH="50%" HEIGHT=20>
<P><B><I>Column head</I></B></TD>
<TD WIDTH="50%" HEIGHT=20>
<P><B><I>Column head</I></B></TD></TR>
<P><TR><TD WIDTH="50%" HEIGHT=20>data</TD>
<TD WIDTH="50%" HEIGHT=20>
<P>data</TD></TR>
</TABLE>
</P>

<P>Tables are useful in HTML because they let you create columns of text:</P>
<TABLE BORDER=0 CELLSPACING=4 CELLPADDING=20 WIDTH=700>
<TR><TD WIDTH="50%" HEIGHT=40>
<P>Here you place the text and graphics for the left-hand column of the screen. If you eliminate the grid lines, your reader sees only the text and graphics. </TD>
<TD WIDTH="50%" VALIGN="TOP" HEIGHT=40>
<P>Here you place the text and graphics for the right-hand column of the screen.</TD>
</TR>
</TABLE>

<P>To help chunk information, use a <I>horizontal rule,</I> which you see here.</P>
<hr>

</BODY>
</HTML>
```

■ **Figure 21.5 The HTML Code for the Page in Figure 21.4**

Formatting graphics for use on the Web is a little trickier, because you sometimes have to open your graphics file in a graphics program, such as Photoshop®, then save it in one of the correct formats. Currently, ".jpeg" and ".gif" are the two most common formats used on the Web. In addition, graphics often have to be sized and compressed to ensure that the file size is as small as possible, so they will download rapidly.

The online tutorials listed on page 614 provide instruction on coding information for the Web.

Revising and Testing the Site

When the page looks the way you want it to, test it as you would a print document to make sure it accomplishes your purposes. Can readers understand the main point? Can they understand how the different pages that make up the site work together? Is the wording of the hyperlinks clear and informative?

Sites such as Web Site Garage (http://websitegarage .netscape.com/) do many diagnostic tests on your site for free.

You will also need to determine whether the technical aspects of the site work correctly. For instance, does the home page load correctly when you enter the URL on your browser? Do all the links work? Does the email form for contacting you work?

You also have to test the site with different kinds of computers, monitors, and browsers. If you find, for example, that one browser displays your tables as masses of meaningless, unformatted numbers, you will have to figure out a different way to display that information. However, the use of XHTML should minimize this problem.

Launching the Site

Ordinarily, you create your Web site on your own computer. Once you have created the site and tested it, you have to transfer the files to an Internet server, a computer that is connected to the Internet and has special software. If you are using a local or national Internet service provider (an ISP), that provider will show you how to use File Transfer Protocol (FTP) to transport your files. If the Internet server is within your own organization, you might use FTP or even carry a disk down the hall.

For more about FTP, see Ch. 4, p. 81.

Once you get the site up on the Web, test it again to make sure it is professional and attractive and that its technical features work.

Registering the Site with Search Engines

Once your site is up and running, you will want to publicize it. In part, you can do this in the traditional way: add the URL to all your product information and advertising. However, you also want to make it easy for people to find the site through search engines on the Web. Although some search engines have programs that roam the Web and automatically record the addresses of new sites, you should formally notify search engines that you have

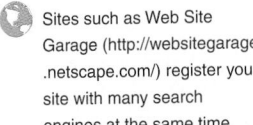

Sites such as Web Site Garage (http://websitegarage .netscape.com/) register your site with many search engines at the same time.

launched a site. When you look at the site of a search engine such as AltaVista, you will see that it has a link to a page where you can register your site by listing its keywords, describing its subject matter, and entering its URL. There are even sites that disseminate this kind of information to a number of search engines, so you do not have to repeat the information for each one. Search for "site registration."

Maintaining the Site

Your goal in creating a Web site is to have people visit it not just once, but often. To encourage visits, you need to maintain the site.

GUIDELINES

Keeping Your Site Current

- ▶ *Add new information.* Many sites have a "what's new" box on the home page that directs readers to new information.
- ▶ *Delete old information.* Few mistakes undercut your credibility more than a page describing an upcoming event from last year.
- ▶ *Test for link rot.* When you link to a site that no longer exists or that has moved to a new address, you have *link rot*. Web-editor programs and a number of Web sites help you check that your links work. But you can check yourself by visiting your own site and trying out the links.
- ▶ *Solicit comments from users.* Ask readers to email you about any features that are not working and with suggestions for adding content to your site. When readers send you such email, be sure to reply with a thank-you note.

DESIGNING EFFECTIVE SITES AND PAGES

For more about designing documents, see Ch. 13.

Most of the principles of good Web page design are similar to the principles of good page design for printed documents. For instance, you should start with a page grid, use white space liberally, and use typography effectively. However, because reading a Web page is different from reading a page in a book, you need to apply these principles a little differently in designing your site.

This section covers seven design principles.

To design effective sites and pages:
Aim for simplicity.
Make the text easy to read and understand.
Create informative headers and footers.

Help readers navigate the site.

Create clear, graceful links.

Avoid Web clichés.

Include extra features your readers might need.

Aim for Simplicity

When you create a site, it doesn't cost anything to use all the colors in the rainbow, to add sound effects and animation, to make whole paragraphs blink on and off. Sometimes these effects can be useful in communicating information to readers. Often, however, all they do is slow the download and annoy the reader. If the effects serve no useful function, avoid them.

GUIDELINES

Designing a Simple Site

▶ *Use simple backgrounds.* If you think a plain white background is ineffective, use a pale pastel color or, at most, a muted background pattern. Avoid loud patterns that distract the reader from the words and graphics of the text. You don't want readers to "see" the background. Here is an example of what can go wrong:

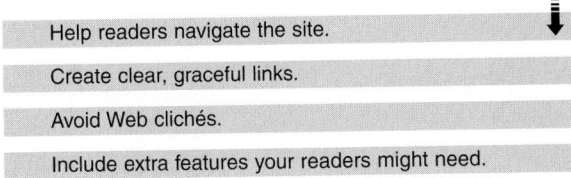

▶ *Use conservative color combinations to increase text legibility.* The greater the contrast between the color of the text and the color of the background, the more legible the text. The most legible text is black against a white background. Avoid garish combinations, such as red text against a purple background. Poor contrast affects legibility.

▶ *Avoid decorative graphics.* Don't waste space by using decorative graphics, which don't convey useful information. Think hard before you use clip art.

▶ *Use thumbnail graphics.* Instead of a large graphic, which takes a long time to download, use a thumbnail graphic. Then tag the thumbnail so that readers who click on it are linked to a large version of the graphic.

Make the Text Easy to Read and Understand

Web pages are harder to read than paper documents because the resolution on the screen is much less sharp and the screen is oriented horizontally rather than vertically.

GUIDELINES

Designing Easy-to-Read Text

▶ *Keep the text short.* The relatively poor resolution on a screen makes reading long stretches of text difficult. In general, pages should contain no more than two or three screens of information.

▶ *Chunk information.* When you write a paper document, you chunk information to help readers understand it better. When you write for the Web, you need to do the same thing. Use frequent headings and brief paragraphs. Lists also help you chunk information.

▶ *Make the text as simple as possible.* Use common words and short sentences to make the information as simple as the subject allows.

Create Informative Headers and Footers

To view Figures 21.6 and 21.7 in context on the Web, click on Links Library for Ch. 21 on TechComm Web (www.bedfordstmartins.com/techcomm).

Readers rely on the information you provide in the header and footer to understand and navigate your site. In addition, headers and footers help establish your credibility. Remember, a lot of information on the Web is not credible. Make sure your readers know they are visiting the official site of your organization, and that it was created by professionals.

Figure 21.6 (Corel, 2001a) shows a typical Web site header. Figure 21.7 (Corel, 2001b) shows a typical Web site footer.

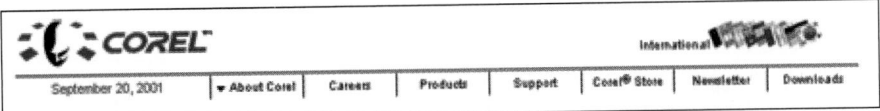

■ **Figure 21.6   Header**

This header on the Corel home page displays the date, along with links to the major contents of the site. Headers should always contain a link to the home page; on this page, the logo and the word Corel *in the upper left link to the home page.*

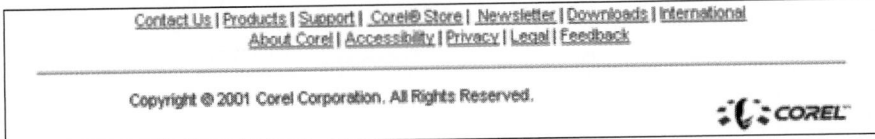

■ **Figure 21.7   Footer**

A footer usually includes a copyright notice. This footer also includes links to legal information, accessibility information, the privacy policy, the Webmaster's email, and several other areas of the site. Note that because the links in the footer are presented as text, they will be visible to visitors with handicaps and to those who have turned off the graphics.

Help Readers Navigate the Site

Unlike readers of a paper document, readers of a Web site cannot hold the Web page in their hands. All they can see is the page on the screen. Therefore, each page should help readers determine where they are in the site and get where they want to go.

A consistent visual Web site design offers readers a predictible way to navigate your site: make sure the header and footer on each page are structured consistently; use the same background color or pattern on all the pages, unless you have a good reason to vary it; use consistent typography — typeface, size, and color — on every page; and put navigational links in the same place on every page (in a column along the left side of the screen, for example, or across the top).

GUIDELINES

Making Your Site Easy to Navigate

▶ *Include a site map or index.* A site map, which lists the pages that make up a site, can take the form of a graphic or a list of the pages, classified according to logical categories. An index is an alphabetized list of the pages. Figure 21.8 on page 622 is a section of the National Science Foundation (1999) site map.

 To view Figure 21.8 in context on the Web, follow the links in Chapter 21 on TechComm Web (www .bedfordstmartins.com /techcomm).

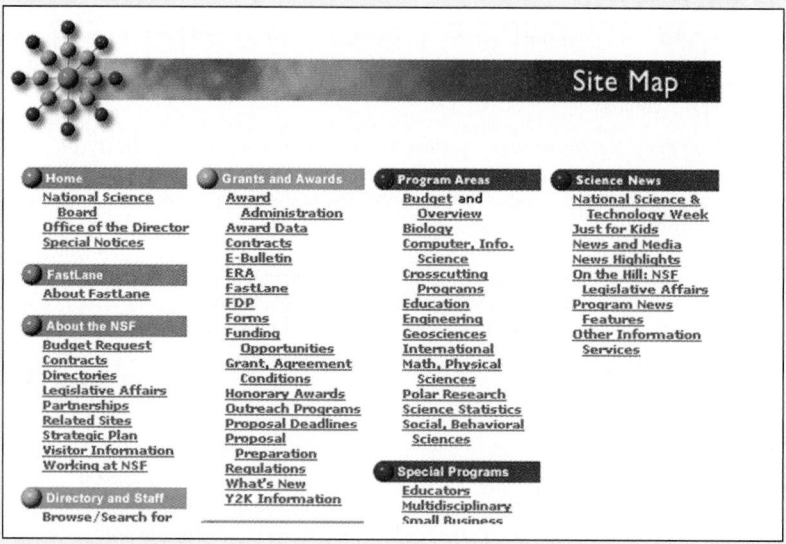

■ **Figure 21.8 Site Map**

Always include a link to the site map on the home page.

▶ *Use a table of contents at the top of long pages.* If your page extends over more than a couple of screens, a table of contents — a set of links to the items on that page — saves your readers from having to scroll down to find the topic they want. Tables of contents can link one page to information further down on the same page or on separate pages. Figure 21.9 (Nolo.com, 1999) shows a table of contents at the top of a page.

▶ *Help readers get back to the top of long pages.* If a page is long enough to justify a table of contents, include a "<u>Back to top</u>" link (a textual link or a button or icon) before the start of each new chunk of information.

▶ *On every page, link to the home page.* This link can be a simple "<u>Back to home page</u>" textual link, a button, or an icon.

▶ *Include redundant textual navigational links at the bottom of the page.* If you are using a button or icon for navigational links on your pages, include textual versions of those links at the bottom of the page. Some of your readers might have turned off the images to speed up the download, and they won't be able to understand the graphical link (unless you have added an alt tag — a tag that instructs the browser to display a word or phrase defining the graphic). In addition, readers with vision impairment might be using special software that

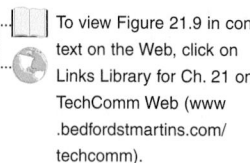

To view Figure 21.9 in context on the Web, click on Links Library for Ch. 21 on TechComm Web (www .bedfordstmartins.com/ techcomm).

Patent, Copyright & Trademark

TABLE OF CONTENTS | FAQ Table of Contents ▼ |

Qualifying For a Patent

- What types of inventions can be patented?
- What types of inventions are not eligible for patent protection?
- Can computer software qualify for patent protection?
- Is it possible to obtain a patent on forms of life?
- What makes an invention novel?
- When is an invention considered nonobvious?
- What makes an invention useful?

Obtaining Your Own Patent Should Be Easy

- What information is typically included in a patent application?
- What happens if there are multiple applications for the same invention?

■ **Figure 21.9   Table of Contents**

These links are classified according to the topics "qualifying for a patent" and "obtaining your own patent should be easy."

reads the information on the screen. This software interprets text only, not graphics. Figure 21.10 (Society for Technical Communication, 2001) shows redundant links at the bottom of a page.

To view Figure 21.10 in context on the Web, click on Links Library for Ch. 21 on TechComm Web (www .bedfordstmartins.com /techcomm).

Membership | Publications | Education | Recognition | For Leaders | Members Only
About Us | Contacts | Jobs Database | For the Press | Search

■ **Figure 21.10   Redundant Textual Links**

Create Clear, Informative Links

Well-phrased links are easy to read and understand. By clearly telling the reader what kind of information the linked site will provide, they help the reader decide whether to follow the link. The following guidelines are based on Sun Microsystems' "Guide to Web Style" (Sun, 1999).

Writing Clear, Informative Links

▶ *Structure your sentences as if there were no links in your text.*

AWKWARD	<u>Click here</u> to go to the Rehabilitation Center page, which includes numerous links to research centers across the nation.
SMOOTH	The <u>Rehabilitation Center</u> page includes numerous links to research centers across the nation.

▶ *Indicate what information the linked page contains.* Readers get frustrated if they wait for a file to download and then discover that it doesn't contain the information they expected.

UNINFORMATIVE	See the <u>Rehabilitation Center</u>.
INFORMATIVE	See the <u>Rehabilitation Center</u> for hours of operation.

▶ *Don't change the colors of the text links.* Readers are used to two common colors: blue for links that have not yet been clicked, and purple for links that have already been clicked.

Avoid Web Clichés

Despite its status as a new medium, the Web has already developed its share of clichés. Like all kinds of writing, text intended for a Web site should be drafted and revised with care. Empty words or phrases can obscure the site's purpose and give readers the impression that they are wasting their time. The following Web clichés are particularly annoying because they insult visitors' intelligence by stating the obvious.

- Don't tell visitors to "check out" your site. If the information looks interesting and useful, they will.
- Don't say your site is "under construction." If the site is a mess, don't launch it. If you want to tell visitors that you update the contents periodically, state when the site was last revised.
- Don't call anything on your site "cool." Very uncool.
- Don't invite visitors to "come back often." If their visit was worth it, they will. If it wasn't, they won't.

Include Extra Features
Your Readers Might Need

Because it is more than likely that readers with a wide range of interests and needs will visit your Web site, you may want to add one or all of the following five features:

- *An FAQ.* A list of frequently asked questions helps new visitors by providing basic information, explaining how to use the site, and directing them to more detailed discussions. Figure 21.11 (Selective Service System, 1999) is an excerpt from an FAQ page.

- *A search page.* A search page lets readers enter a keyword or phrase and find all the pages on the site that contain it. A search page is especially valuable on large sites.

- *Resource links.* Central to many sites is a collection of links to other sources of information. If the purpose of your site is solely to advertise and sell your company's products, you probably wouldn't want to link to other sites. But if your purpose includes educating readers, you should provide links.

- *A printable version of your site.* A Web site is structured to be viewed on the screen, not read on a page. It might consist of dozens of relatively small files that would be difficult to print and assemble. Or it might use yellow text on a dark blue background; unless readers changed their printer settings to make the document print in black and white, the document would print in color, using up a lot of toner or ink and producing results that would be difficult to read. Therefore, you might plan to create a printable version of your site with black text on a white background and all the text and necessary graphics consolidated into one big file.

- *A text-only version of your site.* Many readers with slow connections to the Internet set their browsers to view text only. In addition, as is discussed more fully in the next section, many readers with impaired vision rely on text because their specialized software cannot interpret graphics. Therefore, you might want to create a text-only version of your site and link to it from your home page.

To view Figure 21.11 in context on the Web, click on Links Library for Ch. 21 on TechComm Web (www .bedfordstmartins.com/ techcomm).

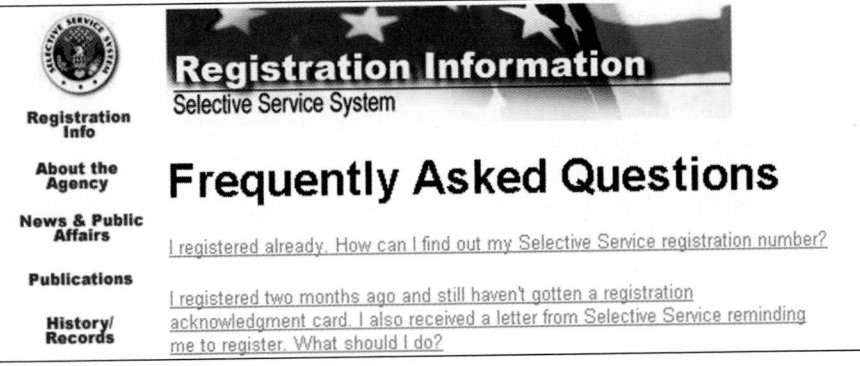

■ Figure 21.11   Excerpt from an FAQ Page

Each question here links to a separate page containing the answer. On other sites, the questions and answers are presented on the same page.

DESIGNING SITES FOR READERS WITH DISABILITIES

The Internet has proved to be a terrific technology for people with disabilities because it brings a world of information to their desktops, allowing them to work from home and participate in virtual communities. However, as sites have become more sophisticated over the last few years, many people with disabilities have found the Internet harder to use. In 1996, a court ruled that the Americans with Disabilities Act covered commercial Web sites, which must now be accessible to people with disabilities. Over the next few years, more effort will go into making hardware and software to help people with disabilities use the Internet.

See the Web Content Accessibility Guidelines (www.w3.org/WAI/), from the World Wide Web Consortium, for a detailed look at accessibility.

The following discussion highlights a few of the ways you can make your site easier for people with disabilities to use. In doing so, you will increase the number of people who can use your site and also make it easier for all your readers to use it.

Consider three main types of disabilities as you design your site:

A site called Bobby (www.cast.org/bobby/?) will check your site for free, to evaluate its adherence to accessibility options.

- *Vision impairment.* People who cannot see, or cannot see well, rely on text-to-speech conversion programs. Be sure to provide either a text-only version of the site or textual equivalents of all your graphics. Use the alt (alternate) tag in your coding to ensure that a textual label pops up when the reader places the mouse over the graphic.

  In addition, do not rely on color or graphics alone to communicate information. For example, if you use a red icon to signal a warning, use the word *warning* as well. If you use tables to create columns on the screen, label the top of each column clearly using a textual label rather than an image.

  Don't select type smaller than the default size of 12 point. Use relative type sizes, not absolute sizes. (*Relative sizing* means that you define size by using plus and minus sizes relative to the default size.) Relative sizing enables readers to enlarge the default size and yet retain the correct proportions.

  Provide audio feedback, for example, by having a button beep when the reader presses it.

- *Hearing impairment.* If you use video, provide captions and (if the video includes sound) a volume control. Also use visual feedback techniques; for example, make a button flash when the reader presses it.

- *Mobility impairment.* Some people with mobility impairments find it easier to use the keyboard than a mouse. Therefore, build in keyboard shortcuts wherever possible. If readers have to click on an area of the screen using a pointing device, make the area large so that it is easy to see and click.

DESIGNING SITES FOR MULTICULTURAL READERS

More than half of the world's Web users are nonnative speakers of English (Global Reach, 2001). Therefore, it makes sense to plan your site as if many of your visitors will not be proficient in English.

Planning for a multicultural Web site is similar to planning for a multicultural paper document. As discussed in Chapter 5, page 109, use short sentences and paragraphs, as well as common words. Avoid idioms, both verbal and visual, that might be confusing. For instance, don't use sports metaphors, such as "full-court press," or a graphic of an American-style mailbox to suggest an email link.

If a large percentage of your readers speak a language other than English, consider creating a version of your site in that language. The expense can be considerable, but so can the benefits.

See the World Wide Web Consortium's internationalization page (www.w3.org /International/) to learn more about the challenges of creating markup languages that meet the needs of international users.

ETHICS, COPYRIGHT LAW, AND THE WEB

The Web is unlike other publishing media in that the content is remarkably easy to steal. Users can download digital versions of the text and graphics and manipulate them in any way they want. An unethical user could republish graphics without permission from the person who owns the copyright; or a user could modify text from a Web site and claim it as his or her own. These practices are illegal as well as unethical, for digital material is covered by the same copyright laws that apply to printed material. All text and graphics on the Web are legally protected, regardless of whether the person who put the material on the site included a copyright symbol. Unless the copyright owner specifically says that you may use the material, you must seek written permission, just as you would for printed material.

However, saying that Web sites are protected like other kinds of documents is just the beginning of the discussion, not the end. Benedict O'Mahoney (2001) has written a thoughtful essay on some of the complicated issues involved in interpreting copyright law. Here are just four of the puzzles he addresses:

For more about copyright law, see Ch. 2, p. 24.

- *Is the design of a Web page protected by copyright law?* Some would say no, because what readers see on the screen is a function of their hardware and software. In addition, readers can customize the image. However, the design of a Web page is an original work, created by a person, and thus should be protected, regardless of how readers might change it after it is transmitted.

- *Are lists of links protected?* Is each link protected by copyright? No. Is the whole list of links protected? Probably, if the person showed some originality in creating the list. Courts have held that phone books are not protected because alphabetized lists of names, addresses, and phone numbers do not show originality. But a list of links to resources for agriculture students, for example, probably is protected, because the author did some original thinking in creating categories for the individual links. Copying another site's list of links and putting it on your own site is probably illegal (and certainly unethical).

- *May you link to anyone else's Web site? May anyone link to yours?* The Web was initially envisioned as an open environment in which anyone can link to anyone else. In theory, this is a sensible idea. But what if some organization doesn't want you to link to its site because it doesn't want the extra traffic on its server? Or what if you discover that a hate group has written a complimentary paragraph about you and linked to your site? Is it your responsibility to find out who has linked to you, or should the site that wishes to link to you notify you and get permission to do so?

- *Are composite Web pages legal?* A composite Web page is a page made up of links to elements on other organizations' sites. For example, a composite page can consist of links to graphics on one organization's site, sound samples on a second, and text on a third. Common sense would suggest that this composite Web page is illegal; unless formal permission has been obtained for each of these elements, it violates copyright. But the fair-use provision of the copyright law permits copying for some purposes, such as commentary or education, within some limits. Therefore, some composite pages might be legal under some interpretations.

As this brief discussion suggests, questions of ethics and legality are extremely complex and will remain unresolved for many years. Over the next decade, the courts will be hearing many cases in which copyright law has to be reinterpreted in light of the unique technical, economic, and social implications of electronic media.

An excellent source of information, news, and analysis of these debates is the World Wide Web Consortium (www.w3.org).

GUIDELINES

Creating an Ethical Site

- ▶ *Don't plagiarize.* If you want to publish text or graphics you found on the Internet, you need to receive written permission from the copyright owner.

- ▶ *Ask permission to link.* Notify an organization if you wish to link to its site, then abide by its wishes. Don't link if the organization says no.

- ▶ *Link to the home page.* Do not link to a lower-level page, even if that page contains the information you want. Organizations have a right to ask people to come in the front door, where advertisements, trademarks, and other legal information are displayed.

- ▶ *Don't misuse meta tags.* If you look at the code of a typical Web page, you will see a "meta" tag near the top. This is the place where you put keywords that describe the contents of your site. If you are a Ford dealer, you list "Ford," "dealer," and the names of Ford models. This way, readers who enter "Ford," or related words, in their search engines will be directed to your site. It is unethical and, according to some intellectual-property attorneys, illegal to list "Chevrolet" in your meta tag as a way to get potential Chevrolet customers to come to your Ford-dealership site.

A LOOK AT SAMPLE WEB PAGES

The best way to learn about designing Web sites and their pages is to study them. Figures 21.12 through 21.14 offer three examples of good Web page design, with a couple of suggestions for improvement.

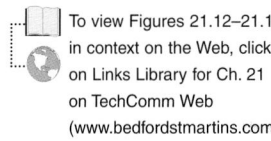 To view Figures 21.12–21.14 in context on the Web, click on Links Library for Ch. 21 on TechComm Web (www.bedfordstmartins.com/techcomm).

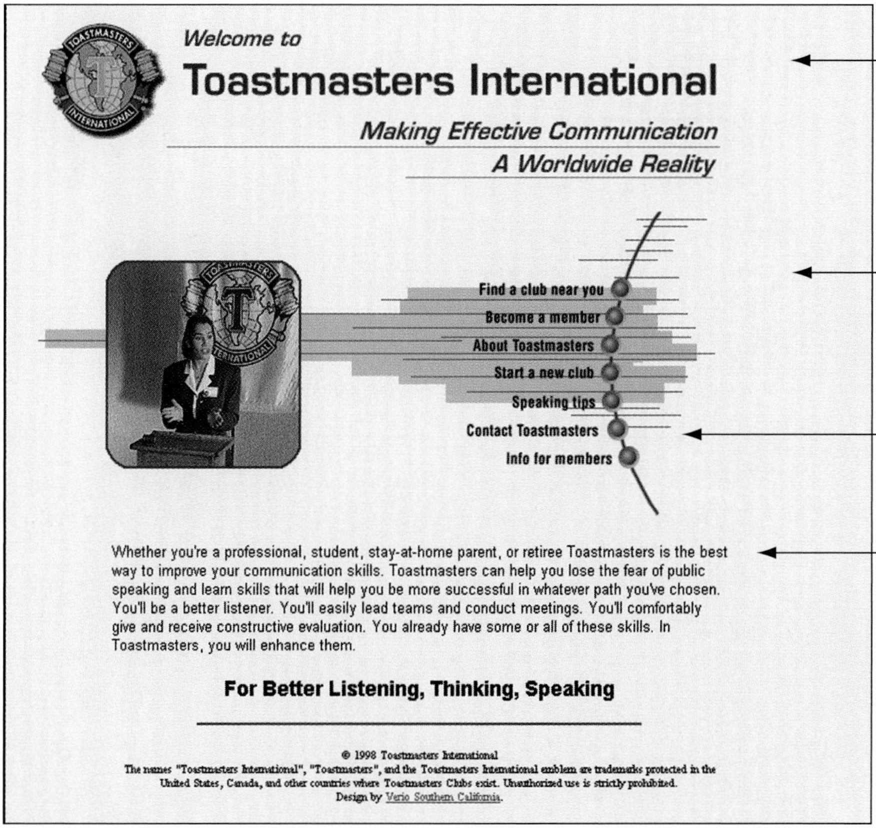

The header includes the organization's logo and a statement of its purpose.

The photograph of the speaker reinforces the purpose of the organization (and includes the organization's logo).

The links are clearly displayed to the right.

The paragraph describing the organization's purpose in detail is easy to read because of the simple typeface and the clear contrast with the pale background.

The page should include redundant textual links for people with vision impairment or people who have turned off the graphics.

■ **Figure 21.12**
The Toastmasters International Home Page (Toastmasters, 1999)

This page is simple and attractive, with a clear purpose and effective organization.

The main navigation bar to the NSF home page and other main pages consists of small graphics that add visual interest to the page and download quickly.

The links to individual news stories are easy to read, and the organization of the links is clear and simple.

Textual links to the main pages make the page accessible to people with disabilities, as well as to people with slow modems who have turned off the graphics.

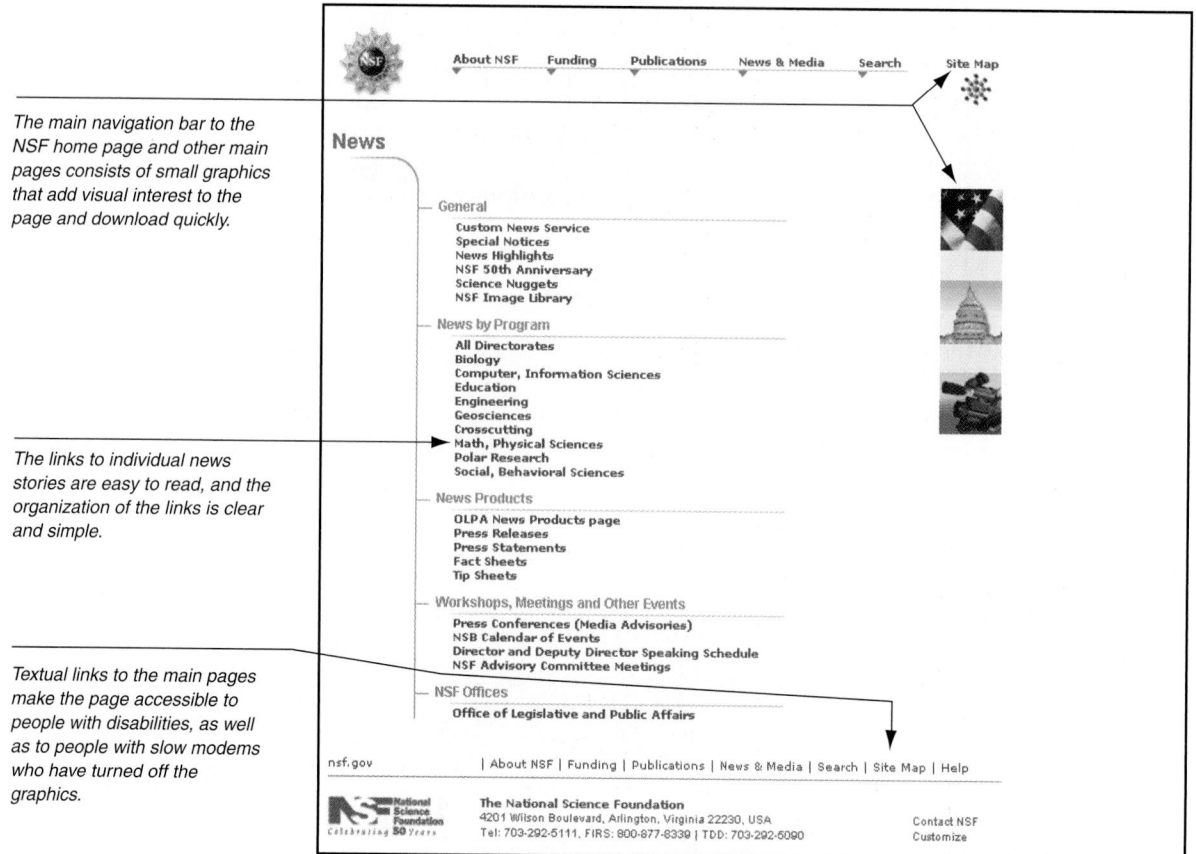

■ **Figure 21.13   News page from the National Science Foundation Web site.**

This page is effective because it is attractive, clear, and compact.

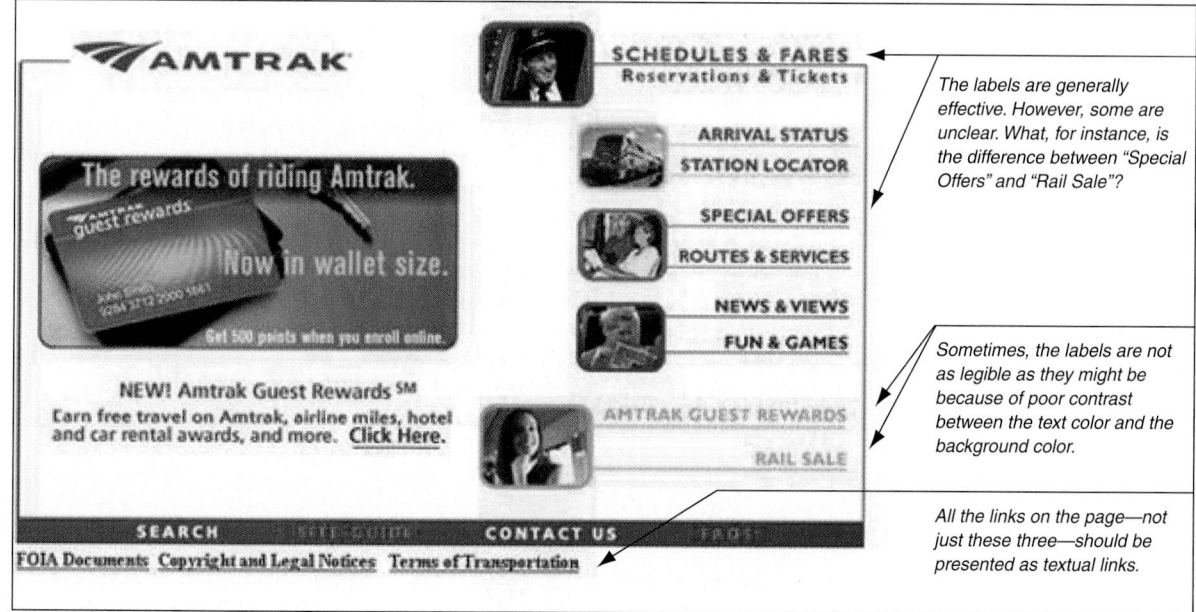

The labels are generally effective. However, some are unclear. What, for instance, is the difference between "Special Offers" and "Rail Sale"?

Sometimes, the labels are not as legible as they might be because of poor contrast between the text color and the background color.

All the links on the page—not just these three—should be presented as textual links.

■ Figure 21.14   Amtrak Home Page (National Railroad Passenger Corporation, 2001)

This page downloads quickly because most of the photographs are relatively small. A large portion of the screen is devoted to white space, giving the page an open, uncluttered feel, and helping readers focus on the text and graphics (see page 618). However, clearer labels would help readers navigate the site, and more textual links would better serve people with disabilities.

    Revision Checklist

In designing the site, did you

- ❑ analyze your audience and purpose before planning your site?
- ❑ use a plain, simple background?
- ❑ allow for effective contrast between the background color and the text color?
- ❑ avoid decorative graphics?
- ❑ use thumbnail graphics rather than large ones?
- ❑ make the text easy to read by using brief chunks of text?
- ❑ use simple language and short sentences?
- ❑ create informative headers and footers?
- ❑ include a site map or index?
- ❑ use a table of contents at the top of long pages?
- ❑ link to the top of long pages?
- ❑ link to the home page on every page?
- ❑ include textual navigational links at the bottom of the page?

❑ create clear and informative links?

❑ avoid Web clichés?

❑ include extra features your readers might need, such as an FAQ, a list of links, a printable version of the site, and a text-only version of the site?

❑ design the site so that it is easy for people with vision, hearing, and movement disabilities to use?

❑ design the site to accommodate the needs of multicultural readers?

❑ get permission to publish any information that you did not generate?

❑ ask permission to link?

❑ link to another site's home page rather than a secondary page?

❑ avoid including misleading information in the meta tags?

❑ revise and test the information?

❑ get the files to an Internet server?

❑ register the site with search engines?

Exercises

1. Find the sites of three manufacturers within a single industry, such as personal watercraft, cars, computers, or medical equipment. Study the three sites, focusing on one of these aspects of site design:

   • use of color

   • effectiveness of the graphics in communicating information

   • quality of the writing

   • quality of the site map or index

   • navigation, including the clarity and placement of links to other pages in the site

   • use of Web clichés

   • accommodation of multicultural readers

   • accommodation of people with disabilities

   • the phrasing of the links

   Which of the three sites is most effective? Which is least effective? Why? Compare and contrast the three sites in terms of their effectiveness.

2. Using a search engine, find a site that serves the needs of people with a physical disability: for example, the Glaucoma Foundation (www.glaucoma-foundation.org/info/). What attempts have the designers made to accommodate the needs of people with glaucoma? How effective are those attempts?

3. You are a native speaker of Spanish who knows some English. You wish to buy a personal computer from a manufacturer who sells online. A friend suggests that you study the Web sites of Dell (www.dell.com), Gateway (www.gateway.com), and Compaq (www.compaq.com). Of these three sites, which does the best job in meeting your needs? Explain.

Research Projects

Some of the following projects ask you to write a memo. See Chapter 15, page 430, for a discussion of writing memos.

4. Using a search engine, find five tutorials on making Web pages. For each site, determine the level of expertise of the intended audience and analyze the strengths and weaknesses of the site. Present your results in a memo to your instructor.

5. Form small groups and describe and evaluate your college's or university's Web site. A different member of the group might carry out each of the following tasks:

- Email the site's Webmaster to ask about the process of creating the site. For example, how involved with the content and design of the site was the Webmaster? What is the Webmaster's role in maintaining the site?

- Analyze the kinds of information the site contains and determine whether the site is intended primarily for faculty, students, alumni, or prospective students.

- Determine the overlap between the information on the site and the information in printed documents published by the school. In those cases in which they overlap, is the information on the site merely a duplication of the printed information, or has it been revised to take advantage of the unique capabilities of the Web?

In a memo to your instructor, present your conclusions and recommendations for improving the site.

6. Form groups according to major and analyze the site of a major professional organization in your field. (For example, IEEE is the major professional organization for electrical engineers; the American Bar Association, for attorneys; the Society for Technical Communication, for technical communicators.) Each member of the group might investigate one of the following questions:

- Who is responsible for creating and maintaining the site?

- What are the major kinds of information provided on the site?

- What is the major function of the organization? Whom does it serve? What services does it provide? What activities does it sponsor?

- What is the role of students in the organization? Are there special activities or opportunities for students? Can you think of ways to increase the role of students in the organization?

Present your findings to the appropriate officer at the organization and to your instructor.

C A S E
Creating a Web Site

Create a Web site for your major, student organization, or community group. (You do not actually have to put the site on the Web; you can create it and test it on a personal computer that has a browser.) Follow these steps:

1. Analyze your audience and purpose. Who would view the site, and what would you hope to achieve with the site?

2. Gather or generate your text and other information. If you do not have a sufficient amount of information available in digital format, you can still proceed with building the site.

3. Design the site. What sort of design is appropriate for the information you wish to communicate and the needs of your audience?

4. Convert the text and graphics. What sorts of tools are most useful for the conversion process?

5. Revise and test the information. Other students in the class might help you in testing the site.

Keep a log of the activities each group member has carried out. Be prepared to describe your experiences in an oral presentation to the class. Which tasks were easy to accomplish? Which were difficult? What surprised you about the process? How satisfied are you with the finished product?

 For advice about preparing oral presentations, see Chapter 22.

22 Making Oral Presentations

Vincent Kuraitis, a consultant, on how to take advantage of nervousness:

I look at nervousness as an opportunity to try to channel that adrenaline, to use it as something that will help me do my best rather than detract from my ability to be persuasive. I can get a little bit more alert, a little bit more in touch with my audience. And I look at the adrenaline rush that comes with speaking in front of an audience as something to look forward to because it makes me do my best.

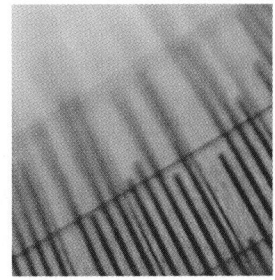

So far, this book has focused on written documents. In this chapter, the focus shifts to oral presentations. The techniques for preparing both are similar: analyze your audience and purpose, gather information, organize that information, and create graphics. The big difference, of course, is the form of delivery.

 See TechComm Web (www .bedfordstmartins.com /techcomm) for guidelines, additional examples, and links related to topics in this chapter.

There are four basic types of presentations:

- *Impromptu presentations.* You deliver an impromptu presentation without advance notice. For instance, at a meeting, your supervisor calls on you to speak for a few minutes about a project you are working on. You did not know you were going to be asked.

- *Extemporaneous presentations.* In an extemporaneous presentation, you might refer to notes or an outline, but you actually make up the sentences as you go along. Regardless of how much you have planned and rehearsed the presentation, you create it as you speak. At its best, an extemporaneous presentation is clear and sounds spontaneous. If you can think well on your feet, the presentation will have a naturalness that will help your audience concentrate on what you are saying.

- *Scripted presentations.* In a scripted presentation, you read a text written out completely in advance (by you or someone else). You sacrifice naturalness for increased clarity and precision.

- *Memorized presentations.* In a memorized presentation, you speak without notes or script. Memorized presentations are not appropriate for most technical subjects because of the difficulty of remembering technical data. In addition, few people other than trained actors can memorize presentations of more than a few minutes.

635

This chapter discusses extemporaneous and scripted presentations. It explains how to prepare an outline or speaking script, how to prepare graphics, and how to rehearse. Then it considers how to give the presentation and how to respond to questions afterward.

UNDERSTANDING THE ROLE OF ORAL PRESENTATIONS

In some ways, an oral presentation is inefficient. For the speaker, preparing and rehearsing generally take more time than writing a document. For the audience, physical conditions during the presentation — noise, poor lighting or acoustics, or an uncomfortable room temperature — can make concentration difficult.

Yet an oral presentation has one big advantage over a written one: it permits a dialogue between the speaker and the audience. Listeners can offer alternative explanations and viewpoints or simply ask questions that help the speaker clarify points of information. And the speaker and listeners can converse before and after the presentation.

Oral presentations are therefore common in technical communication. You can expect to give oral presentations to four types of audiences:

- *Clients and customers.* Whether you are trying to interest clients in a silicon chip or a bulldozer, you will present its features and its advantages compared to the competition. Then, after the sale, you might provide oral operating instructions and maintenance tips to users.

- *Colleagues in your organization.* If you are the resident expert on a mechanism, procedure, or technical subject, you will instruct fellow workers, both technical and nontechnical. After you return from an important conference or an out-of-town project, your supervisors will want a briefing — an oral report. If you have an idea for improving operations at your organization, you will probably write an informal proposal and then present the idea orally to a small group of managers. Your presentation will help them determine whether it is prudent to devote resources to studying the idea.

- *Fellow professionals at technical conferences.* You might speak about your own research project or about a team project carried out at your organization. Or you may be invited to speak to professionals in other fields. If you are an economist, for example, you might be invited to speak to real-estate agents about interest rates.

- *The general public.* As you assume greater prominence in your field, you will receive more invitations to speak to civic organizations and governmental bodies. Your organization will probably encourage these presentations, because they reflect positively on the organization.

You might not have had much experience in public speaking, and perhaps your few attempts have been difficult. Natural speakers who can talk off the cuff are rare. For most of us, an oral presentation requires deliberate and careful preparation.

PREPARING AN ORAL PRESENTATION

Professional speakers make presentations look easy. But when you see an excellent 20-minute presentation, you are seeing only the last 20 minutes of a process that took many hours. How much time should you devote to preparing an oral presentation? Experts recommend devoting 20 to 60 minutes for each minute of the presentation (Smith, 1991, p. 6). At an average of 40 minutes, you would need more than 13 hours to prepare a 20-minute presentation. Obviously there are many variables, including your knowledge of the subject, experience in creating presentation graphics, and experience in giving presentations on that subject. But the point is that good presentations don't just happen.

Preparing an oral presentation requires five steps.

The Virtual Presentation Assistant (www.ukans.edu /cwis/units/coms2/vpa/vpa8 .htm) gives advice and links to text and videos of speeches.

To prepare an oral presentation:
Assess the speaking situation.
Prepare an outline or note cards.
Prepare the graphics.
Choose effective language.
Rehearse the presentation.

Assessing the Speaking Situation

The first step in assessing the speaking situation is to analyze your audience and purpose and then determine how much information you can deliver in the allotted time.

Online Technical Writing (www.io.com/~hcexres /tcm1603/acchtml/oral .html) includes advice and transcripts of several student presentations.

Analyzing Your Audience and Purpose

In planning an oral presentation, consider audience and purpose, just as you would in writing a document.

- *Audience.* How much does the audience know about your subject? Answering this question helps you determine the level of technical vocabulary and concepts you will use, as well as the types of graphics. Speaking over an audience's head puzzles them; oversimplifying can make you appear condescending and insulting. What do audience members want to accomplish in listening to your presentation? Are they likely to be hostile, enthusiastic, or neutral? A presentation on the virtues of free trade, for instance, will be received one way by conservative economists and another way by U.S. steelworkers.

- *Purpose.* Are you attempting to inform, or to both inform and persuade? If you are explaining how windmills can be used to generate power, you might describe the process. If you are explaining why your windmills are an economical means of generating power, you might compare their results with those of other power sources.

Your analysis of your audience and purpose will affect the strategy — the content and the form — of your presentation. For example, you might have to emphasize some aspects of your subject and ignore others altogether. Or you might have to arrange topics so as to accommodate a particular audience's needs.

Budgeting Your Time

At most professional meetings and conferences, organizers clearly state a maximum time, such as 20 minutes, for each speaker. If the question-and-answer period is part of your allotted time, plan accordingly. Even at an informal presentation, you will probably have to work within an unstated time limit that you must determine from the speaking situation. Taking more than your share of an audience's time is rude and egotistical, and eventually your listeners will start to resent you or simply stop paying attention.

For a 20-minute presentation, the time allotment shown in Table 22.1 is typical. For scripted presentations, most speakers need a little over a minute to deliver a double-spaced page of text effectively.

■ **Table 22.1 Time Allotment in a Presentation**

Task	Time (minutes)
• Introduction	2
• Body	
– First Major Point	4
– Second Major Point	4
– Third Major Point	4
• Conclusion	2
• Questions	4

Preparing an Outline or Note Cards

After assessing your audience, purpose, and strategy, prepare an outline or a set of note cards. Some speakers prepare both. They prepare the outline when they are planning the presentation, just as they would if they were writing a document. Then, when they are ready to begin rehearsing, they prepare the note cards they will use in making the presentation. During rehearsals, they revise the notes cards as they consider more effective ways of presenting their information.

Figure 22.1 on page 640 shows an outline for a presentation. The speaker is a specialist in waste-treatment facilities. The audience is a group of civil engineers interested in gaining a general understanding of new developments in industrial-waste disposal. The speaker's purpose is to provide this information and to suggest that his company is a leader in the field.

Notice that this writer uses a problem-methods-solution pattern in developing the presentation. The introduction describes the problem — new environmental regulations will mean cities have to develop new methods of waste treatment and disposal. The following sections present different methods of solving urban waste-management problems.

See Ch. 8, p. 186, for a discussion of organizational patterns.

In preparing note cards to bring to the presentation, your command of the facts — and your ability to remember them under stress — will determine how specific and detailed you make them. Figure 22.2 on page 641 shows one such note card.

You can also make notes using presentation-graphics software. One advantage of software is that revising is easy. As you prepare the notes, you also see the graphics, just as your audience will, and might discover opportunities to improve the organization and development of your presentation.

Preparing Presentation Graphics

Graphics fulfill the same purpose in an oral presentation that they do in a written one: they clarify or highlight important ideas or facts. Statistical data, in particular, lend themselves to graphical presentation, as do descriptions of equipment or processes. Research reported by Smith (1991) indicates that presentations that include transparencies are judged more professional, persuasive, and credible than those that do not, and that audiences remember information better if it is accompanied by graphics. Smith (1991, p. 58) offers these figures:

Writing Guidelines for Engineering and Science Students (http://fbox.vt.edu:10021/eng/mech/writing/) includes advice and sample presentations graphics.

	Retention after	
	3 hours	3 days
Without graphics	70%	10%
With graphics	85%	65%

Purpose: to describe, to a group of civil engineers, a new method of treatment and disposal of industrial waste.

1. Introduction
   1.1 The recent Resource Conservation Recovery Act places stringent restrictions on plant engineers.
   1.2 With neutralization, precipitation, and filtration no longer available, plant engineers will have to turn to more sophisticated treatment and disposal techniques.

2. The Principle behind the New Techniques
   2.1 Waste has to be converted into a cementitious load-supporting material with a low permeability coefficient.
   2.2 Conversion Dynamics, Inc., has devised a new technique to accomplish this.
   2.3 The technique is to combine pozzolan stabilization technology with the traditional treatment and disposal techniques.

3. The Applications of the New Technique
   3.1 For new low-volume-generators, there are two options.
       3.1.1 Discussion of the San Diego plant.
       3.1.2 Discussion of the Boston plant.
   3.2 For existing low-volume generators, Conversion Dynamics offers a range of portable disposal facilities.
       3.2.1 Discussion of the Montreal plant.
       3.2.2 Discussion of the Albany plant.
   3.3 For new high-volume generators, Conversion Dynamics designs, constructs, and operates complete waste-disposal management facilities.
       3.3.1 The Chicago plant now processes up to 1.5 million tons per year.
       3.3.2 The Atlanta plant now processes up to 1.75 million tons per year.
   3.4 For existing high-volume generators, Conversion Dynamics offers add-on facilities.
       3.4.1 The Roanoke plant already complies with the new RCRA requirements.
       3.4.2 The Houston plant will be in compliance within six months.

4. Conclusion
   The Resource Conservation Recovery Act will necessitate substantial capital expenditures over the next decade.

■ **Figure 22.1**
Outline Used to Organize an Oral Presentation

Principle Behind New Technique
— reduce permeability of waste
— use pozzolan stabilization
technology

■ **Figure 22.2**
**Note Card for an Oral
Presentation**

*If you are concerned that you
will forget important facts or
concepts during the presentation, include more of them
on the note cards.*

Characteristics of an Effective Graphic

Effective graphics have five characteristics:

- *Visibility.* The most common problem with presentation graphics is that they are too small. Many speakers mistakenly try to transfer information from an 8.5 × 11-inch page to a slide or transparency. As a general rule, text has to be in 24-point type or larger to be visible on the screen.

   To save space, compress sentences into brief phrases:

For more about creating graphics, see Ch. 14.

TEXT IN A DOCUMENT	The current system has three problems: • It is expensive to maintain. • It requires nonstandard components. • It is not compliant with the new MILSPEC.
SAME TEXT ON A SCREEN	Three Problems: • Expensive Maintenance • Nonstandard Components • Noncompliant with MILSPEC

- *Legibility.* Use clear, legible lines for drawings and diagrams: black on white works best for transparencies. Use legible typefaces for text; a bold-faced sans-serif typeface such as Arial or Helvetica is effective because it reproduces clearly on a screen. Avoid shadowed and outlined letters.

See Chapter 13, p. 345, for more about typefaces. See Chapter 14, p. 372, for more about using color in graphics.

- *Simplicity.* Both text and drawings must be simple. Each graphic should present only one idea. Remember that your listeners have not seen the graphic before and will not be able to linger over it.

- *Clarity.* Of course, the information has to make sense to your audience. In cutting words and simplifying concepts and visual representations, make sure the point of the graphic remains clear.

- *Correctness.* Rare is the presentation that does not contain at least one graphic with a typo or some other error. Everyone makes mistakes, but mistakes are particularly embarrassing when they are 10 inches tall on a screen.

Two points from Chapter 14 also bear repeating here: be careful when you use graphics templates in your software. Some of the templates violate basic principles of design. And use clip art only if it helps you communicate. Don't use clip art just to fill blank space on a transparency or slide.

Graphics and the Speaking Situation

To plan your graphics, analyze four aspects of the speaking situation:

- *Length of the presentation.* How many graphics should you have? A guideline is to have a different graphic for every 30 seconds of the presentation. It is far better to have a series of simple graphics than to have one complicated one that stays on the screen for 10 minutes.
- *Audience aptitude and experience.* What kinds of graphics can your audience understand easily? You don't want to present scatter graphs if your listeners have no experience in interpreting them.
- *Size and layout of the room.* Graphics to be used in a small meeting room differ from those to be used in a 500-seat auditorium. Think first about the size of the images, then about the layout of the room. For instance, will a window create glare that you will have to consider as you plan the type or placement of the graphics?
- *Equipment.* Find out ahead of time what kind of equipment is available in the presentation room. Inquire about backups in case of equipment failure. If possible, bring your own equipment. Some speakers bring graphics in two media just in case; that is, they have slides but they also have transparencies of the same graphics.

When experienced speakers make presentations away from the office, they often bring a set of supplies with them just in case something goes wrong. The following list, based on Smith (1991, pp. 148–49), specifies some of these items:

- electrical plug adapter
- extension cord
- spare bulbs for overhead projector and slide projector
- chalk and eraser
- transparency pens
- blank transparency sheets
- transparent tape
- scissors

Using Graphics to Signal the Organization of the Presentation

Used effectively, graphics can help you communicate the organizational plan of your presentation. For example, you can use the transition from one graphic to the next to indicate the transition from one point to the next.

Figure 22.3 shows seven slides for a presentation that accompanied the report in Chapter 19 on hiring a Webmaster (see page 563). The writers chose the background and accent colors, typeface, and other design elements on a *master slide*, then applied all the elements to all the slides in the presentation.

To view Figure 22.3 as slides, click on Additional Sample Documents for Ch. 22 on TechComm Web (www.bedfordstmartins.com /techcomm).

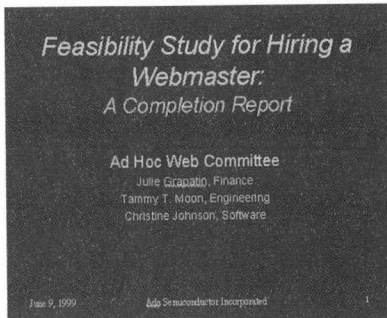

The first slide — the title slide — shows the title of the presentation and the names and affiliation of the speakers. At the bottom of each slide is a footer with the date, the name of the organization, and the number of the slide.

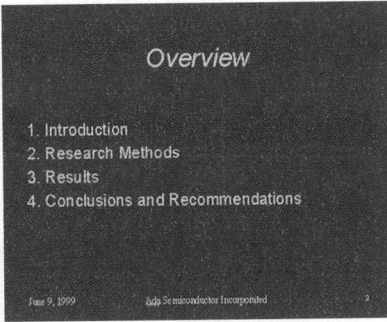

Next, present an overview, which outlines the presentation.

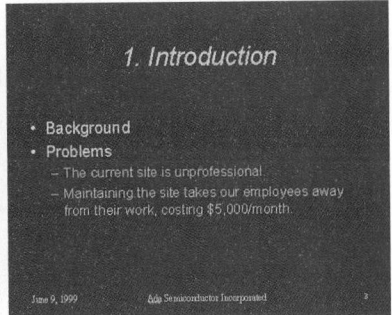

Notice that the heading on the third slide is numbered according to the list introduced in the previous slide. Such cues help readers understand the structure of your presentation.

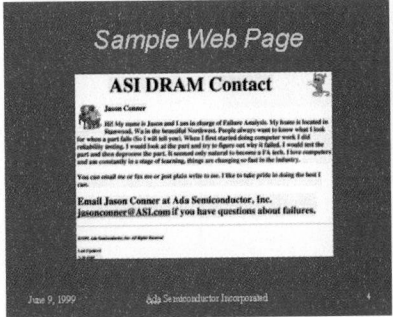

You can insert an image in a slide. This image is the nonstandard Web page the speakers are using to explain part of the problem with the current situation.

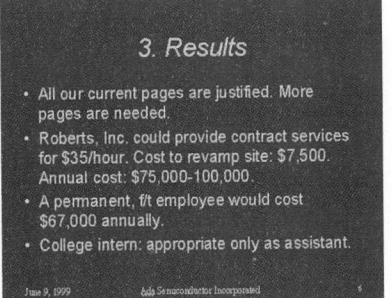

Notice that the bullet list on this slide uses parallel structure: all the phrases begin with the past tense of the verb. If you are projecting your presentation graphics from a computer, you can set the software so that each bullet item appears only after you click the mouse. This way, the audience will not read ahead; you control when the next bullet item appears on the screen.

To save space, the speakers sometimes compress their sentences into phrases.

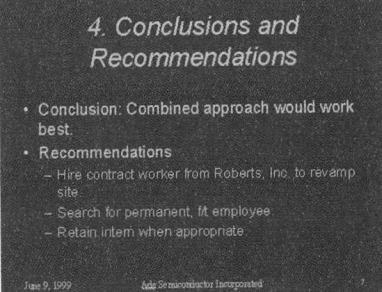

■ **Figure 22.3**
Slides for a Brief
Presentation

For your last graphic, consider a summary of your main points or a brief set of questions that restate your main points and prompt the audience to synthesize the information you have presented.

With presentation software, it is easy to create two other kinds of documents: *speaking notes* and *handouts*. Figure 22.4 shows a page of speaking notes. Figure 22.5 shows a handouts page.

To create speaking notes for each slide, type the notes in the empty box under the picture of the slide, then print the notes pages.

You can choose to print the slides on your notes page either in color or in black and white.

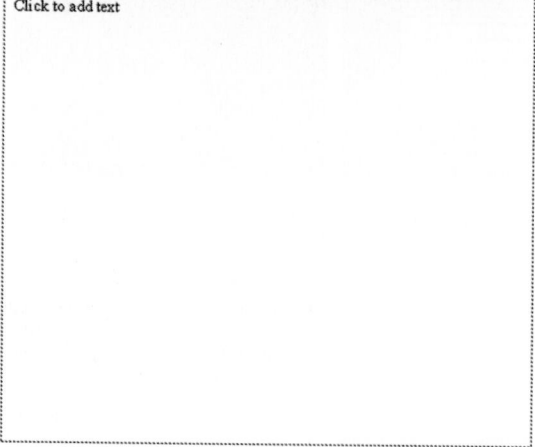

■ **Figure 22.4   Speaking Notes**

Feasibility Study for Hiring a Webmaster:
A Completion Report

Ad Hoc Web Committee
Julie Grapatin, Finance
Tammy T. Moon, Engineering
Christine Johnson, Software

June 9, 1999   Ada Semiconductor Incorporated   1

Overview

1. Introduction
2. Research Methods
3. Results
4. Conclusions and Recommendations

June 9, 1999   Ada Semiconductor Incorporated   2

You can print multiple slides on a page, in color or in black and white. The advantage of using handouts is that if you announce at the start of the presentation that you will make them available at the end, your audience will concentrate on what you are saying and not be distracted by trying to take notes.

---

1. Introduction

• Background
• Problems
  – The current site is unprofessional.
  – Maintaining the site takes our employees away from their work, costing $5,000/month.

June 9, 1999   Ada Semiconductor Incorporated   3

Sample Web Page

ASI DRAM Contact

Jason Conner

Hi! My name is Jason and I am in charge of failure analysis. My home is located in Somewhere, WA in the beautiful Northwest. People always want to know what I look for when a part fails (So I edit all you). When I first started doing computer work I did reliability testing. I would look at the part and try to figure out why it failed. I would test the part and then determine the part. If needed only interest to become a FA tech. I love computers and am constantly in a state of learning; things are changing so fast in the industry.

You can email me or fax me or just plain write to me. I like to take pride in doing the best I can.

Email Jason Conner at Ada Semiconductor, Inc.
jasonconner@ASI.com if you have questions about failures.

June 9, 1999   Ada Semiconductor Incorporated   4

---

2. Research Methods

• Distributed questionnaires to ASI employees involved with the site.
• Interviewed employees who maintain pages.
• Met with Paul Mitchell, HR supervisor.
• Met with internship coordinator at CSU.
• Analyzed the data.
• Wrote and submitted completion report.

June 9, 1999   Ada Semiconductor Incorporated   5

---

3. Results

• All our current pages are justified. More pages are needed.
• Roberts, Inc. could provide contract services for $35/hour. Cost to revamp site: $7,500. Annual cost: $75,000-100,000.
• A permanent, f/t employee would cost $67,000 annually.
• College intern: appropriate only as assistant.

June 9, 1999   Ada Semiconductor Incorporated   6

■ **Figure 22.5   Handouts Page**

Different Media Used for Graphics

Table 22.2 describes the basic media for graphics.

▧ **Table 22.2 Basic Media for Oral Presentations**

Medium	Advantages	Disadvantages
Computer presentations: images are projected from a computer to a screen.	• Very professional appearance. • You can produce any combination of static or dynamic images, from simple graphs to sophisticated, three-dimensional animations. • You can control the rate at which the images change.	• The equipment is not available every-where. • Preparing the graphics can be time consuming. • Presentations prepared using one piece of software might not run on all other systems.
Slide projector: projects previously pre-pared slides onto a screen.	• Very professional appearance. • Versatile — can handle photographs or artwork, color or black-and-white. • With a second projector, you can eliminate the pause between slides. • During the presentation, you can easily advance and reverse the slides. • Graphics software lets you create small paper copies of your slides to distribute to the audience after the presentation.	• Slides can be expensive to produce. • Room has to be kept relatively dark during the slide presentation.
Overhead projector: projects trans-parencies onto a screen.	• Transparencies are inexpensive and easy to create. • You can draw transparencies "live." • You can create overlays by placing one transparency over another. • Lights can remain on during the pres-entation. • You can face the audience. • Graphics software lets you create small paper copies of your trans-parencies to distribute to the audi-ence after the presentation.	• Not as professional-looking as slides. • Each transparency must be loaded separately by hand.
Opaque projector: projects images on paper onto a screen.	• You can project single sheets or pages in a bound volume. • Requires no expense or advance preparation.	• Room has to be kept dark during the presentation. • Cannot magnify sufficiently for a large auditorium. • Each page must be loaded separately by hand. • The projector is noisy.
Poster: a graphic drawn on oak tag or another paper product.	• Inexpensive. • Posters can be drawn or modified "live."	• Too small for large rooms.

⬇

■ Table 22.2 *(Continued)*

Medium	Advantages	Disadvantages
Flip chart: a series of posters bound together at the top like a loose-leaf binder; generally placed on an easel.	• Relatively inexpensive. • You can easily flip backward or forward. • Flip charts can be drawn or modified "live."	• Too small for large rooms.
Chalkboard or other hard writing surface.	• Almost universally available. • You have complete control — can add, delete, or modify the graphic easily.	• Complicated or extensive graphics are difficult to create. • Ineffective in large rooms. • Very informal appearance.
Objects: models or samples of material that can be held up or passed around through the audience.	• Interesting for the audience. • They provide a close look at the object.	• Audience members might not be listening while they are looking at the object. • It can take a long while to pass an object around a large room. • The object might not survive intact.
Handouts: photocopies of written material given to each audience member.	• Much material can fit on the page. • Audience members can write on their copies and keep them.	• Audience members might read the handout rather than listen to the speaker.

If you are using presentation-graphics software, keep in mind that many of the templates provided with the software are unnecessarily ornate, full of fancy shading and designs and colors. Choose a simple template, then modify it for your situation. You want to focus on delivering information, not on the complex design of the graphic.

One other note about projecting presentation graphics from a computer: set the software so that you use the mouse to control the rate of advance from one graphic to the next. Do not set it so that it advances automatically at a specified interval, such as 60 seconds. You will only be distracted by having to speed up or slow down your presentation to keep up with the graphics.

Choosing Effective Language

Delivering an oral presentation is more challenging than writing a document for two reasons:

- Listeners can't go back to listen again to something they didn't understand.
- Because you are speaking live, you must maintain your listeners' attention, even if they are hungry or tired or the room is too hot. (Readers can be distracted, too, but they have more freedom to deal with the distractions.)

Using language effectively helps you meet these two challenges.

Using Language to Signal Advance Organizers, Summaries, and Transitions

Even if you use graphics effectively, listeners cannot "see" the organization of a presentation as well as readers can. For this reason, use language to alert your listeners to advance organizers, summaries, and transitions.

- *Advance organizers.* An advance organizer is a statement that tells the listener what you are about to say. You will use an advance organizer in the introduction when you tell your audience your purpose, scope, main points, and organization. In addition, you will use advance organizers when you introduce main ideas in the body of the presentation. Advance organizers have to be explicit:

  In the next 20 minutes, I'd like to discuss the implications of the new RCRA regulations on the long-range waste-management strategy for Radnor Township. I want to make three major points. First, that.... Second, that.... And third, that.... After the presentation, I'll be happy to answer your questions.

  Notice that the speaker numbers his points. He can use this numbering system throughout the presentation to help listeners follow him.

- *Summaries.* The major summary is in the conclusion, but you might also summarize at strategic points in the body of the presentation. For instance, after a three- to four-minute discussion of a major point, you might summarize it in one sentence before going on to the next major point. Here is a sample summary from a conclusion:

  Let me conclude by summarizing my three main points about the implications of the new RCRA regulations on the long-range waste-management strategy for Radnor Township. The first point: ... The second point: ... The third point: ... I hope this presentation will give you some ideas as you think about the RCRA. If you have any questions, I'd be happy to try to answer them at this time.

- *Transitions.* As you move from one point to the next, signal the transition clearly. Summarize the previous point, then announce that you are moving to the next point:

  It is clear, then, that the federal government has issued regulations without indicating how it wants county governments to comply with them. I'd like to turn now to my second main point: ...

Using Memorable Language

If people doze off while reading a document you have written, you probably won't know it, at least until they complain or you discover that the document failed to accomplish your purpose. But if they doze off while you are speaking to them, you'll know it right away. Effective presentations require memorable language.

GUIDELINES

Using Memorable Language in Oral Presentations

Plan to draw on these three techniques to help make a lasting impression on your audience.

▶ *Involve the audience.* People are more interested in their own concerns than in yours. Talk to the audience about their problems and their solutions. In the introduction, establish a link between your topic and the audience's interests. For instance, the presentation to the Radnor Township Council about waste management might begin like this:

> Picture yourself on the Radnor Township Council two years from now. After exhaustive hearings, proposals, and feasibility studies, you still don't have a waste-management plan that meets federal regulations. What you do have is a mounting debt: the township is being fined $1,000 per day until you implement an acceptable plan.

▶ *Refer to people, not to abstractions.* People remember specifics; they forget abstractions. To make a point memorable, describe it in human terms:

> What could you do with that $365,000 every year? You could buy more than 250 personal computers; that's a computer for almost every classroom in every elementary school in Radnor Township. You could expand your school lunch program to feed every needy child in the township. You could extend your after-school programs to cover an additional 3,000 students.

▶ *Use interesting facts, figures, and quotations.* Do your research and find interesting information about your subject. For instance, you might find a brief quotation from an authoritative figure in the field or a famous person not generally associated with the field (for example, Abraham Lincoln on waste management).

A note about humor: only a few hundred people in the United States make a good living being funny. Don't plan to tell a joke. If something happens in the context of the presentation that provides an opening for a witty remark, and you are good at making witty remarks, fine. But don't *prepare* to be funny.

Rehearsing the Presentation

Even the most gifted speakers need to rehearse. It is a good idea to set aside enough time to rehearse your speech thoroughly.

Rehearsing an Extemporaneous Presentation

Rehearse your extemporaneous presentation at least three times.

- *First rehearsal.* Don't worry about posture or voice projection. Just try to compose your presentation out loud with your outline or notes before you. Your goal is to see if the speech makes sense — if you can explain all the points you have listed and can forge effective transitions from point to point. If you have any trouble, stop and try to figure out the problem. If you need more information, get it. If you need a better transition, create one. You are likely to find that you need to revise the order of your outline or notes. Pick up where you left off and continue the rehearsal, stopping again where necessary to revise. When you have finished your first rehearsal, put the outline or notes away and do something else.

- *Second rehearsal.* Once you are rested, try the presentation again. This time, it should flow more easily. Make any necessary revisions to the outline or notes. When you have complete control over the organization and flow, check to see if you are within the time limits.

- *Third rehearsal.* After a satisfactory second rehearsal, try the presentation again, under more realistic circumstances — if possible, in front of people. The listeners might offer constructive advice about sections they don't understand or about your speaking style. If people aren't available, tape-record or videotape the presentation and then evaluate your own delivery. If you can visit the site of the presentation to get the feel of the room and rehearse there, you will find giving the actual speech a little easier.

Rehearse again until you are satisfied with your presentation. Then stop. Don't attempt to memorize it. If you do, you will probably panic the first time you forget a phrase. During the presentation, you should be thinking of your subject, not trying to remember the words you used during the rehearsals.

Rehearsing a Scripted Presentation

Rehearsing a scripted presentation is a combination of revising the text and rehearsing it. As you revise, read the script out loud to hear how it sounds. Once you think the presentation says what you want it to say, try reading it into a tape recorder. Revise it until you are satisfied, and then rehearse in front of people. Again, do not memorize. There is no need to: you will have your script in front of you on the podium.

GIVING THE ORAL PRESENTATION

After all your preparation, the time to give the presentation finally arrives. In giving the presentation, you will concentrate on what you have to say. However, you will have three additional concerns.

In giving the oral presentation:
Calm your nerves.
Use your voice effectively.
Use your body effectively.

Calming Your Nerves

Most professional actors freely admit to being nervous before a performance, so it is no wonder that most technical speakers are also nervous. You might well fear that you will forget everything or that people at the back of the room will not be able to hear you. These fears are common. But keep in mind three facts about nervousness:

- *You are much more aware of your nervousness than the audience is.* They are farther away from your trembling hands.
- *Nervousness gives you energy and enthusiasm.* Without energy and enthusiasm, your presentation will be flat. If you seem bored and listless, your audience will be bored and listless.
- *After a few minutes, your nervousness will pass.* You will be able to relax and concentrate on the subject.

This advice, however, is unlikely to make you feel much better if you are distracted by nerves as you wait to give your presentation. Experienced speakers suggest a few points to keep in mind when you find yourself getting nervous before a presentation:

- *Realize that you are prepared.* If you have done your homework, prepared speaking notes, and rehearsed the presentation, you'll be fine. You are in control of the presentation.
- *Realize that the audience is there to hear you, not to judge you.* Your listeners want to hear what you have to say. They are much less interested in your nervousness than you are.
- *Realize that your audience is made up of individual people who happen to be sitting in the same room.* If you tell yourself that the audience members are individuals like yourself who also get nervous before making presentations, you'll feel better.

GUIDELINES

Releasing Nervous Energy

Experienced speakers suggest the following strategies for dealing with nervousness before a presentation.

▶ *Walk around.* A brisk walk of a minute or two can calm you by dissipating some of your nervous energy.

▶ *Go off by yourself for a few minutes.* Some people find that getting away for a moment helps them compose their thoughts and realize that they can handle the nervousness.

> ▸ *Talk with someone for a few minutes.* For some speakers, distraction works best. Talk with someone who has come to the presentation a few minutes early.
>
> ▸ *Take several deep breaths, exhaling slowly.* Doing so will help you control your nerves.

When it is time to begin, don't jump up to the lectern and start speaking quickly. Walk up slowly and arrange your text, outline, or note cards before you. If water is available, take a sip. Look out at the audience for a few seconds before you begin. It is polite to begin formal presentations with "Good morning" (or "Good afternoon," or "Good evening") and to refer to the officers and dignitaries present. If you have not been introduced, introduce yourself. In less formal contexts, just begin your presentation.

So that the audience will listen to you and have confidence in what you say, try to project the same attitude that you would in a job interview: restrained self-confidence. Show interest in your topic and knowledge about your subject. You can convey this sense of control through your voice and your body.

Using Your Voice Effectively

Inexperienced speakers often encounter problems with five aspects of vocalizing.

- *Volume.* Because acoustics vary greatly from room to room, you won't know how well your voice will carry in a room until you have heard someone speaking there. In some well-constructed auditoriums, speakers can use a conversational volume. Other rooms require greater voice projection, and some have an annoying echo. These circumstances aside, more people speak too softly than too loudly. After your first few sentences, ask if the people in the back of the room can hear you. When people speak into microphones, they tend to bend down toward the microphone and end up speaking too loudly. Glance at your audience to see if you are having volume problems. The body language of audience members will be clear.

- *Speed.* Nervousness makes people speak quickly. Even if you think you are speaking at the right rate, you might be going a little too fast for some members of your audience. Remember: you know what you are going to say, but your listeners are trying to understand new information. For particularly difficult points, slow down for emphasis. After finishing one major point, pause before introducing the next point.

- *Pitch.* In an effort to control their voices, many speakers end up flattening their pitch. The resulting monotone is boring and, for some listeners, actually distracting. Try to let the pitch of your voice go up or down as it would in a normal conversation. In fact, experienced speakers often exaggerate pitch variations slightly.

- *Articulation.* The nervousness that goes along with an oral presentation tends to accentuate sloppy pronunciation. If you want to say *environment*, don't say *envirament*. Don't drop final *g*s. Say *trying*, not *tryin'*. A related pronunciation problem involves technical words and phrases, especially the important ones. When a speaker uses a phrase over and over, it tends to get clipped and becomes difficult to understand. Unless you articulate carefully, *Scanlon Plan* will end up as *Scanluhplah*.

- *Nonfluencies.* Avoid meaningless fillers like *you know*, *like*, *okay*, *right*, *uh*, and *um*. They do not disguise the fact that you aren't saying anything; they call attention to it. A thoughtful pause is better than an annoying verbal tic.

Using Your Body Effectively

In addition to listening to what you say, the audience will be looking at you. Effective speakers make use of their body language to help listeners follow the presentation.

GUIDELINES

Facing an Audience

As you give a presentation, keep in mind four guidelines about physical movement.

▶ *Maintain eye contact.* It is only polite to look at your audience. This is called *eye contact.* For small groups, look at each listener randomly; for larger groups, look at each segment of the audience frequently during your speech. Do not stare at your notes, at the floor, at the ceiling, or out the window. Eye contact helps you see how the audience is receiving the presentation. You will see, for instance, if listeners in the back are having trouble hearing you.

▶ *Use natural gestures.* When people talk, they tend to gesture with their hands. Most of the time, these gestures make the presentation look natural and improve listeners' comprehension. You can supplement your natural gestures by using your arms and hands to signal pauses and to emphasize important points. When referring to graphics, walk to the screen and point to direct the audience's attention. Avoid mannerisms — physical gestures that serve no useful purpose. Don't play with your glasses or your jewelry or the coins in your pocket. These nervous gestures can quickly distract an audience from what you are saying. Don't pace back and forth. Like verbal mannerisms, physical mannerisms are often unconscious. Constructive criticism from friends can help you pinpoint them.

▶ *Don't block the audience's view of the screen.* Don't stand at the overhead projector if doing so blocks some people's view of the screen. After you put on a transparency, step back to the side of the screen. Use a pointer to indicate key words or images on the screen.

▶ *Control the audience's attention.* People will listen to and look at anything that is interesting. Don't lose the audience's attention. If you hand out photocopies at the start of the presentation, some people will start to read them and stop listening to you. If you leave an image on the screen after you finish talking about it, some people will keep looking at it instead of listening to you. When you want the audience to look at you and listen to you, remove the graphics or turn off the projector.

ANSWERING QUESTIONS AFTER THE PRESENTATION

In some presentations, particularly informal ones, audience members ask questions throughout the presentation. This method of questioning helps you clarify points as you go along, but it can make it difficult for you to stay on track.

More often, an oral presentation is followed by a question-and-answer period. When you invite questions, don't abruptly say, "Any questions?" This phrasing suggests that you don't really want to take any questions. Instead, say something like this: "If you have any questions, I'll be happy to try to answer them now." If invited politely, people will be much more likely to pose questions; in that way, you will be more likely to communicate your information effectively.

When you respond to questions, you might encounter any of these five situations:

• *You're not sure everyone heard the question.* If there is no moderator to ask if the question was audible, ask if people have heard the question. If they haven't, repeat or paraphrase it, perhaps as an introduction to your response: "Your question is about the efficiency of these three techniques. . . ." Some speakers always repeat the question; that way, they are sure everyone hears it, and they get an extra moment to think about their answer.

- *You don't understand the question.* Ask for a clarification. After responding, ask if you have answered the question adequately.

- *You don't know the answer to the question.* Tell the truth. Only novices believe that they ought to know all the answers. If you have some ideas about how to find out the answer — by checking a certain reference source, for example — share them. If the question is obviously important to the person who asked it, you might offer to meet with him or her after the question-and-answer period to discuss ways for you to give a more complete response, perhaps by email.

- *You get a question that you have already answered in the presentation.* Restate the answer politely. Begin your answer with a phrase such as the following: "I'm sorry I didn't make that point clear in my talk. I wanted to explain how. . . ." Never insult the person by pointing out that you already covered that: "I already answered that question in my talk, but let me repeat it for you. . . ."

- *A belligerent member of the audience rejects your response and insists on restating his or her original point.* Politely offer to discuss the matter further after the session. This strategy will prevent the person from boring or annoying the rest of the audience.

If it is appropriate to stay after the session to talk individually with members of the audience, offer to do so. Remember to thank them for their courtesy in listening to you.

SAMPLE EVALUATION FORM

Figure 22.6 on page 656 is a list of questions that can help you focus your thoughts as you watch and listen to a presentation.

To download this form in an electronic format, see Forms for Technical Communication on TechComm Web (www .bedfordstmartins.com /techcomm).

Evaluation of a Presentation

Speaker: _____
Topic: _____
Date: _____

To the left of each of the following statements, write a number from 1 to 5, with 5 signifying strong agreement and 1 signifying strong disagreement.

( ) 1. In the introduction, the speaker made an attempt to relate the topic to my concerns.
( ) 2. In the introduction, the speaker explained the main points he or she wanted to make in the presentation.
( ) 3. In the introduction, the speaker explained the organization of the presentation.
( ) 4. The speaker used interesting, clear language to get the points across.
( ) 5. The speaker made the transitions from one point to the next clearly.
( ) 6. The speaker used clear and distinct enunciation.
( ) 7. The speaker exhibited no distracting vocal mannerisms.
( ) 8. The speaker exhibited no distracting physical mannerisms.
( ) 9. The speaker used graphics effectively to reinforce and explain the main points.
( ) 10. The speaker summarized the main points effectively in the conclusion.
( ) 11. The graphics helped me understand the organization of the presentation.
( ) 12. Throughout the presentation, the speaker paid attention to the audience.
( ) 13. The speaker seemed to be enthusiastic throughout the presentation.
( ) 14. The speaker used the allotted time effectively.
( ) 15. The speaker invited questions politely.
( ) 16. The speaker answered questions effectively.

Answer the following two questions on the reverse side of this page.
17. What did you particularly like about this presentation?
18. What would you have done differently if you had been the speaker?

■ **Figure 22.6   Sample Evaluation Form**

Speaker's Preparation Checklist

1. Did you assess the speaking situation — the audience and purpose of the presentation?
2. Did you determine how much information you can communicate in your allotted time?
3. Did you outline your information?
4. Did you prepare graphics that are
   - ❏ visible?
   - ❏ legible?
   - ❏ simple?
   - ❏ clear?
   - ❏ correct?
5. In planning your graphics, did you consider your audience's aptitude and experience, the size and layout of the room, and the equipment?
6. Did you plan your graphics to help the audience understand the organization of your presentation?
7. Did you choose appropriate media for your graphics?
8. Did you make sure that the presentation room will have the necessary equipment for the graphics?
9. Do you have appropriate supplies with you in case you need them?
10. Did you choose language to signal advance organizers, summaries, and transitions?
11. Did you choose language that is vivid and memorable?
12. Did you rehearse your presentation several times with a tape recorder or a live audience?

Exercises

1. Learn some of the basic functions of a presentation-graphics software program. For instance, modify a template, create your own original design, add footer information to a master slide, insert a graphic onto a slide, and set the animation feature to make each bullet item appear only after a mouse click.

2. Using presentation-graphics software, create a design to be used for the master slide of a computer presentation. Then, for the same information, create a design to be used in a transparency made on a black-and-white photocopier.

Research Projects

3. Prepare a five-minute presentation, including graphics, on one of the topics listed here. For each presentation, your audience consists of the other students in your class, and your purpose is to introduce them to an aspect of your academic field.

a. Define a key term or concept in your field.

b. Describe how a particular piece of equipment is used in your field.

c. Describe how to carry out a procedure common in your field.

The instructor and the other students will evaluate the presentation by filling out the form in Figure 22.6.

4. Prepare a five-minute presentation based either on your proposal for a research-report topic or on your completion report. Your audience consists of the other students in your class, and your purpose is to introduce them to your topic. The instructor and the other students will evaluate the presentation by filling out the form in Figure 22.6. If your instructor wishes, this assignment can be done collaboratively.

5. Write a 500-word memo to your instructor in which you describe and evaluate a recent oral presentation of a guest speaker at your college or a politician on television. See Chapter 15, page 430, for a discussion of writing memos.

6. Many speeches of different kinds can be found on the Internet. Some of them are audio files, some are just text. These speeches are from the public sector (see www.access.gpo.gov to link to federal government sites) as well as the private sector (find the site of a large organization, such as Exxon — www.exxon.com; or AT&T — www.att.com; or Boeing — www.boeing.com). Find a speech on a topic that interests you. Download the text, then do the following tasks:

a. Create a set of presentation graphics in the form of bulleted text that might accompany the speech.

b. Describe and evaluate the speech. Consider such questions as the following: What is the audience of the presentation? In what ways has the speaker attempted to meet that audience's needs and goals? How is the speech organized? How has the speaker used language? On the basis of the text, how effective do you think the presentation was?

c. Present your findings in a memo to your instructor. Attach the transcript of the presentation and a printout of your graphics.

For additional cases, click on Case of the Month and Archive on TechComm Web (www.bedfordstmartins.com /techcomm).

C A S E
Publicizing the School Newspaper

As the editor of your school's student newspaper, you want to publicize the newspaper on campus. You have contacted a number of teachers, who have agreed to give you 10 minutes of class time to make a presentation to their students about the newspaper. Prepare and deliver a presentation — complete with graphics — on one of the following topics:

- the role of the newspaper in campus life
- five features most students don't know your newspaper provides
- proposed changes to the contents or design of the newspaper
- writing for the newspaper as a means of professional development
- how the newspaper is produced

Appendix:
Reference Handbook

Documentation identifies the sources of the ideas and the quotations in your document. Integrated throughout your document, documentation consists of citations in the text and a reference list (or list of works cited) at the back of your document. Documentation serves three basic functions:

- *To help you acknowledge your debt to your sources.* Complete and accurate documentation is a professional obligation, a matter of ethics. Failure to document a source, whether intentional or unintentional, is plagiarism. At most colleges and universities, plagiarism means automatic failure of the course and, in some instances, suspension or expulsion. In many companies, it is grounds for immediate dismissal.

- *To help you establish credibility.* Effective documentation helps you place your document within the general context of continuing research and to define it as a responsible contribution to knowledge in the field. Knowing how to use existing research is one mark of a professional.

- *To help your readers find your source in case they want to read more about a particular subject.*

For more about quoting and paraphrasing sources, see Ch. 7, pp. 160 and 162.

Three kinds of material should always be documented:

- *Any quotation from a written source or an interview, even if it is only a few words.*

- *A paraphrased idea, concept, or opinion gathered from your reading.* There is one exception. An idea or concept so well known that it has become general knowledge, such as Einstein's theory of relativity, needs no citation. If you are unsure about whether an item is general knowledge, document it, just to be safe.

- *Any graphic from a written or electronic source.* Cite the source for a graphic next to the graphic or in the reference list. For an online source, be sure to include a retrieval statement in the bibliographic entry. If you are publishing your work, you must also request permission to use any graphic protected by copyright.

For more about using graphics from other sources, see Ch. 14, p. 370.

Just as organizations have their own preferences for formatting and punctuation, many organizations also have their own documentation style. The documentation systems included in this section of the appendix are based on the following style manuals:

- *Publication Manual of the American Psychological Association*, 5th ed. (Washington, DC: APA, 2001). This system, often referred to as APA, is used widely in the social sciences.

- *Scientific Style and Format: The CBE Manual for Authors, Editors, and Publishers* (New York: Cambridge University Press, 1994). This system, known as CBE, is from the Council of Science Editors (formerly the Council of Biology Editors) and is widely used in the natural sciences.

- *MLA Handbook for Writers of Research Papers*, 5th ed. (New York: MLA, 1999). This system, from the Modern Language Association, is used widely in the humanities.

Other organizations use other published style guides, such as the *U.S. Government Printing Office Style Manual,* the *American Chemical Society's Handbook for Authors*, or the *Chicago Manual of Style*. Find out what your organization's style is and abide by it. And check with your instructor to see which documentation system to use in the documents you write for class.

Click on the link to Diana Hacker's Research and Documentation Online on TechComm Web (www.bedfordstmartins.com/techcomm) for links to advice about documentation styles.

APA STYLE

APA style consists of two elements: the citation in the text and the references at the end of the document.

APA Textual Citations

In APA style, a textual citation typically includes the name of the source's author and the date of its publication. Textual citations will vary depending on the type of information cited, the number of authors, and the context of the citation. The following models illustrate a variety of common textual citations; for additional examples, consult the *Publication Manual of the American Psychological Association*.

APA Style for Textual Citations

1. Summarized or Paraphrased Material
2. Quoted Material or Specific Fact
3. Source with Multiple Authors
4. Source Issued by an Organization
5. Source with an Unknown Author
6. Multiple Sources in One Citation
7. Multiple Authors with the Same Last Name
8. Personal Communication
9. Electronic Document

1. Summarized or Paraphrased Material

For material or ideas that you have summarized or paraphrased, include the author's name and publication date in parentheses immediately following the borrowed information.

This phenomenon was identified more than fifty years ago (Wilkinson, 1948).

If your sentence already includes the source's name, do not repeat it in the parenthetical notation.

Wilkinson (1948) identified this phenomenon more than fifty years ago.

2. Quoted Material or Specific Fact

If the reference is to a specific fact, idea, or quotation, add the page number(s) of the source to your citation.

This phenomenon was identified more than fifty years ago (Wilkinson, 1948, p. 36).

Wilkinson (1948) identified this phenomenon more than fifty years ago (p. 36).

3. Source with Multiple Authors

For a source written by two authors, cite both names. Use an ampersand (&) in the citation itself, but use the word *and* in regular text.

(Allman & Jones, 1987)

As Allman and Jones (1987) suggested, . . .

For a source written by three, four, or five authors, include all the names the first time you cite the reference; after that, include only the last name of the first author followed by "et al."

FOR THE FIRST REFERENCE:

Bradley, Edmunds, and Soto (1995) argued . . .

FOR SUBSEQUENT REFERENCES:

Bradley et al. (1995) found . . .

For a source written by six or more authors, use only the first author's name followed by "et al."

(Smith et al., 1987)

4. Source Issued by an Organization

If the author is an organization rather than a person, use the name of the organization.

The causes of narcolepsy are discussed in a recent booklet (Association of Sleep Disorders, 1999).

In a recent booklet, the Association of Sleep Disorders (1999) discusses the causes of narcolepsy.

5. Source with an Unknown Author

If the source does not identify an author, use a shortened name of the title in your parenthetical citation.

This trend has been evident in American society since the beginning of the twentieth century ("Modernism," 1999).

6. Multiple Sources in One Citation

When you refer to two sources or more in one citation, present the sources in alphabetical order, separated by a semicolon.

> This phenomenon was identified more than fifty years ago (Betts, 1949; Wilkinson, 1948).

7. Multiple Authors with the Same Last Name

Use first initials if two or more sources have authors with the same last name.

> This phenomenon was identified more than fifty years ago (B. Wilkinson, 1948).

8. Personal Communication

Include the words "personal communication" and the date of the communication when citing personal interviews, phone calls, and email.

> C. Connors (personal communication, October 6, 1999) argued that . . .

9. Electronic Document

Cite the author and date of the source as you would for other kinds of documents. If the date is unknown, use "n.d." for "no date."

> Resnick (2000) discusses usability testing and the Palm Beach County election ballots.

If the document is posted as a PDF file, include a page number in the citation. If page numbers are not available, but the source contains paragraph numbers, give the paragraph number: (Tong, 2001, ¶ 4) or (Tong, 2001, para. 4). If no paragraph or page numbers are available and the source has headings, cite the appropriate heading and paragraph.

> Vidoli (2000) warns against using "jargon because it may be misinterpreted" (Writing Naturally section, para. 2).

The APA Reference List

A reference list provides the information your readers will need in order to find each source you have cited in the text. Note that the reference list includes only those sources that you actually used in researching and preparing your document; it should not include background reading. Following are some guidelines for an APA-style reference list.

 Click on the link to Diana Hacker's Research and Documentation Online on TechComm Web (www.bedfordstmartins.com /techcomm) for help with formatting an APA reference list.

- *Arranging Entries.* The individual entries in the reference list are arranged alphabetically by author's last name. Two or more works by the same author are arranged by date, earliest to latest; two or more works by the same author in the same year should also include a lowercase letter: Smith 1999a, Smith 1999b, etc. Works by an organization are alphabetized by the first significant word in the name of the organization.

- *Book Titles.* Titles of books should be italicized. The first word of the book's title and subtitle are capitalized, but all other words (except for proper nouns) should be in lowercase.

- *Publication Information.* Give the publisher's full name or consult your style guide for the preferred abbreviations. Include both the publisher's city and state or country, unless the city is well known (such as New York, Boston, or London).

- *Periodical Titles.* Titles of periodicals should be italicized, and all major words should be capitalized.

- *Article Titles.* Titles of articles should not be italicized or placed in quotations. The first word of the article's title and subtitle are capitalized, but all other words (except for proper nouns) should be in lowercase.

- *Electronic Sources.* Include as much information as you can about electronic sources, such as date of publication, identifying numbers, and retrieval information. Also, be sure to record the date you retrieved the information, because electronic information changes frequently.

- *Indenting.* APA style recommends using a hanging indent, with the second and subsequent lines of each entry indented 5 to 7 spaces:

See Electronic Reference Formats Recommended by the American Psychological Association (www.apa.org /journals/webref.html) for more information.

Chapman, D. L. (1995, June 12). Detroit makes a big comeback. *Motorist's Metronome, 12*, 17–26.`

Paragraph indents, in which the first line of each entry is indented 5 to 7 spaces, may be preferred by your instructor:

Chapman, D. L. (1995, June 12). Detroit makes a big comeback. *Motorist's Metronome, 12*, 17–26.

- *Spacing.* Double space the entire reference list; do not add extra spacing between entries.

- *Page Numbers.* When citing a range of page numbers for articles, always give the complete numbers (for example, 121–124, not 121–24). If an article continues on subsequent pages interrupted by other articles or advertisements, use a comma to separate the page numbers. Use the abbreviation *p.* or *pp.* only with articles in newspapers, chapters in edited books, and proceedings.

- *Dates.* For a reference list, follow this format: year, month, day (2000, October 31).

The following are models of reference list entries for a variety of sources. For further examples of APA-style citations, consult the *Publication Manual of the American Psychological Association*.

APA Style for Reference List Entries

1. Book by One Author
2. Book by Multiple Authors
3. Book Issued by an Organization
4. Book Compiled by an Editor or Issued under an Editor's Name
5. Book in Edition Other Than First
6. Magazine or Journal Article
7. Newspaper Article
8. Article Included in a Book
9. Article from a Volume of Proceedings
10. Unsigned Article
11. Technical Report
12. Government Document
13. Nonperiodical Web Document
14. Article from an Online Periodical
15. Article from a Database
16. Email Message
17. Online Posting
18. Computer Software

1. Book by One Author

Begin with the author's last name, followed by the first initial or initials. If the author has two first initials, put a space between the initials. Place the date of publication in parentheses, followed by a period. Place the location and name of the publisher last, and end the citation with a period.

> Cunningham, W. S. (1980). *Crisis at Three Mile Island: The aftermath of a near melt-down.* New York: Madison.

2. Book by Multiple Authors

To cite two or more authors, use the ampersand (&) instead of *and* between their names. Use a comma to separate the authors' names.

> Bingham, C., & Withers, S. (1999). *Neural networks and fuzzy logic.* New York: IEEE.

To cite more than six authors, list only the first six followed by "et al."

3. Book Issued by an Organization

Use the full name of the organization in place of an author's name.

> American Psychological Association. (1994). *Publication manual of the American Psychological Association* (4th ed.). Washington: American Psychological Association.

4. Book Compiled by an Editor or Issued under an Editor's Name

Place the abbreviation "Ed." in parentheses after the name. For sources with more than one editor, use the abbreviation "Eds."

> Morgan, D. E. (Ed.). (2000). *Readings in alternative energies.* Boston: Smith-Howell.

5. Book in Edition Other Than First

Include the edition number in parentheses following the title.

> Schonberg, N. (1997). *Solid state physics* (3rd ed.). London: Paragon.

6. Magazine or Journal Article

Follow the author's name and date with the article title, then give the periodical title. If the journal issue is identified by a word such as "Fall," include that word in the date. Volume numbers should be italicized, but issue and page numbers should not. Include the volume number, set off by commas, after the journal name.

> Hastings, W. (1990, Fall). The space shuttle debate. *The Modern Inquirer, 13,* 311–318.

In citing an article from a journal paginated by issue, put the issue number (not italicized) in parentheses immediately after the volume number.

> Tran, X. (1999). Report: Breeding pandas in captivity. *Nature, 14*(4), 10–17.

7. Newspaper Article

The exact date of publication follows the year.

> Eberstadt, A. (1995, July 31). Carpal tunnel syndrome. *Morristown Mirror and Telegraph,* p. B3.

8. Article Included in a Book

Present the book editor's initials and name—unlike those of the article authors—in normal order. The pages on which the article appears are given after the book title.

May, B., & Deacon, J. (1995). Amplification systems. In A. Kooper (Ed.), *Advances in electronics* (pp. 101–114). Miami, FL: Valley Press.

9. Article from a Volume of Proceedings

After the proceedings title, give the page numbers on which the article appears.

Carlson, C. T. (1995). Advanced organizers in manuals. In K. Rainey (Ed.), *Proceedings of the 45th International Technical Communication Conference* (pp. RT56–58). Fairfax, VA: Society for Technical Communication.

10. Unsigned Article

If the author of a periodical article is not indicated, alphabetize it by title, ignoring the initial articles *the, a,* and *an.*

The state of the art in microcomputers. (2000, Fall). *Newscene, 56,* 406–421.

11. Technical Report

Include identifying numbers after the name of the technical report. After the location and the name of the publisher, add the name of the service you used to locate the item.

Birnest, A. J., & Hill, G. (1996). *Early identification of children with ATD.* (Report No. 43-8759). State College, PA: Pennsylvania State University College of Education. (ERIC Document Reproduction Service No. ED186389).

12. Government Document

For most government agencies, use the abbreviation *U.S.* instead of spelling out *United States.* Include identifying document numbers in parentheses after the publication title.

U.S. Department of Energy. (1998). *The energy situation in the next decade.* (Technical Publication 11346-53). Washington, DC: U.S. Government Printing Office.

13. Nonperiodical Web Document

To cite a nonperiodical Web document, provide as much of the following information as possible: author's name, date of publication (use "n.d." if there is no date), title of document (in italics), retrieval date, and the URL for the document.

 See the APA style site (www.apastyle.org) for more information.

McMurrey, D. A. (n.d.). *Online technical writing: Audience analysis.* Retrieved August 28, 2001, from http://www.io.com/~hcexres/tcm1603/acchtml/aud.html

If the author of a document is not identified, begin the reference with the title of the document.

> *Judgment against spammer.* (1997, November 11). Retrieved November 6, 2001, from
> http://www. matrix. net/company/news/19971111_spam.html

If the document is from a university program's Web site, identify the host institution and program or department followed by a colon and the URL for the document.

> Hooser, Sue. (1997, October 2). *Tips for speakers.* Retrieved September 12, 2001,
> from the New Mexico State University, Techprof Web site: http://www.nmsu.edu
> /techprof/backgrnd/sueback.html

For a chapter or section in a Web document, give the author, year of publication, and chapter or section title, followed by "In" and the document title. Identify a chapter or section, instead of page numbers, in parentheses, followed by a retrieval date and URL for the chapter or section.

> Ware, W. H. (1998). Key elements of a solution approach. In *The cyber-posture of the
> national information infrastructure* (chap. 4). Retrieved August 27, 2001, from
> http://www.rand.org/publications/MR/MR976/mr976.html#chap4

14. Article from an Online Periodical

To cite online articles, provide the same information you would for print articles (see item 6 on page 666). If an identical version of the article also appears in print, do not include a URL; instead, follow the title of the article with the words "Electronic version" in brackets.

> Quesenbery, W. (2001, May). On beyond help: Meeting user needs for useful online
> information [Electronic version]. *Technical Communication, 48*(2), 182–188.

If there is no print version, or if you are citing an online article that is different from the print version (for example, the format is different or the online version contains additional material, such as animations), include the date you retrieved the article and its URL.

> Miller, D. (2001, February). Moving from documentation to usability: The Dangerfield
> effect. *The Willamette Galley, 4*(1). Retrieved August 28, 2001, from http://www
> .stcwvc.org/galley/jan01_newsletter/Moving.htm

If you are citing an article you retrieved from a searchable Web site, such as for a newspaper, give the URL for the site as opposed to the specific source.

> Bredemeier, K. (2001, June 18). Ethical dilemmas call for careful approaches. *The
> Washington Post.* Retrieved November 7, 2001, from http://www.washingtonpost
> .com

15. Article from a Database

To cite an article from an electronic database, provide the publication information followed by the access date, the name of the database, and the item number, if any.

> Johanek, J. (2000, January). Readability: Rule one in magazine design. *Folio: The Magazine for Magazine Management, 29(*1), 154. Retrieved August 28, 2001, from Business Index ASAP database (A58836541).

16. Email Message

Email messages are not cited in the reference list. Instead they should be cited in the text as personal communications.

17. Online Posting

If an online posting is not archived, and therefore not retrievable, cite it as a personal communication and do not include it in the reference list. If the posting can be retrieved from an archive, provide the author's name or screen name, exact date of the posting, the title or subject line, and any identifier in brackets, then finish with "Message posted to" followed by the address.

> Graham, T. (2001, August 27). Simple font size question [Msg. 7815]. Message posted to http://groups.yahoo.com/group/wwp-users/message/7815

18. Computer Software

Include any identifying information, such as the version number, in parentheses after the proper name of the software.

> Block, K. (1999). Planner (Release 3.1) [Computer software]. New York: Global Software.

If the software has no author, begin the entry with the name of the program. Use standard capitalization; do not italicize or underline the program title.

> Tools for Drafting (Version 2.3) [Computer software]. (1999). San Jose, CA: Software International.

Sample APA Reference List

The following is a sample reference list using the APA citation system:

References

Andress, K. (1999, July 12). More on the Scanlon Plan. *Online Journal of Accounting Studies*. Retrieved May 13, 2000, from http://www.ojac.org/journals/9907/andress.htm

Daly, P. H. (1993). Selecting and designing a group insurance plan. *Personnel Journal, 54*, 322–323.

Flanders, A. (1999). Measured daywork and collective bargaining. *British Journal of Industrial Relations, 9*, 368–392.

Goodman, R. K., Wakely, J., & Ruh, K. (1995). What employees think of the Scanlon Plan. *Personnel, 6*, 22–29.

Trencher, P., & Coughlin, C. (1988). *Recent trends in labor-management relations*. New York: Westly.

Zwicker, D. (1995, August). More on the Scanlon Plan: A response. *Scanlon News*. Retrieved May 12, 2000, from http://www.aiiu.com/pubs/scnews/95Aug/zwicker.htm

CBE STYLE

CBE style consists of two elements: the citation in the text and the list of references at the end of the document.

CBE Textual Citations

For CBE-style textual citations, you may use either the *citation-sequence system* or the *name-year system*. Find out which method your instructor or organization prefers.

- *The Citation-Sequence System.* In this method, superscripted or parenthetical numbers are inserted into the text to indicate borrowed material.

  . . . travels at the speed of light[1], but still others contend that gravity is responsible for the phenomenon[2-4].

  or

. . . travels at the speed of light (1), but still others contend that gravity is responsible for the phenomenon (2–4).

Later textual references to the same source repeat the numbers already used.

. . . as experiments have shown[23]. If the velocity theory[1,6] is to be taken . . .

The list of references at the end of the document includes each of the cited sources in numerical order.

- *The Name-Year System.* In this method, the author's last name(s) and year of publication are mentioned either in the text or in a parenthetical citation immediately following the borrowed material.

   . . . travels at the speed of light (Posdevna 1999), but still others contend that gravity is responsible for the phenomenon (Walters 1984; Chang 1999; Rivera 2000).

   . . . as experiments have shown (Rao and Leschley 1998). If Posdevna's velocity theory (1999) is to be taken . . .

   . . . the most recent study on this topic was inconclusive (Matthews and others 2000).

The list of references at the end of the document lists each of the cited sources alphabetically by author's last name.

The CBE Reference List

Whether you use the citation-sequence system or the number-year system in the body of your paper, you will also need to prepare a list of references at the end. The following guidelines will help you prepare a CBE-style reference list.

 Click on the link to Diana Hacker's Research and Documentation Online on TechComm Web (www.bedfordstmartins.com /techcomm) for help with formatting a CBE reference list.

- *Arranging Entries: Citation-Sequence System.* The entries are arranged in numerical order, each entry having a number that corresponds to a number in the text of the paper. Sources are not repeated in the list of references, even if one is referred to many times in the text.
- *Arranging Entries: The Name-Year System.* The individual entries in the reference list are arranged alphabetically by the author's last name. Two or more works by the same author are arranged by date, earliest to latest; two or more works by the same author in the same year should add lowercase letters: Smith 1999a, Smith 1999b, etc. Works by an organization are alphabetized by the first significant word in the name of the organization.
- *Book Titles.* Book titles should not be underlined, italicized, or placed in quotation marks. The first word of the book's title is capitalized, but all other words (except proper nouns) should be lowercase.
- *Publication Information.* Give the publisher's full name or consult your style guide for the preferred abbreviation. Give the publisher's city, and

include the state or country in parentheses, unless the city is well known (such as New York, Boston, or London).

- *Periodical Titles.* Titles of periodicals should be abbreviated according to CBE style. The abbreviated titles should be capitalized, but not underlined, italicized, or placed in quotations.

- *Article Titles.* Titles of articles should not be underlined, italicized, or placed in quotations. The first word of the article title is capitalized, but all other words (except proper nouns) should be lowercase.

- *Electronic Sources.* Include as much information about electronic sources as you can, such as date of publication, identification numbers, and retrieval information. Also, be sure to record the date you retrieved the information, because electronic information changes frequently.

- *Spacing.* Double space the entire reference list and leave no extra spacing between entries.

- *Page Numbers.* Give the total number of pages for a book entry, followed by the abbreviation *p* with no period (for example, 298 p). If you are giving a range of pages for specific articles in books and periodicals, use the abbreviation *p* and the last digit of the second number, but only if the previous digits are identical (for example, p 151–3, not 151–153).

- *Dates.* Follow this format: year, month, day, with no periods or commas (2000 Oct 31). Use only the first three letters of each month.

- *Additional References.* A reference list includes only the sources you actually cite in your document. You may include other sources you used in researching and preparing your document (but didn't cite) in a separate alphabetical list entitled "Additional References."

The models in this section show CBE's citation-sequence system. In the name-year system, the reference list would be alphabetized rather than numbered, and the publication year would immediately follow the name of the author(s). In addition, the list would follow the hanging indent style, as shown here.

Cunningham W. S. 1980. Crisis at Three Mile Island: the aftermath of a near meltdown. New York: Madison. 342 p.

For further examples of both the citation-sequence system and the number-year system, consult the CBE style manual, *Scientific Style and Format.* The documentation models for online sources in this section are based on recommendations in *Online!* (Harnack & Kleppinger, 2000).

CBE Style for Reference List Entries

1. Book by One Author	10. Unsigned Article
2. Book by Multiple Authors	11. Web Site
3. Book Issued by an Organization	12. Online Book
4. Book in Edition Other Than First	13. Online Article
5. Journal Article	14. Online Abstract
6. Magazine Article	15. Email Message
7. Newspaper Article	16. Listserv or Newsgroup Message
8. Article Included in a Book	17. Synchronous Communication
9. Article from a Volume of Proceedings	

1. Book by One Author

Include the author's last name and initials (not separated by a comma), followed by the book title, the location and name of the publisher, the year of publication, and the number of pages in the book.

> [1]Cunningham W. S. Crisis at Three Mile Island: the aftermath of a near meltdown. New York: Madison; 1980. 342 p.

2. Book by Multiple Authors

List all names in reverse order. Do not use the word *and* between names.

> [2]Bingham C, Withers S. Neural networks and fuzzy logic. New York: IEEE; 1999. 276 p.

3. Book Issued by an Organization

The organization takes the position of the author.

> [3]National Commission on Corrections. The future of incarceration. Publication 11346-53. St. Louis, MO: Liberty; 1998. 112 p.

In the name-year system, an abbreviated form of the organization name may be used in both the text and the reference list. The entry should be alphabetized according to the abbreviation, not the full name of the organization.

> [NCC] National Commission on Corrections. 1998. The future of incarceration. Publication 11346-53. St. Louis, MO: Liberty. 112 p.

4. Book in Edition Other Than First

The edition number follows the title of the book.

> [4]Schonberg N. Solid state physics. 3rd ed. London: Paragon; 1997. 354 p.

5. Journal Article

List the author's name, the article title, and the abbreviated journal title followed by the year, month, volume number, and page number(s). If the journal is continuously paginated by volume, include only the volume number after the year.

6Hastings W. The space shuttle debate. Mod Inq 1990;13:311–8.

If the journal paginates each issue separately, include the issue number in parentheses after the volume number.

7Juneja G. Asynchronous transmission techniques. Video Q 1994;6(2):11–9.

6. Magazine Article

List the author's name, the article title, and the abbreviated magazine title, followed by the issue date and page number(s).

8Newman D. Passive restraint systems. Car Lore 1995 Dec 12:41–6.

7. Newspaper Article

List the author's name, the article title, and the newspaper title, followed by the issue date and section, page, and column number(s). If the newspaper does not use section numbers, use a colon between the date and the page number.

9Felder M. Smokeless tobacco: New danger signs. New York Post 1995 May 4; Sect A:13(col 2).

8. Article Included in a Book

Give the author and name of the article first, followed by the word *In* with a colon and the book editor and title. Then give the publication information for the book and the page numbers on which the article appears.

5Deacon M. Amplification systems. In: Kooper A, editor. Advances in electronics. Miami, FL: Valley Pr; 1995. p 34–51.

9. Article from a Volume of Proceedings

List the author's name, the article title, and the word *In* followed by the proceedings editor(s) and title, the date and place of the conference, and publication information, including any identifying numbers. Because conference titles are proper nouns, they should be capitalized.

11Carlson CT. Advanced organizers in manuals. In: Rainey K, editor. Proceedings of the 45th International Technical Communication Conference; 1995 Sep 23–26; Boston. Fairfax (VA): Society for Technical Communication; 1995. p RT56–8.

10. Unsigned Article

If the author of an article is not indicated, insert the word *Anonymous* in brackets.

[10][Anonymous]. The state of the art in microcomputers. Newscene 2000;56:406–21.

11. Web Site

Begin with the author name and date of publication. Next give the title of the document (if applicable) and the name of the site. Also provide the URL in angle brackets and your access date.

[14]Hendl KB. 2000 Jan 1. Internet resources for nursing students. The nursing home page. <http://www.carney.edu/nursing/index.html>. Accessed 2000 May 10.

12. Online Book

Begin by giving the author name and original publication year, followed by the book title. Also provide the URL in angle brackets. End with the date of access.

[13]Rawlins, GJE. 1997. Moths to the flame. <http://mitpress.mit.edu/e-books/Moths/>. Accessed 2000 Feb 10.

13. Online Article

Begin by giving the author name, date of publication, article title, and abbreviated title of the periodical. Then give any volume, issue, and page numbers. Include the URL in angle brackets and end with the date of access.

[12]Webster L. 1999. New hope for Alzheimer's sufferers. Ann Neur Online 13(2). <http://www.annals.com/neurology/issues.99spr>. Accessed 1999 May 27.

14. Online Abstract

Include "abstract" in square brackets following the title.

[1]Kaufman L, Kaufman JH. 2000 Jan 4. Explaining the moon illusion [abstract]. In Proc Natl Acad Sci 97(1):500–5. <http://www.pnas.org/cgi/content/abstract/97/1/500>. Accessed 2000 Feb 8.

15. Email Message

List the author's name and the date of the message, the subject line (if any), the type of communication, and the date the email was accessed.

[15]Nelworth KC. 1999 May 7. Flat-panel displays. [Personal email]. Accessed 1999 May 8.

16. Listserv or Newsgroup Message

Give the author's name, date of message, subject line (if any), listserv or newsgroup address, and date of access.

> [16]Rajiv CV. 1998 May 25. Portable document formats. <techwr-l@listserv.okstate.edu>. Accessed 1998 Jun 10.

17. Synchronous Communication

To cite a synchronous discussion from a MOO, a MUD, or an IRC, give the name of the speaker (if known) or the name of the site. Then provide the date and name of the discussion, the type of communication, the URL (in angle brackets) or command line instructions, and the date of access.

> [17]Prisley L. 1997 Oct 5. Seminar discussion on FTP. [Group discussion]. telnet moo.ku.edu/port=9999. Accessed 1997 Nov 19.

Sample CBE Reference List

Following is a sample list of references using the CBE citation-sequence system.

References

[1]Goodman RK, Wakely J, Ruh K. What employees think of the Scanlon Plan. Personnel 1995;6:22–9.

[2]Zwicker D. 1999 Oct 6. More on the Scanlon Plan: a response. Insurance fundamentals. <http://www.swisu.edu/ec201/index.html>. Accessed 1999 Nov 16.

[3]Daly PH. Selecting and designing a group insurance plan. Personnel J 1996 Nov;54:322–3.

[4]Flanders A. Measured daywork and collective bargaining. British J Industrial Relations 1997;9(4):368–92.

[5]Trencher P, Coughlin C. Recent trends in labor-management relations. New York: Westly; 1995. 422 p.

MLA STYLE

MLA style consists of two elements: the citation in the text and the list of works cited at the end of the document.

MLA Textual Citations

In MLA style, the textual citation typically includes the name of the source's author and the number of the page being referred to. Textual citations will vary according to the type of information cited, the authors' names, and the context of the citation. The following models illustrate a variety of common textual citations; for additional examples, consult the *MLA Handbook for Writers of Research Papers*.

MLA Style for Textual Citations

1. Citing Specific Pages
2. Citing Entire Works
3. Multiple Sources by Same Author
4. Source with Multiple Authors
5. Source Quoted within Another Source
6. Source Issued by an Organization
7. Source with Unknown Author
8. Multiple Sources in One Citation
9. Multiple Authors with Same Last Name
10. Electronic Sources

---

1. Citing Specific Pages

Immediately following the borrowed material, include a parenthetical reference with the author's name and the page number(s) being referred to. Do not use a comma between the name and the page number, and do not use the abbreviation "p." or "pp."

> This phenomenon was identified more than fifty years ago (Wilkinson 134).

If your sentence already includes the author's name, include only the page number in the parenthetical notation.

> Wilkinson identified this phenomenon more than fifty years ago (134).

2. Citing Entire Works

If you are referring to the whole source, not to a particular page or pages, use only the author's name.

> This phenomenon was identified more than fifty years ago (Wilkinson).
>
> Wilkinson identified this phenomenon more than fifty years ago.

3. Multiple Sources by Same Author

If you cite two or more sources by the same author, either include the full source title in the text or add a shortened title after the author's name in the parenthetical citation to prevent confusion.

> Wilkinson identified this phenomenon more than fifty years ago in his book *Particle Physics* (36–37).

Later technologies were able to further prove the early genetic theories (Wilkinson, "Ascent" 11).

4. Source with Multiple Authors

For a source written by two or three authors, cite all names.

. . . as early as 1974 (Allman and Jones 15–34).

Finn, Crenshaw, and Zander contend that . . .

For a source written by four or more authors, list only the first author followed by "et al" (Latin for "and others").

Bradley et al. argued . . .

5. Source Quoted within Another Source

Give the name of the author of the quotation in the text. In the parenthetical citation, give the author and page number of the source in which you found the quotation.

According to Hanson, multimedia will be the dominant mode of instruction in colleges by 2005 (qtd. in Ortiz 211).

Note that only the source by Ortiz will appear in the list of works cited.

6. Source Issued by an Organization

If the author is an organization rather than a person, use the name of the organization.

The causes of narcolepsy are discussed in a recent booklet (Association of Sleep Disorders, 2–3).

In a recent booklet, the Association of Sleep Disorders (2–3) discusses the causes of narcolepsy.

7. Source with Unknown Author

If the source does not identify an author, use a shortened form of the title in your parenthetical citation.

This trend has been evident in American society since the beginning of the twentieth century ("Modernism" 179).

8. Multiple Sources in One Citation

To refer to two sources at the same point, separate the sources in parentheses with a semicolon.

This phenomenon was identified more than fifty years ago (Betts 29; Wilkinson 134).

9. Multiple Authors with Same Last Name

If two or more sources have authors with the same last name, spell out the first names of those authors in your textual citations.

> This phenomenon was identified more than fifty years ago (Brian Wilkinson 134).
>
> Renee Wilkinson has suggested two possible explanations for the digression (81).

10. Electronic Sources

Follow the same rules as for print sources when citing electronic sources in your document, providing author names and page numbers if available. If an author's name is not given, use either the full title of the source in the text or a shortened version of the title in the parenthetical citation. If no page numbers are used, include any given identifying numbers, such as paragraph or section numbers, abbreviated "par." or "sec." Otherwise, use no number at all. Include URLs in the works cited list but not in the text.

> Twenty million books were in print by the early sixteenth century (Rawlins, ch. 3, sec. 2).

The MLA List of Works Cited

A list of works cited provides the information your readers will need to find each source you used. Note that the reference list includes only those items that you actually used in researching and preparing your document and that you cite in your document; it should not include your background reading materials. Following are some guidelines for creating an MLA-style list of works cited.

Click on the link to Diana Hacker's Research and Documentation Online on TechComm Web (www.bedfordstmartins.com /techcomm) for help with formatting an MLA list of works cited.

- *Arranging Entries.* The individual entries in the reference list are arranged alphabetically by the author's last name. Two or more works by the same author are arranged by title. Works by an organization are alphabetized by the first significant word in the name of the organization.

- *Book Titles.* Titles of books should follow standard capitalization rules and should be either underlined or italicized; use a consistent method throughout your document.

- *Publication Information.* Give the publisher's full name or consult your style guide for the preferred abbreviations. Include both the publisher's city and state or country, unless the city is well known (such as New York, Boston, or London).

- *Periodical Titles.* Titles of periodicals should be either underlined or italicized (consistently), and they should be capitalized normally.

- *Article Titles.* Titles of articles and other short works should be placed in quotation marks and should follow standard capitalization rules (all major words should be capitalized).

- *Electronic Sources.* Include as much information as you can about electronic sources, such as date of publication, identifying numbers, and

See the MLA Web site (www.mla.org/style/sources .htm) for their statement on documenting Web sources.

retrieval information. Also be sure to record the date you retrieved the information, because electronic information changes frequently. Titles of entire Web sites should be either underlined or italicized (consistently); titles of works within Web sites, such as articles and video clips, should be treated normally.

- *Indenting.* Use a hanging indent, with second and subsequent lines of each entry indented one-half inch.

- *Spacing.* Double space the entire reference list, with no extra spacing between entries.

- *Page Numbers.* Do not use the abbreviation "p." or "pp." when giving page numbers in MLA citations. For a range of pages, give only the last two digits of the second number if the previous digits are identical (for example, 243–47, not 243–247). Use a plus sign (+) to indicate that an article continues on subsequent pages interrupted by other articles or advertisements.

- *Dates.* Follow this format: day, month, year, with no commas (20 Feb. 1999). Spell out *May, June,* and *July;* abbreviate all other months by using the first three letters, with the exception of *Sept.*

The following models of reference list entries include a variety of sources. For further examples of MLA-style citations, consult the *MLA Handbook for Writers of Research Papers.*

MLA Style for Works Cited Entries

1. Book by One Author
2. Book by Multiple Authors
3. Multiple Books by the Same Author
4. Book Issued by an Organization
5. Book Compiled by an Editor or Issued under an Editor's Name
6. Book in Edition Other Than First
7. Journal Article
8. Magazine Article
9. Newspaper Article
10. Article Included in a Book
11. Article from a Volume of Proceedings
12. Unsigned Article
13. Government Document
14. Interview You Conducted
15. Map or Chart
16. Lecture
17. Web Site
18. Online Book
19. Online Article
20. Online Government Publication
21. Document from an Online Subscription Service
22. Email Message
23. Listserv or Newsgroup Message
24. Synchronous Communication

1. Book by One Author

Include the author's full name, in reverse order, followed by the book title. Next, give the location and name of the publisher, followed by the year of publication.

Cunningham, Walter. *Crisis at Three Mile Island: The Aftermath of a Near Meltdown.* New York: Madison, 1980.

2. Book by Multiple Authors

For a book by two or three authors, present the names in the sequence in which they appear on the title page. Only the name of the first author is presented in reverse order. A comma separates the names of the authors.

Bingham, Christine, and Stephen Withers. *Neural Networks and Fuzzy Logic.* New York: IEEE, 1999.

For a book by four or more authors, you may use the abbreviation *et al.* ("and others") after the first author's name.

Foster, Glenn, et al. *The American Renaissance.* Binghamton: Archive, 1995.

3. Multiple Books by the Same Author

For second and subsequent entries by the same author, use three hyphens followed by a period. Arrange the entries alphabetically by title.

Smith, Louis. *International Standards.* New York: IEEE, 1995.

– – –. *Wave-Propagation Technologies.* Berkeley: Stallings, 1994.

4. Book Issued by an Organization

The organization takes the position of the author.

National Commission on Corrections. *The Future of Incarceration.* Publication 11346-53. St. Louis, MO: Liberty, 1998.

5. Book Compiled by an Editor or Issued under an Editor's Name

The book editor's name(s) followed by *ed.* or *eds.* is used in place of an author's name.

Morgan, Donald E., ed. *Readings in Alternative Energies.* Boston: Smith-Howell, 1995.

6. Book in Edition Other Than First

The edition number follows the title of the book.

Schonberg, Nathan. *Solid State Physics.* 3rd ed. London: Paragon, 1997.

7. Journal Article

List the author's name, the article title (in quotes), and the magazine title (in italics), followed by the volume number, year, and page number(s). If the journal is paginated continuously throughout a volume, include only the volume number after the journal title.

> Hastings, Wendy. "The Space Shuttle Debate." *The Modern Inquirer* 13 (1990): 311–18.

If the journal paginates each issue separately, include the issue number after the volume number, separated by a period.

> Juneja, Gupta. "Asynchronous Transmission Techniques." *Video Quarterly* 6.2 (1994): 11–19.

8. Magazine Article

List the author's name, the article title (in quotes), and the magazine title (in italics) followed by the issue date and page number(s).

> Newman, Daniel. "Passive Restraint Systems." *Car Lore* 12 Dec. 1995: 41+.

If the article has no author listed, alphabetize it by title.

> "Passive Restraint Systems." *Car Lore* 12 Dec. 1995: 41+.

9. Newspaper Article

List the author's name, the article title (in quotes), and the newspaper title (in italics), followed by the issue date and page number(s). If the newspaper appears in more than one edition, cite the edition.

> Felder, Melissa. "Smokeless Tobacco: New Danger Signs." *New York Post* 4 May 1995, morning ed.: 13.

10. Article Included in a Book

Give the name of the author and the title of the article first, followed by the book title and the editor. Present the editor's name in normal order.

> May, Bruce, and James Deacon. "Amplification Systems." *Advances in Electronics.* Ed. Alvin Kooper. Miami: Valley Press, 1995.

11. Article from a Volume of Proceedings

List the author's name, the article title, and the proceedings volume title and editor name, followed by the publication information, including any identifying numbers.

> Carlson, Carl T., "Advanced Organizers in Manuals." *Proceedings of the 45th International Technical Communication Conference.* Ed. Ken Rainey. Fairfax: Society for Technical Communication, 1995. RT56–58.

12. Unsigned Article

If the author of an article is not indicated, alphabetize it by title, ignoring the articles *the, a,* and *an.*

> "The State of the Art in Microcomputers." *Newscene* 56 (fall 1990): 406–21.

13. Government Document

Give the government name and agency as the author name, followed by the publication title, edition or identifying number (if any), place, and date.

> United States. Department of Energy. *The Energy Situation in the Next Decade.* Technical Publication 11346-53. Washington: GPO, 1998.

14. Interview You Conducted

List the interviewee's name, the words *Personal interview,* and the date.

> Gangloff, Richard. Personal interview. 24 Jan. 1995.

15. Map or Chart

Give the title of the map or chart, followed by the label "Map" or "Chart."

> *Central Orange County.* Map. Mill Creek, WA: King of the Road, 1999.

16. Lecture

Give the speaker's name and lecture title, followed by the location and the date on which the lecture was given.

> Robbins, Bruce. "Trends in Secondary Education." Lecture in E382. Boise State University. 2 May 1999.

17. Web Site

If you are citing an entire Web site, begin with the author, title or description of the site, and publication date or most recent update. Then give the name of the sponsoring institution or organization (if any), your access date, and the URL in angle brackets.

 See the MLA Web site (www.mla.org/style/sources .htm) for their statement on documenting Web sources.

> Hendl, Kevin B. *The Nursing Page.* 1 Jan. 2000. Northwest Consortium of Medical Professionals. 10 May 2000 <http://www.carney.edu/nursing/index.html>.

18. Online Book

Begin with the author's name and the title of the work, along with any available information about the print source. If the book has not been published before, include the online publication date and publisher. End with your access date and the URL in angle brackets.

Rawlins, Gregory J. E. *Moths to the Flame.* Cambridge, MA: MIT P, 1997. 10 Feb. 2000
    <http://mitpress.mit.edu/e-books/Moths/>.

19. Online Article

Include both the title of the document and the periodical, along with the date of publication. If the periodical is a scholarly journal, include relevant identifying numbers, such as volume, issue, and page numbers. For abstracts of articles, include the word *Abstract*, followed by a period. End with your access date, and the URL in angle brackets.

Carnevale, Dan. "University Uses New Format to Send Televised Courses by Computer." *The Chronicle of Higher Education* 11 Feb. 2000: A45. 8 Feb. 2000
    <http://chronicle.com/weekly/v46/i23/23a04502.htm>.

20. Online Government Publication

Begin with the name of the country and the government agency (abbreviated if appropriate). Next give the title of the document and the relevant publication information: author (if known) and date of publication. End with your access date, and the URL in angle brackets.

United States. Federal Communications Commission. *FCC Consumer Alert on
    Telephone Slamming.* Jan. 1999. 8 Feb. 2000 <http://www.fcc.gov/Bureaus
    /Common_Carrier/Factsheets/slamming.html>.

21. Document from an Online Subscription Service

If you are citing a document retrieved from an online subscription service, and you have not been given a URL for the document, record the name of the service, the library, and your access date. End with the URL of the service in angle brackets.

Siegfried, Tom. "A Way to Get the Message without Using Up Energy." *The Dallas
    Morning News* 1 July 1996: 7D. Electric Lib. O'Neill Lib., Boston College, Chestnut
    Hill, MA. 8 Feb. 2000 <http://www.elibrary.com>.

22. Email Message

List the author's name, the subject line (if any), the word *Email,* and the date the email was sent.

Nelworth, Karen C. "Flat-panel Displays." Email to the author. 7 May 1997.

23. Listserv or Newsgroup Message

List the author's name, the subject line (if any), the posting date, the words *Online posting,* the name of the listserv or newsgroup, any identifying number of the posting, the date you accessed the post, and the address of the list-serv or newsgroup.

> Rajiv, Chris V. "Portable Document Formats." 25 May 1998. Online posting. TECHWR-L. 10 June 1998 <techwr-l@listserv.okstate.edu>.

24. Synchronous Communication

To cite a synchronous discussion from a MOO, a MUD, or an IRC, give the name of the speaker, the name and date of the discussion, the forum title (if any), the date you accessed the information, and the URL. If an archival version of the communication is unavailable, include the telnet address.

> Prisley, Lauren. "Seminar Discussion on FTP." 5 Oct. 1997. Tech MOOspace. 19 Nov. 1997. telnet moo.ku.edu/port=9999.

Sample MLA List of Works Cited

The following sample list of works cited illustrates the MLA citation system.

Works Cited

Daly, Peter H. "Selecting and Designing a Group Insurance Plan." *Personnel Journal* 54 (Nov. 1996): 322–23.

Flanders, Andrew. "Measured Daywork and Collective Bargaining." *British Journal of Industrial Relations* 9.4 (1997): 368–92.

Goodman, Raymond K., Janice Wakely, and Kathleen Ruh. "What Employees Think of the Scanlon Plan." *Personnel* 6 (1995): 22–29.

Trencher, Patricia, and Christopher Coughlin. *Recent Trends in Labor-Management Relations*. New York: Westly, 1995.

Zwicker, David. "More on the Scanlon Plan: A Response." *Insurance Fundamentals*. 6 Oct. 1999. 16 Nov. 1999 <http://www.swisu.edu/ec201/index.html>.

This part of the handbook contains advice on editing your documents for grammar, punctuation, and mechanics. The final section provides a concise guide to some of the most challenging aspects of English for nonnative speakers.

If your organization or professional field has a style guide with different recommendations about grammar and usage, you should of course follow those guidelines.

For a full-length treatment of the topics discussed here, refer to one of the handbooks listed in the Selected Bibliography (page 727).

GRAMMATICAL SENTENCES

Avoid Sentence Fragments

`frag`

Click on the link to Exercise Central on TechComm Web (www.bedfordstmartins.com /techcomm) for online exercises covering these grammar skills.

A sentence fragment is an incomplete sentence, an error that occurs when a sentence is missing either a verb or an independent clause. To correct a sentence fragment, use one of the following two strategies:

1. Introduce a verb.

FRAGMENT The pressure loss caused by a worn gasket.

This example is a fragment because it lacks a verb. (The word *caused* does not function as a verb here; rather, it introduces a phrase that describes the pressure loss.)

COMPLETE
SENTENCE The pressure loss was caused by a worn gasket.

Pressure loss has a verb: *was caused.*

COMPLETE
SENTENCE We identified the pressure loss caused by a worn gasket.

Pressure loss becomes the object in a new main clause: *We identified the pressure loss.*

FRAGMENT A plotting program with clipboard plotting, 3D animation, and FFTs.

COMPLETE
SENTENCE It is a plotting program with clipboard plotting, 3D animation, and FFTs.

COMPLETE
SENTENCE A plotting program with clipboard plotting, 3D animation, and FFTs will be released today.

2. Link the fragment (a dependent element) to an independent clause.

FRAGMENT The article was not accepted for publication. Because the data could not be verified.

Because the data could not be verified is a fragment because it lacks an independent clause: a clause that has a subject and a verb and could stand alone as a sentence. To be complete, it needs more information.

COMPLETE SENTENCE	The article was not accepted for publication because the data could not be verified.

The dependent element is joined to the independent clause that precedes it.

COMPLETE SENTENCE	Because the data could not be verified, the article was not accepted for publication.

The dependent element is followed by the independent clause.

FRAGMENT	Delivering over 150 horsepower. The two-passenger coupe will cost over $32,000.
COMPLETE SENTENCE	Delivering over 150 horsepower, the two-passenger coupe will cost over $32,000.
COMPLETE SENTENCE	The two-passenger coupe will deliver over 150 horsepower and cost over $32,000.

Avoid Comma Splices

CS

A comma splice is an error that occurs when two independent clauses are joined, or spliced together, by a comma. Independent clauses in a comma splice can be linked correctly in three ways:

1. **Use a comma and a coordinating conjunction (*and, or, nor, but, for, so, and yet*).**

COMMA SPLICE	The 909 printer is our most popular model, it offers an unequaled blend of power and versatility.
CORRECT	The 909 printer is our most popular model, for it offers an unequaled blend of power and versatility.

The coordinating conjunction *for* explicitly states the relationship between the two clauses.

2. **Use a semicolon.**

COMMA SPLICE	The 909 printer is our most popular model, it offers an unequaled blend of power and versatility.
CORRECT	The 909 printer is our most popular model; it offers an unequaled blend of power and versatility.

The semicolon creates a somewhat more distant relationship between the two clauses than the comma-and-coordinating-conjunction link; the link remains implicit.

3. Use a period or another form of terminal punctuation.

COMMA
SPLICE   The 909 printer is our most popular model, it offers an unequaled blend of power and versatility.

CORRECT   The 909 printer is our most popular model. It offers an unequaled blend of power and versatility.

The two independent clauses are separate sentences. Of the three ways to punctuate the two clauses correctly, this punctuation suggests the most distant relationship between them.

Avoid Run-On Sentences

run

In a run-on sentence (sometimes called a *fused sentence*), two independent clauses appear together with no punctuation between them. A run-on sentence can be corrected in the same three ways as a comma splice:

1. Use a comma and a coordinating conjunction (*and, or, nor, but, for, so, and yet*).

RUN-ON
SENTENCE   The 909 printer is our most popular model it offers an unequaled blend of power and versatility.

CORRECT   The 909 printer is our most popular model, for it offers an unequaled blend of power and versatility.

2. Use a semicolon.

RUN-ON
SENTENCE   The 909 printer is our most popular model it offers an unequaled blend of power and versatility.

CORRECT   The 909 printer is our most popular model; it offers an unequaled blend of power and versatility.

3. Use a period or another form of terminal punctuation.

RUN-ON
SENTENCE   The 909 printer is our most popular model it offers an unequaled blend of power and versatility.

CORRECT   The 909 printer is our most popular model. It offers an unequaled blend of power and versatility.

Avoid Ambiguous Pronoun References

`ref`

Pronouns must refer clearly to their antecedents—the words or phrases they replace. To correct ambiguous pronoun references, try one of these four strategies:

1. **Clarify the pronoun's antecedent.**

   UNCLEAR   Remove the cell cluster from the medium and analyze it.

   Analyze what: the cell cluster or the medium?

   CLEAR   Analyze the cell cluster after removing it from the medium.

   CLEAR   Analyze the medium after removing the cell cluster from it.

   CLEAR   Remove the cell cluster from the medium. Then analyze the cell cluster.

   CLEAR   Remove the cell cluster from the medium. Then analyze the medium.

2. **Clarify the relative pronoun, such as *which*, introducing a dependent clause.**

   UNCLEAR   She decided to evaluate the program, which would take five months.

   What would take five months: the program or the evaluation?

   CLEAR   She decided to evaluate the program, a process that would take five months.

   By replacing *which* with *a process that*, the writer clearly indicates that it is the evaluation that will take five months.

   CLEAR   She decided to evaluate the five-month program.

   By using the adjective *five-month*, the writer clearly indicates that it is the program that will take five months.

3. **Clarify the subordinating conjunction, such as *where*, introducing a dependent clause.**

   UNCLEAR   This procedure will increase the handling of toxic materials outside the plant, where adequate safety measures can be taken.

   Where can adequate safety measures be taken: inside the plant or outside?

   CLEAR   This procedure will increase the handling of toxic materials outside the plant. Because adequate safety measures can be taken only in the plant, the procedure poses risks.

   CLEAR   This procedure will increase the handling of toxic materials outside the plant. Because adequate safety measures can be taken only outside the plant, the procedure will decrease safety risks.

Sometimes the best way to clarify an unclear reference is to split the sentence in two, drop the subordinating conjunction, and add clarifying information.

4. Clarify the ambiguous pronoun that begins a sentence.

UNCLEAR   Allophanate linkages are among the most important structural components of polyurethane elastomers. They act as cross-linking sites.

What act as cross-linking sites: allophanate linkages or polyurethane elastomers?

CLEAR   Allophanate linkages, which are among the most important structural components of polyurethane elastomers, act as cross-linking sites.

The writer has rewritten part of the first sentence to add a clear nonrestrictive modifier and combined it with the second sentence.

If you begin a sentence with a pronoun that might be unclear to the reader, be sure to follow it immediately with a noun that clarifies the reference.

UNCLEAR   The new parking regulations require that all employees pay for parking permits. These are on the agenda for the next senate meeting.

What are on the agenda: the regulations or the permits?

CLEAR   The new parking regulations require that all employees pay for parking permits. These regulations are on the agenda for the next senate meeting.

Compare Items Clearly

comp

When comparing or contrasting items, make sure your sentence communicates their relationship clearly. A simple comparison between two items often causes no problems: "The X3000 has more storage than the X2500." Simple comparisons, however, can sometimes result in ambiguous statements:

AMBIGUOUS   Trout eat more than minnows.

Do trout eat minnows in addition to other food, or do trout eat more than minnows eat?

CLEAR   Trout eat more than minnows do.

If you are introducing three items, make sure the reader can tell which two are being compared:

AMBIGUOUS   Trout eat more algae than minnows.

CLEAR   Trout eat more algae than they do minnows.

CLEAR   Trout eat more algae than minnows do.

Beware of comparisons in which different aspects of the two items are compared:

ILLOGICAL   The resistance of the copper wiring is lower than the tin wiring.

LOGICAL   The resistance of the copper wiring is lower than that of the tin wiring.

Resistance cannot be logically compared with *tin wiring*. In the revision, the pronoun *that* substitutes for *resistance* in the second part of the comparison.

Use Adjectives Clearly

adj

In general, adjectives are placed before the nouns that they modify: *the plastic washer*. In technical communication, however, writers often need to use clusters of adjectives. To prevent confusion in technical communication, follow two guidelines.

1. **Use commas to separate coordinate adjectives.**

   Adjectives that describe different aspects of the same noun are known as coordinate adjectives.

   portable, programmable CD player

   adjustable, removable housings

   The comma is used instead of the word *and*.

   Sometimes an adjective is considered part of the noun it describes: *electric drill*. When one adjective modifies *electric drill*, no comma is required: *a reversible electric drill*. The addition of two or more adjectives, however, creates the traditional coordinate construction: *a two-speed, reversible electric drill*.

2. **Use hyphens to link compound adjectives.**

   A compound adjective is made up of two or more words. Use hyphens to link these elements when compound adjectives precede nouns.

   a *variable-angle* accessory

   increased *cost-of-living* raises

   The hyphens prevent *increased* from being read as an adjective modifying *cost*.

   A long string of compound adjectives can be confusing even if you use hyphens appropriately. To ensure clarity, turn the adjectives into a clause or a phrase following the noun.

   UNCLEAR   an *operator-initiated default-prevention* technique

   CLEAR   a technique *initiated by the operator to prevent default*

Maintain Subject-Verb Agreement

agr s/v

The subject and verb of a sentence must agree in number, even when a prepositional phrase comes between them. The object of the preposition may be plural in a singular sentence.

INCORRECT	The *result* of the tests *are* promising.
CORRECT	The *result* of the tests *is* promising.

The object of the preposition may be singular in a plural sentence.

INCORRECT	The *results* of the test *is* promising.
CORRECT	The *results* of the test *are* promising.

Don't be misled by the fact that the object of the preposition and the verb don't sound natural together, as in *tests is* or *test are*. Here, the noun *test(s)* precedes the verb, but it is not the subject of the verb. As long as the subject and verb agree, the sentence is correct.

Maintain Pronoun-Antecedent Agreement

agr p/a

A pronoun and its antecedent (the word or phrase being replaced by the pronoun) must agree in number. Often an error occurs when the antecedent is a collective noun—one that can be interpreted as either singular or plural, depending on its usage.

INCORRECT	The *company* is proud to announce a new stock option plan for *their* employees.
CORRECT	The *company* is proud to announce a new stock option plan for *its* employees.

Company acts as a single unit; therefore, the singular pronoun is appropriate.

When the individual members of a collective noun are emphasized, however, plural pronouns are appropriate.

CORRECT	The inspection team have prepared their reports.

The use of *their* emphasizes that the team members have prepared their own reports.

Use Tenses Correctly

t

Two verb tenses are commonly used in technical communication: the present tense and the past perfect tense. It is important to understand the specific purpose of each.

1. The **present tense** is used to describe scientific principles and recurring events.

   INCORRECT   In 1992, McKay and his coauthors argued that the atmosphere of Mars *was* salmon pink.

   CORRECT   In 1992, McKay and his coauthors argued that the atmosphere of Mars *is* salmon pink.

   Although the argument was made in the historical past—1992—the point is expressed in the present tense, because the atmosphere of Mars continues to be salmon pink.

   When the date of the argument is omitted, some writers express the entire sentence in the present tense.

   CORRECT   McKay and his coauthors *argue* that the atmosphere of Mars *is* salmon pink.

2. The **past perfect tense** is used to describe the earlier of two events that occurred in the past.

   CORRECT   We *had begun* excavation when the foreman *discovered* the burial remains.

   Had begun is the past perfect tense. The excavation began before the burial remains were discovered.

   CORRECT   The seminar *had concluded* before I *got* a chance to talk with Dr. Tran.

PUNCTUATION

Commas

The comma is the most frequently used punctuation mark, as well as the one about whose usage writers most often disagree. Examples of common misuses of the comma are noted within the following guidelines. This section concludes with advice about editing for unnecessary commas.

 Click on the link to Exercise Central on TechComm Web (www.bedfordstmartins.com /techcomm) for online exercises covering punctuation.

1. **Use a comma in a compound sentence, to separate two independent clauses linked by a coordinating conjunction (*and, or, nor, but, so, for, yet*).**

   INCORRECT   The mixture was prepared from the two premixes and the remaining ingredients were then combined.

   CORRECT   The mixture was prepared from the two premixes, and the remaining ingredients were then combined.

2. **Use a comma to separate items in a series composed of three or more elements:**

The manager of spare parts is responsible for ordering, stocking, and disbursing all spare parts for the entire plant.

Despite the presence of the conjunction *and*, most technical-communication style manuals require a comma after the second-to-last item. The comma clarifies the separation and prevents misreading.

CONFUSING	The report will be distributed to Operations, Research and Development and Accounting.
CLEAR	The report will be distributed to Operations, Research and Development, and Accounting.

3. **Use a comma to separate introductory words, phrases, and clauses from the main clause of the sentence:**

However, we will have to calculate the effect of the wind.

To facilitate trade, the government holds a yearly international conference.

In the following example, the comma actually prevents misreading:

Just as we finished eating, the rats discovered the treadmill.

NOTE: Writers sometimes make errors by omitting commas following introductory words, phrases, or clauses. A comma is optional only if the introductory text is brief and cannot be misread.

CORRECT	First, let's take care of the introductions.
CORRECT	First let's take care of the introductions.
INCORRECT	As the researchers sat down to eat the laboratory rats awakened.
CORRECT	As the researchers sat down to eat, the laboratory rats awakened.

4. **Use a comma to separate a dependent clause from the main clause:**

Although most of the executive council saw nothing wrong with it, the advertising campaign was canceled.

Most PCs use green technology, even though it is relatively expensive.

For more about restrictive and nonrestrictive modifiers, see Ch. 11, p. 282.

5. **Use commas to separate nonrestrictive modifiers (parenthetical clarifications) from the rest of the sentence:**

Jones, the temporary chairman, called the meeting to order.

NOTE: Writers sometimes introduce an error by dropping one of the commas around a nonrestrictive modifier.

INCORRECT	The phone line, which was installed two weeks ago had to be disconnected.
CORRECT	The phone line, which was installed two weeks ago, had to be disconnected.

6. **Use a comma to separate interjections and transitional elements from the rest of the sentence:**

    Yes, I admit that your findings are correct.

    Their plans, however, have great potential.

    NOTE: Writers sometimes introduce an error by dropping one of the commas around an interjection or a transitional element.

    INCORRECT   Our new statistician, however used to work for Konaire, Inc.

    CORRECT   Our new statistician, however, used to work for Konaire, Inc.

7. **Use a comma to separate coordinate adjectives:**

    The finished product was a sleek, comfortable cruiser.

    The heavy, awkward trains are still being used.

    The comma here takes the place of the conjunction *and*.

    If the adjectives are not coordinate—that is, if one of the adjectives modifies the combined adjective and noun—do not use a comma:

    They decided to go to the first general meeting.

    For more about coordinate adjectives, see page 691.

8. **Use a comma to signal that a word or phrase has been omitted from a sentence because it is implied:**

    Smithers is in charge of the accounting; Harlen, the data management; Demarest, the publicity.

    The commas after *Harlen* and *Demarest* show that the phrase *is in charge* has not been repeated.

9. **Use a comma to separate a proper noun from the rest of the sentence in direct address:**

    John, have you seen the purchase order from United?

    What I'd like to know, Betty, is why we didn't see this problem coming.

10. **Use a comma to introduce most quotations:**

    He asked, "What time were they expected?"

11. **Use a comma to separate towns, states, and countries:**

    Bethlehem, Pennsylvania, is the home of Lehigh University.

    He attended Lehigh University in Bethlehem, Pennsylvania, and the University of California at Berkeley.

    Note that a comma precedes and follows *Pennsylvania*.

12. **Use a comma to set off the year in a date:**

   August 1, 2003, is the anticipated completion date.

   If the month separates the date and the year, you do not need to use commas because the numbers are not next to each other:

   The anticipated completion date is 1 August 2003.

13. **Use a comma to clarify numbers:**

   12,013,104

   NOTE: European practice is to reverse the use of commas and periods in writing numbers: periods signify thousands, and commas signify decimals.

14. **Use a comma to separate names from professional or academic titles:**

   Harold Clayton, Ph.D.
   Marion Fewick, CLU
   Joyce Carnone, P.E.

   NOTE: The comma also follows the title in a sentence:

   Harold Clayton, Ph.D., is the featured speaker.

UNNECESSARY COMMAS

Writers often introduce errors by using unnecessary commas. Do not insert commas in the following situations:

- Commas are not used to link two independent clauses without a coordinating conjunction (known as a "comma splice"):

   INCORRECT   All the motors were cleaned and dried after the water had entered, had they not been, additional damage would have occurred.

   CORRECT   All the motors were cleaned and dried after the water had entered; had they not been, additional damage would have occurred.

   CORRECT   All the motors were cleaned and dried after the water had entered. Had they not been, additional damage would have occurred.

   For more about comma splices, see page 687.

- Commas are not used to separate the subject from the verb in a sentence:

   INCORRECT   Another of the many possibilities, is to use a "first in, first out" sequence.

   CORRECT   Another of the many possibilities is to use a "first in, first out" sequence.

- Commas are not used to separate the verb from its complement:

INCORRECT   The schedules that have to be updated every month are, numbers 14, 16, 21, 22, 27, and 31.

CORRECT   The schedules that have to be updated every month are numbers 14, 16, 21, 22, 27, and 31.

- Commas are not used with a restrictive modifier:

INCORRECT   New and old employees who use the processed order form, do not completely understand the basis of the system.

The phrase *who use the processed order form* is a restrictive modifier necessary to the meaning: it defines which employees do not understand the system.

CORRECT   New and old employees who use the processed order form do not completely understand the basis of the system.

INCORRECT   A company, that has grown so big, no longer finds an informal evaluation procedure effective.

The clause *that has grown so big* is a restrictive modifier.

CORRECT   A company that has grown so big no longer finds an informal evaluation procedure effective.

- Commas are not used to separate two elements in a compound subject:

INCORRECT   Recent studies, and reports by other firms confirm our experience.

CORRECT   Recent studies and reports by other firms confirm our experience.

Semicolons

Semicolons are used in the following instances.

1. **Use a semicolon to separate independent clauses not linked by a coordinating conjunction:**

   The second edition of the handbook is more up-to-date; however, it is also more expensive.

2. **Use a semicolon to separate items in a series that already contains commas:**

   The members elected three officers: Jack Resnick, president; Carol Wayshum, vice president; Ahmed Jamoogian, recording secretary.

   Here the semicolon acts as a "supercomma," grouping each name with the correct title.

MISUSE OF SEMICOLONS

Sometimes writers incorrectly use a semicolon when a colon is called for:

INCORRECT   We still need one ingredient; luck.

CORRECT   We still need one ingredient: luck.

Colons

■

Colons are used in the following instances.

1. **Use a colon to introduce a word, phrase, or clause that amplifies, illustrates, or explains a general statement:**

   The project team lacked one crucial member: a project leader.

   Here is the client's request: we are to provide the preliminary proposal by November 13.

   We found three substances in excessive quantities: potassium, cyanide, and asbestos.

   The week was productive: fourteen projects were completed and another dozen were initiated.

   NOTE: The text preceding a colon should be able to stand on its own as a sentence:

   INCORRECT   We found: potassium, cyanide, and asbestos.

   CORRECT   We found the following: potassium, cyanide, and asbestos.

   CORRECT   We found potassium, cyanide, and asbestos.

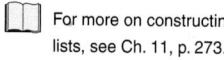 For more on constructing lists, see Ch. 11, p. 273.

2. **Use a colon to introduce items in a vertical list if the sense of the introductory text would be incomplete without the list:**

   We found the following:

   potassium

   cyanide

   asbestos

3. **Use a colon to introduce long or formal quotations:**

   The president began: "In the last year . . ."

MISUSE OF COLONS

Writers sometimes incorrectly use a colon to separate a verb from its complement:

INCORRECT   The tools we need are: a plane, a level, and a T-square.

CORRECT   The tools we need are a plane, a level, and a T-square.

CORRECT   We need three tools: a plane, a level, and a T-square.

Periods

Periods are used in the following instances.

1. **Use a period at the end of sentences that do not ask questions or express strong emotion:**

   The lateral stress still needs to be calculated.

2. **Use a period after some abbreviations:**

   M.D.

   U.S.A.

   etc.

   For more about abbreviations, see page 710.

3. **Use a period with decimal fractions:**

   4.056

   $6.75

   75.6 percent

Exclamation Points

The exclamation point is used at the end of a sentence that expresses strong emotion, such as surprise or doubt.

   The nuclear plant, which was originally expected to cost $1.6 billion, eventually cost more than $8 billion!

In technical documents, which require objectivity and a calm, understated tone, exclamation points are rarely used.

Question Marks

The question mark is used at the end of a sentence that asks a direct question.

What did the commission say about effluents?

NOTE: When a question mark is used within quotation marks, no other end punctuation is required.

She asked, "What did the commission say about effluents?"

MISUSE OF QUESTION MARKS

Do not use a question mark at the end of a sentence that asks an indirect question.

He wanted to know whether the procedure had been approved for use.

Dashes

To make a dash, use two uninterrupted hyphens (—). Do not add spaces before or after the dash. Some word-processing programs turn two hyphens into a dash, but with most, you have to use a special character to make a dash; there is no dash key on the keyboard.

Dashes are used in the following instances.

1. **Use a dash to set off a sudden change in thought or tone:**

   The committee found—can you believe this?—that the company bore full responsibility for the accident.

   That's what she said—if I remember correctly.

2. **Use a dash to emphasize a parenthetical element:**

   The managers' reports—all 10 of them—recommend production cutbacks for the coming year.

   Arlene Kregman—the first woman elected to the board of directors—is the next scheduled speaker.

3. **Use a dash to set off an introductory series from its explanation:**

   Wetsuits, weight belts, tanks—everything will have to be shipped in.

   NOTE: When a series follows the general statement, a colon replaces the dash.

   Everything will have to be shipped in: wetsuits, weight belts, and tanks.

MISUSES OF DASHES

Sometimes writers incorrectly use a dash as a substitute for other punctuation marks:

INCORRECT	The regulations—which were issued yesterday—had been anticipated for months.
CORRECT	The regulations, which were issued yesterday, had been anticipated for months.
INCORRECT	Many candidates applied—however, only one was chosen.
CORRECT	Many candidates applied; however, only one was chosen.

Parentheses

Parentheses are used in the following instances.

1. Use parentheses to set off incidental information:

Please call me (x3104) when you get the information.

Galileo (1546–1642) is often considered the father of modern astronomy.

The cure rate for lung cancer has almost doubled in the last thirty years (Capron 1999).

2. Use parentheses to enclose numbers and letters that label items listed in a sentence:

To transfer a call within the office, (1) place the party on HOLD, (2) press TRANSFER, (3) press the extension number, and (4) hang up.

Use both a left and a right parenthesis—not just a right parenthesis—in this situation.

MISUSE OF PARENTHESES

Sometimes writers incorrectly use parentheses instead of brackets to enclose their insertion within a quotation.

INCORRECT	He said, "The new manager (Farnham) is due in next week."
CORRECT	He said, "The new manager [Farnham] is due in next week."

For more about square brackets, see page 705.

Apostrophes

Apostrophes are used in the following instances.

1. Use an apostrophe to indicate possession:

the manager's goals	the employee's credit union
the workers' lounge	Charles's T-square

For joint possession, add an apostrophe and an *s* only to the last noun or proper noun:

Watson and Crick's discovery

For separate possession, add an apostrophe and an *s* to each of the nouns or pronouns:

Newton's and Galileo's theories

NOTE: Do not add an apostrophe or an *s* to possessive pronouns: *his, hers, its, ours, yours, theirs.*

2. **Use an apostrophe to indicate possession when a noun modifies a gerund:**

We were all looking forward to Bill's joining the company.

The gerund *joining* is modified by the proper noun *Bill.*

3. **Use an apostrophe to form contractions:**

I've　　　shouldn't

can't　　　it's

The apostrophe usually indicates an omitted letter or letters:

can(no)t = can't

it (i)s = it's

NOTE: Some organizations discourage the use of contractions; others have no preference. Find out the policy your organization follows.

4. **Use an apostrophe to indicate special plurals:**

three 9's

two different JCL's

the why's and how's of the problem

NOTE: For plurals of numbers and abbreviations, some style guides omit the apostrophe: *9s, JCLs.* Because usage varies considerably, check with your organization.

MISUSE OF APOSTROPHES

Writers sometimes incorrectly use the contraction *it's* in place of the possessive pronoun *its.*

INCORRECT　　The company does not feel that the problem is it's responsibility.

CORRECT　　The company does not feel that the problem is its responsibility.

Quotation Marks

" "

Quotation marks are used in the following instances.

1. **Use quotation marks to indicate titles of short works, such as articles, essays, or chapters:**

   Smith's essay "Solar Heating Alternatives" was short but informative.

2. **Use quotation marks to call attention to a word or phrase used in an unusual way or in an unusual context:**

   A proposal is "wired" if the sponsoring agency has already decided who will be granted the contract.

   Do not use quotation marks as a means of excusing poor word choice:

   The new director has been a real "pain."

3. **Use quotation marks to indicate a direct quotation.**

For more about quoting sources, see Ch. 7, p. 162.

   "In the future," he said, "check with me before authorizing any large purchases."

   As Breyer wrote, "Morale *is* productivity."

   NOTE: Quotation marks are not used with indirect quotations.

INCORRECT	He said that "third-quarter profits will be up."
CORRECT	He said that third-quarter profits will be up.
CORRECT	He said, "Third-quarter profits will be up."

   Also note that quotation marks are not used with quotations that are longer than four lines; instead, set the quotation in block format. In a word-processed manuscript, a block quotation is usually

   - indented one-half inch from the left-hand margin
   - typed without quotation marks
   - introduced by a complete sentence followed by a colon

   Different style manuals recommend variations on these basic rules; the following example illustrates APA style.

   McFarland (1997) writes:

   > The extent to which organisms adapt to their environment is still being charted. Many animals, we have recently learned, respond to a dry winter with an automatic birth control chemical that limits the number of young to be born that spring. This prevents mass starvation among the species in that locale (p. 49).

Hollins (1999) concurs. She writes, "Biological adaptation will be a major research area during the next decade" (p. 2).

USING QUOTATION MARKS WITH OTHER PUNCTUATION

- If the sentence contains a *tag*—a phrase identifying the speaker or writer—a comma separates it from the quotation:

  Wilson replied, "I'll try to fly out there tomorrow."

  "I'll try to fly out there tomorrow," Wilson replied.

  Informal and brief quotations require no punctuation before a quotation mark:

  She asked herself "Why?" several times a day.

- In the United States (unlike most other English-speaking nations), commas and periods at the end of quotations are placed within the quotation marks:

  The project engineer reported, "A new factor has been added."

  "A new factor has been added," the project engineer reported.

- Question marks, dashes, and exclamation points are placed inside quotation marks when they are part of the quoted material:

  He asked, "Did the shipment come in yet?"

  When question marks, dashes, and exclamation points apply to the whole sentence, they are placed outside the quotation marks:

  Did he say, "This is the limit"?

- When a punctuation mark appears inside a quotation mark at the end of a sentence, do not add another punctuation mark.

INCORRECT	Did she say, "What time is it?"?
CORRECT	Did she say, "What time is it?"

Ellipses

Ellipses (three spaced periods) indicate the omission of material from a direct quotation.

SOURCE	My team will need three extra months for market research and quality assurance testing to successfully complete the job.
QUOTE	She responded, "My team will need three extra months . . . to successfully complete the job.

Insert an ellipsis after a period if you are omitting entire sentences that follow:

> Larkin refers to the project as "an attempt . . . to clarify the issue of compulsory arbitration. . . . We do not foresee an end to the legal wrangling . . . but perhaps the report can serve as a definition of the areas of contention."

The writer has omitted words from the source after *attempt* and after *wrangling*. After *arbitration*, the writer has inserted an ellipsis after a period to indicate that a sentence has been omitted.

NOTE: MLA style recommends that writers insert brackets around an ellipsis that is introduced in a quotation.

> Larkin refers to the project as "an attempt [. . .] to clarify the issue of compulsory arbitration."

Square Brackets

[ ]

Square brackets are used in the following instances.

1. **Use square brackets around words added to a quotation:**

   > As noted in the minutes of the meeting, "He [Pearson] spoke out against the proposal."

   A better approach would be to shorten the quotation:

   > The minutes of the meeting note that Pearson "spoke out against the proposal."

2. **Use square brackets to indicate parenthetical information within parentheses:**

   > (For further information, see Charles Houghton's *Civil Engineering Today* [1997].)

MECHANICS

Italics

ital

Although italics are generally preferred, you may use underlining in place of italics. Whichever method you choose, be consistent throughout your document. Italics (or underlining) are used in the following instances.

 Click on the link to Exercise Central on TechComm Web (www.bedfordstmartins.com /techcomm) for online exercises covering mechanics.

1. **Use italics for words used as words:**

   > In this report, the word *operator* will refer to any individual who is actually in charge of the equipment, regardless of that individual's certification.

2. **Use italics to indicate titles of long works (books, manuals, and so on), periodicals and newspapers, long films, long plays, and long musical works:**

See Houghton's *Civil Engineering Today.*

We subscribe to the *Wall Street Journal.*

Note that *the* is not italicized or capitalized when the title is used in a sentence.

NOTE: The MLA style guide recommends that the names of Web sites be italicized.

The Library of Congress maintains *Thomas*, an excellent site for legislative information.

3. **Use italics to indicate the names of ships, trains, and airplanes:**

The shipment is expected to arrive next week on the *Penguin.*

4. **Use italics to set off foreign expressions that have not become fully assimilated into English:**

The speaker was guilty of *ad hominem* arguments.

Check a dictionary to determine whether a foreign expression has become assimilated.

5. **Use italics to emphasize words or phrases:**

Do not press the red button.

Angle Brackets

Many style guides now advocate using angle brackets around URLs in print documents to set them off from the text.

Our survey included a close look at three online news sites: *The New York Times on the Web* <http://www.nytimes.com>, *The Washington Post* <http://www.washingtonpost.com>, and *CNN Interactive* <http://www.cnn.com>.

You may want to check with your instructor or organization before following this recommendation.

Hyphens

Hyphens are used in the following instances.

1. **Use hyphens to form compound adjectives that precede nouns:**

general-purpose register

meat-eating dinosaur

chain-driven saw

NOTE: Hyphens are not used after adverbs that end in *-ly*.

newly acquired terminal

Also note that hyphens are not used when the compound adjective follows the noun:

The Woodchuck saw is chain driven.

Many organizations have their own preferences about hyphenating compound adjectives. Check to see if your organization has a preference.

For more about compound adjectives, see page 691.

2. **Use hyphens to form some compound nouns:**

once-over

go-between

NOTE: There is a trend away from hyphenating compound nouns (*vice president, photomicroscope, drawbridge*); check your dictionary for proper spelling.

3. **Use hyphens to form fractions and compound numbers:**

one-half

fifty-six

4. **Use hyphens to attach some prefixes and suffixes:**

post-1945

president-elect

5. **Use hyphens to divide a word at the end of a line:**

We will meet in the pavilion in one hour.

Whenever possible, however, avoid such line breaks; they slow the reader down. When you do use them, check the dictionary to make sure you have divided the word between syllables. If you need to break a URL at the end of a line, do not add a hyphen. Instead, break it before a slash or a period:

http://www.bedfordstmartins.com
/techcomm

Numbers

num

Ways of handling numbers vary considerably. Therefore, in choosing between words and numerals, consult your organization's style guide. Many organizations observe the following guidelines.

1. **Technical quantities of any amount are expressed in numerals, especially if a unit of measurement is included:**

   3 feet   43,219 square miles

   12 grams   36 hectares

2. **Nontechnical quantities of fewer than 10 are expressed in words:**

   three persons

   six whales

3. **Nontechnical quantities of 10 or more are expressed in numerals:**

   300 persons

   12 whales

   35 percent increase

4. **Approximations are written out:**

   approximately ten thousand people

   about two million trees

5. **Round numbers over nine million are expressed in both words and numerals:**

   14 million light-years

   $64 billion

6. **Decimals are expressed in numerals:**

   3.14

   1,013.065

   Decimals of less than one should be preceded by a zero:

   0.146

   0.006

7. **Fractions are written out, unless they are linked to technical units:**

   two-thirds of the members

   3 1/2 hp

8. **Time of day is expressed in numerals if A.M. or P.M. is used; otherwise, it is written out:**

    6:10 A.M.

    six o'clock

    the nine-thirty train

9. **Page numbers and titles of figures and tables are expressed in numerals:**

    Figure 1

    Table 13

    page 261

10. **Back-to-back numbers are written using both words and numerals:**

    six 3-inch screws

    fourteen 12-foot ladders

    3,012 five-piece starter units

    In general, the technical unit should be expressed with the numeral. If the nontechnical quantity would be cumbersome in words, use the numeral for it instead.

11. **Numbers in legal contracts or in documents intended for international readers should be represented in both words and numerals:**

    thirty-seven thousand dollars ($37,000)

    five (5) relays

12. **Street addresses may require both words and numerals:**

    3801 Fifteenth Street

SPECIAL CASES

- A number at the beginning of a sentence should be spelled out:

    Thirty-seven acres was the size of the lot.

    Many writers would revise the sentence to avoid this problem:

    The lot was 37 acres.

- Within a sentence, the same unit of measurement should be expressed consistently in either numerals or words:

    INCORRECT   On Tuesday the attendance was 13; on Wednesday, eight.

    CORRECT   On Tuesday the attendance was 13; on Wednesday, 8.

    CORRECT   On Tuesday the attendance was thirteen; on Wednesday, eight.

- In general, months should not be expressed as numbers. In the United States, 3/7/99 means March 7, 1999; in many other countries, it means July 3, 1999. The following forms, in which the months are written out, are preferable:

March 7, 1999

7 March 1999

Abbreviations

abbr

Abbreviations save time and space, but you should use them carefully because your readers may not understand them. Many companies and professional organizations provide lists of approved abbreviations.

Analyze your audience to determine whether and how to abbreviate. If your readers include a general audience unfamiliar with your field, either write out the technical terms or attach a list of abbreviations. If you are new to an organization or are publishing in a field for the first time, find out which abbreviations are commonly used. If for any reason you are unsure about a term, write it out.

The following are general guidelines about abbreviations:

1. **When an unfamiliar abbreviation is introduced for the first time, the full term should be given, followed by the abbreviation in parentheses. In subsequent references, the abbreviation may be used alone. For long works, the full term and its abbreviation may be written out at the start of major units, such as chapters.**

   The heart of the new system is the self-loading cartridge (slc).

   The cathode-ray tube (CRT) is your control center.

2. **To form the plural of an abbreviation, an *s* is added, either with or without an apostrophe, depending on the style of your organization:**

   GNPs

   PhD's

   Most unit-of-measurement abbreviations do not take plurals:

   10 in

   3 qt

3. **Most abbreviations in scientific writing are not followed by periods:**

   lb

   cos

   dc

If the abbreviation can be confused with another word, however, a period should be used:

in.

Fig.

4. **If no number is used with a measurement, an abbreviation should not be used.**

INCORRECT   How many sq meters is the site?

CORRECT   How many square meters is the site?

Capitalization

cap

For the most part, the conventions of capitalization in general writing apply in technical communication.

1. **Proper nouns, titles, trade names, places, languages, religions, and organizations should be capitalized:**

William Rusham

Director of Personnel

Quick-Fix Erasers

Bethesda, Maryland

Italian

Methodism

Society for Technical Communication

In some organizations, job titles are not capitalized unless they refer to specific persons.

Alfred Loggins, Director of Personnel, is interested in being considered for vice president of marketing.

2. **Headings and labels should be capitalized:**

A Proposal to Implement the Wilkins Conversion System

Mitosis

Table 3

Section One

The Problem

Rate of Inflation, 1995–2005

Figure 6

GUIDELINES FOR SPEAKERS OF ENGLISH AS A SECOND LANGUAGE

Basic Characteristics of a Sentence

sent

A sentence has five characteristics.

1. **It starts with an uppercase letter and ends with a period, a question mark, or (rarely) an exclamation point.**

2. **It has a subject, usually a noun. The subject is what the sentence is about.**

3. **It has a verb, which tells the reader what happens to the subject or what the subject does.**

4. **It has a standard word order.**

   The most common sequence in English is subject-verb-object:

   subject   verb   object

   We hired a consulting firm.

   You can add information to the start of the sentence:

   Yesterday we hired a consulting firm.

   or to the end of the sentence:

   Yesterday we hired a consulting firm: *Sanderson & Associates*.

   or in the middle:

   Yesterday we signed a *nontransferable contract* with a consulting firm: Sanderson & Associates.

   In fact, any element of a sentence can be expanded.

5. **It has an independent clause (a subject and verb that can stand alone).**

   The following is a sentence because it can stand alone:

   The pump failed because of improper maintenance.

   The following is also a sentence:

   The pump failed.

   But the following is a phrase, not a sentence, because the thought is incomplete:

Because of improper maintenance.

An independent clause is required to complete this sentence:

Because of improper maintenance, the pump failed.

Linking Ideas by Coordination

coor

One way to connect ideas in a sentence is by coordination. Coordination means that ideas in the sentence are roughly equal in importance. There are three main ways to coordinate ideas.

1. **Use a semicolon (;) to coordinate ideas that are independent clauses:**

   The information for bid was published last week; the proposal is due in less than a month.

2. **Use a comma and a coordinating conjunction (*and, but, or, nor, so, for,* and *yet*) to coordinate two independent clauses:**

   The information for bid was published last week, but the proposal is due in less than a month.

   In this example, *but* clarifies the relationship between the two clauses: the writer hasn't been given enough time to write the proposal.

3. **Use transitional words and phrases to coordinate two independent clauses:**

   The Pentium II chip has already been replaced; *as a result*, it is hard to find a Pentium II in a new computer.

   The Pentium II chip has already been replaced. *As a result*, it is hard to find a Pentium II in a new computer.

For more about transitional words and phrases, see Ch. 10, p. 260.

Linking Ideas by Subordination

sub

Two ideas can also be linked by subordination, that is, by deemphasizing one of them. There are two basic ways to use subordination.

1. **Use a subordinating word or phrase to turn one idea into a subordinate clause.**

   Subordinating words and phrases:

after	as	because	even though	since
although	as if	before	if	so that

Subordinating words and phrases, continued:

that	until	where	while	whom
unless	when	which	who	whose

Start with two independent clauses:

The bridge was completed last year. The bridge already needs repairs.

Then choose a subordinating word and combine the clauses:

Although the bridge was completed last year, it already needs repairs.

Although subordinates the first clause, leaving *it already needs repairs* as the independent clause.

Note that the order of the ideas could be reversed:

The bridge already needs repairs *even though* it was completed last year.

> For more about restrictive and nonrestrictive modifiers, see Ch. 11, p. 282.

Another way to subordinate one idea is to turn it into a nonrestrictive clause using the subordinate word *which*:

The bridge, which was completed last year, already needs repairs.

This version deemphasizes *was completed last year* by turning it into a nonrestrictive clause and emphasizes *already needs repairs* by leaving it as the independent clause.

2. Turn one of the ideas into a phrase modifying the other.

Completed last year, the bridge already needs repairs.

Completed last year was turned into a phrase by dropping the subject and verb from the independent clause. Here the phrase is used to modify *the bridge*.

Verb Tenses

vb

Verb tenses in English can be complicated, but in general there are four kinds of time relationships that you need to understand.

1. Simple past, present, and future

PAST — Yesterday we *subscribed* to a new ecology journal.

PRESENT — We *subscribe* to three ecology journals every year.

Meaning: We regularly subscribe to the three journals.

FUTURE — We *will subscribe* to the new ecology journal next year.

Or: We *are going to subscribe* to the new ecology journal next year.

2. **An action in progress at a known time (progressive)**

PAST PROGRESSIVE

We *were updating* our directory when the power failure occurred.

PRESENT PROGRESSIVE

We *are updating* our directory now.

FUTURE PROGRESSIVE

We *will be updating* our directory tomorrow when you arrive.

3. **An action completed before a known time (perfect)**

PAST PERFECT   We *had* already *started* to write the proposal when we got your call.

The writing began before the call.

PRESENT PERFECT   We *have started* to write the proposal.

FUTURE PERFECT   We *will have started* to write the proposal by the time you arrive.

Both events are in the future, but in both cases, the writing begins before the arrival.

4. **An action in progress until a known time (perfect progressive)**

PAST PERFECT PROGRESSIVE

We *had been working* on the reorganization when the news of the merger was publicized.

The work was in progress when the news of the merger was publicized.

PRESENT PERFECT PROGRESSIVE

We *have been working* on the reorganization for over a year.

The work will continue into the future.

FUTURE PERFECT PROGRESSIVE

We *will have been working* on the reorganization for two years by the time it occurs.

Forming Verbs with *-ing*

-ing

English uses the *-ing* form of verbs in three major ways.

1. **As part of a progressive or perfect progressive verb (see numbers 2 and 4 above):**

We are *shipping* the materials by UPS.

We have been *waiting* for approval since January.

2. **As a present participle, which functions as an adjective either by itself:**

the *leaking* pipe

or as part of a participial phrase:

Analyzing the sample, we discovered two anomalies.
The sample *containing* the anomalies appears on Slide 14.

3. **As a gerund, which functions as a noun either by itself:**

Writing is the best way to learn to write.

or as part of a gerund phrase:

The designer tried *inserting* the graphics by hand.

Infinitives

inf

Infinitives consist of the word *to* plus the base form of the verb (*to write, to understand*). An infinitive can be used in three main ways.

1. **As a noun:**

The editor's goal for the next year is *to publish* the journal on schedule.

2. **As an adjective:**

The company requested the right *to subcontract* the project.

3. **As an adverb:**

We established the schedule ahead of time *to prevent* the kind of mistake we made last time.

Helping Verbs and Main Verbs

help

There are a number of helping verbs in English. The following discussion explains four categories of helping verbs.

1. **Modals**

There are nine modal verbs: *can, could, may, might, must, shall, should, will,* and *would*. After a modal verb, use the base form of the verb (the form of the verb used after *to* in the infinitive).

base form
↓
The system *must meet* all applicable codes.

2. Forms of *do*

After a helping verb that is a form of *do*—*do, does,* and *did*—use the base form of the verb.

base form
↓

Do we *need* to include the figures for the recovery rate?

3. Forms of *have* plus the past participle

To form one of the perfect tenses (past, present, or future), use a form of *have* plus the past participle of the verb (usually the -*ed* or -*en* form of the verb).

PAST PERFECT

We *had written* the proposal before learning of the new RFP.

PRESENT PERFECT

We *have written* the proposal according to the instructions in the RFP.

FUTURE PERFECT

We *will have written* the proposal by the end of the week.

4. Forms of *be*

To describe an action in progress, use a form of *be* (*be, am, is, are, was, were, being, been*) and the present participle (the -*ing* form of the verb).

For more about active and passive voice, see Ch. 11, p. 286.

We *are testing* the new graphics tablet.

The company *is considering* flextime.

To create the passive voice, use a form of *be* and the past participle.

The piping *was installed* by the plumbing contractor.

Agreement of Subject and Verb

ESL agr s/v

The subject and the verb in a clause or sentence must agree in number:

The new *valve is* installed according to the manufacturer's specifications.

The new *valves are* installed according to the manufacturer's specifications.

When you edit your document for subject-verb agreement, keep in mind the following guidelines.

1. Make sure the subject and verb agree when information comes between the subject and the verb.

The *result* of the tests *is* included in Appendix C.

The *results* of the test *are* included in Appendix C.

2. **Certain pronouns and quantifiers always require singular verbs. Pronouns that end in -*body* or -*one*—such as *everyone, everybody, someone, somebody, anyone, anybody, no one,* and *nobody*—are singular. In addition, quantifiers such as *something, each,* and *every* are singular.**

SINGULAR   *Everybody is* invited to the preproposal meeting.

SINGULAR   *Each* of the members *is* asked to submit billable hours by the end of the month.

3. **When the clause or sentence contains a compound subject, the verb must be plural.**

COMPOUND
SUBJECT   *The contractor and the subcontractor want* to meet to resolve the difficulties.

4. **When a relative pronoun such as *who, that,* or *which* begins a clause, make sure the verb agrees in number with the noun that the relative pronoun refers to.**

The *numbers* that *are* used in the formula do not agree with the ones we were given at the site.

Numbers is plural, so the verb in the *that* clause (*are*) is also plural.

The *number* that *is* used in the formula does not agree with the one we were given at the site.

Number is singular, so the verb in the *that* clause (*is*) is also singular.

Conditions

cond

Four main types of condition are used with the word *if* in English.

1. **Conditions of fact**

   For conditions of fact, use the same verb tense in both clauses. In most cases, use a form of the present tense:

   If you *see* "Unrecoverable Application Error," the program *has crashed.*

   If rats *are* allowed to eat as much as they want, they *become* obese.

2. **Future prediction**

   For prediction, use the present tense in the *if* clause. Use a modal (*can, could, may, might, must, shall, should, will,* or *would*) plus the base form of the verb in the independent clause.

   If we *win* this contract, we *will have* to add three more engineers.

   If this weather *keeps* up, we *may need* to postpone the launch.

3. **Present-future speculation**

The present-future speculation usage suggests a condition contrary to fact. Use *were* in the *if* clause. Use a modal plus the base form of the verb in the independent clause.

If I *were* president of the company, I *would be* much more aggressive.

This sentence implies that you are not president of the company.

4. **Past speculation**

Use *had* plus the past participle in the *if* clause. Use a modal plus *have* in the independent clause.

If we *had won* this contract, we *would have* had to add three engineers.

This sentence implies that the condition is contrary to fact: the contract wasn't won, so the engineers were not added.

Articles

art

Few aspects of English can be as frustrating to the nonnative speaker of English as the correct usage of the simple words *a, an,* and *the.* Although there are a few rules that you should try to learn, remember that there are many exceptions and special cases. Here are three general guidelines.

1. **Singular proper nouns—those that name specific persons, places, and things—do not usually take an article:**

Taiwan

James Allenby

But plural proper nouns often do take the article *the:*

the United States

the Allenbys

2. **Countable common nouns (persons, places, or things that can be counted) take an article:**

the microscope

a desk

an electron

Uncountable common nouns generally do not take an article:

overtime   integrity   information

equipment   research

3. **Common nouns can be referred to as either specific or nonspecific. The specific form takes *the*; the nonspecific form takes either *a* or *an*.**

Our department received funding for *an experiment*. The experiment will take six months to complete.

In the first sentence in this example, *experiment* is a nonspecific noun; in the second sentence, *experiment* is a specific noun because it has been identified in the previous sentence.

Adjectives

adj

Keep in mind three main points about adjectives in English.

1. **Adjectives do not take a plural form.**

   a complex project

   two complex projects

2. **Adjectives can be placed either before the nouns they modify or later in the sentence.**

   The *critical* need is to reduce the drag coefficient.

   The need to reduce the drag coefficient is *critical*.

3. **Adjectives of one or two syllables take special endings to create the comparative and superlative forms.**

Positive	Comparative	Superlative
big	bigger	biggest
heavy	heavier	heaviest

   Adjectives of three or more syllables take the word *more* for the comparative form and the words *the most* for the superlative form.

Positive	Comparative	Superlative
qualified	more qualified	the most qualified
feasible	more feasible	the most feasible

Adverbs

adv

Like adjectives, adverbs are modifiers, but their placement in the sentence is somewhat more complex. Remember five points about adverbs.

1. **Adverbs can modify verbs.**

   Management terminated the project *reluctantly*.

2. **Adverbs can modify adjectives.**

   The executive summary was *conspicuously* absent.

3. **Adverbs can modify other adverbs.**

   The project is going *very* well.

4. **Adverbs that describe how an action takes place can appear in different locations in the sentence.**

   Carefully the inspector examined the welds.

   The inspector *carefully* examined the welds.

   The inspector examined the welds *carefully*.

   NOTE: The adverb should not be placed between the verb and the direct object.

   INCORRECT   The inspector examined *carefully* the welds.

5. **Adverbs that describe the whole sentence can also be placed in different locations in the sentence.**

   Apparently, the inspection was successful.

   The inspection was *apparently* successful.

   The inspection was successful, *apparently*.

Omitted Words

omit

Except for imperative sentences, in which the subject *you* is understood (*Get the correct figures*), all sentences in English require a subject.

   The company has a policy on conflict of interest.

Do not omit the expletives *there* or *it*.

For more about expletives, see Ch. 11, p. 279.

   INCORRECT   Are three reasons for us to pursue this issue.

   CORRECT   *There* are three reasons for us to pursue this issue.

   INCORRECT   Is important that we seek his advice.

   CORRECT   *It* is important that we seek his advice.

Repeated Words

rep

1. **Do not repeat the subject of a sentence.**

   INCORRECT   The company we are buying from *it* does not permit us to change our order.

   CORRECT   The company we are buying from does not permit us to change our order.

2. **In an adjective clause, do not repeat an object.**

   INCORRECT   The technical communicator does not use the same software that we were writing in *it*.

   CORRECT   The technical communicator does not use the same software that we were writing in.

3. **In an adjective clause, do not use a second adverb.**

   INCORRECT   The lab where we did the testing *there* is an excellent facility.

   CORRECT   The lab where we did the testing is an excellent facility.

PART C: COMMONLY MISUSED WORDS

The list below explains the correct usage of words commonly confused and misused. Most examples include a sentence showing the words being defined in italics.

accept, except   *Accept* means to receive, while *except* means excluding or to exclude. "We will not *accept* delivery of any items *except* those we have ordered."

adapt, adopt   *Adapt* means to adjust or to modify; *adopt* means to accept. "Management decided to *adapt* the quality-circle plan rather than *adopt* it as is."

affect, effect   *Affect* is a verb: "How will the news *affect* him?" *Effect* is most commonly a noun: "What will be the *effect* of the increase in allowable limits?" *Effect* is also (rarely) a verb meaning to bring about or cause to happen: "The new plant is expected to *effect* a change in our marketing strategy."

already, all ready   "The report had *already* been sent to the printer when the writer discovered that it was not *all ready*."

alright, all right   *Alright* is a misspelling of *all right* and should not be used.

among, between   In general, *among* is used for relationships of more than two items, *between* for only two items. "The collaboration *among* the writer, the illustrator, and the printer," but "the agreement *between* the two companies."

amount, number   *Amount* is used for noncounting items; *number* refers to counting items: "the *amount* of concrete," but "the *number* of bags of concrete."

assure, ensure, insure   To *assure* means to put someone's mind at ease: "let me *assure* you." To *ensure* and to *insure* both mean to make sure: "the new plan will *ensure* [or *insure*] good results." Some writers prefer to use *insure* only when referring to insurance: "to *insure* against fire loss."

can, may, might   *Can* refers to ability: "We *can* produce 300 chips per hour." *May* refers to permission: "*May* I telephone your references?" *Might* refers to possibility: "We *might* see declines in prices this year."

compliment, complement   A *compliment* is a statement of praise: "The owner offered a gracious *compliment* to the architect on his design." *Compliment* is also a verb: "The owner graciously *complimented* the architect." A *complement* is something that fills something up or makes it complete, or something that is an appropriate counterpart: "The design is a perfect *complement* to the landscape." *Complement* is also a verb: "The design *complements* the landscape perfectly."

could of   This is not a correct phrase; it is a corruption of *could've*, the contraction of *could have:* "She *could have* mentioned the abrasion problem in the report."

criteria, criterion   *Criteria,* meaning standards against which something will be measured, is plural; *criterion* is singular.

data, datum   *Data* is plural; *datum* is singular. This distinction is fading in popular usage, although not in some scientific and engineering applications. Check to see how it is spelled in your company or field.

discreet, discrete   *Discreet* means careful and prudent: "She is a very *discreet* manager; you can confide in her." *Discrete* means separate or distinct: "The company will soon split into three *discrete* divisions."

effective, efficient   *Effective* means that the item does what it is meant to do; *efficient* also carries the sense of accomplishing the goal without using more resources than necessary. "Air Force One is an *effective* way to transport the president, but it is not *efficient*; it costs some $40,000 per hour to fly."

farther, further   *Farther* refers to distance: "one mile *farther* down the road." *Further* means greater in quantity, time, or extent: "Are there any *further* questions?"

feedback   Many writers will not use *feedback* to refer to a response by a person: "Let me have your *feedback* by Friday." They limit the term to its original meaning, dealing with electricity, because a human response involves thinking (or should, anyway).

fewer, less   *Fewer* is used for counting items: "*fewer* bags of cement"; *less* is used for noncounting items: "*less* cement." It's the same distinction as between *number* and *amount*.

foreword, forward   A *foreword* is a preface, usually written by someone other than the author, introducing a book. *Forward* refers to being in advance: "The company decided to move *forward* with the project."

i.e., e.g.   *I.e.*, Latin for *id est*, means *that is*. *E.g.*, Latin for *exempli gratia*, means *for example*. Writers often confuse them. That's why I recommend using the English versions. Also, add commas around them: "Use the main entrance, *that is*, the one on Broadway."

imply, infer   The writer or speaker *implies*; the reader or listener *infers*.

input   People who don't like to give their *feedback* also don't like to offer their *input*.

its, it's   *Its* is the possessive pronoun: "The lab rat can't make up *its* mind." *It's* is the contraction of *it is:* "*It's* too late to apply for this year's grant." Why do people mix up these two words? Because they remember learning that possessives take apostrophes — "*Bob's* computer" — when they use the possessive form of *it*, they add the apostrophe. But *its* is a possessive pronoun — like *his, hers, theirs, ours,* and *yours* — a word specifically created to indicate possession; it does not take an apostrophe.

-ize   Many legitimate words end in *-ize*, such as *harmonize* and *sterilize*, but many writers and readers can't stand new ones (such as *prioritize*) when there are perfectly fine words already (such as *rank*).

lay, lie   *Lay* is a transitive verb meaning *to place:* "*Lay* the equipment on the table." *Lie* is an intransitive verb meaning *to recline:* "*Lie* down on the couch." The complete conjugation of *lay* is *lay, laid, laid, laying*; of *lie*, it is *lie, lay, lain, lying*.

lead, led   *Lead* is the infinitive verb: "We want to *lead* the industry." *Led* is the past tense: "Last year we *led* the industry."

parameter   This is a mathematical term referring to a constant whose value can vary according to its application. Many writers object to the nonmathematical uses of the term, including such concepts as *perimeter, scope, outline,* and *limit.* (You guessed it: the same people who don't provide *input* or *feedback* don't use *parameter* very much either.)

phenomena, phenomenon   *Phenomena* is plural; *phenomenon* is singular.

plain, plane   *Plain* means simple and unadorned: "The new company created a very *plain* logo." *Plane* has several meanings: an airplane, the act of smoothing a surface, the tool used to smooth a surface, and the flat surface itself.

precede, proceed   *Precede* means to come before: "Should Figure 1 *precede* Figure 2?" *Proceed* means to move forward: "We decided to *proceed* with the project despite the setback."

shall, will   *Shall* suggests a legal obligation, particularly in a formal specification or contract: "The contractor *shall* remove all existing debris." *Will* suggests intent; it does not suggest a legal obligation: "We *will* get in touch with you as soon as possible to schedule the job."

sight, site, cite   *Sight* refers to vision; *site* is a place; *cite* is a verb meaning to document a reference.

than, then   *Than* is a conjunction used in comparisons: "Plan A works better *than* Plan B." *Then* is an adverb referring to time: "First we went to the warehouse. *Then* we went to the plant."

their, there, they're   *Their* is the possessive pronoun: "They brought *their* equipment with them." *There* refers to a place—"We went *there* yesterday"—or in expletive expressions—"*There* are three problems we have to solve." *They're* is the contraction of *they are.*

to, too, two   *To* is used in infinitive verbs ("*to* buy a new microscope") and in expressions referring to direction ("go *to* Detroit"). *Too* means excessively: "The refrigerator is *too* big for the lab." *Two* is the number 2.

viable   This is a fine Latin word meaning able to sustain life: "*viable* cell culture" and "*viable* fetus." Many writers avoid such clichés as *viable alternative* (while they're avoiding *input* and *feedback*).

weather, whether   *Weather* refers to sunshine and temperature. *Whether* refers to alternatives. "The demonstration will be held outdoors *whether* or not the *weather* cooperates."

who's, whose   *Who's* is the contraction of *who is. Whose* is the possessive case of *who:* "*Whose* printer are we using?"

-wise   Don't say, "*RAMwise,* the computer has 128 MB." Instead, say "The computer has 128 MB of RAM."

Xerox   The people at Xerox become unhappy when writers ask for a *xerox* copy. The correct word is *photocopy; Xerox* is a copyrighted term.

your, you're   *Your* is the possessive pronoun: "Bring *your* calculator to the meeting." *You're* is the contraction of *you are.*

Mark In Margin	Instructions	Mark on Manuscript	Corrected Type
e	Delete	$10 billion dollars	$10 billion
∧	Insert	enviroment	environment
(stet)	Let stand	let it stand	let it stand
(cap)	Capitalize	the english language	the English language
(lc)	Make lowercase	the English Language	the English language
—	Italicize	Technical Communication	*Technical Communication*
(tr)	Transpose	recieve	receive
⌒	Close up space	electric lawn mower	electric lawnmower
(sp)	Spell out	Pres Smithers	President Smithers
#	Insert space	3amp light	3 amp light
¶	Start paragraph	. . . the results. These results	. . . the results. These results
run in	No paragraph	. . . the results. For this reason,	. . . the results. For this reason,
(sc)	Set in small capitals	Needle-nosed pliers	NEEDLE-NOSED PLIERS
(bf)	Set in boldface	Needle-nosed pliers	**Needle-nosed pliers**
⊙	Insert period	Fig 21	Fig. 21
⋏	Insert comma	the plant which was built	the plant, which was built
=	Insert hyphen	menu driven software	menu-driven software
⊙	Insert colon	Add the following ⊙	Add the following:
⋎	Insert semicolon	. . . the plan however the committee	. . . the plan; however the committee
⋎	Insert apostrophe	the users preference	the user's preference
⋎/⋎	Insert quotation marks	Furthermore, she said . . .	"Furthermore," she said . . .
(/)	Insert parenthese	Write to us at the Newark office	Write to us (at the Newark office)
[/]	Insert brackets	President John Smithers	President [John] Smithers
$\frac{1}{N}$	Insert en dash	1984 2001	1984–2001
$\frac{1}{M}$	Insert em dash	Our goal victory	Our goal—victory
⋎	Insert superscript	4,000 ft2	f,000 ft^2
⋏	Insert subscript	H2O	H$_2$O
//	Align	$123.05// $86.95 //	$123.05 $86.95
[	Move to the left	[PVC piping	PVC piping
]	Move to the right	PVC piping]	PVC piping
⌐	Move up	[PVC piping]	PVC piping
⌣	Move down	[PVC piping]	PVC piping

SELECTED BIBLIOGRAPHY

Technical Communication

Allen, O. J., & Deming, L. H. (1994). *Publications management: Essays for professional communicators*. Amityville, NY: Baywood.

Beer, D. F. (1992). *Writing and speaking in the technology professions: A practical guide*. New York: IEEE.

Blicq, R. S., & Moretto, L. A. (1995). *Writing reports to get results: Quick, effective results using the pyramid method*. Piscataway, NJ: IEEE.

Boiarsky, C. R., & Soven, M. K. (1995). *Writings from the workplace: Documents, models, cases*. Needham Heights, MA: Allyn & Bacon.

Brusaw, C. T., Alred, G. J., & Oliu, W. E. (1997). *Handbook of technical writing* (5th ed.). New York: St. Martin's.

Day, R. A. (1995). *Scientific English: A guide for scientists and other professionals* (2nd ed.). Phoenix, AZ: Oryx.

Dombrowski, P. M. (Ed.). (1994). *Humanistic aspects of technical communication*. Amityville, NY: Baywood.

Dragga, S., & Gong, G. (1989). *Editing: The design of rhetoric*. Amityville, NY: Baywood.

Hoft, N. L. (1995). *International technical communication: How to export information about high technology*. New York: Wiley.

Markel, M. (1994). *Writing in the technical fields: A step-by-step guide for engineers, scientists, and technicians*. New York: IEEE.

Mathes, J. C., & Stevenson, D. W. (1991). *Designing technical reports: Writing for audiences in organizations* (2nd ed.). Indianapolis, IN: Bobbs-Merrill.

Pickett, N. A., & Laster, A. A. (1996). *Technical English: Writing, reading, and speaking* (7th ed.). New York: HarperCollins.

Sides, C. H. (1999). *How to write and present technical information* (3rd ed.). Phoenix: Oryx.

Also see the following journals:
IEEE Transactions on Professional Communication
Journal of Business and Technical Communication
Journal of Technical Writing and Communication
Technical Communication
Technical Communication Quarterly

Ethics

Beauchamp, T. L., & Bowie, N. E. (1997). *Ethical theory and business* (5th ed.). Englewood Cliffs, NJ: Prentice-Hall.

Brockmann, R. J., & Rook, F. (Eds.). (1989). *Technical communication and ethics*. Washington, DC: Society for Technical Communication.

Galler, B. A. (1995). *Software and intellectual property protection.* Westport, CT: Quorum.

Markel, M. (2001). *Ethics and technical communication: A synthesis and critique.* Westport, CT: Ablex.

Velasquez, M. G. (1998). *Business ethics: Concepts and cases* (4th ed.). Englewood Cliffs, NJ: Prentice-Hall.

Collaborative Writing

Blyler, N. R., & Thralls, C. (Eds.). (1993). *Professional communication: The social perspective.* Newbury Park, NY: Sage.

Cross, G. A. (1993). *Collaboration and conflict: A contextual exploration of group writing and positive emphasis.* Cresskill, NJ: Hampton.

Ede, L., & Lunsford, A. (1990). *Singular texts/plural authors: Perspectives on collaborative writing.* Carbondale, IL: Southern Illinois University.

Forman, J. (Ed.). (1992). *New visions of collaborative writing.* Portsmouth, NH: Boynton/Cook.

Lay, M. M., and Karis, W. M. (1991). *Collaborative writing in industry: Investigations in theory and practice.* Amityville, NY: Baywood.

Research Techniques

Berkman, R. I. (1995). *Find it online!* New York: Windcrest/McGraw-Hill.

Berkman, R. I. (1997). *Find it fast: How to uncover expert information on any subject* (4th ed.). New York: HarperPerennial.

Levin, J. (1995). *The federal Internet source* (2nd ed.). Washington, DC: National Journal Inc. and NetWeek L.L.C.

Mount, E., & Kovacs, B. (1991). *Using science and technology information sources.* Phoenix, AZ: Oryx.

Zimmerman, D. E., & Muraski, M. L. (1995). *The elements of information gathering: A guide for technical communicators, scientists, and engineers.* Phoenix, AZ: Oryx.

Also see the following journal:

How to access the federal government on the Internet. Washington, DC: Congressional Quarterly.

Usage and General Writing

Burchfield, R. W., & Fowler, H. W. (Eds.). (1996). *The new Fowler's dictionary of modern English usage* (3rd ed.). New York: Oxford University Press.

Bush, D. W., & Campbell, C. P. (1994). *How to edit technical documents.* Phoenix, AZ: Oryx.

Corbett, E. P. J. (1999). *Classical rhetoric for the modern student* (4th ed.). New York: Oxford University Press.

Maggio, R. (1997). *Talking about people: A guide to fair and accurate language.* Phoenix, AZ: Oryx.

Partridge, E. (1995). *Usage and abusage: A guide to good English* (New ed.) (J. Whitcut, ed.). New York: Norton.

Strunk, W., & White, E. B. (1999). *The elements of style* (4th ed.). Boston: Allyn & Bacon.

Williams, J. (1999). *Style: Ten lessons in clarity and grace* (6th ed.). New York: Longman.

Handbooks for Grammar and Style

Hacker, D. (1998). *The Bedford handbook* (5th ed.). Boston: Bedford/St. Martin's.

Lunsford, A., & Connors, R. (1999). *The new St. Martin's handbook.* Boston: Bedford/St. Martin's.

Style Manuals

American National Standards, Inc. (1979). *American National Standard for the preparation of scientific papers for written or oral presentation* (ANSI Z39.16–1972). New York: American National Standards Institute.

The Chicago manual of style. (1993). (14th ed.). Chicago: University of Chicago.

Dodd, J. S. (Ed.). (1997). *The ACS style guide: A manual for authors and editors* (2nd ed.). Washington, DC: American Chemical Society.

Gibaldi, J. (1999). *MLA handbook for writers of research papers* (5th ed.). New York: Modern Language Association.

Li, X., & Crane, N. B. (1996). *Electronic style: A guide to citing electronic information* (Rev. ed.). Westport, CT: Meckler.

Nagle, J. (1995). *Handbook for preparing engineering documents: From concept to completion.* New York: IEEE.

Pollack, G. (1977). *Handbook for ASM editors.* Washington, DC: American Society for Microbiology.

Publication manual of the American Psychological Association (1994). (4th ed.). Washington, DC: American Psychological Association.

Rubens, P. (1994). *Science and technical writing: A manual of style.* New York: Henry Holt.

Style Manual Committee, Council of Biology Editors. (1994). *Scientific style and format: The CBE manual for authors, editors, and publishers* (6th ed.). Chicago: Cambridge University Press.

U.S. Government Printing Office style manual. (1988). (Rev. ed.). New York: Outlet.

Many private corporations, such as John Deere, DuPont, Ford Motor Company, General Electric, Microsoft, and Westinghouse, have their own style manuals.

Graphics, Design, and Multimedia

Baird, R. N., et al. (1992). *The graphics of communication: Methods, media, and technology* (6th ed.). Fort Worth, TX: Harcourt Brace Jovanovich.

Campbell, K. S. (1995). *Coherence, continuity, and cohesion: Theoretical foundations for document design.* Hillsdale, NJ: Erlbaum.

Cleveland, W. S. (1994). *Elements of graphing data.* Summit, NJ: Hobart.

Dillon, P. M., & Leonard, D. C. (1998). *Multimedia and the Web from a to z* (2nd ed.). Phoenix, AZ: Oryx.

Harris, R. L. (1999). *Information graphics: A comprehensive illustrated reference* (2nd ed.). Atlanta, GA: Management Graphics.

Horton, W. (1991a). *The icon book: Visual symbols for computer systems and documentation.* New York: Wiley.

Horton, W. (1991b). *Illustrating computer documentation: The art of presenting information graphically on paper and online.* New York: Wiley.

Horton, W., Taylor, L., Ignacio, A., & Hoft, N. L. (1996). *The Web page design cookbook.* New York: Wiley.

Johnson, S. (1996). *Electronic publishing construction kit*. New York: Wiley.

Jones, G. E. (1995). *How to lie with charts*. Alameda, CA: Sybex.

Kelvin, G. V. (1992). *Illustrating for science*. New York: Watson-Guptill.

Kosslyn, S. M. (1994). *Elements of graph design*. New York: W. H. Freeman.

Nielsen, J. (1995). *Multimedia and hypertext: The Internet and beyond*. Boston: AP Professional.

Parker, R. C., & Berry, P. (1998). *Looking good in print* (4th ed.). Albany, NY: Coriolis Group.

Shushan, R., & Wright, D. (1996). *Desktop publishing by design: Everyone's guide to PageMaker 6* (4th ed.). Redmond, WA: Microsoft.

Tufte, E. R. (1983). *The visual display of quantitative information*. Cheshire, CT: Graphics Press.

Tufte, E. R. (1990). *Envisioning information*. Cheshire, CT: Graphics Press.

Tufte, E. R. (1997). *Visualizing explanations*. Cheshire, CT: Graphics Press.

Vaughan, T. (1998). *Multimedia: Making it work* (4th ed.). Berkeley, CA: Osborne-McGraw-Hill.

Wheildon, C. (1995). *Type & layout: How typography and design can get your message across — or get in the way*. Berkeley, CA: Strathmoor.

White, J. V. (1990). *Color for the electronic age*. New York: Watson-Guptill.

Also see the following journals:
Graphic Arts Monthly
Graphics: USA

The Internet

Cady, G. H., & McGregor, P. (1996). *Mastering the Internet* (2nd ed.). San Francisco: Sybex.

Crumlish, C. (1998). *The Internet for busy people* (3rd ed.). Berkeley, CA: Osborne/McGraw-Hill.

Ernst, T., & Engst, A. (1995). *Create your own home page*. Indianapolis, IN: Hayden.

Lemay, L. (1997). *Teach yourself Web publishing with HTML 4 in a week* (4th ed.). Indianapolis, IN: Sams.

Manger, J. J. (1995). *The essential Internet information guide*. New York: McGraw-Hill.

Williams, R., & Tollett, J. (1998). *The non-designer's Web book: An easy guide to creating, designing, and posting your own Web site*. Berkeley, CA: Peachpit.

Also see the following journals:
InfoWorld
Internet Week
Internet World
Web Week
Wired

Technical Manuals

Barker, T. (1998). *Writing software documentation: A task-oriented approach*. Boston: Allyn & Bacon.

Bias, R. G., & Mayhew, D. J. (Eds.). (1994). *Cost-justifying usability*. Boston: AP Professional.

Brockmann, R. J. (1990). *Writing better computer documentation: From paper to hypertext. Version 2: The First Quarter Century of Service.* New York: Wiley-Interscience.

Crown, J. (1992). *Effective computer user documentation.* New York: Van Nostrand Reinhold.

Dumas, J. S., & Redish, J. C. (1993). *A practical guide to usability testing.* Norwood, NJ: Ablex.

Forbes, M. (1993). *Writing technical articles, speeches, and manuals* (2nd ed.). Malabar, FL: Krieger.

Hackos, J. T. (1994). *Managing your documentation projects.* New York: Wiley.

Haydon, L. M. (1995). *The complete guide to writing and producing technical manuals.* New York: Wiley.

Horton, W. (1994). *Designing and writing online documentation: Hypermedia for self-supporting products* (2nd ed.). New York: Wiley.

Lanyi, G. (1994). *Managing documentation projects in an imperfect world.* Columbus, OH: Battelle.

Nielsen, J., & Mack, R. L. (Eds.). (1994). *Usability inspection methods.* New York: Wiley.

Price, J., & Korman, H. (1993). *How to communicate technical information: A handbook of software and hardware documentation.* Redwood City, CA: Benjamin/Cummings.

Rubin, J. (1994). *Handbook of usability testing: How to plan, design, and conduct effective tests.* New York: Wiley.

Slatkin, E. (1991). *How to write a manual.* Berkeley, CA: Ten Speed Press.

Steehouder, M., Jansen, C., van der Poort, P., & Verheijen, R. (Eds.). (1994). *Quality of technical documentation.* Amsterdam: Editions Rodopi B. V.

Velotta, C. (1995). *Practical approaches to usability testing for technical documentation.* Arlington, VA: Society for Technical Communication.

Weiss, E. H. (1992). *How to write usable user documentation* (2nd ed.). Phoenix, AZ: Oryx.

Whitaker, K. (1995). *A guide to publishing user manuals.* New York: Wiley.

Wieringa, D., Moore, C., & Barnes, V. (1998). *Procedure writing: Principles and practices* (2nd ed.). Columbus, OH: Battelle.

Woolever, K. R., & Loeb, H. M. (1994). *Writing for the computer industry.* Englewood Cliffs, NJ: Prentice-Hall.

Oral Presentations

Anholt, R. H. (1994). *Dazzle 'em with style: The art of oral scientific presentation.* New York: W. H. Freeman.

D'Arcy, J. (1998). *Technically speaking: A guide for communicating complex information.* Columbus, OH: Battelle.

Gurak, L. J. (2000). *Oral presentations for technical communication.* Boston: Allyn & Bacon.

Smith, T. C. (1991). *Making successful presentations: A self-teaching guide* (2nd ed.). New York: Wiley.

Proposals

Bowman, J. P., & Branchaw, B. P. (1992). *How to write proposals that produce.* Phoenix, AZ: Oryx.

Freed, R. C., Freed, S., & Romano, J. (1995). *Writing winning business proposals.* New York: McGraw-Hill.

Hill, J. W., & Whalen, T. (1993). *How to create and present successful government proposals*. New York: IEEE.

Miner, L. E., Miner, J. T., & Griffith, J. (1998). *Proposal planning and writing* (2nd ed.). Phoenix, AZ: Oryx.

Stewart, R. D., & Stewart, A. L. (1992). *Proposal preparation* (2nd ed.). New York: Wiley-Interscience.

REFERENCES

Chapter 1   Introduction to Technical Communication

Anderson, P. V. (1985). What survey research tells us about writing at work. In L. Odell & D. Goswami (Eds.), *Writing in nonacademic settings*. New York: Guilford.

Beer, D., & McMurrey, D. (1997). *A guide to writing as an engineer*. New York: Wiley.

Canon Corporation. (1999). The Clean Earth Campaign. *Canon Corporation Web site*. Retrieved July 20, 1999, from http: // www.usa.canon.com/cleanearth/index.html

Kiggins, C. (1999). Companies spending $300,000 on communication skills training. *CareerMag Web site*. Retrieved September 15, 1999, from http: // www.careermag .com

MIT Industrial Liaison Program. (1984). Communication skills: A top priority for engineers and scientists (Report No. 32929). Cambridge, MA: MIT.

Motorola, Inc. (2001). *Home page*. Retrieved June 5, 2001, from http://www.motorola .com

Scholz, J. (1996, October). Adding value by developing information for online customer support services. *Intercom:* 36–43.

Spretnak, C. M. (1982). A survey of the frequency and importance of technical communication in an engineering career. *The Technical Writing Teacher, 9* (3), 133–136.

Technical Communication. (1990). *37* (4), 385.

Chapter 2   Understanding Ethical and Legal Considerations

Beauchamp, T. L., & Bowie, N. E. (1993). *Ethical theory and business* (4th ed.). Upper Saddle River, NJ: Prentice-Hall.

Bryan, J. (1992). Down the slippery slope: Ethics and the technical writer as marketer. *Technical Communication Quarterly, 1* (1), 73–88.

De George, R. T. (1995). *Business ethics* (4th ed.). Upper Saddle River, NJ: Prentice-Hall.

Donaldson, T. (1989). *The ethics of international business*. New York: Oxford University Press.

Helyar, P. S. (1992). Products liability: Meeting legal standards for adequate instructions. *Journal of Technical Writing and Communication, 22* (2), 125–147.

Maggio, R. (1991). *The dictionary of bias-free usage: A guide to nondiscriminatory language*. Phoenix, AZ: Oryx.

Murphy, P. (1995). Corporate ethics statements: Current status and future prospects. *Journal of Business Ethics, 14,* 727–740.

Texas Instruments. (1999). TI ethics quick test. Retrieved July 20, 1999, from http://www.ti.com/corp/docs/ethics/quicktest.htm

Velasquez, M. G. (1998). *Business ethics: Concepts and cases* (4th ed.). Upper Saddle River, NJ: Prentice-Hall.

Chapter 3   Understanding the Writing Process

Cormier, R. A. (1997). One last look: The final quality control review. *The Editorial Eye.* Retrieved July 21, 1997, from http://www.eeicom.com/eye/qc-lead.html

Dumas, J. S., & Redish, J. C. (1993). *A practical guide to usability testing.* Norwood, NJ: Ablex.

Kantner, L. (1994). The art of managing usability tests. *IEEE Transactions on Professional Communication, 37,* 143–148.

Rubin, J. (1994). *Handbook of usability testing: How to plan, design, and conduct effective tests.* New York: Wiley.

Chapter 4   Writing Collaboratively

About.com. (2001). Overpopulation and population issues. Delphi Forums Web site. Retrieved September 20, 2001, from http://forums.delphi.com/overpopulation/messages/?msg=551.1

Allen, N., Atkinson, D., Morgan, M., Moore, T., & Snow, C. (1987). What experienced collaborators say about collaborative writing. *Iowa State Journal of Business and Technical Communication, 1* (2), 70–90.

Bolton, R. (1979). *People skills.* New York: Simon & Schuster.

Borisoff, D., & Merrill, L. (1987). Teaching the college course in gender differences as barriers to conflict resolution. In L. B. Nadler, M. K. Nadler, & W. R. Todd-Mancillas (Eds.), *Advances in gender and communication research* (pp. 351–361). Lanham, MD: University Press of America.

Bosley, D. (1993). Cross-cultural collaboration: Whose culture is it, anyway? *Technical Communication Quarterly, 2* (1), 51–62.

Chodorow, N. (1978). *The reproduction of mothering: Psychoanalysis and the sociology of gender.* Berkeley, CA: University of California Press.

Couture, B., & Rymer, J. (1989). Interactive writing on the job: Definitions and implications of collaboration. In M. Kogen (Ed.), *Writing in the business professions* (pp. 73–93). Urbana, IL: National Council of Teachers of English.

Duin, A. H., Jorn, L. A., & DeBower, M. S. (1991). Collaborative writing — Courseware and telecommunications. In M. M. Lay & W. M. Karis (Eds.), *Collaborative writing in industry: Investigations in theory and practice* (pp. 146–169). Amityville, NY: Baywood.

Eddy, W. (1983). Qtd. in A. G. Sargent (1983), *The androgynous manager.* In J. Stewart (1990), *Bridges not walls: A book about interpersonal communication* (5th ed.) (pp. 274–282). New York: McGraw-Hill.

Ede, L., & Lunsford, A. (1990). *Singular texts/plural authors: Perspectives on collaborative writing.* Carbondale, IL: Southern Illinois University Press.

Faigley, L., & Miller, T. P. (1982). What we learn from writing on the job. *College English, 44*, 557–569.

Hulbert, J. (1989). Barriers to effective listening. *Bulletin of the Association of Business Communication, 52* (2), 3–5.

Killingsworth, M. J., & Jones, B. G. (1989). Division of labor or integrated teams: A crux in the management of technical communication? *Technical Communication, 36* (3), 210–221.

Lay, M. M. (1994). The value of gender studies to professional communication research. *Journal of Business and Technical Communication, 8* (1), 58–90.

Lustig, M. W., & Koester, J. (1993). *Intercultural competence.* New York: HarperCollins.

Matson, R. (1996, April). The seven sins of deadly meetings. *Fast Company.* Retrieved July 22, 1999, from http://www.fastcompany.com/online/02/meetings.html

McMillan, J. R., Clifton, A. K., McGrath, D., & Gale, W. S. (1977). Women's language: Uncertainty or interpersonal sensitivity and emotionality? *Sex Roles, 3*, 545–549.

PictureTel Corporation. (1998). Intel® TeamStation™ System. *PictureTel Corporation.* Retrieved October 20, 1999, from http: // www.picturetel.com/teamstation/

Tannen, D. (1990). *You just don't understand.* New York: William Morrow.

Thiederman, S. (1991). *Profiting in America's multicultural marketplace.* New York: Macmillan.

Chapter 5 Analyzing Your Audience and Purpose

Bathon, G. (1999, May). Eat the way your mama taught you. *Intercom:* 22–24.

Ferraro, G. P. (1990). *The cultural dimensions of international business.* Englewood Cliffs, NJ: Prentice-Hall.

Hoft, N. L. (1995). *International technical communication: How to export information about high technology.* New York: Wiley.

Jet Propulsion Lab. (1997a). Mars Climate Orbiter flight system description. *Jet Propulsion Lab Web site.* Retrieved October 5, 1999, from http: // mars.jpl.nasa.gov/msp98 /orbiter/bus.html

Jet Propulsion Lab. (1997b). Mars Climate Orbiter science goals. *Jet Propulsion Lab Web site.* Retrieved October 5, 1999, from http: // mars.jpl.nasa.gov/msp98/orbiter /science.html

Jet Propulsion Lab. (1999a). Mars Climate Orbiter/Mars Polar Lander Mission Overview. *Jet Propulsion Lab Web site.* Retrieved October 5, 1999, from http: // mars.jpl.nasa.gov/msp98/mission_overview.html

Jet Propulsion Lab. (1999b). Mars, water and life. *Jet Propulsion Lab Web site.* Retrieved October 5, 1999, from http: // mars.jpl.nasa.gov/msp98/why.html

Limaye, M. R., & Victor, D. A. (1991). Cross-cultural business communication research: State of the art and hypotheses for the 1990s. *Journal of Business Communication, 28* (3), 277–299.

Lustig, M. W., & Koester, J. (1999). *Intercultural competence: Interpersonal communication across cultures* (3rd ed.). New York: Longman.

Schriver, K. A. (1997). *Dynamics in document design: Creating text for readers*. New York: Wiley.

U.S. Bureau of the Census. (2001). *Statistical abstract of the United States: 1999*. Washington, DC: U.S. Government Printing Office.

Chapter 6   Communicating Persuasively

Bowman, J. P. (1999). Understanding persuasion. *Business Communication: Managing Information and Relationships*. Retrieved July 22, 1999, from http: // spider.hcob .wmich.edu / bis / faculty / bowman / persuade.html

Death of a salesman: The road warrior breed. (1999, July). *SC Magazine*. Retrieved June 30, 1999, from http: // www.infosecnews.com

Gateway, Inc. (1999). Gateway Solo5150. *Gateway Web site*. Retrieved October 11, 1999, from http: // www.gateway.com / prod / ed_s5150_Matrix.shtml

General Electric Company. GE in the Community. *GE Web site*. Retrieved September 20, 2001, from http://www.ge.com/annual00/community/index.html

Mayberry, K. J. (1999). *For argument's sake: A guide to writing effective arguments* (3rd ed.). New York: Longman.

Poverty accounts for gap in IQ scores between blacks and whites. (1996). *Northwestern News*. Retrieved June 28, 1999, from http: // nuinfo.nwu.edu / univ-relations / media / news-releases / *archives / *soc-policy / duncan.html

Princess Cruises. (2001). *Princess Cruises home page*. Retrieved September 20, 2001, from http://www.princess.com/home.jsp

Tenet, G. (1999). Statement of the director of Central Intelligence George J. Tenet on diversity. *Central Intelligence Agency*. Retrieved June 28, 1999, from http: // www .odci.gov / cia / public_affairs / press_release / ps020199.html

Vanguard Group. (2000). SEC to investors: Don't chase performance. *Vanguard Group Web site*. Retrieved February 20, 2000, from http: // www.vanguard.com / cgi-bin / NewsPrint / 949429335

Chapter 7   Researching Your Subject

Bowman, J. P. (1999). Human relations: Conversations and interviews. *Business Communication: Managing Information and Relationships*. Retrieved on July 22, 1999, from http: // spider.hcob.wmich.edu / bis / faculty / bowman / dyads.html

Chemical Abstracts Service. (1999). CA on CD quick start tips. *Chemical Abstracts Web site*. Retrieved on June 2, 1999, from http: // www.cas.org / ONLINE / CD / CACD / QUICKSTART / author.html

Harris, C. (1999). Using the Internet for research (FAQ). *Pure Fiction Web site*. Retrieved October 14, 1999, from http: // www.purefiction.com / pages / res1.htm

Lovgren, J. (1994). How to choose good metaphors. *IEEE Software, 11* (3), 86–88.

McComb, G. (1991). *Troubleshooting and repairing VCRs* (2nd ed.). Blue Ridge Summit, PA: TAB/McGraw-Hill.

Zakon, R. H. (2001). Hobbes' Internet Timeline 5.3. Retrieved on June 5, 2001, from http://www.zakon.org/robert/internet/timeline/

Chapter 8   Organizing Your Information

Bernard, B. P. (Ed.). (1997). Neck Musculoskeletal Disorders: Evidence for work-relatedness. Chap. 2 in *Musculoskeletal Disorders (MSDs) and Workplace Factors: A Critical Review of Epidemiologic Evidence for Work-Related Musculoskeletal Disorders of the Neck, Upper Extremity, and Low Back.* U.S. Department of Health and Human Services. Retrieved July 8, 1999, from http: //www.cdc.gov/niosh/ergtxt2.html

Boeing Corporation. (1999). A brief history. *Boeing Corporation.* Retrieved July 8, 1999, from http: //www.boeing.com/companyoffices/history/boeing/chr6_future _3.html

Brusaw, C. T., Alred, G. J., & Oliu, W. E. (1997). *Handbook of technical writing* (5th ed.). New York: St. Martin's.

Hsiao, A. (1998). Introducing Linux, Part 2: Considering Linux. *About.com.* Issue 36: 17 June 1998. Retrieved July 9, 1999, from http: //linux.about.com/library/weekly /aa061798.htm

Larson, D. E. (1990). *Mayo Clinic family health book.* New York: William Morrow.

United States Agency for International Development. (1999). *U.S. International Food Assistance Report 1998.* Chapter 2. Retrieved July 8, 1999, from http: //www .info.usaid.gov/hum_response/farpt1998/chapter2.htm

University of Texas at Austin. (1999). Campus overview map. *University of Texas.* Retrieved July 8, 1999, from http: //www.utexas.edu/maps/main/overview

U.S. Chemical Safety and Hazard Investigation Board. (1999). Investigation report, propane tank explosion, Herrig Brothers Feather Creek Farm, Albert City, Iowa, April 9, 1998, Report No. 98-007-I-IA. Retrieved July 8, 1999, from http: //www.chemsafety.gov/reports/1999/herrig/herrig05.htm#4.0

U.S. Department of Labor. (1997). Methylene chloride. OSHA 3144. Retrieved July 8, 1999, from http: //www.oshaslc.gov/Publications/Osha3144.pdf

Chapter 9   Drafting and Revising Definitions and Descriptions

Eisenberg, A. (1992). *Effective technical communication* (2nd ed.). New York: McGraw-Hill.

Farkas, D. (2000). How digital cameras work. *CNN.com Web site.* June 15, 2000. Retrieved September 20, 2001, from http://www.cnn.com/2000/TECH/computing/06/15/digital.camera.idg/index.html

National Institutes of Health. (1993). Don't lose sight of glaucoma: Information for people at risk. NIH Publication No. 91-3251. Retrieved July 12, 1999, from http: //www.pueblo.gsa.gov/cic_text/health/glaucoma/glaucoma.htm

Praxis, Inc. (1997). SSULI brochure. *Praxis.* Retrieved July 12, 1999, from http: //www .pxi.com/public/brochures/ssuli/index.html

Roblee, C. L., & McKechnie, A. J. (1981). *The investigation of fires*. Englewood Cliffs, NJ: Prentice-Hall.

U.S. Congress, Office of Technology Assessment. (1995a). *Bringing health care online: The role of information technologies* (OTA-ITC-624). Washington, DC: U.S. Government Printing Office.

U.S. Congress, Office of Technology Assessment. (1995b). *Renewing our energy future* (OTA-ETI-614). Washington, DC: U.S. Government Printing Office.

U.S. Environmental Protection Agency. (1991). *Building air quality: A guide for building owners and facility managers*. Washington, DC: U.S. Environmental Protection Agency.

Wilson, J. R. (1964). *The mind*. New York: Time, Inc.

Chapter 10   Drafting and Revising Coherent Documents

Benson, P. (1985). Writing visually: Design considerations in technical publications. *Technical Communication, 32,* 35–39.

Cohen, S., & Grace, D. (1994). Engineers and social responsibility: An obligation to do good. *IEEE Technology and Society, 13,* 12–19.

Darling, C. (1999). Coherence: Transitions between ideas. *Guide to grammar and writing*. Retrieved August 8, 1999, from http://webster.commnet.edu/hp/pages/darling/grammar/transitions.htm

Snyder, J. D. (1993). Off-the-shelf bugs hungrily gobble our nastiest pollutants. *Smithsonian, 24,* 66.

U.S. Department of Health and Human Services. (1997, March 31). Pneumonia: More patients may be treated at home: Research findings for consumers. *Agency for Health Care Policy and Research*. Retrieved August 5, 1999, from http://www.ahcpr.gov/consumer/pneucons.htm

Chapter 11   Drafting and Revising Effective Sentences

Carliner, S. (1987). Lists: The ultimate organizer for engineering writing. *IEEE Transactions on Professional Communication, 30,* 218–221.

Fuchsberg, G. (1990, December 7). Well, at least "terminated with extreme prejudice" wasn't cited. *Wall Street Journal,* p. B1.

Horn, R. E. (1985). Results with structured writing using the Information Mapping® writing service standards. In T. M. Duffy and R. Waller (Eds.), *Designing Usable Texts*. Orlando, FL: Academic Press.

National Organization on Disability. (2001). Disability etiquette tips. *National Organization on Disability Web site*. Retrieved June 13, 2001, from http://www.nod.org/etiquette.html

Peterson, D. A. T. (1990). Developing a Simplified English vocabulary. *Technical Communication, 37,* 130–133.

Strunk, W. (1918). *The elements of style*. Ithaca, NY: (Privately printed). Retrieved November 1, 1999, from http://www.bartleby.com/141/strunk.html#13

Thomas, M., Jaffe, G., Kincaid, J. P., & Stees, Y. (1992). Learning to use Simplified English: A preliminary study. *Technical Communication, 39*, 69–73.

Williams, J. (1997). *Style: Ten lessons in clarity & grace* (5th ed.). New York: Harper-Collins.

Chapter 12   Drafting and Revising Front and Back Matter

Bonura, L. S. (1994). *The art of indexing*. New York: Wiley.

Chicago manual of style (14th ed.). (1993). Chicago: University of Chicago Press.

Crowe, B. (1985). Design of a radio-based system for distribution automation. Unpublished manuscript, Drexel University.

Gibaldi, J. (1995). *MLA handbook for writers of research papers* (4th ed.). New York: Modern Language Association.

Ruiu, D. (1994). Testing ATM systems. *IEEE Spectrum, 31* (6), 25.

U.S. Congress, Office of Technology Assessment. (1995). *Bringing health care online: The role of information technologies* (OTA-ITC-624). Washington, DC: U.S. Government Printing Office.

Vacca, R., in Rubens, P. (Ed.). (1992). *Science and technical writing: A manual of style*. New York: Henry Holt.

Chapter 13   Designing the Document

Berry, R., Mobley, K., & Turk, K. (1994). Preparing to document an object-oriented project. *Technical Communication, 41*, 643–652.

Biggs, J. R. (1980). *Basic typography*. New York: Watson-Guptill.

Bonneville Power Administration. (1993). *Resource programs: Final environmental impact statement: Vol. 1. Environmental analysis*. Washington, DC: U.S. Department of Energy.

Felker, D. B., Pickering, F., Charrow, V. R., Holland, V. M., & Redish, J. C. (1981). *Guidelines for document designers*. Washington, DC: American Institutes for Research.

Fischer, J. M. (1992). *Inside DesignCAD*. Carmel, IN: New Riders Publishing.

Haley, A. (1991). All caps: A typographic oxymoron. *U&lc, 18* (3), 14–15.

Hewlett-Packard. (1997). *HP OpenView Desktop Administrator desktop configuration user's guide. Hewlett-Packard*. Retrieved on August 1, 1999, from http://www.openview.hp.com/pdfs/48.pdf

Keyes, E. (1993). Typography, color, and information structure. *Technical Communication, 40*, 638–654.

Kostelnick, C., & Roberts, D. D. (1998). *Designing visual language: Strategies for professional communicators*. Needham Heights, MA: Allyn & Bacon.

National Commission on Libraries and Information Science. (1999). *Moving Toward More Effective Public Internet Access: The 1998 National Survey of Public Library Outlet Internet Connectivity. National Commission on Libraries and Information Science.* Retrieved August 2, 1999, from http://www.nclis.gov/what/1998plo.pdf

Poulton, E. (1968). Rate of comprehension of an existing teleprinter output and of possible alternatives. *Journal of Applied Psychology, 52,* 16–21.

Texas Instruments. (1999). Planning for PCS: The TI strategy for upbanded GSM. *Texas Instruments.* Retrieved August 3, 1999, from http://www.ti.com/sc/docs/wireless/gsmweb.pdf

United States Environmental Protection Agency. (1999, July). EPA oil spill update, Vol. 2, No. 4, p. 7. *United States Environmental Protection Agency.* Retrieved August 3, 1999, from http://www.epa.gov/oilspill/docs/epaupd8.pdf

White, J. V. (1990). *Great pages: A common-sense approach to effective desktop design.* El Segundo, CA: Serif Publishing.

Williams, T., & Spyridakis, J. (1992). Visual discriminability of headings in text. *IEEE Transactions on Professional Communication, 35,* 64–70.

Chapter 14   Creating Graphics

Barnum, C. M., & Carliner, S. (1993). *Techniques for technical communicators.* New York: Macmillan.

Brockmann, R. J. (1990). *Writing better computer user documentation: From paper to hypertext.* New York: Wiley.

Curtis, H., & Barnes, N. S. (1989). *Biology* (5th ed.). New York: Worth.

Dean, R. S., & Kulhavy, R. W. (1981). Influence of spatial organization in prose learning. *Journal of Educational Psychology, 73,* 57–64.

Gatlin, P. L. (1988). Visuals and prose in manuals: The effective combination. In *Proceedings of the 35th International Technical Communication Conference* (pp. RET 113–115). Arlington, VA: Society for Technical Communication.

Grimstead, D. (1987). Quality graphics: Writers draw the line. In *Proceedings of the 34th International Technical Communication Conference* (pp. VC 66–69). Arlington, VA: Society for Technical Communication.

Horton, W. (1991). *Illustrating computer documentation: The art of presenting information graphically on paper and online.* New York: Wiley.

Horton, W. (1993). The almost universal language: Graphics for international documents. *Technical Communication, 40,* 682–693.

Labuz, R. (1984). *How to typeset from a word processor: An interfacing guide.* New York: R. R. Bowker.

Levie, W. H., & Lentz, R. (1982). Effects of text illustrations: A review of research. *Journal of Educational Psychology, 73,* 195–232.

Mankiw, N. G. (1997). *Macroeconomics* (3rd ed.). New York: Worth.

McComb, G. (1991). *Troubleshooting and repairing VCRs* (2nd ed.). Blue Ridge Summit, PA: TAB/McGraw-Hill.

McGuire, M., & Brighton, P. (1990). Translating text into graphics. Paper presented at the meeting of the 37th International Technical Communication Conference. Dallas, TX.

Morrison, C., & Jimmerson, W. (1989, July). Business presentations for the 1990s. *Video Manager, 4,* 18.

Software602, Inc. (2001) Software 602Pro PLUS PACK. Retrieved September 20, 2001 from http://www.software602.com/products/pp/features.html#spell

Tufte, E. R. (1983). *The visual display of quantitative information.* Cheshire, CT: Graphics Press.

Tufte, E. R. (1999). *The visual display of quantitative information.* Adapted by Saul Greenberg. Retrieved August 2, 1999, from http://www.cpsc.ucalgary.ca/projects/grouplab/699/vis_display.html

U.S. Bureau of the Census. (2001). *Statistical abstract of the United States: 1999.* Washington, DC: U.S. Government Printing Office. Retrieved June 26, 2001, from http://www.census.gov/prod/2001pubs/statab/sec03.pdf

U.S. Congress, Office of Technology Assessment. (1993). *Making government work: Electronic delivery of federal services* (OTA-TCT-578). Washington, DC: U.S. Government Printing Office.

U.S. Consumer Product Safety Commission. (1999). Your used crib could be deadly. *Consumer Product Safety Commission.* Retrieved August 3, 1999, from http://www.cpsc.gov/cpscpub/pubs/usedcrib.pdf

U.S. Department of Health and Human Services. (1990). *Health status of the disadvantaged chartbook* (DHHS Publication HRS-P-DV-90-1). Washington, DC: U.S. Department of Health and Human Services.

U.S. Department of the Treasury. (1999). Business use of your home. Publication 587, p. 4. *Internal Revenue Service.* Retrieved August 3, 1999, from http://ftp.fedworld.gov/pub/irs-pdf/p587.pdf

U.S. Environmental Protection Agency. (1991). *Building air quality: A guide for building owners and facility managers.* Washington, DC: U.S. Government Printing Office.

White, J. V. (1984). *Using charts and graphs: 1000 ideas for visual persuasion.* New York: R. R. Bowker.

White, J. V. (1990). *Color for the electronic age.* New York: Watson-Guptill.

Chapter 15 Writing Letters, Memos, and Emails

Bowman, J. P. (1999). Writing persuasive messages. *Business communication: Managing information and relationships.* Retrieved September 21, 1999, from http://spider.hcob.wmich.edu/bis/faculty/bowman/persuade2.html

Crowe, E. P. (1994). *The electronic traveller: Exploring alternative online systems.* New York: Windcrest/McGraw-Hill.

ePolicy Institute. (2001). What are your employees up to? *ePolicy Web site.* Retrieved July 5, 2001, from http://www.epolicyinstitute.com

Chapter 16   Writing Job-Application Materials

Bowman, J. P. (1999). Selling yourself. *Business communication: Managing information and relationships*. Retrieved September 18, 1999, from http://spider.hcob .wmich.edu/bis/faculty/bowman/job2.html

Dumas, M. (2001). Résumé consultation & writing. *Distinctive Documents Web site*. Retrieved July 9, 2001, from http://www.distinctiveweb.com/writing.htm

Harcourt, J., Krizan, A. C., & Merrier, P. (1991). Teaching résumé content: Hiring officials' preferences versus college recruiters' preferences. *Business Education Forum, 45*(7), 13–17.

Isaacs, K. (2001). Résumé Center FAQ. *Monster.com Web site*. Retrieved July 9, 2001, from http://resume.monster.com/faq/

Parker, Y. (1999). Hot tips for résumé-writing from expert Yana Parker. *Mary Ellen Guffey's Communication@Work*. Retrieved September 18, 1999, from http://www .westwords.com/GUFFEY/parker.html

Robinson, K. (1997, March 24). Job search engine. *Idaho Statesman*, p. D1.

Chapter 17   Writing Proposals

CBDNET. (1999). *Commerce Business Daily Online*. Retrieved August 9, 1999, from http://frwebgate.access.gpo.gov/cgi-bin/CBDbrowse.cgi?system=162.140.64.30 &file=AG001364.645

Grapatin, J., Moon, T., & Johnson, C. (1999). Proposal for a feasibility study for hiring a Webmaster. Unpublished document.

Reid, A. N. T. (1998). A practical guide for writing proposals. Retrieved August 24, 1999, from http://www.members.dca.net/areid/proposal.htm

SmartDraw.com. (2001). SmartDraw cool examples. *SmartDraw.com Web site*. Retrieved September 20, 2001, from http://www.smartdraw.com/resources/examples /business/gantt15.htm

U.S. Department of Commerce. (2001). *Statistical abstract of the United States, 2000*. Retrieved July 9, 2001, from http://www.census.gov/prod/2001pubs/statab /sec11.pdf

Wells, M., Tommack, D., & Tuck, B. (1997). Proposal to research whether Central State University should recognize American Sign Language as a foreign language. Unpublished document.

Chapter 18   Writing Informal Reports

Beer, D., & McMurrey, D. (1997). *A guide to writing as an engineer*. New York: Wiley.

Grapatin, J., Moon, T., & Johnson, C. (1999). Progress report on proposal for a feasibility study for hiring a Webmaster. Unpublished document.

Robert, H. M., & Patnode, D. (1994). *Robert's rules of order*. New York: Berkley.

Sabin, W. (1999). *Gregg reference manual* (7th ed.). Lake Forest, IL: Glencoe/McGraw-Hill.

Chapter 19   Writing Formal Reports

Corder, Z. (1993). QA Lab SEM evaluation: Project completion report. Unpublished document.

Grapatin, J., Moon, T., & Johnson, C. (1999). Completion report on proposal for a feasibility study for hiring a Webmaster. Unpublished document.

Mathes, J. C., & Stevenson, D. W. (1991). *Designing technical reports: Writing for audiences in organizations* (2nd ed.). New York: Macmillan.

Chapter 20   Writing Instructions and Manuals

Befuddled PC users flood help lines, and no question seems to be too basic. (1994, March 1). *Wall Street Journal*, sec. p. B1.

Group 1 Software, Inc. (1994). *AccuMail user's guide*. Lanham, MD: Group 1 Software, Inc.

HCL, Inc. (1999). Hazard communication labels. *HCL, Inc. Web site*. Retrieved September 25, 1999, from http: //www.hclco.com/hazard.htm

Hewlett-Packard Company. (1998). *HP LaserJet 4000, 4000 T, 4000 N, and 4000 TN printers*. Boise, ID: Hewlett-Packard Company.

Learn2.com. (1999a). Jump-start a car. *Learn2.com Web site*. Retrieved September 26, 1999, from http: //www.learn2.com/05/0508/05082.html

Learn2.com. (1999b). Jump-start a car. *Learn2.com Web site*. Retrieved September 26, 1999, from http: //www.learn2.com/05/0508/05081.html

Mead, J. (1998). Measuring the value added by technical documentation: A review of research and practice. *Technical Communication, 45*, 353–379.

Microsoft Corporation. (1998). *Getting started: Microsoft® Windows® 98*. Redmond, WA: Microsoft Corporation.

Sanderlin, S. (1988). Preparing instruction manuals for non-English readers. *Technical Communication, 35*, 96–100.

U.S. Department of Labor, Occupational Safety and Health Administration. (2000). *OSHA Regulations (Standards — 29 CFR)*. Retrieved February 20, 2000, from http://www.osha-slc.gov/OshStd_toc/OSHA_Std_toc.html

Chapter 21   Creating Web Sites

Computer Industry Almanac, Inc. (2001). U.S. has 33% share of Internet users worldwide year-end 2000. *Computer Industry Almanac, Inc. Web site*. Retrieved July 11, 2001, from http://www.c-i-a.com/200103iu.htm

Corel (2001a). *Corel Web page*. Retrieved on September 20, 2001, from http://www3.corel.com/cgi-bin/gx.cgi/AppLogic+FTContentServer?pagename=Corel/Index

Corel (2001b). *Corel Web page*. Retrieved on September 20, 2001, from http://www3.corel.com/cgi-bin/gx.cgi/AppLogic+FTContentServer?pagename=Corel/Index

Gahran, A. (1999). Get with the plan! Setting goals for your site. *Contentious*. Retrieved on November 30, 1999, from http://www.content-exchange.com/cx/html/newsletter/1-16/vt1-16htm

Global Reach. (2001). Global Internet statistics (by language). *Global Reach Web site*. Retrieved July 12, 2001, from http://www.euromktg.com/globstats/

Goodstein, A. (1999, April 1). People with disabilities reach for web access. *PC World*. Retrieved September 11, 1999, from http://www.pcworld.com/pcwtoday/article/0,1510,10362+1+0,00.html

Lynch, P. J., & Horton, S. (1997). *Web style guide*. Retrieved on September 12, 1999, from http://info.med.yale.edu/caim/manual/contents.html

Microsoft Corporation. (1999a). Corporate information. *Microsoft Web page*. Retrieved on September 9, 1999, from http://www.microsoft.com/mscorp

Microsoft Corporation. (1999b). *Microsoft.com Home*. Retrieved on September 9, 1999, from http://www.microsoft.com/ms.htm

National Railroad Passenger Corporation. (2001). *Amtrak Web site*. Retrieved June 25, 2001, from http://www.amtrak.com/index1.html

National Science Foundation. (1999). Site map. Retrieved September 9, 1999, from http://www.nsf.gov/home/help/sitemap.htm

National Science Foundation. (2001). News. *National Science Foundation Web site*. Retrieved July 3, 2001, from http://www.nsf.gov/home/menus/news.htm

Nolo.com. (1999). Encyclopedia. *Nolo.com Web Site*. Retrieved September 9, 1999, from http://www.nolo.com/encyclopedia/faqs/pct_ency.html

O'Mahoney, B. (1999). *The copyright Web site*. Retrieved November 15, 1999, from http://www.benedict.com

Selective Service System. (1999). Registration information. FAQs. *Selective Service System Web site*. Retrieved on September 12, 1999, from http://www.sss.gov/qa.htm

Shape Up America. (1999). Fitness. *Shape Up American Web site*. Retrieved on September 10, 1999, from http://www.shapeup.org/fitness/index.htm

Society for Technical Communication. (2001). *Society for Technical Communication Web site*. Retrieved July 13, 2001, from http://www.stc.org

Sun Microsystems, Inc. (2001). Site index. *Sun Microsystems Web site*. Retrieved July 5, 2001, from http://www.sun.com/siteindex/

Toastmasters International. (1999). *Toastmasters International Web site*. Retrieved September 12, 1999, from http://www.toastmasters.org/

Zakon, R. (2000). Hobbes' Internet Timeline v. 5.0. Retrieved February 20, 2000, from http://info.isoc.org/guest/zakon/Internet/History/HIT.html

Chapter 22   Making Oral Presentations

Smith, T. C. (1991). *Making successful presentations: A self-teaching guide*. New York: Wiley.

Appendix. Reference Handbook

Hacker, D. (1999). *A writer's reference* (4th ed.). Boston: Bedford/St. Martin's.

Harnack, A., & Kleppinger, E. (2000). *Online! A reference guide to using Internet sources.* Boston: Bedford/St. Martin's.

Raimes, A. (1992). *Grammar troublespots: An editing guide for students* (2nd ed.). New York: St. Martin's Press.

Figures 9.5 and 9.6: Excerpts from "Special Sensor Ultraviolet Limb Imager" brochure. Reprinted by permission of Praxis, Inc., Alexandria, VA, <http://www.pxi.com>.

Chapter 9, exercise 5: Digital camera based on Farkas, 2000. Source: Farkas Graphic Resources Web site <www.farkas.com>. © David F. Farkas 1998. Courtesy of Farkas Graphic Resources.

Figure 11.1: Information Mapping®. From Robert Horn et al., "Results with Structured Writing Using the Information Mapping Writing Service Standards," from T. M. Duffy and R. Waller, eds., *Designing Usable Texts* (Orlando: Academic Press, 1985). Reprinted with permission from Information Mapping, Inc. Copyright Information Mapping, Inc., 1985. All rights reserved.

Figure 13.14: A page from *HP Openview Desktop Administrator.* Source: Hewlett-Packard Web site. Copyright © 1999 Hewlett-Packard Company. Reproduced with permission.

Figure 13.16: "Planning for PCS: The TI Strategy for Upbanded GSM." Source: Texas Instruments Web site. Copyright © 1999 by Texas Instruments. All rights reserved. Reprinted by permission.

Chapter 13, Case 2: "Join the Club!" Scrabble® brochure. SCRABBLE®, and the distinctive game letter tiles and game board, are the trademarks of Hasbro in the U.S. and Canada. © 2000 Hasbro. All rights reserved. Used with permission.

Figure 14.4: Color used to establish patterns. Source: From p. 532 in *Biology,* 5/e, by Helena Curtis and N. Sue Barnes. Worth Publishers, New York (1989). Reprinted with permission.

Figure 14.16: Pie chart. Source: "The Three Groups of Population," p. 37 from *Macroeconomics,* 3/e, by N. Gregory Mankiw. Worth Publishers, New York (1997). Reprinted with permission.

Figure 14.24: Closed-system flowchart. Source: From p. 251 in *Biology,* 5/e, by Helena Curtis and N. Sue Barnes. Worth Publishers, New York (1989). Reprinted with permission.

Figure 20.2: Safety label. Source: HCL, Inc., Web site. Used by permission.

Figure 20.3: Safety information on machinery. Source: John Deere operator's manual. Reproduced by permission of Deere & Company, © 1995. Deere & Company. All rights reserved.

Figures 20.4 and 20.5: "Learn2 Jump-Start a Car: Intro/Before You Begin" and "Learn2 Jump-Start a Car: Step 1." Source: Learn2.com Web site. Reprinted by permission.

Figure 20.6: A set of instructions from a Hewlett-Packard manual. Source: *HP Laserjet 4000, 4000T, 4000N, and 4000TN Printers,* copyright © 1999 Hewlett-Packard Company. Reproduced with permission.

Figures 21.06 & 21.07: Header and Footer screen shots on Corel home page. Source: Corel Web site <www.corel.com>. Copyright © 2001 Corel Corporation. All Rights Reserved. Reprinted by permission.

Figure 21.9: Table of Contents for "Frequently Asked Patent, Copyright & Trademark Questions." Source: Nolo.com (at <http://www.nolo.com>). Used by permission.

Figure 21.10: Patent, Copyright & Trademark Table of Contents screen shot. Source: Society for Technical Communication <www.stc.org>. © 2001 Society for Technical Communication.

Figure 21.12: The Toastmasters International home page. Source: Toastmasters International Web site. Copyright © 1998 Toastmasters International. Reprinted by permission.

Figure 21.14: Amtrak home page. Source: AMTRAK Corp. <www.amtrak.com>. Courtesy of Amtrak.

INDEX

ABBREVIATIONS IN THE REFERENCE HANDBOOK

Your instructor might use the following abbreviations to refer you to specific topics in the appendix, "A Reference Handbook."

Abbreviation	Topic	Page Number	Abbreviation	Topic	Page Number
abbr	abbreviation	710	MLA	MLA Style	676
adj	adjective	691, 720	num	number	708
adv	adverb	720	omit	omitted word or words	721
agr p/a	pronoun/ante-cedent agreement	692	ref	ambiguous pronoun reference	689
agr s/v	subject/verb agreement	692, 717	rep	repeated word (ESL)	722
APA	APA style	661	run	run-on sentence	688
art	article (*a, an, the*)	719	sent	sentence part (ESL)	712
cap	capitalization	711	sub	subordinating clause (ESL)	713
CBE	CBE style	670	t	verb tense	692
comp	comparison of items	690	vb	verb tense (ESL)	714
cond	conditional sentence (ESL)	718	.	period	699
coor	coordinating clause (ESL)	713	!	exclamation point	699
cs	comma splice	687	?	question mark	699
ESL agr/sv	subject-verb agreement (ESL)	717	,	comma	693
frag	sentence fragment	686	;	semicolon	697
help	helping verb and main verb	716	:	colon	698
inf	infinitive form of the verb	716	—	dash	700
-ing	*-ing* form of the verb	715	( )	parentheses	701
ital	italics (underlining)	705	-	hyphen	706
			,	apostrophe	701
			" "	quotation marks	703
			...	ellipses	704
			< >	angle brackets	706
			[ ]	square brackets	705